Advances in Intelligent Systems and Computing

Volume 398

Series editor

Janusz Kacprzyk, Polish Academy of Sciences, Warsaw, Poland
e-mail: kacprzyk@ibspan.waw.pl

About this Series

The series "Advances in Intelligent Systems and Computing" contains publications on theory, applications, and design methods of Intelligent Systems and Intelligent Computing. Virtually all disciplines such as engineering, natural sciences, computer and information science, ICT, economics, business, e-commerce, environment, healthcare, life science are covered. The list of topics spans all the areas of modern intelligent systems and computing.

The publications within "Advances in Intelligent Systems and Computing" are primarily textbooks and proceedings of important conferences, symposia and congresses. They cover significant recent developments in the field, both of a foundational and applicable character. An important characteristic feature of the series is the short publication time and world-wide distribution. This permits a rapid and broad dissemination of research results.

More information about this series at http://www.springer.com/series/11156

L. Padma Suresh · Bijaya Ketan Panigrahi
Editors

Proceedings of the International Conference on Soft Computing Systems

ICSCS 2015, Volume 2

Springer

Editors
L. Padma Suresh
Noorul Islam Centre for Higher Education
Kumaracoil, Tamil Nadu
India

Bijaya Ketan Panigrahi
IIT Delhi
New Delhi
India

ISSN 2194-5357 ISSN 2194-5365 (electronic)
Advances in Intelligent Systems and Computing
ISBN 978-81-322-2672-7 ISBN 978-81-322-2674-1 (eBook)
DOI 10.1007/978-81-322-2674-1

Library of Congress Control Number: 2015953797

Springer New Delhi Heidelberg New York Dordrecht London

Printed on acid-free paper

Springer (India) Pvt. Ltd. is part of Springer Science+Business Media (www.springer.com)

Preface

The volumes contain the papers presented at the *International Conference On Soft Computing Systems* (ICSCS) held during 20 and 21 April 2015 at Noorul Islam Centre for Higher Education, Noorul Islam University, Kumaracoil. ICSCS 2015 received 504 paper submissions from various countries across the globe. After a rigorous peer-review process, 163 full-length articles were accepted for oral presentation at the conference. This corresponds to an acceptance rate of 36 % and is intended for maintaining the high standards of the conference proceedings. The papers included in these AISC volumes cover a wide range of topics in Genetic Algorithms, Evolutionary Programming and Evolution Strategies such as AIS, DE, PSO, ACO, BFA, HS, SFLA, Artificial Bees and Fireflies Algorithm, Neural Network Theory and Models, Self-organization in Swarms, Swarm Robotics and Autonomous Robot, Estimation of Distribution Algorithms, Stochastic Diffusion Search, Adaptation in Evolutionary Systems, Parallel Computation, Membrane, Grid, Cloud, DNA, Quantum, Nano, Mobile Computing, Computer Networks and Security, Data Structures and Algorithms, Data Compression, Data Encryption, Data Mining, Digital Signal Processing, Digital Image Processing, Watermarking, Security and Cryptography, Bioinformatics and Scientific Computing, Machine Vision, AI methods in Telemedicine and eHealth, Document Classification and Information Retrieval, Optimization Techniques and their applications for solving problems in these areas.

In the conference separate sessions are arranged for delivering the keynote address by eminent members from various academic institutions and industries. Three keynote lectures were given in two different venues as parallel sessions on 20 and 21 April 2015. In the first session, Dr. Pradip K. Das, Associate Professor, Indian Institute of Technology, Guwahati gave a talk on "Trends in Speech Processing" in Venue 1 and Prof. D. Thukaram, Professor, Department of EEE, IISC, Bangalore, gave his lecture in "Recent Trends in Power System" and covered various evolutionary algorithms and its applications to Power Systems in Venue 2. In the second session, Dr. Willjuice Iruthayarajan, Professor, Department of EEE, National Engineering College, Kovilpatti, gave his talk on "Evolutionary

Computational Algorithm for Controllers" at Venue 1. All these lectures generated great interest among the participants of ICSCS 2015 in paying more attention to these important topics in their research work. Four sessions are arranged for electrical-related papers, five sessions for electronics-related papers and nine sessions for computer-related papers.

We take this opportunity to thank the authors of all the submitted papers for their hard work, adherence to the deadlines and suitably incorporating the changes suggested by the reviewers. The quality of a refereed volume depends mainly on the expertise and dedication of the reviewers. We are indebted to the Programme Committee members for their guidance and co-ordination in organizing the review process.

We would also like to thank our sponsors for providing all the support and financial assistance. We are indebted to our Chairman, Vice Chancellor, Advisors, Pro-Vice-Chancellor, Registrar, faculty members and administrative personnel of Noorul Islam Centre for Higher Education, Noorul Islam University, Kumaracoil, for supporting our cause and encouraging us to organize the conference in a grand scale. We would like to express our heartfelt thanks to Prof. B.K. Panigrahi and Prof. S.S. Dash for providing valuable guidelines and suggestions in the conduct of the various parallel sessions in the conference. We would also like to thank all the participants for their interest and enthusiastic involvement. Finally, we would like to thank all the volunteers, whose tireless efforts in meeting the deadlines and arranging every detail meticulously, made sure that the conference could run smoothly. We hope the readers of these proceedings find the papers useful, inspiring and enjoyable.

L. Padma Suresh
Bijaya Ketan Panigrahi

Organization Committee

Chief Patron

Dr. A.P. Majeed Khan, President, NICHE
Dr. A.E. Muthunayagam, Advisor, NICHE
Mr. M.S. Faizal Khan, Vice President, NICHE

Patron

Dr. R. Perumal Swamy, Vice Chancellor, NICHE
Dr. S. Manickam, Registrar, NICHE
Dr. N. Chandra Sekar, Pro-Vice Chancellor, NICHE
Dr. K.A. Janardhanan, Director (HRM), NICHE

General Chair

Dr. B.K. Panigrahi, IIT, Delhi

Programme Chair

Dr. L. Padma Suresh, Professor and Head, EEE

Programme Co-chair

Dr. I. Jacob Raglend, Professor, EEE
Dr. G. Glan Devadhas, Professor, EIE
Dr. R. Rajesh, Professor, Mech
Dr. V. Dharun, Professor, Biomedical

Co-ordinator

Prof. V.S. Bindhu, NICHE
Prof. K. Muthuvel, NICHE

Organizing Secretary

Dr. D.M. Mary Synthia Regis Prabha, NICHE
Prof. K. Bharathi Kannan, NICHE

Executive Members

Dr. S. Gopalakrishnan, Controller of Exams, NICHE
Dr. A. Shajin Nargunam, Director (Academic Affairs), NICHE
Dr. B. Krishnan, Director (Admissions), NICHE
Dr. Amar Pratap Singh, Director (Administration), NICHE
Dr. M.K. Jeyakumar, Additional Controller, NICHE
Prof. S.K. Pillai, NICHE
Dr. J. Jeya Kumari, NICHE
Prof. F. Shamila, NICHE
Prof. K. Subramanian, NICHE
Dr. P. Sujatha Therese, NICHE
Prof. J. Arul Linsely, NICHE
Dr. D.M. Mary Synthia Regis Prabha, NICHE
Prof. H. Vennila, NICHE
Prof. M.P. Flower Queen, NICHE
Prof. S. Ben John Stephen, NICHE
Prof. V.A. Tibbie Pon Symon, NICHE
Prof. Breesha, NICHE
Prof. R. Rajesh, NICHE

Contents

About the Editors

Dr. L. Padma Suresh obtained his doctorate from MS University and Dr. M.G.R. University, respectively. He is presently working as a Professor and Head in Department of Electrical and Electronics Engineering, Noorul Islam University, Kumaracoil, India. Dr. Suresh is well known for his contributions to the field in both research and education contributing over 75 research articles in journal and conferences. He is the editorial member of International Journal of Advanced Electrical and Computer Engineering and also served as a reviewer for various reputed journals. He has been a life member of the Indian Society for Technical Education. He also served in many committees as Convener, Chair, and Advisory member for various external agencies. His research is currently focused on Artificial Intelligence, Power Electronics, Evolutionary Algorithms, Image Processing and Control Systems.

Dr. Bijaya Ketan Panigrahi is an associate professor in Electrical and Electronics Engineering Department in Indian Institute of Technology Delhi, India. He received his Ph.D. degree from Sambalpur University. He is serving as a chief editor in the International Journal of Power and Energy Conversion. He has contributed more than 200 research papers in well-reputed indexed journals. His interests include Power Quality, FACTS Devices, Power System Protection, and AI Application to Power System.

A Hybrid Ant Colony Tabu Search Algorithm for Solving Task Assignment Problem in Heterogeneous Processors

M. Poongothai and A. Rajeswari

Abstract Assigning real-time tasks in a heterogeneous parallel and distributed computing environment is a challenging problem, in general, to be NP hard. This paper addresses the problem of finding a solution for real-time task assignment to heterogeneous processors without exceeding the processor capacity and fulfilling the deadline constraints. The proposed Hybrid Max–Min Ant System (HACO-TS) makes use of the merits of Max–Min ant system with Tabu search algorithm for assigning tasks efficiently than various metaheuristic approaches. The Tabu search is used to intensify the search by the MMAS method. The performance of the proposed HACO-TS algorithm has been tested on consistent and inconsistent heterogeneous multiprocessor systems. Experimental comparisons with existing Modified BPSO algorithms demonstrate the effectiveness of the proposed HACO-TS algorithm.

Keywords ACO · Real-time task assignment · Heterogeneous processors · Metaheuristic algorithm · Parallel computing

1 Introduction

Power consumption is increased due to the complexity of multiprocessor designs. Effective task mapping is an important problem in a parallel and distributed computing environment which provides better utilization of the parallelism and enhances the system performance.

The complexity of the system increases if it includes heterogeneous multiprocessors. The heterogeneous multiprocessor system includes two or more processors with different executing speeds [1]. It has the potential to process multiple tasks simultaneously, which is widely used for executing computationally intensive

M. Poongothai (✉) · A. Rajeswari
Department of Electronics and Communication Engineering, Coimbatore Institute of Technology, Coimbatore 641014, India
e-mail: gothaikathirvel@gmail.com

L.P. Suresh and B.K. Panigrahi (eds.), *Proceedings of the International Conference on Soft Computing Systems*, Advances in Intelligent Systems and Computing 398, DOI 10.1007/978-81-322-2674-1_1

applications. Multiple processing environments will improve the system performance and fulfill the increase in energy consumption. Although the effective assignment of real-time tasks in heterogeneous multiprocessor systems to achieve high performance is said to be an NP-hard problem [2], the problem can be solved by applying approximate or metaheuristics algorithms to obtain suboptimal results within a reasonable time. Several metaheuristic algorithms [1, 3–7] have been proposed in the literature for handling the task assignment problem. The most used metaheuristics are particle swarm optimization (PSO), genetic algorithm (GA), ant colony optimization (ACO), and Tabu search (TS). Among these algorithms, a hybrid algorithm provides good results for the task assignment problem. Hybrid algorithms are formed by combining two metaheuristic algorithms. In this paper, a hybrid metaheuristic task assignment algorithm is proposed with the objective of resource utilization and energy consumption, which combines the advantages of the MAX–MIN ant system and Tabu search algorithm. Max–Min Ant System (MMAS) is extended with a Tabu search heuristic to improve the task assignment solution. The performance of MMAS algorithm can be significantly improved by adding a Tabu search algorithm called hybridization (HACO-TS) in which some or all ants are allowed to improve their solutions by Tabu search algorithm.

In this paper, Real-time heterogeneous multiprocessor systems is considered based on task heterogeneity, processor heterogeneity, and consistency. The system is said to be a consistent (C) system if a particular processor always runs at the same speed for the entire task (i.e., processor speed does not depend on task characteristics). While the inconsistent system (IC) is one where a particular processor runs at different speeds for different tasks (i.e., processor speed depends on task characteristics). The proposed algorithm is applied and compared for both consistency and inconsistency systems. From the results obtained it has been found that the proposed algorithm outperforms for the inconsistent problem instances compared to consistent problem instances. Finally, the performance of the proposed HACO-TS algorithm for inconsistent problem instances is compared with the existing algorithm Modified BPSO.

2 System Model and Problem Statement

The heterogeneous multiprocessor environment with m pre-emptive processors $\{P_1, P_2 \ldots P_m\}$ based on CMOS technology is considered [1, 4, 8]. The processors in the heterogeneous environment are operated at different speeds and one instruction per cycle is limited to execute in each processor at variable speed. The energy consumption is calculated as

$$\text{Energy}_{i,j} = \text{Power}_{i,j} \cdot e_{i,j} = \left(C_{\text{ef}} \cdot s_{ij}^3/k^2\right) \cdot e_{i,j} = \left(C_{\text{ef}}/k^2\right) \cdot c_i \cdot s_{ij}^2 \tag{1}$$

where C_{ef} is the effective switching capacitance related to tasks, k is the constant, $e_{i,j}$ is the execution time for task T_i on processor P_j , $s_{i,j}$ is the speed of P_j for task T_i, and c_i is the number of clock cycles to execute a task T_i. From Eq. (1), it is understood that energy consumption is directly proportional to $c_i \times s_{i,j}^2$ [8]. This equation is significant because the processors operate at different speeds.

2.1 Construction Graph and Constraints

Scheduling algorithms include the problem of task assignment and scheduling [9]. The task assignment problem determines the processor that executes a given task. The scheduling problem decides when and in which order the tasks must be executed. The scheduling of real-time tasks can be done by partitioned scheme or global scheme [1]. In the partitioned scheme, all the instances of the task are statically assigned to a specific processor, as shown in Fig. 1. It converts the multiprocessor scheduling problem into a set of uniprocessor ones. This helps to implement uniprocessor scheduling algorithms on each processor.

To solve the task assignment problem, the utilization matrix $u_{i,j}$ is given as input to the proposed algorithm. The utilization matrix $u_{i,j}$ is an nxm matrix in which m is the number of heterogeneous processors and n is the number of tasks. The row of utilization matrix represents the estimated utilization value for a specified task on each heterogeneous processor. Similarly, the column of utilization matrix represents the estimated utilization value of a specified processor for each task. The utilization matrix U_{n*m} holds the real numbers in (0, 1) and infinity. If $U_{ij} = \infty$ means, the particular task is not suitable to execute on a specified processor P_j. For a given set of heterogeneous multiprocessor and task set, the artificial ant stochastically assigns the tasks to the processor until the specific processor without exceeding its computing capacity.

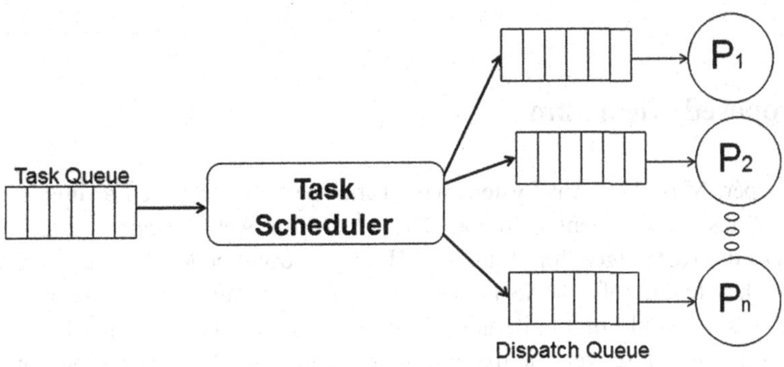

Fig. 1 Model for proposed partitioned task scheduler scheme

An artificial ant constructs the task assignment solution based on the constraints given by Eq. (2)

$$\sum_{i=1}^{n} u_{i,j} \cdot \Omega^{s}_{i,j} \leq 1 \quad j = 1,2\ldots m; \quad s = \text{solution} \tag{2}$$

$$\Omega^{s}_{i,j} = \begin{cases} 1 & \text{if square } (T_i, P_j) \text{ is visited in solution} \\ 0 & \text{otherwise} \end{cases} \tag{3}$$

The main aim of our proposed algorithm is to maximize the number of tasks assigned as well as to minimize the energy consumption of all assigned tasks by applying Hybrid Max–Min Ant Colony optimization algorithm (HACO-TS) in heterogeneous processor.

$f(s)$ is the objective function which measures the quality of the solution and is given as

$$f(s) = \mathrm{TA}(s) + \left(1 - \frac{EC(s)}{\max EC}\right) \tag{4}$$

where

$$EC(s) = \sum_{j=1}^{m} \left(\sum_{i=1}^{n} E_{i,j} \times \Omega^{s}_{i,j} \right) \tag{5}$$

$$\mathrm{Max}\, EC(s) = \sum_{i=1}^{n} c_i \times \max\left(s^{s}_{i,j}\right) \tag{6}$$

where TA(s) is equal to the number of assigned tasks in the solution s. $E_{i,j}$ is the energy utilized by task T_i on processor P_j; c_i is the execution cycle of task T_i; $s_{i,j}$ is the speed at which task T_i is executed on processor P_j.

3 Proposed Algorithm

In this paper, Max–Min Ant System with Tabu search is proposed to improve the quality of task assignment solution. The Max–Min Ant System is one of the variants of the ACO algorithm. Stützle & Hoos proposed the Max–Min Ant System [10]. The key feature of MMAS is that the pheromone trails are updated with only one ant; this ant could be the iteration-best ant or the global-best ant which found the best solution. Moreover, the maximum and minimum values of the pheromones are limited to certain values to escape getting stuck at local solutions. In order to avoid stagnation, the amount of pheromone is restricted to a range [$\tau_{\min}$, $\tau_{\max}$] [10]. The ants are situated at random places initially. At each construction step, ant k applies a probabilistic action choice rule to choose a next city to visit until a complete solution has been built. Finally, all of the solutions are evaluated and the pheromone updating rule was applied until all the ants have built a complete solution. After all ants have constructed a tour, pheromones are updated by applying evaporation as in Ant System as given in Eq. (7).

$$\tau(i,j) = (1 - \rho)\tau(i,j)\forall(i,j) \in N(s) \tag{7}$$

where $0 < \rho < 1$ is the pheromone evaporation rate and $\tau\,(i, j)$ is the pheromone trail. This is followed by the deposit of new pheromone as follows:

$$\tau(i,j) = \tau(i,j) + \Delta\tau(i,j)^{\text{best}} \tag{8}$$

where $\Delta\tau(i,j)^{\text{best}} = 1/f(s)$ best; $f(s)$ best as a solution cost of iteration-best sbest = s_{ib}.

The pheromone limits are calculated as

$$\tau_{\max} = f\left(s^{\text{best}}\right)/\rho \tag{9}$$

$$\tau_{\min} = \tau_{\max}/(\omega \times \ln(\theta + 1)) \tag{10}$$

where θ is the sequential number of the current iteration starting with 1, ω is a constant, and $\omega \geq 1$. Here ρ is the evaporation rate of pheromone trails and sbest denotes the iteration best solution.

3.1 Solution Construction

An artificial ant increases the pheromone value $\tau(i, j)$ at the edge between T_i and P_j which represents the possibility of assigning task T_i to processor P_j. The pheromone values of the ant are initialized the same as for solution construction. Each ant builds a tour from a starting pair of tasks and processors [1]. The probability of selecting the next pair of task and processor is given by Eq. (11).

In Eq. (11), $N(s)$ denotes the set of eligible pairs of (task, processor) obtained; $\tau(i, j)$ denotes the pheromone trial of (T_i, P_j).

$$p(s,i,j)(t) = \begin{cases} \frac{\tau(i,j)(t)}{\sum_{(i',j')\in N(s)}^{n} \tau(i',j')(t)} & \text{if} \quad (i,j) \in N(s) \\ 0 & \text{otherwise} \end{cases} \tag{11}$$

Step 1: Set k=1. Select S_1 and set $S_0=S_1$

Step 2 : Select S_c from $N(S_k)$.

Remove a task from one processor and assign it to the neighborhood processor;

$U_{NEW} = (\Sigma^m{}_{j=1}Uj)/m$

If $U_{NEW}<U_{AVG}$; New assignment is updated;

If $U_{NEW}>U_{AVG}$; Old assignment is retained;

(i) If the move $S_k \rightarrow S_c$ is on the tabu list set $S_{k+1}= S_k$ and go to 3.
(ii) If $S_k \rightarrow S_c$ is not on the tabu list set $S_{k+1}=S_c$.
Add the reverse move to the top of the tabu list and delete the entry at the bottom.
If $f(S_c)>f(S_0)$, set $S_0=S_c$
Step 3: Set k=k+1.
Stop if stopping criteria are satisfied;
otherwise go to 2.

Fig. 2 Algorithm for Tabu search

3.2 Tabu Search Algorithm

Tabu search has been widely applied to solve combinatorial optimization problems [11, 12]. Tabu search is a metaheuristic algorithm which directs a local heuristic search procedure for exploring the solution space in order to avoid being stuck in local optima. The important feature is the tabu-list maintained throughout the search [12].

The tabu energy search and tabu resource search algorithms are applied at different stages of the proposed algorithm. Both the algorithms follow the same procedures shown in Fig. 2. Tabu resource search will search for a solution that needs the smallest fraction of the overall computing capacity in heterogeneous multiprocessors. In tabu energy search, the quality of the solution s is inversely proportional to the energy consumed by all tasks in it. The proposed HACO-TS algorithm performs tabu energy search algorithm for every feasible solution and tabu resource search algorithm for every infeasible solution until no improvement can be made [12]. The pseudo code of the proposed HACO-TS algorithm is shown in Fig. 3.

4 Results and Discussion

To test the proposed algorithm HACO-TS, experiments are performed on an Intel core i3 CPU processor running at 2.27 GHZ with 1.87 GB RAM. The operating system is MS Windows 7, 64 bit running the MATLAB R2011b environment.

```
Input: Random problem instances
Output:Set of tasks mapped to processor j
Procedure: HACO-TS
{
  Set parameters, Initialize pheromone trail to τmax, I = 1,2... number of ants
  While (termination condition not met)
   {
     do
     {
     for each ant i;
     {
       Construct Solution Si under the condition U<=1;
      if (S is a partial solution)
        {
         move ant i one step further;
        }
    else if (S is a feasible solution)
     {
      Apply Tabu search_ Energy algorithm to improve the solution S;
      Calculate quality for each solution using equation (4)
        }
       else if (S is an infeasible solution)
        {
          Apply Tabu search_ Resource algorithm;
             if( s becomes a partial solution )
                  {
                   move ant i one step further;
                  }
                 else
                  {
                Calculate quality for each solution using equation (4)
                  }
       }// end else if

      find unassigned task and assign it randomly with EDF bound;
       }//end For
    Sib ={ Sj:f(Sj) = maxi=1 no of Ants (f(Sj) };
    Choose the Ant with the best fitness value of all Ants as the gbest
       if f (sib ) > f (sgb ) then f (sgb ) = f (sib );
       Update pheromone trails of only the gbest solution using equation (7)&(8)
          }// end for
       } // end do
   }//end- while
}//end – procedure
```

Fig. 3 Algorithm for proposed HACO-TS

4.1 Data Set Description

In the experiments, the utilization matrix is generated based on task heterogeneity, processor heterogeneity, and consistency for evaluating the performance of the proposed and existing algorithms. The utilization matrix is generated as in [1, 8]. The steps are given below:

1. A *nx1* **clock cycle matrix** *C* is generated, the number of cycles to execute task T_i is a random number between [100, 1000].
2. A *nx1* **task frequency matrix** T_B is generated, the task frequency of T_i is a random number between [*1*, Φ_T], here Φ_T is task heterogeneity. It may be high task heterogeneity (HT; [$\Phi_T = 100$]) or low task heterogeneity (LT; [$\Phi_T = 5$]).
3. A *1xm* **speed vector** is generated for each $T_B(i)$, the speed to execute task T_i on P_j that is $S_i(j)$ to a random number between [Φ_T, $\Phi_T \cdot \Phi_p$], here Φ_p is processor heterogeneity; it may be high processor heterogeneity (HP) or low processor heterogeneity (LP). For high processor heterogeneity, $\Phi_p = 20$ and for low processor heterogeneity, $\Phi_p = 5$.
4. An *nxm* **utilization matrix** $U_{i,j}$ is generated by $T_B(i)/S_i(j)$ and $U_{i,j} \in [1/(\Phi_T \cdot \Phi_p), 1]$
5. The utilization matrix is said to be consistent (*C*) if each speed vector value is sorted by descending with processor P_0 always being the fastest and processor $P_{(m-1)}$ the slowest. But the inconsistent matrix (IC) holds unsorted speed vector values that are random state as they are generated.

Figure 4 shows the comparison of the quality of the solution obtained by the proposed HACO-TS algorithm for different consistent and inconsistent problem instances. It is interesting that the proposed HACO-TS algorithm performs better

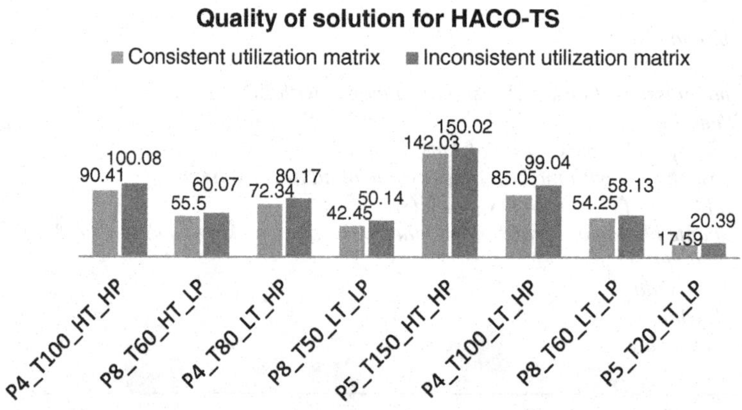

Fig. 4 Comparison of the quality solution for real-time consistent and inconsistent problem instances for proposed HACO-TS algorithm

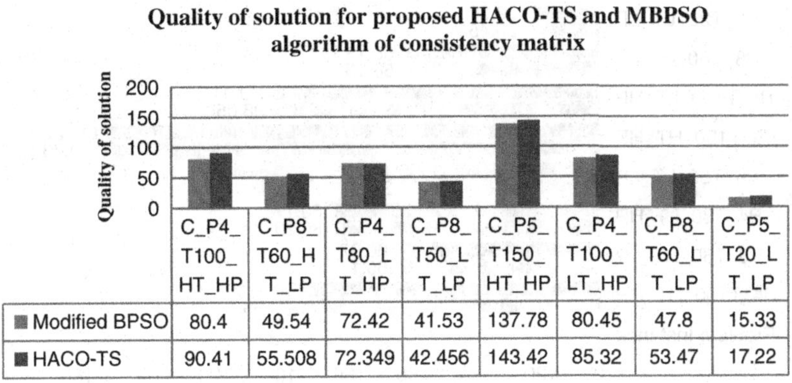

Fig. 5 Comparison of the quality of solution by the proposed HACO-TS and Modified BPSO algorithm for consistent problem instances

for inconsistent problem instances than consistent problem instances. From Fig. 4 it can be observed that the proposed HACO-TS has proved to be the best algorithm for the task assignment optimization problem in the inconsistent heterogeneous multiprocessor systems. Hence, the investigations has started against the existing algorithm Modified BPSO for consistent and inconsistent problem instances [13]. Figure 5 shows the comparison of the results obtained by the proposed HACO-TS and Modified BPSO [13] algorithms for consistent problem instances. The proposed HACO-TS algorithm is achieved more tasks than the Modified BPSO algorithm for consistent problem instances. As a result, HACO-TS algorithm gets more favorable achievements in all cases because Tabu search trics to achieve minimized utilization by removing the task from one processor and assigning to the other processor and thereby helping to add more tasks and also to minimize average normalized energy consumption.

Table 1 shows the quality of the solutions obtained by proposed HACO-TS and Modified BPSO [13] algorithms. It can be observed that the proposed algorithm performs better than existing Modified BPSO for inconsistent problem instances.

Table 1 Comparison of quality of solutions obtained by proposed HACO-TS and Modified BPSO [13] algorithms for inconsistent problem instances

Problem set	Size	Modified BPSO [13]	Proposed HACO-TS
IC_HT_HP	U4*100	98.1	100.08
IC_HT_LP	U8*60	60.1	60.07
IC_LT_HP	U4*80	80.29	80.17
IC_LT_LP	U8*50	50.23	50.14
IC_HT_HP	U5*150	149.03	150.02
IC_LT_HP	U4*100	98.05	99.04
IC_LT_LP	U8*60	58.15	58.13
IC_LT_LP	U5*20	20.45	20.39

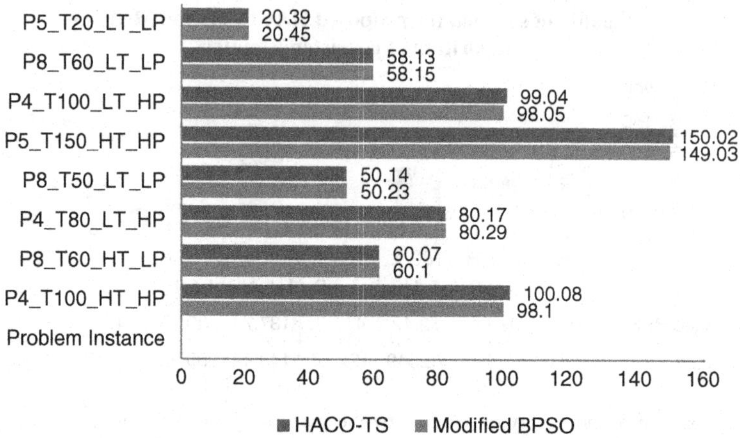

Fig. 6 Comparison of quality of solutions obtained by proposed HACO-TS and Modified BPSO algorithms for inconsistent problem instances

The reason is that the proposed algorithm includes Tabu search heuristic at different stages, which enriched MMAS performance by improving the quality of task assignment solution. The inclusion of Tabu search has its influence on the energy part of the solution [11, 12]. The comparison of quality of solutions obtained by HACO-TS and Modified BPSO algorithms for inconsistent problem instances is shown in Fig. 6.

Table 2 gives the comparison of average CPU time (s) consumed by proposed HACO-TS and existing Modified BPSO algorithm. HACO-TS needs less execution time compared to Modified BPSO or inconsistent problem instances, because the HACO-TS requires less number of iterations to obtain the best quality of solution.

Table 2 Comparison of average CPU time (s) by proposed HACO-TS and Modified BPSO algorithms for inconsistent problem instances

Problem set	Size	Proposed HACO-TS	Modified BPSO [13]
IC_HT_HP	U4*100	2.25	2.886
IC_HT_LP	U8*60	1.985	9.52
IC_LT_HP	U4*80	1.84	31.81
IC_LT_LP	U8*50	1.549	16.03
IC_HT_HP	U5*150	3.43	51.367
IC_LT_HP	U4*100	1.789	1.962
IC_LT_LP	U8*60	2.15	14.609
IC_LT_LP	U5*20	0.478	0.187
Average CPU time		1.933	16.04
		Proposed HACO-TS/ Modified BPSO	87.9 %

5 Conclusion

In this paper, a hybrid metaheuristic algorithm is proposed for the task assignment problem in the heterogeneous multiprocessor system. The proposed algorithm consists of the Max–Min Ant System which manages the task assignment problem constraints and is extended with Tabu search algorithm for improving the quality of the solutions found. The Tabu search has resulted in maximizing the number of tasks assigned as well as minimizing the energy consumption. The computational results obtained by the proposed algorithm are superior to the existing Modified BPSO algorithm. It has been found that the proposed HACO-TS performs better for inconsistent problem instances than consistent problem instances in terms of quality of solution. The test performed shows that HACO-TS algorithm is an option for handling the task assignment problem of inconsistent heterogeneous multiprocessor system.

References

1. Chen H, Cheng AMK, Kuo YW (2011) Assigning real-time tasks to heterogeneous processors by applying ant colony optimization. J Parallel Distrib Comput 71(1):132–142
2. Garey MR, Johnson DS (1979) Computers and intractability: a guide to the theory of NP-completeness. Freeman & Co, San Francisco
3. Srikanth UG, Maheswari VU, Shanthi P, Siromoney A (2012) Tasks scheduling using ant colony optimization. J Comput Sci 8(8):1314–1320
4. Prescilla K, Selvakumar AI (2013) Modified binary particle swarm optimization algorithm application to real-time task assignment in heterogeneous multiprocessor. Microprocess Microsyst 37(6):583–589
5. Abdelhalim MB (2008) Task assignment for heterogeneous multiprocessors using re-excited particle swarm optimization. In: Proceedings of the IEEE international conference on computer and electrical Engineering, pp 23–27
6. Braun TD, Siegel HJ, Beck N, Bölöni L, Maheswaran M, Reuther AI, Robertsong JP, Mitchell DT, Bin Y, Debra H, Freund RF (2001) A comparison of eleven static heuristics for mapping a class of independent tasks onto heterogeneous distributed computing systems. J Parallel Distrib Comput 61(56):810–837
7. Kang Q, He H (2013) Honeybee mating optimization algorithm for task assignment in heterogeneous computing systems. Intell Autom Soft Comput 19(1):69–84
8. Poongothai M, Rajeswari A, Kanishkan V (2014) A heuristic based real time task assignment algorithm for the heterogeneous multiprocessors. IEICE Electron Express 11(3):1–9
9. Krishna CM, Shin KG (2010) Real-time systems. Tata MacGraw-Hill Edition
10. Stützle T, Hoos H (1997) The MAX–MIN ant system and local search for the traveling salesman problem. In: Proceedings of the IEEE international conference on evolutionary computation, pp 309–314
11. Thamilselvan R, Balasubramanie P (2012) Integration of genetic algorithm with tabu search for job shop scheduling with unordered subsequence exchange crossover. J Comput Sci 8 (5):681–693
12. Ho SL, Yang S, Ni G, Machado JM (2006) A modified ant colony optimization algorithm modeled on tabu-search methods. IEEE Trans Magn 42(4):1195–1198
13. Prescilla K, Selvakumar AI (2013) Comparative study of task assignment on consistent and inconsistent heterogeneous multiprocessor system. In: Proceedings of the IEEE international conference on advanced computing and communication systems (ICACCS), pp 1–6

Study of Chunking Algorithm in Data Deduplication

A. Venish and K. Siva Sankar

Abstract Data deduplication is an emerging technology that introduces reduction of storage utilization and an efficient way of handling data replication in the backup environment. In cloud data storage, the deduplication technology plays a major role in the virtual machine framework, data sharing network, and structured and unstructured data handling by social media and, also, disaster recovery. In the deduplication technology, data are broken down into multiple pieces called "chunks" and every chunk is identified with a unique hash identifier. These identifiers are used to compare the chunks with previously stored chunks and verified for duplication. Since the chunking algorithm is the first step involved in getting efficient data deduplication ratio and throughput, it is very important in the deduplication scenario. In this paper, we discuss different chunking models and algorithms with a comparison of their performances.

Keywords Chunking algorithm · Data deduplication · Fixed and variable chunking method

1 Introduction

With the enormous growth of digital data in the cloud storage and enterprise organization, backup and disaster recovery have become crucial to the data center environment. Based on the latest forecast by Cisco, data center traffic on a global scale will grow at a 23 % CAGR, while cloud data center traffic will grow at a faster rate (32 % CAGR) or 3.9-fold growth from 2013 to 2018 [1]. To handle this

A. Venish (✉)
Computer Science and Engineering, Noorul Islam University, Kumarakoil, Tamil Nadu, India
e-mail: Venish07@gmail.com

K. Siva Sankar
Information Technology, Noorul Islam University, Kumarakoil, Tamil Nadu, India
e-mail: sivasankarniu@gmail.com

L.P. Suresh and B.K. Panigrahi (eds.), *Proceedings of the International Conference on Soft Computing Systems*, Advances in Intelligent Systems and Computing 398, DOI 10.1007/978-81-322-2674-1_2

massive data growth in storage environment and balance the resource consumption, such as RAM/CPU/disk I/O effectively, the data deduplication concept has been accepted worldwide.

There are four major steps involved in data deduplication. The chunking method splits the incoming large data into small portions or 'chunks'. Using hash algorithm, a unique identifier value is assigned to each chunk. The new incoming chunks are compared with the existing stored chunk using the unique identifier. If the identifier matches, the redundant chunk will be removed and will be given the reference point; otherwise, new chunk will be stored. The key objective of the chunking algorithm is to divide the data object into small fragments. The data object may be a file, a data stream, or some other form of data. There are different chunking algorithms for deduplication including file-level chunking, block-level chunking, content-based chunking, sliding window chunking, and TTTD chunking.

2 Chunking Methods

Efficient chunking is one of the key elements that decide the overall deduplication performance. There are a number of methodologies to detect duplicate chunk of data using fixed-level chunking [2] and fixed-level chunking using rolling checksums [3, 4].

As described by Won et al., chunking is one of the main challenges in the deduplication system [5, 6]. Policroniades et al. [7] and Kulkarni et al. [8] analyzed the various factors that affect the overall deduplication performance using file-level chunking, fixed-size chunking, and content-aware chunking. Also, various studies including Bimodal chunking [9] and Two Threshold Two Denominators (TTTD) [10] emerged to optimize chunking algorithms.

Figure 1 explains the different chunking module structures in detail. The chunking module contains file-level, fixed-size, variable-size, and content-aware chunking. Input files come to the deduplication system and then the files are moved to the chunking module where the file is divided into small chunks based on the chunking method used. After the chunks division, the unique hash value is assigned to each chunk. In this section, we discuss different chunking modules in detail.

2.1 *File-Level Chunking*

File-level chunking or whole file chunking considers an entire file as a chunk, rather than breaking files into multiple chunks. In this method, only one index is created for the complete file and the same is compared with the already stored whole file indexes. As it creates one index for the whole file, this approach stores less number of index values, which in turn saves space and helps store more index values compared to other approaches. It avoids maximum metadata lookup overhead and

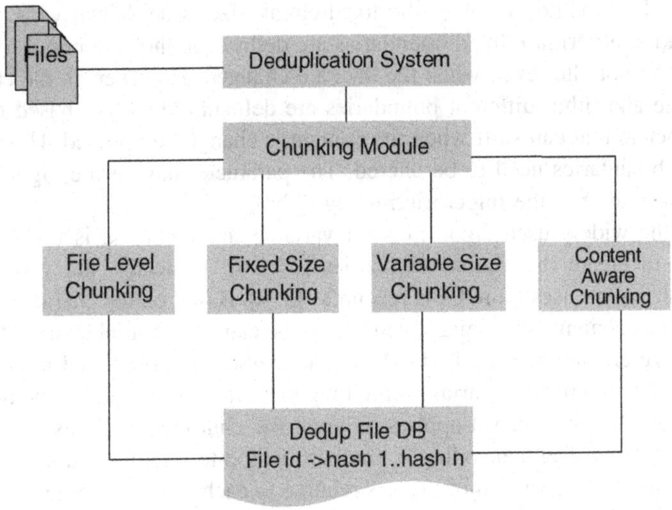

Fig. 1 Different chunking module structure

CPU usage. Also, it reduces the index lookup process as well as the I/O operation for each chunk. However, this method fails when a small portion of the file is changed. Instead of computing the index for the changed parts, it calculates the index for the entire file and moves it to the backup location. Hence, it affects the throughput of the deduplication system. Especially for backup systems and large files that change regularly, this approach is not suitable.

2.2 *Block Level Chunking*

Fixed-Size Chunking. Fixed-size chunking method splits files into equally sized chunks. The chunk boundaries are based on offsets like 4, 8, 16 kB, etc. [11, 12]. This method effectively solve issues of the file-level chunking method: If a huge file is altered in only a few bytes, only the changed chunks must be reindexed and moved to the backup location. However, this method creates more chunks for larger file which requires extra space to store the metadata and the time for lookup of metadata is more. As it splits the file into fixed size, byte shifting problem occurs for the altered file. If the bytes are inserted or deleted on the file, it changes all subsequent chunk position which results in duplicate index values. Hash collision is likely to happen on chunking method by creating same hash value for different chunks. This can be eliminated by using bit-by-bit comparison which is more accurate, but requires more time to compare the files.

Variable-Size Chunking. The files can be broken into multiple chunks of variable sizes by breaking them up based on the content rather than on the fixed size of the files. This method resolves the fixed chunk size issue. When working on a fixed chunking algorithm, fixed boundaries are defined on the data based on chunk size which do not alter even when the data are changed. However, in the case of a variable-size algorithm different boundaries are defined, which are based on multiple parameters that can shift when the content is changed or deleted. Hence, only less-chunk boundaries need to be altered. The parameter having the highest effect on the performance is the fingerprinting algorithm.

One of the widely used algorithms for variable-size chunking is Rabin's algorithm [13] to create the chunk boundaries [14–16]. Basically, the variable-size chunking algorithm uses more CPU resources [5, 6]. Based on the characteristics of the file such as content, size, image, color, etc., we can apply variable-size chunking and fixed-size chunking. An ADMAD scheme method proposed by Liu et al. [17] applies not only the fixed or variable chunking method, but is based on the metadata of individual files on which it applies different file chunking methods.

Delta Encoding. Delta encoding is another method to find the boundaries of the file, which prevents repetition of objects relative to each other retaining an existing version of the object with the same name [18]. This method eliminates an object completely in some of the cases, and so the necessary presence of the basic versions, for computing a delta, will be challenging.

Compared to variable-size and fixed-size chunking, Delta encoding gives a good deduplication ratio in desktop applications such as Word, Excel, Zip, Photos, though the problem in delta encoding creates more fingerprints than other methods which yield more space utilization. Bolosky et al. [19] found a file granularity method to detect duplicate data instead of using chunk granularity; on the other hand, Deep Store [20] is planned to work with not only one chunking method but can accommodate various chunking approaches.

Basic Sliding Window. The other approach is the basic sliding window (BSW) approach [21]: it applies break condition logic and marks the boundary of file. Chunk boundary is computed based on the fingerprint algorithm. File boundary is marked based on break condition. The problem with this approach is the chunk size. The size of the chunk cannot be predicted with this approach, but it is possible to predict the probability of getting a larger chunk or a smaller one. This probability is defined based on the probability of getting a particular fingerprint. A divisor D and the sliding window size define if the probability is bigger or smaller.

Two Threshold Two Divisor (TTTD). Another most frequently used variable-length chunking algorithm is TTTD [9, 10, 13, 22]. The Two Threshold Two Divisor (TTTD) chunking method [10] ascertains that chunks smaller than a particular size are not produced. However, it has a drawback that the chunks produced might escape duplicate detection as larger chunks are more likely to be related than smaller ones. To achieve this, the method ignores the chunk boundaries after a minimum size is reached. TTTD applies two techniques where two divisors (D, the regular divisor and D0, the backup divisor) are used. By applying these divisors, TTTD guarantees a minimum and maximum chunk size. A minimum and a maximum size limit is used for splitting a file into chunks for searching for duplicates.

TTTD-S algorithm. This is the method proposed by Moh et al. [10] which overcomes the drawback of the TTTD algorithm by introducing TTTD-S algorithm. It computes the average threshold as a new parameter of the main and the backup divisors and uses it as a benchmark for the chunking algorithm. When the average threshold is hit, both primary and secondary divisors shrink to half. The primary and secondary divisors get restored after chunking is completed. This method helps in avoiding a lot of unnecessary computation.

The algorithm by Kruus et al. [9] is based on two fundamental assumptions. The first one is during long runs, where there is a higher probability of getting a large set of unknown data; the second one is during long runs where breaking of large chunks into smaller chunks improves efficiency.

Content or Application Aware-Based Chunking. The chunking methods discussed previously do not consider file format of the file of the underlying data, such as file characteristics. If the chunking method understands the data stream of the file (format of the file), the deduplication method can provide the best redundancy detection ratio compared than the fixed and other variable-size chunking methods. The fixed and the variable-size chunking methods set the chunk boundaries based on the parameter or some predefined condition, whereas the file-type-aware chunking understands the structure of the original file data and defines the boundaries based on the file type such as audio, video, or document.

The advantage of the content-aware chunking method is to facilitate more space savings because this algorithm is aware of the file format and sets the boundaries more natural than other algorithm methods. If the file format is known, it also provides the option to change the chunk size depending on the section of the file. Some formats or file section will not change anytime, by this method it can assume how often the file or section is going to change in the future and the size of the chunk accordingly. For example, potential chunk boundaries include the number of pages in a word document or slides in a PowerPoint presentation.

Douglis et al. [23] used different approach to handle the content-aware deduplication. In this method the data is considered as an object. Incoming data is converted into the object and the same has been compared with the already stored objects for finding the duplicate data in effectively. Using of the Byte level comparison and the knowledge of the content of the data, the input file is split into large data segments. The splitted data segments are compared with the already stored segments and similar segments are determined. At end the altered bytes are saved.

As corroborated by Meister and Brinkmann [24], the understanding of the compression formats such as TAR or ZIP can have consequential space savings. They also exploited the characteristic of the file format and analyzed the deduplication ratio and the scaling efficiency under different chunking methods. They found another interesting thing that playing with Meta data and the payload we can yield more savings space. The knowledge of the archive structure and the matching of the chunk boundaries with the payload boundaries can result in the invalidation of a lot of small size chunks and can save a lot of space by avoiding duplicate data.

However, the mechanism might not be realistic as to grasp all possible file formats and then using them for breaking the files based on their content is a bit unrealistic for any computational method. Another important issue with file-type chunking is that the algorithm might not be able to detect duplicate data if it comes across in a different file format. This second problem can rather adversely impact the overall performance of this computational approach than improving it. While this approach holds the most promise as far as the space conservation goes, it can utilize more processing power and time if a detailed analysis of all the individual files is done [25, 26].

Figure 2, compares the different deduplicate/undeduplicate size with different chunk size algorithm. File-level and fixed-size chunking occupying more spaces compare than the Rabin chunking algorithm. Rabin chunking algorithm provides good results for all the chunk size such as 8, 16, 32, and 64 kB [27]. Figure 3 compares the different deduplicate/undeduplicate sizes with different deduplication file system (Domain Size) size. As the domain size increases, amount of space reclaimed by the deduplication algorithm raises linearly. Best possible result is achieved in 8 kB Rabin algorithm. While using whole file without chunking yields to the worst deduplication result [27].

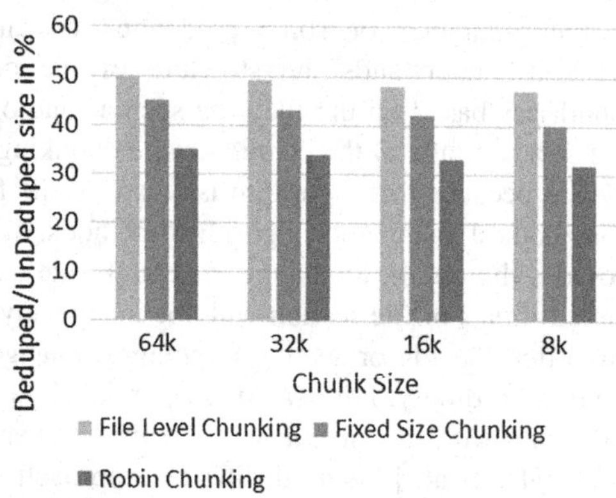

Fig. 2 Comparison of different deduplication versus various chunk size

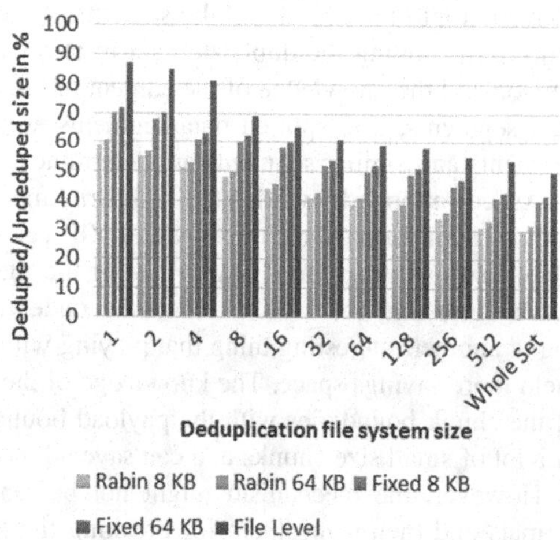

Fig. 3 Deduplication versus deduplication domain size

3 Conclusion

We have summed up different algorithms and methods for the deduplication of data. In the data deduplication process, the first and the most important step is the granular division and subdivision of data. For proper granularity, an effective and efficient chunking algorithm is a must. If the data is chunked accurately, it increases the throughput and the net deduplication performance. The file-level chunking method is efficient for small files deduplication, but not relevant for a big file environment or a backup environment. The problem with the fixed-size chunking method is that it fails to detect redundant data if some bytes are altered. The variable-size chunking methods overcome this fixed file size chunking problem by creating boundaries based on the content of the files. In the second section, we discussed various variable chunking methods. In the current deduplication scenario, the content-based chunking methods provide good throughput as well as decrease the space utilization. While working and managing the multimedia files like images, videos, or audio, the content-aware chunking methods are accepted mostly.

References

1. Cisco Global Cloud Index: Forecast and methodology (2015) white paper. http://www.cisco.com/c/en/us/solutions/collateral/service-provider/global-cloud-index-gci/Cloud_Index_White_Paper.html. Visited last on 02 Apr 2015
2. Quinlan S, Venti SD (2002) A new approach to archival storage. In: Proceedings of the first USENIX conference on file and storage technologies, Monterey, CA
3. Denehy TE, Hsu WW (2003) Reliable and efficient storage of reference data. Technical Report RJ10305, IBM Research, Oct 2003
4. Andrew Tridgell (1999) Efficient algorithms for sorting and synchronization. PhD thesis, Australian National University
5. Won Y, Kim R, Ban J, Hur J, Oh S, Lee J (2008) Prun: eliminating information redundancy for large scale data backup system. In: Proceedings IEEE international conference computational sciences and its applications (ICCSA'08)
6. Won Y, Ban J, Min J, Hur J, Oh S, Lee J (2008) Efficient index lookup for de-duplication backup system. In: Proceedings of IEEE international symposium modeling, analysis and simulation of computers and telecommunication systems (MASCOTS'08), pp 1–3, Sept 2008
7. Kulkarni P, Douglis F, LaVoie J, Tracey J (2004) Redundancy elimination within large collections of files. In: Proceedings of the USENIX annual technical conference, pp 59–72
8. Kruus E, Ungureanu C, Dubnicki C (2010) Bimodal content defined chunking for backup streams. In: Proceedings of the 8th USENIX conference on file and storage technologies. USENIX Association
9. Policroniades C, Pratt I (2004) Alternatives for detecting redundancy in storage systems data. In: Proceedings of the annual conference on USENIX annual technical conference. USENIX Association
10. Eshghi K, Tang HK (2005) A framework for analyzing and improving content-based chunking algorithms
11. Kubiatowicz J et al (2000) Oceanstore: an architecture for global store persistent storage. In: Proceedings of the 9th international conference on architectural support for programming languages and operating systems

12. Quinlan S, Dorwards S (2002) Venti: a new approach to archival storage. In: Proceedings of USENIX conference on file and storage technologies
13. Rabin M (1981) Fingerprinting by random polynomials. Center for Research in Computing Technology, Aiken Computation Laboratory, University
14. Lillibridge M, Eshghi K, Bhagwat D, Deolalikar V, Trezise G, Camble P (2009) Sparse indexing: large scale, inline deduplication using sampling and locality. In: Proceedings of the 7th USENIX conference on file and storage technologies (FAST'09), San Francisco, CA, USA, Feb 2009, pp 111–124
15. Muthitacharoen A, Chen B, Mazi`eres D (2001) A low-bandwidth network file system. SIGOPS Oper Syst Rev 35(5):174–187
16. Zhu B, Li K, Patterson H (2008) Avoiding the disk bottleneck in the data domain deduplication file system. In: FAST'08: Proceedings of the 6th USENIX conference on file and storage technologies, Berkeley, CA, USA, pp 1–14
17. Liu C, Lu Y, Shi C, Lu G, Du D, Wang D (2008) ADMAD: application-driven metadata aware de-duplication archival storage system. In: Proceedings o fifth IEEE international workshop storage network architecture and parallel I/Os (SNAPI'08), pp 29–35
18. Mogul J, Douglis F, Feldmann A, Krishnamurthy B (1997) Potential benefits of delta encoding and data compression for HTTP. In: Proceedings of ACM SIGCOMM'97 conference, pp 181–194, Sept 1997
19. Bolosky WJ, Corbin S, Goebel D, Douceur JR (2000) Single instance storage in windows 2000. In: Proceedings of fourth USENIX windows systems Symposium, pp 13–24
20. You LL, Pollack KT, Long DDE (2005) Deep store: an archival storage system architecture. In: Proceedings of international conference on data engineering (ICDE'05), pp 804–8015
21. Muthitacharoen A, Chen B, Mazieres D (2001) A low-bandwidth network file system. ACM SIGOPS Oper Syst Rev 35(5):174–187
22. Thein NL, Thwel TT (2012) An efficient Indexing Mechanism for data de-duplication. In: Proceedings of the 2009 international conference on the current trends in information technology (CTIT), pp 1–5
23. Bloom BH (1970) Space/time tradeoffs in hash coding with allowable errors. Commun ACM 13(7):422–426
24. Meister D, Brinkmann A (2009) Multi-level comparison of data deduplication in a backup scenario. In: Proceedings of SYSTOR'09: The Israeli experimental systems conference, May 2009, pp 1–12
25. Cannon D (2009) Data deduplication and tivoli storage manager, Mar 2009
26. Data Domain LLC. Deduplication FAQ. url:http://www.datadomain.com/resources/faq.html
27. Meyer DT, Bolosky WJ (2011) A study of practical deduplication. In: Proceedings of 9th USENIX conference on file and storage technologies

Secured Key Sharing in Cloud Storage Using Elliptic Curve Cryptography

M. Breezely George and S. Igni Sabasti Prabu

Abstract Cloud computing provides a standard architecture to share the application as same as the other network sharing. It is providing the online storing facility on several servers, which is organized by a trusted authority rather than the resource provider. Sharing the key and managing plays a vital role in sharing the process over cloud computing. Usual processing of the key cryptosystem lacks in efficient safety method; those keys are produced by the existing system. Existing system, proposes the thoughts about aggregate cryptosystem in that the key generated by several origins of the cipher text class, which properly ties the data and related keys. The aggregate key was produced once after losing the key after which it was not possible to access the data. Therefore, to solve the problem, we proposed a novel technique to utilize the key sharing with proper security such as Triple DES algorithm and Elliptic Curve Cryptography (ECC). The Triple DES algorithm is used for file encryption and decryption process. The ECC transmits the secret key. The secret key will be generated by the one-time password (OTP) method that will verify the real user by the unique password confirmation method after which the user will get the secret key on his/her e-mail ID.

Keywords Cloud storage · Elliptic curve cryptography · Triple data encryption standard · Cipher text · Encryption · Decryption

M. Breezely George (✉) · S. Igni Sabasti Prabu
Sathyabama University, Rajiv Gandhi Road, Jeppiaar Nagar, Sholinganallur, Chennai 600119, Tamil Nadu, India
e-mail: breezemtech@gmail.com

S. Igni Sabasti Prabu
e-mail: igni.prabu@gmail.com

L.P. Suresh and B.K. Panigrahi (eds.), *Proceedings of the International Conference on Soft Computing Systems*, Advances in Intelligent Systems and Computing 398, DOI 10.1007/978-81-322-2674-1_3

1 Introduction

Nowadays, cloud storage is famous in storing process. Third party is doing maintenance for storing the data in physical storage. The third party or authenticator is saving the digital data in physical storage or in the logical hub by using several servers. The data availability and accessibility in the physical environment and its protection is the third parties' responsibility. So, it is better to save the data in cloud storage, which can be accessible from anywhere and anytime, instead of saving in local storage disk or any hard drive. It decreases labor to carry the data. The user has right to use the stored data in the cloud storage from any computer by using the Internet, and it provides the facility to access information from any computer; not only from the same computer through which it was stored [1].

While considering about data security, it is tough to trust the traditional method for authentication, because of unpredicted dispensation appreciation that could explore all the data.

The solution is encrypting the data with user key. The sharing of data is the important functionality in cloud storage, as the user could share the data to anyone from anywhere at any time. For instance, an organization could allow for accessing the sensitive part of any data for their employees. But the difficult part is that how to share the encrypted data. By the traditional way it was easy to share data, by downloading and decrypting the data, but use of cloud storage did not allow it anymore. Cryptography techniques could be applied in two major ways: they are Symmetric key encryption and Asymmetric key encryption.

In symmetric key encryption process, for both decryption and encryption processes only one key was used [2]. By comparison, in asymmetric key encryption, different keys are used; private key and public key for decryption and encryption, respectively. In our approach, asymmetric key encryption is flexible. This can be discussed by the given example. Let us consider that Alice saved her entire data on Box.com and she does not want to share any data with anyone. Because of the possibilities of data leakage, she does not believe on the facility which is provided by Box.com, so she encrypts all the data before uploading. If she wants to share some data with Bob, she can share the data with Bob by utilizing shared function of Box.com. But the main issue is that how could she share the encrypted data. There are two techniques to share the encrypted data

1. Alice can encrypt data with one secret key and share the private key with bob, because by using this private key only bob will be able to decrypt the encrypted data.
2. Alice can encrypt the data with different key and send the related key to Bob via a channel after which Bob can access the data.

Both these technique are insufficient for users. In the second technique, the number of secret keys is as much as the shared files, which may not be possible to decrypt the data easily.

Hence, the suitable solution for this problem is that Alice has to encrypt all the data with different keys, but should send only decryption key to Bob, for using the data, at a constant size. While decryption, the key is sent through the protected channel and it is difficult to keep the small size key secure. For designing an enhanced public key, the encryption system is suitable for allocation, if any part of the cipher text is able to decrypt by constant size [3].

The rest of this research work is organized as follows: Section 2 states the related work of our research. Section 3 presents the detailed description of our proposed work. Section 4 presents the results and discussion part. Section 5 concludes with the future work of our proposed work.

2 Related Work

In the research work [4], author Benaloh et al. had proposed a scheme that is really presented for concisely transferring the huge number of secret keys in the scenario [5]. The creation is easy and we evaluate the key origin method for the actual description of the popular characteristics that the users want to gain. The key statement for the class set is as follows. A complex modulus is chosen where p, q are large prime numbers. A Master Secret Key (MSK) is selected at arbitrary. Each and every class is connected with different key. The entire primary keys could be kept in the system limit. A steady key for the set could be produced for the assigned access rights. Though, a steady key is designed for the symmetric key settings. The data providers need to acquire the secret keys for encrypting the data, it will not be suitable for the other applications, because this technique is used to produce a value related to a pair of keys, but it is not clear that how to apply this method for the public key encryption system. At last, we note that there are different schemes that try to decrease the size of the key for attaining verification in symmetric key encryption scheme [6]. Moreover, the distribution of decryption control is not an anxiety in these systems.

Identity-based encryption (IBE) [7–9], is a public key encryption where the user could make an identity-string. The private key generator (PKG) in IBE holds a private key and generates a secret key for every user with the proper identity. The authors Guo et al. [10, 11] had tried to make IBE with the secret key aggregation. In this method, the key aggregation controlled the sense that every key aggregated must have to validate from the distinct location. When there is an exponential number of identities and respective private keys, then aggregate a polynomial number only. These particularly increment the costs of storing and transferring the cipher text which are unfeasible with more conditions like sharing the data from cloud storage. This way is applied to Hash functions in the string for denoting the class, and continue with the function until obtaining the output of the hash function. Our methods state the characteristics constant for the cipher text size, and security in the model. In research work [12], the fuzzy IBE, a single secret key can decrypt cipher text encryption under many more identities which are near to the assured metric gap or space, but not for the identities arbitrary and then does not match with the key aggregation idea. Attribute-based encryption (ABE) [13, 14] allows every cipher text for associating the attribute, and the secret key holder could easily explore the information from the decrypted cipher text policy-based encryption. For instance, with the private key for the policy (1 ∨ 3 ∨ 6 ∨ 8), one can decrypt an encrypted text (cipher text), tagged with class 8, 6, 3, or 1. Though, the aim issue is in which ABE is collision-resistant. In fact, the key size is frequently increasing with the number of attributes that were included, or the cipher text size is not a fixed value [15].

3 Proposed Work

3.1 Overview

The preliminary functions of our proposed framework are shown in Fig. 1. Our proposed system implements a secured aggregation mechanism using Elliptic Curve Cryptography (ECC) and Triple Data Encryption Standard (DES) Algorithm. The system model in this research work involves three parties: the cloud service provider (CSP), data owner (DO), and new user. Initially, the original user or DO stores his/her own data file in cloud storage. Before storing the data, the user must encrypt the data file using Triple DES. This algorithm improves the privacy of stored data and it is used for secured data sharing. After that, a new user sends an access request to the DO for accessing the data. Then, the DO gives access permission for particular files and provides an aggregate key for that. The secret key is generated by the authenticator which is sent to the users through one-time password (OTP) technique, which keep the secret key secure from malicious users. The OTP process is preventing the leakage of the secret key. The cloud provides secure data sharing in cloud storage by using Key aggregation. Our system utilizes ECC for key aggregation, i.e., secured key sharing. Finally, our proposed system efficiently aggregates the keys and files.

3.2 Proposed Architecture Overview

See Fig. 1.

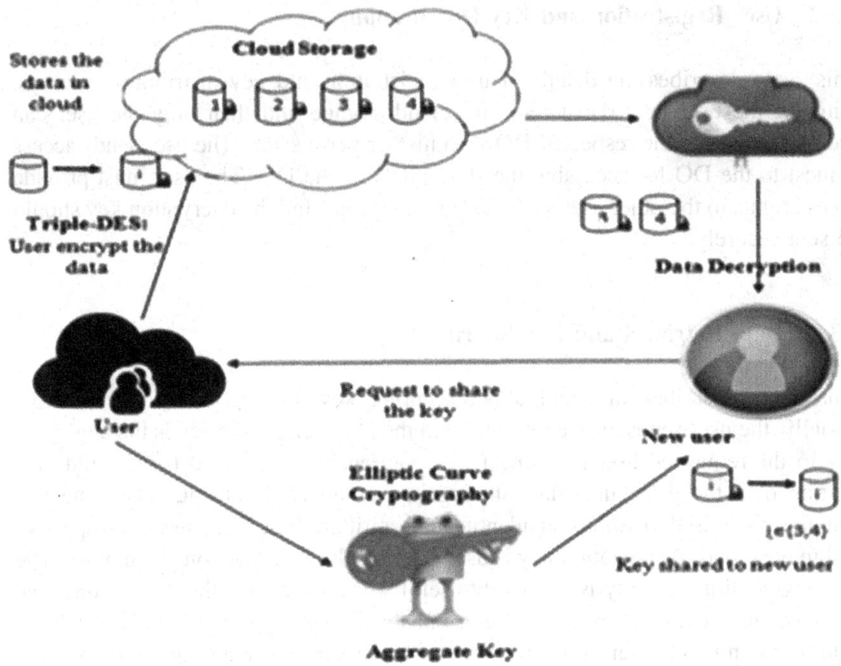

Fig. 1 System architecture

3.3 *Working Methodology*

Our proposed framework consists of three major stages such as

- Cloud Deployment and Data Encryption
- User Registration and Key Distribution
- Data Retrieval and Key sharing

3.3.1 Cloud Deployment and Data Encryption

This stage describes the details about cloud deployment. The cloud contains CSP, DOs, and users. Initially, the DO gets connected with the CSP. The DO directly handles the data. The DO provides the secret key to the requested user. Initially, the DO makes user registration by entering the required details and login to the cloud. DO is an entity who owns his data or message and desires to store them into the exterior data storage node for reliable delivery to users or for easy sharing in the tremendous cloud environments. The DO encrypts the data using Triple DES algorithm. This algorithm is used for secured data sharing.

3.3.2 User Registration and Key Distribution

This stage describes the details of user registration and key distribution process. Initially, the user gets registered with cloud storage and then only the user can access the data of the respected DO with his/her permission. The user sends access request to the DO for accessing the stored data of the DO. The user must provide access rights to the other user as the data is encrypted and the decryption key should be sent securely.

3.3.3 Data Retrieval and Key Sharing

This stage describes in detail about securing key sharing technique or ECC. Initially, the user sends an access request to the DO then he/she sends the aggregate key to the requested user by using ECC. Aggregate key is used for secured data sharing over the distributed data sharing in a cloud environment. Aggregate key contains various derivations of identity and attribute-based classes for respective DO in the cloud. Aggregation key is used to share the data from one to another. The key aggregation property is especially useful when we expect the delegation to be efficient and flexible. Key aggregation enables content provider to share other's data in a confidential and selective way with a fixed and small cipher text expansion, by distributing to each authorized user a single and small aggregate key. By using this aggregate key, the requested user could be able to download and view the requested data not only the remaining data. This algorithm improves the key sharing strategy. The key sharing is done by OTP method, which restricts the other users with malicious behavior from stealing the secret key. The key is being transferred by secure method where the owner sends a password to the related user for confirmation of sharing the key to access the database. Without matching the confirmation password, the authenticator does not generate any key for the specific request. Encryption Standard Algorithm is used to secure data encryption and decryption process.

3.4 Algorithms Used

3.4.1 Triple DES

In our proposed work, we are using Triple DES algorithm for both the data encryption and decryption process. General steps are given below.

Triple DES is used as a "Key Collection" that consist three DES keys such as Key1, Key_2, and Key_3. Each and every key size is 56 bits long without the parity bits.

The triple DES encryption algorithm process is below.

Cipher text = E_{Key3} (D_{Key2} (E_{Key1} (plain text)))

The DES method encrypts the normal text with Key1, then decrypt with Key2, and then again DES method encrypts the text with Key3

The Triple DES decryption is the reverse of encryption process which is shown in below

Plain text = D_{Key1} (E_{Key2} (D_{Key3} (cipher text)))

The DES method decrypts cipher text with Key3 initially, then it encrypts with Key2 and again it decrypts with Key1

Each time 64 bit of data is encrypted by triple encryption method.

The center process overwrites process of first and last in every test case (Fig. 2).

3.4.2 Elliptic Curve Cryptography

ECC works under the principle of elliptic curves. In our work, the equation of an elliptic curve over a field M is considered and it known as,

$$j2 = i3 + ri + s \tag{1}$$

where

i, j co ordinates

r, s elements of *M.*

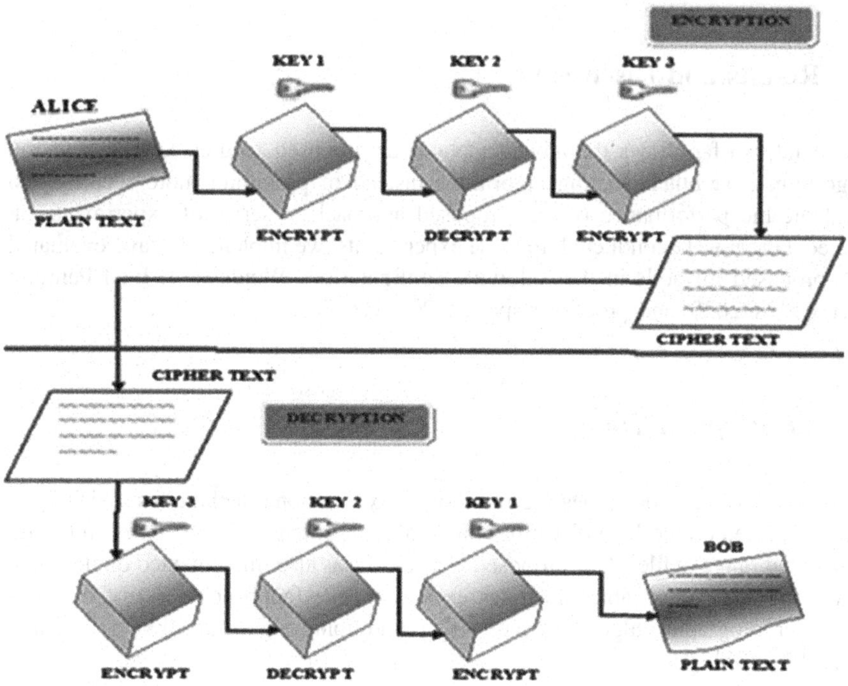

Fig. 2 Symmetric key Triple DES encryption

There is one important process in our proposed aggregation algorithm, i.e., key generation.

A. Key Generation:

The key generation process is a significant factor where the sender generates public keys and secret keys. The sender will encrypt the data by using a public key of the receiver. Moreover, by using the sender's private key, the receiver will decrypt the data.

Now, the sender randomly picks a number '*v*' within the range of '*u*'.

By using the following equation, the user can generate the public key

$$N = v * p \tag{2}$$

where

d is the random number and selects this number within the range (1 to n − 1)
P is the point on the elliptic curve
d is the secret key and
N is the public key.

The DO utilizes this generated aggregate key as an OTP for data decryption process of the receiver. Our proposed system utilizes private key pair for key generation process.

4 Results and Discussion

The study so far shows the advantages of our proposed secured key aggregation algorithm over other existing algorithms instead of providing more security. To evaluate the performance of our proposed approach, a series of experiments on stored dataset were conducted. In these experiments, we implemented and evaluated the proposed methods in the following configurations Windows 7, Intel Pentium (R), CPU G2020, and processor speed 2.90 GHz.

4.1 Encryption Time

This research work describes the traditional cryptographic techniques which have been executed on the Java platform of a number of different content sizes and types of a broad range of files. Our proposed algorithm was also implemented on the same platform. The below table illustrates execution time (approximately) of our proposed cryptographic algorithm with other traditional algorithms like DJSA and NJJSAA (Table 1).

Table 1 Comparison of file encryption between existing and proposed algorithm

S. No	File size (kB)	DJSA	NJJSAA	Proposed algorithm
		Execution time in second (approximately)		
1.	189	18–20	16–19	8–10 (expected)
2.	566	37–40	34–38	25–30 (expected)
3.	1136.64	70–73	65–67	50–55 (expected)
4.	1495.04	97–99	70–72	60–65 (expected)

4.2 Results

4.2.1 Accuracy of Proposed System

Figure 4 illustrates the key aggregation accuracy of our proposed algorithm and other traditional algorithms. Furthermore, ECC has a better accuracy than other common encryption algorithms which are shown in Fig. 3.

4.2.2 Efficiency of Proposed System

Figure 4 illustrates the key aggregation time period for both proposed and existing systems. Preceding more competence will also appear in terms of the total CPU process and clock time for all the three considered algorithms. Our proposed key aggregation algorithm utilizes minimum time duration for key aggregation process while comparing other traditional algorithms like IBE with compact key and Attribute-Based Encryption Algorithm.

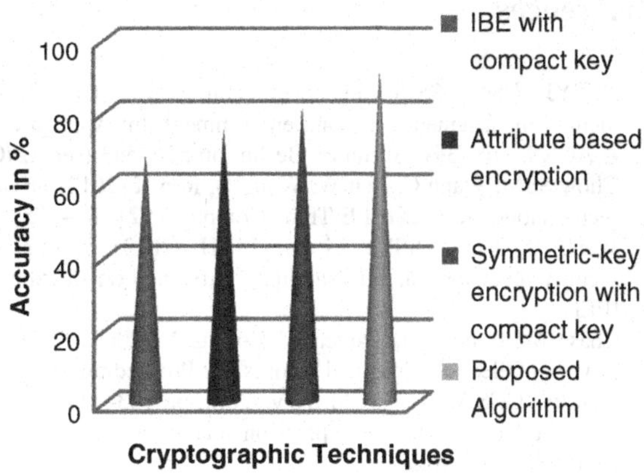

Fig. 3 Accuracy versus number of cryptographic techniques

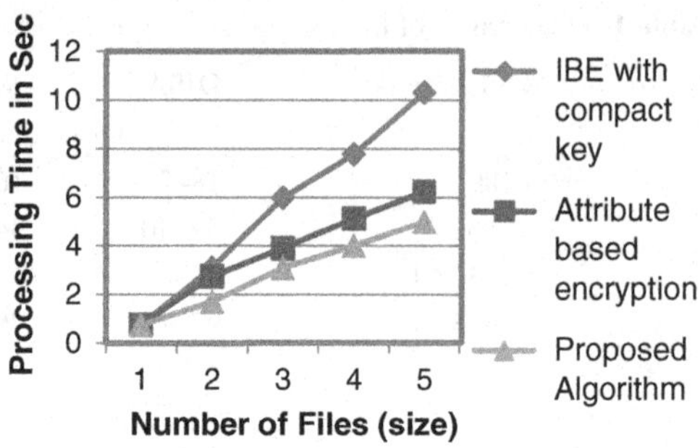

Fig. 4 Processing time versus number of files

5 Conclusions

This research work is proposing a key aggregation mechanism for securing the data by using Triple DES algorithm and ECC. Basically, the encrypted data are stored in the cloud storage by the users. Then, we used ECC algorithm for the safe access and key aggregation. The smooth data sharing is playing a vital role in cloud computing. Users upload their data on cloud storage; data outsourcing is increasing the susceptibility of data leakage. So, encryption is the perfect way to share data securely with the desired users. For secure sharing of the secret key, owners using OTP technique that provides the safest way for sharing the secret or aggregate key. The secure sharing of decryption secret key is acting as an imperative character. The cryptosystems of public key is providing the delegation of secret public key for distinct cipher text in cloud storage. So, finally, our proposed work is capable to gain the files and key values. This makes a way for secure key sharing and ensures files from being misused.

References

1. He YJ, Chow SSM, Yiu S-M, Hui LCK (2012) SPICE—Simple privacy-preserving identity-management for cloud environment. In: Applied cryptography and network security —ACNS, vol 7341. Springer, Berlin, pp 526–543 (ser. LNCS)
2. Chow SSM, Wang C, Lou W, Wang Q, Ren K (2013) Privacy-preserving public auditing for secure cloud storage. IEEE Trans Comput 62(2):362–375
3. Wang B, Chow SSM, Li M, Li H (2013) Storing shared data on the cloud via security-mediator. In: International conference on distributed computing systems—ICDCS. IEEE
4. Chase M, Benaloh J, Lauter K, Horvitz E (2009) Patient controlled encryption: ensuring privacy of electronic medical records. In: Proceedings of ACM workshop on cloud computing security (CCSW '09). ACM, New York, pp 103–114
5. Benaloh J (2009) Key compression and its application to digital fingerprinting. Microsoft Research, Tech. Rep
6. Poovendran R, Alomair B (2009) Information theoretically secure encryption with almost free authentication J UCS 15(15):2937–2956

7. Franklin MK, Boneh D (2001) Identity-based encryption from the weil pairing. In: Proceedings of advances in cryptology—CRYPTO '01, vol 2139. Springer, Berlin, pp 213–229 (ser. LNCS)
8. Waters B, Sahai A (2005) Fuzzy identity-based encryption. In: Proceedings of advances in cryptology, EUROCRYPT '05, vol 3494. Springer, Berlin, pp 457–473 (ser. LNCS)
9. Dodis Y, Chow SSM, Waters B, Rouselakis Y (2010) Practical leakage-resilient identity-based encryption from simple assumptions. In: ACM conference on computer and communications security, pp 152–161
10. Mu Y, Chen Z, Guo F (2007) Identity-based encryption: how to decrypt multiple ciphertexts using a single decryption key. In: Proceedings of pairing-based cryptography (Pairing '07), vol 4575. Springer, Berlin, pp 392–406 (ser. LNCS)
11. Mu Y, Guo F, Xu L, Chen Z (2007) Multi-identity single-key decryption without random oracles. In: Proceedings of information security and cryptology (Inscrypt '07), vol. 4990. Springer, Berlin, pp 384–398 (ser. LNCS)
12. Dodis Y, Chow SSM, Waters B, Rouselakis Y (2010) Practical leakage-resilient identity-based encryption from simple assumptions. In: ACM conference on computer and communications security, pp 152–161
13. Goyal V, Sahai A, Pandey O, Waters B (2006) Attribute-based encryption for fine-grained access control of encrypted data. In: Proceedings of the 13th ACM conference on computer and communications security (CCS '06). ACM, New York, pp 89–98
14. Chow SSM, Chase M (2009) Improving privacy and security in multi-authority attribute-based encryption. In: ACM conference on computer and communications security, pp 121–130
15. Takashima K, Okamoto T (2011) Achieving short cipher texts or short secret-keys for adaptively secure general inner-product encryption. In: Cryptology and network security (CANS '11), pp 138–159

Designing a Customized Testing Tool for Windows Phones Utilizing Background Agents

J. Albert Mayan, R. Julian Menezes and M. Breezely George

Abstract The communication gadgets that are being assembled together these days are well endowed with sophisticated ball grid mother boards and Qualcomm as well as Snapdragon processors that have, what it takes to redeem the required Intel as well as the instruments that form a part and parcel in the life of a common individual. The unique gadgets have their own traits, and aptness which are distinct in their identities. Before emancipating the application to a user, who has an intention of using the application, it is a necessity to fortify that the codes which are declared are triggered and the same is running and the property of the application is willful to provide the required reinforcement to the person, he/she who is wishing to use the application. In order to attain the declaration of the above statements, there exist some global functional testing tools that have received recognition like Anteater, Apodora, Canoo WebTest, Eclipse Jubula, etc., which are used to check the proper functioning of the application. There are several thousands of tools to achieve the end product of testing the app functionality, but the tools that are being declared have the caliber to get fired in an environment which is fit to be on a personal computer. The project explores and speaks of an app that inhibits a caliber to get installed like a (Tom, Jones, and Harry) application over a communication device. When using this application, the stress gets reduced in terms of monitoring that if certain applications are functioning in accordance to the predetermined functionality. The application that is taken under pilot study, has certain traits that gives its distinctiveness, are inspected and apprehended by the Tool like a spider's web on a wireless fidelity network. The Tool which is portable is extremely facile to operate

J. Albert Mayan (✉) · R. Julian Menezes · M. Breezely George
Department of Computer Science and Engineering, Sathyabama University, Rajiv Gandhi Road, Jeppiaar Nagar, Sholinganallur, Chennai 600119, Tamil Nadu, India
e-mail: albertmayan@gmail.com

R. Julian Menezes
e-mail: mtechit2k13@gmail.com

M. Breezely George
e-mail: breezemtech@gmail.com

L.P. Suresh and B.K. Panigrahi (eds.), *Proceedings of the International Conference on Soft Computing Systems*, Advances in Intelligent Systems and Computing 398, DOI 10.1007/978-81-322-2674-1_4

so much that a novice user can use it. The app can be uprooted from one communication device to the other and can be planted on to a phone like a cordial application on a phone, for examining the functionality of other set of codes that build up an application.

Keywords Functionality testing tool · Testing 8 phones · Windows OS phones · Phone windows functional testing · Testing windows operating system phones

1 Introduction

When the functionality and the security testing of a specific application are taken under consideration, the technology that is involved in the development of the application makes a benchmark for the application. When irrefutable ingredients of a particular code are monitored for its conventional functioning, it can be defined as the testing performed in the absence of an external entity aka (Automation). The handheld devices require some unit of charge somewhere between 3 and 7 W(-source: Lawrence Berkeley Lab), which means even if the phone takes 2 h for charging it will just consume 0.006–0.014 units or kWh of electricity to charge; this statement arrives at a conclusion that if a handheld device is charged every day, an end user will get to spend only 2–5 units in the whole year! These devices are hence segregated under low-powered devices, and if any application is developed for such low-powered devices, they are termed as mobile applications or mobile apps [11]. These applications when intended to get triggered on various handheld devices like PDAs, Tablets, Pads, Netbooks, Notebooks, and other electronic devices, a mnemonic terminology is used, it is nothing but a simple mobile app [15]. A particular type of application is more desired if its sole intention is to create an interactive session with the end user; otherwise, to present an application that requires a lot of work as that of a program, as when it is compared to a web URL [13]. A normal blue print of a mobile might vary in several entities; some of them include the condition of the device, the proper functioning when compared to the needs of a user, the entry and the functioning in various environments, and the establishment of the connection to a sophisticated server in so many unique ways [9].

When taking into discussion, the blue print of mobile device inhibits the following: devices, type of connectivity (Wired & Wireless), clients and servers, and the inherent networking infrastructure. The mobile applications can be classified into two forms, where the former runs on the device, and the latter getting executed on the server, and the mobile device is used as a portal to view the application [12, 14]. The execution of the mobile app is influenced by various entities like, deciding the location of the triggering of the application. The local client or the remote server: the functionality of the hardware of a mobile is based on the quality of the structure [2], the ability of the device, and the way the mobile app functions [17].

In our day-to-day life, mobile phones are manufactured with several entities kept in mind like, sophisticated processors ranging from Snapdragon, Qualcomm, etc.; the potentiality of the software, storage space, means of connection like Bluetooth, Xender, etc.; multimedia enhancements like Dual audio speakers, adjusted for both mp4 and mp3 formats, and the same device is utilized for traveling from one destination to the other [18]. The resolution of the camera varies from one model to the other, to capture special moments, and when in terms of security, the front view camera identifies the face to unlock the Mobile, with the concept of facial recognition [3]. End users are attracted to several mobile applications based on their needs. There are several volumes of apps in the store to tailor the needs of an end user. These apps are developed by the professionals in the industry. The applications vary from entertainment to official needs of an end user. The applications are scattered from several portals and through licensed providers, by which the apps are sealed and verified [16]. The Applications getting coded by the developers might work in one model and it might not work on a phone of the same brand. The genuine reason is based on the abilities of a phone to execute the application, which is directly dependent on the hardware components.

The application developers focus their primary attention on the functionalities of the application being developed and not on the hardware configuration of the mobile phone (The Processor, RAM, etc.). Moreover the applications get tested [10] over the inbuilt emulators rather than the actual hardware [8]. The applications tested over the emulators might or might not show the errors; if the latter happens, when the applications get deployed over the phone, a fatal error might happen leading to the burning of microchips residing in the hardware of the phone. The need for the process of testing [1] makes sure the developer to come out with apps that would lead the developer to imprint their foot in the industry. There exist several online tutorials that guide the novice to develop simple but powerful applications [17].

The applications getting developed by the software professionals are based on the environment of the desktop, and definitely not based on the mobile environment. The desktop-based apps have the ability in testing a Windows operating system-based mobile via a personal computer. The weakest link in the chain is present in the entanglement of the applications. The thought flow of the developer is quite different from that of an end user. Therefore the ability to understand the working of a Mobile App is not as easy as it is. To overcome the mentioned scenario, the research project presents the designing of a testing tool that gets initiated over the phone without the need of any special required Hardware, apart from the phone. Utilizing the Phone Application Development software from Microsoft and the Inbuilt Emulator, i.e., Visual Studio 2014 and Windows WVGA 512 MB Emulator, the end user can initialize a secondary application using a primary application, and test the secondary application using the primary application, while both the applications being present on the same hand held device.

2 Related Work

There is an extension which exists in reality from a project which is called as the Apache Jakarta. The extension is a framework which illustrates the collective behavior of the ants. The terminology used to describe the extension is called the Anteater. This framework gives out a facile way to list out the ways, in which an application that is based out on the web or a service based on the Extensible Markup Language functions on the web.

Some of the ways to test certain applications via the Anteater are described below.

In the initial step, a particular request based on http/https is sent to the server based only on the web. Like the functioning of SONAR, when the client gets the reply from the server, it ought to meet certain requirements. The end user can also check for attributes like codes in the reply packet from the server, Hyper Text Transfer Protocol Headers, and added to it, the user may endorse the body part of the reply by the use of some entities like Relaxng, xpath, or xp with rege, provided along with formats in zeros and ones. Inclusion of tests is much easier.

The requests dispersed from a particular URL are in the form of Hyper Text Transfer Protocol and the utilization of the mentioned tool is utilized to listen to the request, when the URL is present on the local host. The user has the ability to check the attributes and, what is contained, as and when a specific request is triggered from a unique Uniform Resource Locator and he/she can send the required response, based on his/her wish. There is one particular entity which is exceptionally good with regard to Anteater is that it waits for the arrival of the Hyper Text Transfer Protocol messages. This unique quality makes the tool noteworthy in terms of constructing, tests for the applications that which uses communication based out of Simple Object Access Protocol. Some of the apps that use SOAP for communications are BizTalk and EBXML.

The tool which is defined as Apodora is open source and it is defined as a framework for performing testing based on web. This tool was introduced in the year 2007, which was conceived into form by Development Services for Software, STARWEST, Aculis by Southern Seth. This tool is based on open source and availed from the Internet under Public License. The organization which formed this tool introduces an end user to the community which is based on open source that illustrates the utilization of the mentioned tool, and pin points the varieties in terms of functionality, concerned with Apodora and other available tools in the industry. A lot of valuable resource like time is not getting wasted that is depicted clearly by Southern Seth. A unique feature that belongs to Apodora defined by Seth, shows the way how, changes would affect the software which is being tested and what sort of scripts need to be maintained.

When new and advanced gadgets and operating systems get released every year, the needs and wants of the users get changed accordingly. The science and technology has grown by leaps and bounds, i.e., by utilizing wifi and 3G a user has the ability to get the required app up and running in his/her gadgets at any point of time.

The users can do it at their own time frame. The application developers, when they code the blueprint for an application, make sure that when it gets released to the user, they are made available through the common download point called "The Store." The users might have any sort of device; it might be a Nokia, Samsung, HTC, or any sort of a device based on any operating system. The coders who are working in these organizations have their own repository for downloading the apps any time. The servers who are holding on to the apps are made available to be online 24/7 to maintain operational stability. Encapsulation is put into action by the app coders, in terms of the app behavior with respect to the concerned pedestal. Software development methods are always dependant on some criteria, in the same way when the abovementioned methods are dependent on the stable growth of an app, it is indistinctively called as Development based on Agile [4, 6]. If in the event there occurs any issues, then the resolution is achieved by a team of cross-functioning experts. To grasp things in a nut shell, the mentioned lines can be understood as a way to bring about a conclavity, in the middle of various cycle of development [9]. Eminent minds came forward to bring about a model rooted on Agile development. There might exist a lot of these people, but some noteworthy laureates have made their mark in history, they are none other than Mr. Sutherland who is a doctorate and Mr. Swaber, and the important aspect with regard to these gentlemen are the organizations that are utilizing the ideas which have been introduced by these eminent personalities [5, 7].

As mentioned in the previous lines along with Mr. Sutherland, one more intellectual, called Beck, came out with an idea called as the Extreme Programming [5]. This concept plays a vital role in Agile process [2], which in turn is used for molding the codes that have unique and defining qualities as, not involving a lot of risks, small extension of a whole new entity, extremely comforting to use, and above all it is purely in terms of science and scientific [3]. Here, the end user gets an opportunity to function with the coders that are developing the required application and are tested on a repetition criteria.

There exist several forms of crystals, in the universe that show their uniqueness in terms of their shapes and their ability to reflect the light into several colors. Likewise, there exists several modus operandi for putting the codes into motion, terminology stating it as "Crystal." There exist several variations, in the crystal in terms like the transparency, the color, by which we can say the qualities are obtained via the number of members in the specific team, and the importance given to the project.

3 Proposed Work

3.1 Architecture

See Fig. 1.

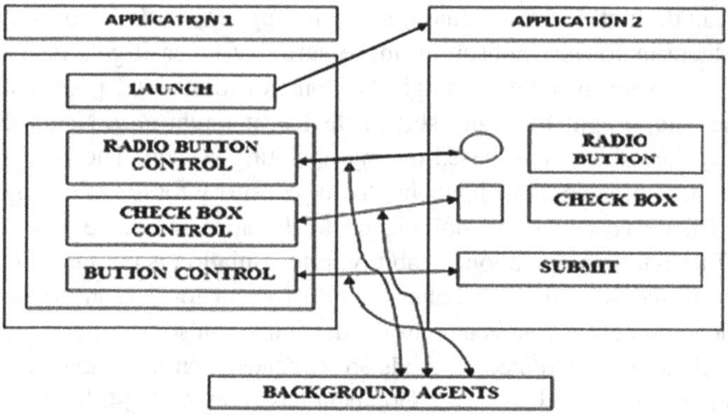

Fig. 1 Architecture of the system being introduced

3.2 Proposed Architecture Overview

The architecture which is discussed, speaks of broken codes that when it gets assembled together in a proper mannerism gets evolved into a Tool meant for testing specific apps that gets triggered on the gadget with an operating system called Windows 8. According to Bill Gates, the definition for a personal computer is as follows: "I think it's fair to say that personal computers have become the most empowering tool we've ever created. They're tools of communication, they're tools of creativity, and they can be shaped by their user." Likewise the certified professionals who are assembling the codes, to give shape to the custom tool, utilize some toolsets which are made available from websites like MSDN as well as Microsoft. The Microsoft Developer Network is a repository, to get all the softwares to satisfy the needs, right from the end user to the certified professionals. The downloaded programs from the website which is being utilized to give an image to the testing instrument, includes The Visual Studio 2013, Express 2012 for Windows Phones, and Inclusion of Emulators within SDK 8.0. When combining the acquired codes, it gives rise to a deadly combination of syllables that spits out a Tool that gets triggered on a gadget which in turn tests a custom app for the working of the functional attributes. To let the codings emerge from the charted blue print, onto a working application it is vital to have a bird's eye view of the working procedure of the instrument getting triggered onto a gadget. From the abovementioned statements, unique applications are created which are two in number. One application forms a pedestal for the other application to get fired upon. The application that forms the pedestal is considered as the background, whilst the application that gets triggered gets considered as foreground application. The underlying technology which is used to achieve the above is termed as Resource Identifier or simply as RI. There exists a lot of ways of testing, say for an instance Black Box Testing, White Box Testing, and likewise the tool, given a shape, comes under functional testing. From the blueprint, involving the conceiving of the tool to the time frame involved, in snap shooting the interface used by the user, there involves a behind the scenes technique termed as BLAs, Back Line Agents or Agents functioning in the Background. In the world of internet, the same is termed as Background Agents (Fig. 1).

3.3 Execution

This module discusses about how the tool is designed, utilizing the tools made to be availed in Microsoft, and functions the way it is supposed to be functioned. The same can be achieved in terms of utilizing certain agents who have the ability to get executed in the background. In any industry, if an electronic gadget gets released, the drivers get released along with it, and not as separate entities. Likewise, for the purpose of identifying the gadget which forms the environment for executing the codes, requires a driver, and it is available from the internet as Zune software application. Before making any major changes on to the personal computer, the registry is backed up and an entry is created for system restore. In the same mannerism, the data residing inside the gadget is backed up for safety precaution. When a gadget gets released on to the industry, it is in lock mode. The famous devices like Apple's IPOD, Nano, and IPHONES are always locked when they get released. The positive approach is that the device is always under warranty and if some issue occurs, the manufacturer can replace with a new device, and the device with the issue goes under refurbishment. Only the apps listed under the stores are allowed to get installed on to the gadget. In android, the same principle is applied, but there exists a loophole to overcome this shortcoming. The concept of rooting is extremely efficient when it comes to installing custom application. In Windows platform, the phone gets unlocked by registering via the Visual Studio Developer Registration. The stage is set for the custom applications to get fired up properly and efficiently.

3.3.1 Emulation Toolset

When a new engine gets manufactured, it does not get installed on a vehicle directly, before that the entire design is scanned digitally, by Mathematical tools like Matlab or Weka and the full working ability is digitalized and the way each and every part is getting worked up is visualized digitally. This concept is also seen under the development of the apps using Windows phone 8. In the real world scenario, the tools like Matlab, etc., are utilized. In the world of testing, tools like Visual Studio 2013 plays a vital role in digitalizing the real working environment of a gadget on to a simulation. When the late 1970s and 1980s had a revolution in gaming, the latest hardware were released and there were 8-bit and 16-bit cartridges that gets inserted on to the machines and the data gets read on to the screen. Likewise, the Windows phone apps development software gets released with an inclusion of program to emulate the working environment of a gadget. The terminology which is used to indicate that process is defined as emulator. The Tool MATLAB is utilized to pin point any failure that may arise in the design part of an engine, and the same is visualized on a screen. The engineers supervise these short comings and work toward rectifying the flaws which had risen, through the visual. In the same way, the certified professionals have a trump card to revisit the errors arising in the emulator when designing an app, and have an opportunity to clear the miswritten codes, and then get the codes to form a good working application.

The forthcoming statements depict the importance of the emulation kit.

- The common notion among the certified professionals is to get their hands on a gadget, unlock it, feel the screen, features and so on, but the simulation toolkit helps to overcome the shortcomings in the previous statement.
- A variety of activities based on testing can be done via the tool used for simulation.
- By the use of a simple click and select, the required emulator can be selected for the use of an end user.

3.4 *The Working Phase*

3.4.1 Triggering of the Pedestal Application

The codes which are available in MSDN as well as in the Microsoft websites are collected together and it is formed into a meaningful set of programming statements and they undergo a step called as compilation and the tool Visual Studio 2012 Express for Windows phones, detects the phone and executes the code inside the gadget to form an Icon on the screen with the name given to the project under Visual Studio 2012 Express.

3.4.2 Initiating the Secondary Application

The steps followed in triggering the pedestal application is followed here, but only the project name alone differs.

3.4.3 Hoisting the Secondary Application via the Pedestal Application

The terminology which is named as URI is utilized in this scenario, for hoisting the secondary application via the pedestal application. The user taps on the pedestal app icon to bring it up on the screen. The screen possesses a button, which when tapped brings up another screen that displays both the applications. The user needs to choose the apps from the given options.

3.4.4 Seizing the Interface Components Utilizing the Agents Functioning in the Background

For the working scenario, a secondary application with example components like the check box, radio buttons, list boxes, and sample buttons are established. The same components are seized by the pedestal application and the testing called as functional is performed.

3.5 Demo Codes Available Under MSDN and Microsoft

```
{Component Initialize (); }
Private sync button_click_0(obj send, RootEveargu)
{System.Window.Launch.Riusync(SystemNew.Ri("phoneapp9:")); }
```

4 Output and Idea Exchange

In order to check the caliber of the tool being proposed, a fair number of trial and error experiments were performed on the apps that were present under the phone that has operating system from windows 8. For testing a particular entity, let it be from any walk of industry, it requires a steady ground work or a foundation. In the same way, the defined configurations are required for checking the caliber of the instrument being introduced. The requirements are tabulated (Table 1), down for a better understanding.

Table 1 Requirements of hardware and software

S. No	Requirements
1.	Windows 8 mobile phone
2.	Processor: Intel(R) Core(TM) i7-3770 K CPU @ 3.90 GHz
	RAM: 8.00 GB hard disk drive: 500 GB
	Operating system: windows 8.1 × 64 (professional)
3.	Microsoft visual studio 2013
	Software development Kit 8
4.	Emulator WVGA 512 MB

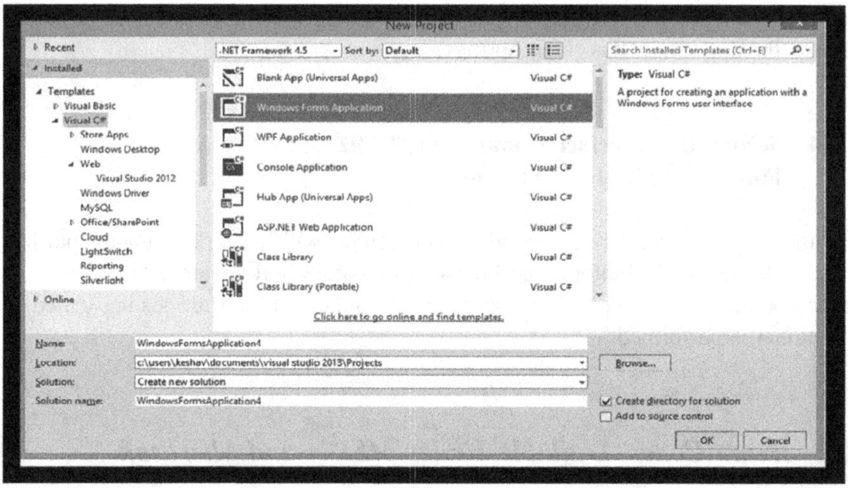

Fig. 2 Compilation of the pedestal application

4.1 Execution Steps

With reference to the requirements table, when the mentioned hardware and software are made available, the tool gets compiled and it gets installed on to the gadget to study the execution of other apps (Fig. 2).

The very first process discusses about the compilation step for the codes acquired from MSDN website as well as from Microsoft. This is achieved by the utilization of the universal serial bus cable attached to the PC and the Gadget, and Zune that runs in the background. The Interface used in this stage is the Visual Studio 2012 Express Edition for phones (Fig. 3).

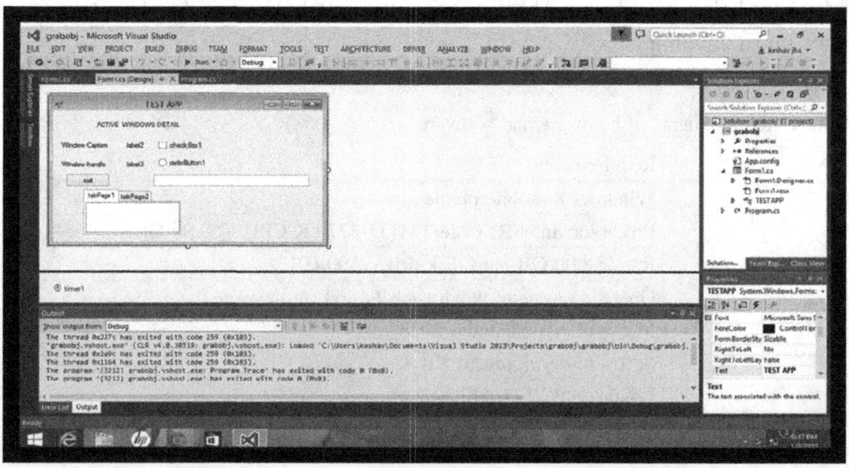

Fig. 3 Pedestal application installation

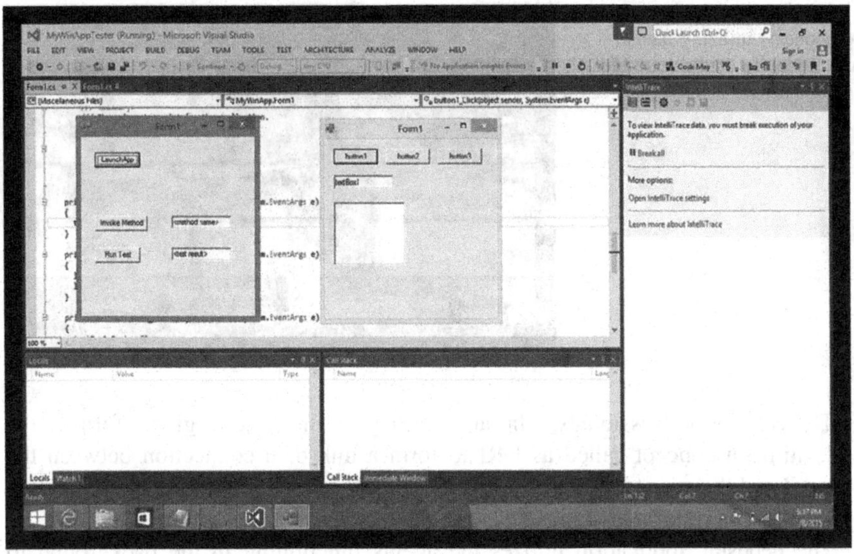

Fig. 4 Secondary application installation

Post compilation, the pedestal application gets seated on to the gadget, which possess Windows 8 OS (Fig. 4).

The compilation and execution steps that were followed for the pedestal application gets followed here for the installation of secondary application (Fig. 5).

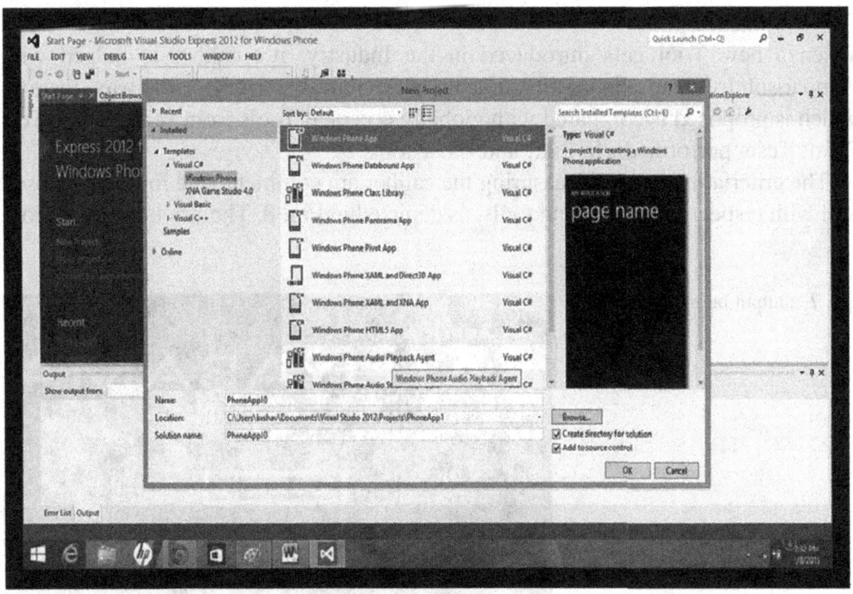

Fig. 5 Execution of the secondary application via the pedestal application

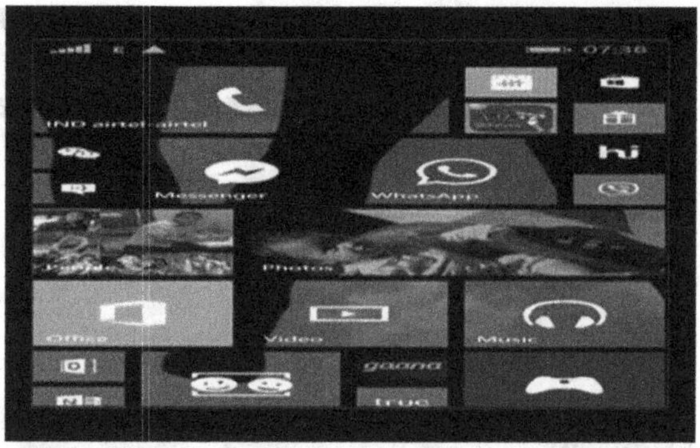

Fig. 6 Seizing interface components utilizing agents functioning in the background

The certified professionals, who are working on the code to give a form to the tool, utilize a concept called as URI to form a link or a connection between the pedestal and the secondary app. The same concept is used to boot up the secondary app via the pedestal app (Fig. 6).

The pedestal application utilizes the agents functioning in the background to seize the Interface object from the secondary app to test the working caliber (Fig. 7).

Post seizure of the Interface Component by the agents working in the background, the IC is used to verify it's own caliber via the pedestal application.

4.2 Caliber Verification

When a new Tool gets introduced in the Industry, it's caliber is judged with comparison to other existing instruments in the industry. In the same way, the tool which is proposed is compared with globally accepted Tools from vendors like Test Plant, Tests performed via Silk, and other tools.

The criteria in terms of measuring the caliber are commendable for the proposed tool with respect to the existing tools, as depicted in Fig. 8. The Tool triggered from

Fig. 7 Output being expected

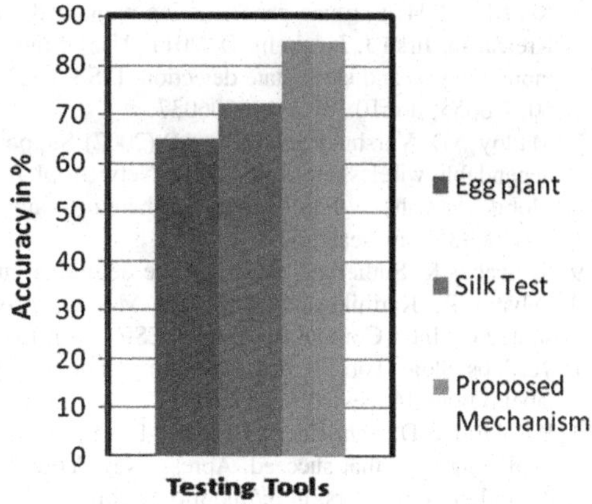

Fig. 8 Test tool accuracy

the gadget is user-friendly and fits like a hand and glove to take care of the need of an organization. The existing tools require an external machine to get installed and then test the required apps. The tool getting introduced requires a desktop only at the time of compilation and then the app gets utilized like a casual store app.

5 Conclusion

The research brings up a low-end Tool utilized for testing, specific apps. As mentioned in the previous statements, the tool requires the use of a machine with processor, RAM, HDD only at the time of compilation. The form for the Tool takes hoisting from apps like Visual Studio Express 2012 and Software Development Kit for Windows Phone 8. The stores are brimming with apps that have the ability to transfer certain files from one gadget to the next gadget at lightning speed. Utilizing one of the apps, the testing app can be transmitted from one gadget to the next gadget and it can be used. Testing a specific work of an app is the underlying term that was kept in mind while developing the mentioned testing app.

References

1. Albert MJ, Ravi T (2015) Structural software testing: hybrid algorithm for optimal test sequence selection during regression testing. Int J Eng Technol (IJET)7(1)
2. Aroul canessane R, Srinivasan S (2013) A framework for analyzing the system quality. IEEE Trans Int Conf Circ Power Comput Technol 1111–1115
3. Eom HE, Lee SW (2013) Human-centered software development methodology in mobile computing environment: agent supported agile approach. Eom Lee EURASIP J Wirel Commun Netw 111. http://jwcn.eurasipjournals.com/content/2013/1/111
4. IEEE (2000) IEEE standard for software test documentation. IEEE Std 829. 2000

5. Kent B (2004) Extreme programming explained, 2nd edn. Addison-Wesley, Boston
6. Krejcar O, Jirka J, Janckulik D (2011) Use of mobile phones as intelligent sensors for sound input analysis and sleep state detection. ISSN 1424-8220; www.mdpi.com/journal/sensors/11, 6037-6055; doi:10.3390/s110606037
7. Malloy AD, Varshney U, Snow AP (2002) Supporting mobile commerce applications using dependable wireless networks. Mob Netw Appl 7(3):225–234
8. Mobile stats. http://www.slideshare.net/vaibhavkubadia75/mobile-web-vs-mobile-apps-27540693?from_search=1
9. Schwaber K, Sutherland J (2011) The definitive guide to scrum: the rules of the game
10. Selvam R, Karthikeyani V (2011) Mobile software testing-automated test case design strategies. Int J Comput Sci Eng (IJCSE) 3(4). ISSN: 0975-3397 (Selvam R et al)
11. Testlabs blog. Top 10 tips for testing iPhone applications. http://blog.testlabs.com/search/label/iPhone. Accessed Sept 2010
12. Wooldridge D, Schneider M (2010) The business of iPhone app development: making and marketing apps that succeed. Apress, New York
13. A guide to emulators. http://mobiforge.com
14. Agile modelling. [Online] Available at: http://en.wikipedia.org/wiki/Agile_Modeling
15. Crystal clear (software development). [Online] Available at: http://en.wikipedia.org/wiki/Crystal_Clear_(software_development)
16. Extreme programming. [Online] Available at: http://www.extremeprogramming.org
17. Feature driven development. [Online] Available at: http://en.wikipedia.org/wiki/Feature_Driven_Development
18. http://en.wikipedia.org/wiki/Agile_software_development

Real-Time Remote Monitoring of Human Vital Signs Using Internet of Things (IoT) and GSM Connectivity

R. Prakash, Siva V. Girish and A. Balaji Ganesh

Abstract The next generation of technology is connecting to internet from a standalone device. The background of Internet of Things (IoT), which covers various areas like ubiquitous positioning, biometrics, energy harvesting technologies, and machine vision, makes researches to make deep study in these individual technologies. The paper presents human vital sign monitoring system which includes measurement of pulse rate and body temperature of patients from remote locations. Standalone device connectivity with Simple Mail Transfer Protocol (SMTP) server can be achieved by implementing a TCP/IP stack and Ethernet interface in a MSP430 controller. An IoT enabled network processor CC3100 and Computational processor MSP430 is interfaced with sensory components and made to communicate emergency signals through Global System for Mobile (GSM) and SMTP server. The sensors are pre-calibrated and the wireless system showed good response in sending messages and email. The physiological signals can be analyzed and data logged to identify history of ailments of particular user/patient. The system possesses the characteristics, such as less intrusiveness to the users and helps the caretakers as well as physicians with comfort in taking care of patients.

Keywords IoT · CC3100 · Simple mail transfer protocol (SMTP) · GSM

R. Prakash (✉) · S.V. Girish · A. Balaji Ganesh
Electronic System Design Laboratory, TIFAC-CORE, Velammal Engineering College, Chennai 600066, India
e-mail: prakash.rama121@gmail.com

S.V. Girish
e-mail: sivagirish1@gmail.com

A. Balaji Ganesh
e-mail: abganesh@live.in

L.P. Suresh and B.K. Panigrahi (eds.), *Proceedings of the International Conference on Soft Computing Systems*, Advances in Intelligent Systems and Computing 398, DOI 10.1007/978-81-322-2674-1_5

1 Introduction

For several decades, vital sign measurement is considered very important as it involves continuous monitoring of patient health and finding out abnormal conditions. In any environment, vital sign monitoring focuses on many parameters like blood pressure, pulse, temperature, etc. Most of the system includes continuous collection, multiple vital sign evaluation, cellular communication in emergency case, and data transfer over internet in normal cases [1]. Also, some several wireless sensor nodes which are capable of sending data to the coordinator node and forward to base station [2].

In general, a hospital environment needs every patient's body parameter to be monitored continuously either in wireless or through manual measured product. Continuous monitoring of vital parameter of a patient is very essential in medical field which gives more attention for caretakers. This leads to on-body propagation issues from multiple sensors [3]. In collaboration with different specialists, ipath was designed to facilitate with data objects. Base class takes care of storing and retrieving data from the database and is handling access control [4]. The current research in area of vital monitoring is Emergency Health Support System (EHSS). The system is based on the platform of Cloud Computing. It not only improves performance, but also cuts down networking resource requirements to a large extent [5].

Wireless sensor network finds much application with health care monitoring. On focusing in vital sign monitoring of patients, it places sensor nodes in a person's body to form a wireless network which can communicate with the patient's health status to the cluster node. Among human vital signals, pulse assessment of many persons has long been a research area of interest in the physiology field, because the pulse reflects state of health [6]; monitoring systems that can measure various bio-signals and provide QRS detection and arrhythmia classification, real-time ECG classification algorithm [7], and heart rate variability measurement [8]. The paper focuses on calibrating temperature and pulse sensor for parameters like temperature and BPM. LCD interface for monitoring real-time data. Global System for Mobile (GSM) modem for primary emergency notification and CC3100 simple link acts a secondary for emergency notification.

2 Experimental Setup

A mixed signal microcontroller, MSP430 with 16-bit RISC architecture, enhanced UART supports, baud rate detection, synchronous SPI, 12-bit analog-to-digital (A/D) converter, has been chosen for data processing and computing. For data communication, two protocols have been identified for sending emergency signals. GSM service acts as a primary emergency communication protocol. Email act as secondary emergency notification. The paper uses CC3100 Wi-Fi-board from Texas instruments in order to connect with TCP/IP stack which fulfills Internet of Things (IoT). The temperature sensor MA100 thermistor from GE, its sensitivity ranges from 0 to 50 °C, size is 0.762 × 9.52 mm, and is created for biomedical applications [9]. Pulse sensor SEN-11574 from Spark fun-Electronics plays here as a vital sign parameter module. The system consists of a 16 × 2 basic LCD module for display with 8 bit data mode programming. The selection of the processor in paper suits with MSP430 as it has ultra low-power consumption on standby mode with 2.1 μA at 3.0 V and shutdown mode with 0.18 18 μA at 3.0 V. Vital parameters can be calibrated with the processing unit in 12 bit ADC mode.

The GSM modem used in this paper is SIM900 which communicates with Quad-Band 850/900/1800/1900 MHz. SIM900 is designed with a very powerful single-chip processor integrating AMR926EJ-S core [10]. Basic set AT commands works with the GSM modem foe emergency messages. The programming of the microcontroller is based on an IDE, code composer studio v6 has been purchased from Texas Instrument and has been tested with various functions.

CC3100 simple link Wi-Fi consists of Wi-Fi Network Processor and Power-Management. Network processor consists of the Internet-On-a-Chip with dedicated ARM microcontroller. The paper uses this CC3100 simple link Wi-Fi board in order to communicate the emergency signal through mail transfer protocol. Email logging through this system can be attained. This emergency email acts as a secondary protocol which gives the patient current BPM in case of emergency. Smart Security Solutions uses this CC3100 simple link Wi-Fi for data logging. The prototype described the provision of accepting inputs from a smart-card reader (RFID reader) or a biometric sensor. Logs are wirelessly transmitted to the using of a Wi-Fi module [11].

3 System Architecture

The proposed implementation on vital sign monitoring system using Wi-Fi focuses on interfacing a network processor CC3100 with a selected computational processor MSP430. The transformation of emergency body parameters with two communication technologies, GSM and mail transfer protocol. The proposed work involves the following module.

- Calibrating Primary sensor for vital sign monitoring with MSP430.
- Configuring CC3100 Wi-Fi network processor with MSP430.
- Implementing TCP/IP stack programming via SPI interface.
- 16 × 2 LCD interface for monitoring real-time vital parameter of patient's body.
- UART communication protocol of MSP430 device uses a SIM900A GSM modem which sends emergency parameter as a message to concern persons.

MA100 senses the body temperature and pulse sensor senses the systolic BPM. MSP430 calibrates both the sensor reading and detection of emergency signal which is identified with the processor. The overall block diagram is shown in Fig. 1.

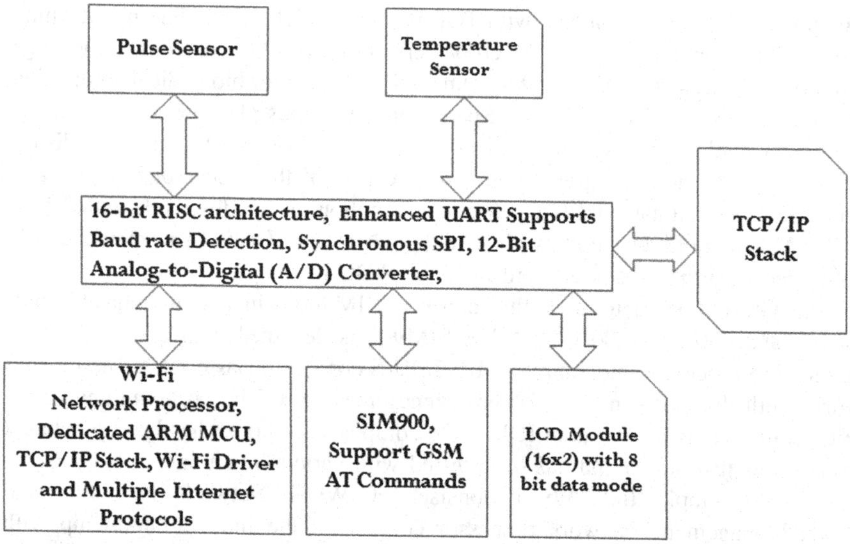

Fig. 1 The overall block diagram of the system is shown in Fig. 1. The MSP430 computes the sensor reading for detection of emergency signals. Real-time data can be monitored in LCD and implementation of TCP/IP stack with Ethernet interface is done in Wi-Fi network processor

As an initial stage of the upcoming technology IoT, we focus on Simple Mail Transfer Protocol (SMTP) from an IoT-enabled device. This makes use of transmitting email messages from our embedded device to a SMTP server. In Internet protocol stack, the application layer protocol, such as SMTP provide us a specialized service such as sending status report, periodic logs, and event alert to monitoring center through any SMTP server. A simple Application Programming Interface (API) has been developed for sending event alert to the caretaker. It supports transmission to multiple addresses so that it can be easy to integrate and configure with the computational processor (Fig. 2).

The analysis of the circuit is translated in Eq. (1), where 3162 represents overall digital value of the sensor at 0 °C [12]. $V_{\text{out}0^\circ\text{C}}$ (mV) corresponds to voltage measurement of MA100 terminal at 0 °C. 30.04 is the constant value that varies for every 1 °C.

$$T = \frac{3162 - V_{\text{out}0^\circ\text{C}}(\text{mV})}{39(\text{mV}/0^\circ\text{C})}{}^{o}C \tag{1}$$

The characteristics of Pulse sensor at various locations have been identified. Normal and emergency parameter signal is differentiated with pulse sensor visualizer. The operating voltage of pulse sensor varies from 3 to 5 V. Depending upon the input voltage and location of BP measurement in the body, the pulse signal characteristics changes in both amplitude and frequency (Figs. 3 and 4).

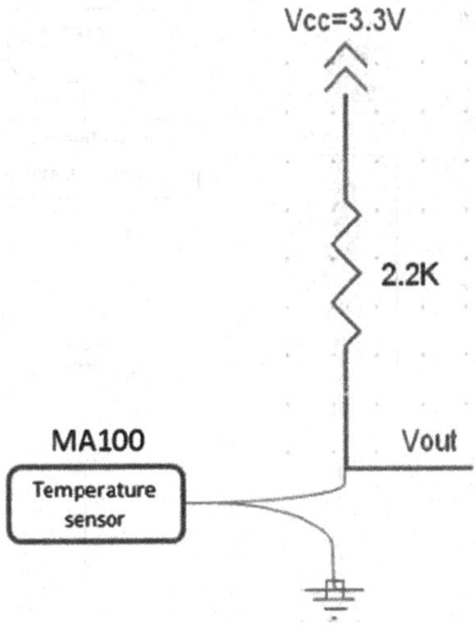

Fig. 2 The temperature sensor (MA100) reference circuit in Fig. 2 shows that the analog input (V_{out}) is connected with GPIO port of the processor (MSP430)

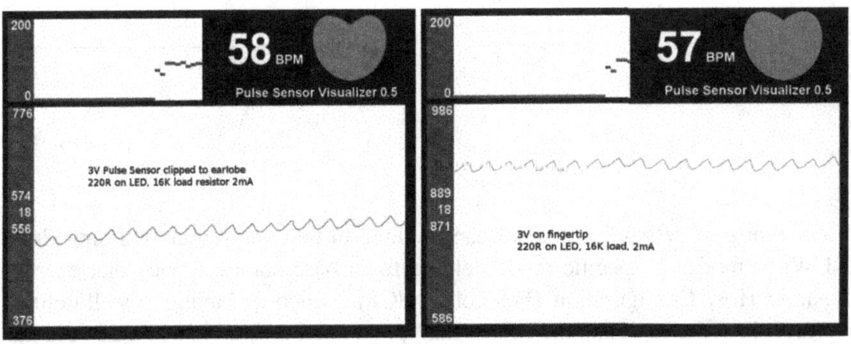

Fig. 3 The figure shows the characteristics of the pulse sensor signal, were it has been powered with 3 V and kept in different locations like earlobe and fingertip

- *Start*: when simple link receives the connect command, it looks for mode configuration parameters. The Wi-Fi mode initiates its own parameters like station selection, set power policy, transmission power, etc. SMTP connection is established with the server.
- *Email initialization*: Detection of emergency messages has been identified and it has been feed into the subject line of the mail.
- *Connection with SMTP server*: A secure socket has been created in order to connect with predefined SMTP server machine.
- *End*: This command stops the mail transfer protocol in stack flow.

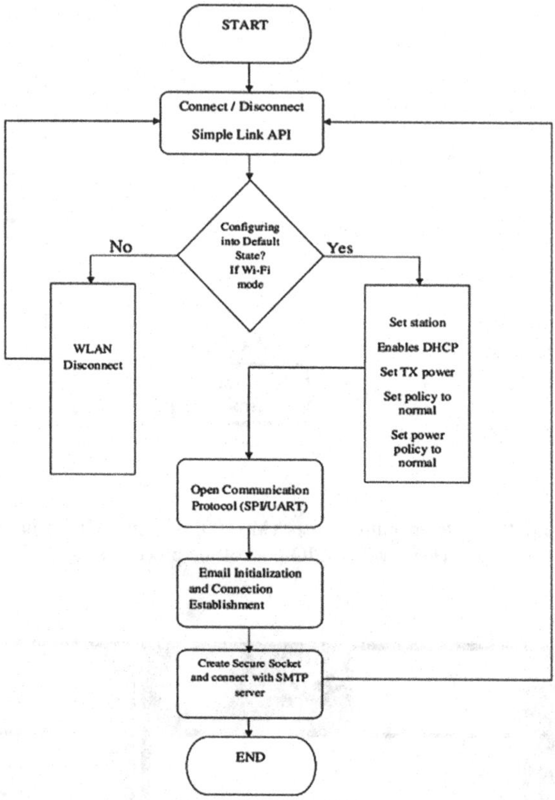

Fig. 4 Flow chart diagram of connection establishment/termination with SMTP server

Accessing of network processor can be done in two ways: Access Point Mode and Wi-Fi mode. A specific mode defined for a base station device that enables Dynamic Host Configuration Protocol (DHCP), which is having a well-defined interface. In order to define the source mail parameter, username and password has been predefined with SMTP server settings. Connection establishment with AP can be fulfilled with required parameters like Service Set Identification (SSID) name, pass key of the access point, continued with length of the pass key.

```
SMTP setting serversetEmail();
retVal = establishConnectionWithAP();
getsensorreading();
retVal = sendEmail(load the obtained sensor reading);
```

The major connectivity of UART communication depends on Transmit and Receive which enables to transmit and receive the data. SIM900 has the capability of accepting the standard attenuation command (AT Commands) from the user. The

IoT-enabled device has the GSM connectivity through one of its communication interface. GSM connection with developed module checks for Subscriber Identification Module (SIM), which is registered in the network. In case of SIM lock, it requires AT + CPIN attenuation command from the processor. Emergency messages are equipped with the following at commands.

GSM Interface
AT;
AT+CMGF=1;
AT+CMGS="MOBILE NUMBER";
CTRL+Z;

Data transfer in Liquid Crystal Display (LCD) can be done in two ways of operation, 4-bit mode and 8-bit mode, respectively. One of the challenges in these types of healthcare devices has to face limited number of output pins on a microprocessor. Obviously, 4-bit mode suits us since it has less data lines. Function set of LCD varies with display data RAM and character generator RAM. Data has been shifted according to the enabled set of the data lines.

4 Results and Discussion

The pulse sensor coupled by a GPIO port of the controller normalizes the wave, approximately V/2. Midpoint of the voltage refers to the mid value of 12 bit ADC. They do changes with light intensity. It is reported that if light intensity remains constant, then the appropriate digital value will be in V/2. The digital calibrated data has been plotted as shown in Fig. 5.

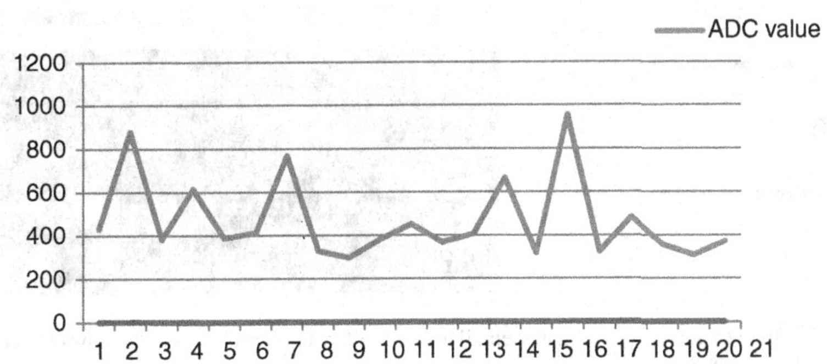

Fig. 5 This figure shows digital output values from the controller which is coupled with pulse sensor. The digital output may vary with different bit ADC of the controller

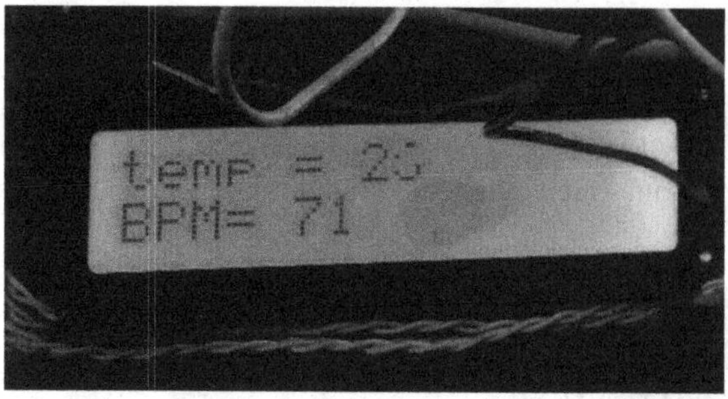

Fig. 6 The figure represents monitoring of real-time data in 16 × 2 LCD module

Body parameters like temperature and pulse are monitored in the attached LCD module with the device. The system acts as a handheld module, where patients need an individual system in order to monitor vital parameters. With the specification of the sensor, they are calibrated in the processor. Pulse sensor timer has been generated in the processor for stabilization of values (Fig. 6).

4.1 Mail Transfer Protocol

CC3100 simple link Wi-Fi board attached with MSP430, where the sending of emergency mail can be performed. Email application runs in Wi-Fi board that collects the sensor data from MSP430 in emergency cases and sends it over mail. The device has been configured in Wi-Fi mode and connected to the nearby Wi-Fi modem. Once the access point has been established, the mail has been sent to the concerned person or caretaker (Fig. 7).

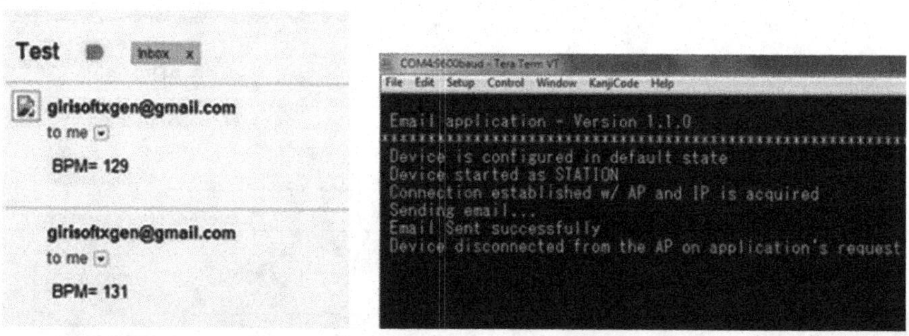

Fig. 7 The figure shows an emergency email and an API virtual terminal which shows the current status of the configuration

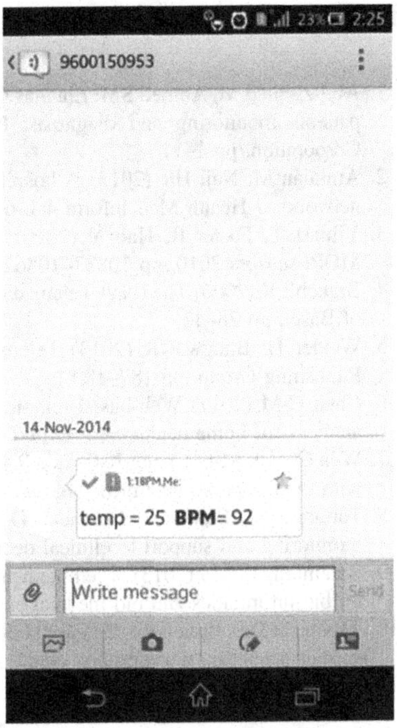

Fig. 8 Emergency message notification

4.2 Emergency Message Notification

GSM modem acts an emergency notification system, independent with mail transfer protocol. It acts as a primary emergency system in the device. The system indicates the current temperature and pulse of the patient as a message (Fig. 8).

5 Conclusion and Future Work

The paper presented a device-based approach for real-time monitoring of individual patients. The sensor outputs such as temperature, pulse signal, beats per minute, and inter beat interval has been calculated. The calibrated value shows good accuracy for detecting emergency signal. This system can be used in hospital environment. Since this paper presents the initial implementation of vital sign monitoring, rest of the activities have been postponed to further work in the near future. The future work of the system leads to wireless sensor network-based approach, where a wearable device sends the emergency signal through the co-coordinator node.

Acknowledgments The authors gratefully acknowledge the financial support from Department of Science and Technology, New Delhi, through Science for Equity Empowerment and Development Division No: SSD/TISN/047/2011-TIE (G) to Velammal Engineering College, Chennai.

References

1. Abo-Zahhad M, Ahmed SM, Elnahas O (2014) A wireless emergency telemedicine system for patients monitoring and diagnosis. Int J Telemed Appl, vol 2014. Hindawi Publishing Corporation, pp 1–11
2. Aminian M, Naji HR (2013) A hospital healthcare monitoring system using wireless sensor networks. J Health Med Inform 4:1–6
3. Yilmaz T, Foster R, Hao Y (2010) Detecting vital signs with wearable wireless sensors. MDPI-sensors 2010, pp 10837–10862
4. Brauchli K (2006) The iPath Telemedicine System. Department of Pathology at the University of Basel, pp 28–42
5. Weider D, Bhagwat R (2013) Telemedicine health support system, Chap 11. IGI Global Publishing Group, pp 187–189
6. Chen C-M (2011) Web-based remote human pulse monitoring system with intelligent data analysis for home health care. Expert Syst Appl 38(3):2011–2019
7. Wen C, Yeh M-F, Chang K-C, Lee R-G (2008) Real- time ECG telemonitoring system design with mobile phone platform. Measurement 41(4):463–470
8. Tartarisco G, Baldus G, Cordaetal D (2012) Personal Health System architecture for stress monitoring and support to clinical decisions. Comput Commun 35(11):1296–1305
9. Balamurugan MS (2013) A remote health monitoring system. International work—conference on bio-informatics and bio medical engineering, pp 195–204, March 2013
10. Technical Description for "SIM900 GSM/GPRS Module". The GSM/GPRS Module for M2M application. http://www.propox.com/download/docs/SIM900.pdf
11. Shah CM, Sangoi VB, Visharia RM (2014) Smart security solutions based on internet of things (IoT). Int J Current Eng Technol 4(5):3401–3404
12. Technical Description for "NTC Type MA" Thermometrics biomedical chip thermistors. http://www.ge-mcs.com/download/temperature/920_321a.pdf

Task Scheduling Using Multi-objective Particle Swarm Optimization with Hamming Inertia Weight

S. Sarathambekai and K. Umamaheswari

Abstract Task scheduling in a distributed environment is an NP-hard problem. A large amount of time is required for solving this NP-hard problem using traditional techniques. Heuristics/meta-heuristics are applied to obtain a near optimal solution within a finite duration. Discrete Particle Swarm Optimization (DPSO) is a newly developed meta-heuristic population-based algorithm. The performance of DPSO is significantly affected by the control parameter such as inertia weight. The new inertia weight based on hamming distance is presented in this paper in order to improve the searching ability of DPSO. Make span, mean flow time, and reliability cost are performance criteria used to assess the effectiveness of the proposed DPSO algorithm for scheduling independent tasks on heterogeneous computing systems. Simulations are carried out based on benchmark ETC instances to evaluate the performance of the algorithm.

Keywords Distributed system · Inertia weight · Particle swarm optimization

1 Introduction

Task Scheduling (TS) is one of the key challenges in Heterogeneous Systems (HS). The most well-known meta-heuristic algorithms are Genetic Algorithm (GA), Ant Colony Optimization (ACO), and Particle Swarm Optimization (PSO).

PSO is an optimization algorithm proposed by Kennedy and Eberhart [1], in 1995. Initially, the development of PSO has intended in continuous search space. Recently, many researchers proposed different conversion techniques for mapping continuous positions of particles in PSO to the discrete values [2]. Kang and He [3]

S. Sarathambekai (✉) · K. Umamaheswari
Department of IT, PSG College of Technology, Coimbatore, Tamil Nadu, India
e-mail: vrs070708@gmail.com

K. Umamaheswari
e-mail: uma@ity.psgtech.ac.in

L.P. Suresh and B.K. Panigrahi (eds.), *Proceedings of the International Conference on Soft Computing Systems*, Advances in Intelligent Systems and Computing 398, DOI 10.1007/978-81-322-2674-1_6

proposed a new position update method for a particle in discrete domain. Dongarra et al. [4] proposed Discrete Particle Swarm Optimization (DPSO) for grid job scheduling problem to minimize Make Span (MS) and Flow Time (FT).

Different metrics may be used to evaluate the effectiveness of optimization algorithms such as MS, FT, Resource Utilization (RU) [2, 3]. All the existing works investigated a number of scheduling algorithms for minimizing MS or FT. The issue of reliability for such an environment needs to be addressed or else an application running on a very large HS may crash because of the hardware failure [4]. The previous research work [5, 6] evaluated the scheduler with the Reliability Cost (RC) using DPSO algorithm. To enhance the reliability of the HS, this paper presents a modified DPSO algorithm.

The traditional way to solve a Multi-Objective Optimization (MOO) problem is to convert the MO into a single objective by using a weighting sum. Kim and de Weck [7] presented an Adaptive Weighted Sum (AWS) method for MOO problems. This paper presents the fitness value of each solution using AWS method that was proposed by Kim and de Weck [7].

The performance of PSO greatly depends on the Inertia Weight (IW). Kennedy and Eberhart [1] developed PSO with no IW. Shi and Eberhart [8] presented the concept of IW with constant values. Xin et al. [9] presented Linearly Decreasing Inertia weight (LDI) for enhancing the efficiency and performance of PSO. Bansal et al. [10] presented a comparative study on 15 strategies to set IW in PSO. Comparative analyses among the four different DPSO variants were addressed in [5]. Suresh et al. [11] proposed an Adaptive IW (AIW) in continuous domain. This paper makes the AIW more suitable for discrete domain.

The previous research work [6] presented DPSO approach to minimize MS, FT, and RC. The MS and FT values are in incomparable ranges and the FT has a higher magnitude order over the MS. Proposed work in this paper extends the DPSO [3] to minimize MS, MFT, and RC using modified DPSO.

The remainder of the paper is organized as follows: The proposed DPSO is presented in Sect. 2. Experimental results are reported in Sect. 3. Finally, Sect. 4 concludes the paper.

2 The HDPSO Algorithm

The LDI was used in the existing DPSO [3, 5, 6]. This inertia factor does not guarantee that the particles have not moved away from the global best. Instead of LDI, the HDPSO uses HIW. Moving the particles based on the hamming distance between the particles and their Gbest that ensures that the particles have not moved away from the Gbest. The HIW is calculated using the Eq. (1), where W_0 the random number between them is 0.5 and 1, H_i is the current hamming distance of ith particle from the Gbest and MDH is the maximum distance of a particle from the Gbest in that generation.

$$W = W_0\left(1 - \frac{H_i}{\mathrm{MDH}}\right) \tag{1}$$

$$H_i = \text{Hamming distance}\left(\mathrm{Gbest}^t, \mathrm{present}_i^t\right) \tag{2}$$

$$\mathrm{MDH} = \mathrm{Max}(H_i) \tag{3}$$

The following sections describe in detail the steps of HDPSO algorithm.

2.1 Particle Representation and Swarm Initialization

The DPSO algorithm [3] starts with a random initial swarm. The problem in this technique is that some processors are busy with processing, while some processors are idle without any processing. To make better utilization of the processors, the HDPSO performs load sharing which assures that no processor is idle initially.

2.2 Particle Evaluation

The MS [6], MFT, and RC [6] are used to assess the efficiency of the HDPSO algorithm. The value of MFT [12] is given in Eq. (4).

$$\mathrm{MFT} = \frac{\sum_{i=1}^{m} M\_\mathrm{Flow}_i}{m} \tag{4}$$

$$M\_\mathrm{Flow}_i = \frac{\sum_{j=1}^{k} F_{ji}}{k} \tag{5}$$

The AWS [7] is used to calculate the fitness value of each solution. This can be estimated using Eq. (6).

$$\mathrm{Fitness} = \alpha_1\alpha_2\mathrm{MS} + (1-\alpha_1)\alpha_2\mathrm{MFT} + (1-\alpha_2)\mathrm{RC} \quad \alpha_i \in [0,1] \tag{6}$$

2.3 Update the Particle's Pbest and Gbest Position

The Pbest and Gbest can be determined based on the fitness value.

2.4 Updating the Particles

This algorithm has taken the redefined particle update methods from [3] and incorporates HIW to update the particles in the swarm.

2.5 VND Heuristic

Local search was not included in DPSO [5, 6] algorithm. Therefore, the algorithm may get struck in local optima. The HDPSO applies VND [3] when there is no appreciable improvement in the Gbest for more than ten iterations.

2.6 Migration Technique

The HDPSO performs this [3] in TS, in order to increase the exploration of the search space.

2.7 Stopping Condition

The above iterative processes will continue until a predefined maximum number of iterations or no significant improvement in the fitness value for sufficiently large number of iterations.

3 Simulation Results and Analysis

All algorithms are coded in Java and executed in Intel i7 processor.

3.1 Brief Description of Benchmark ETC Instances

The computational simulation is presented using a set of benchmark ETC instances [3, 13, 14]. These instances are classified into twelve types based on three metrics: task heterogeneity, machine heterogeneity, and consistency. The instances consist of 512 tasks and 16 processors.

The instances are marked as "a_b_c" as follows:

"a" indicates the type of consistency: c-consistent, i-inconsistent, and s-semi consistent.
"b" says the heterogeneity of the tasks: hi-high and lo-low.
"c" shows the heterogeneity of the machines: hi-high and lo-low.

3.2 Algorithms Comparison

(i) Differential Evolution (DE) [15] (ii) DPSO [3].

3.3 Parameter Setup

The following parameters are initialized for simulating the DE [15], DPSO [3], and HDPSO algorithms.

The population size is set to 32 as recommended in [3]. Number of iteration is set to 1000. The failure rate for each processor is uniformly distributed [16, 17] in the range from 0.95×10^{-6}/h to 1.05×10^{-6}/h.

3.4 Performance Comparisons

The average results of all the algorithms are shown in Table 1 for MS, Table 2 for MFT, and Table 3 for RC. From Tables 1, 2, and Table 3, the first column indicates the ETC instances; the second, third, and fourth columns indicate the values

Table 1 Comparison of MS (in seconds) of all algorithms

ETC instances	DE	DPSO	HDPSO
c_lo_lo	591,107.71	745,933.68	**567,356.89**
c_lo_hi	78,413.01	**60,126.25**	70,649.588
c_hi_lo	279,996.61	299,992.55	**263,979.15**
c_hi_hi	176,838.66	172,331.87	**154,498.19**
i_lo_lo	606,821.01	**450,264.28**	451,509.15
i_lo_hi	732,555.11	726,507.91	**609,668.93**
i_hi_lo	**504,768.31**	618,875.46	896,500.11
i_hi_hi	877,159.02	806,980.29	**452,256.76**
s_lo_lo	354,417.61	378,620.65	**195,943.41**
s_lo_hi	68,716.82	**94,893.557**	212,459.33
s_hi_lo	648,728.21	825,120.63	**500,859.85**
s_hi_hi	402,668.22	462,668.21	**232,549.81**

Table 2 Comparison of MFT (in seconds) of all algorithms

ETC instances	DE	DPSO	HDPSO
c_lo_lo	98,138.46	93,241.71	**70,919.61**
c_lo_hi	15,551.81	**7515.78**	8831.19
c_hi_lo	**20,240.81**	37,499.06	32,997.39
c_hi_hi	19,854.93	21,541.48	**19,312.27**
i_lo_lo	98,352.59	**56,283.03**	56,438.64
i_lo_hi	109,066.18	408,063.49	**76,208.61**
i_hi_lo	204,358.06	204,472.02	**112,062.51**
i_hi_hi	107,144.84	103,372.54	**56,532.09**
s_lo_lo	41,552.14	41,077.58	**24,492.92**
s_lo_hi	**10,339.61**	11,861.69	26,557.41
s_hi_lo	90,591.02	98,140.07	**62,607.48**
s_hi_hi	50,083.52	54,083.52	**29,068.73**

Table 3 Comparison of RC of all algorithms

ETC instances	DE	DPSO	HDPSO
c_lo_lo	1.050639	1.001478	**1.001101**
c_lo_hi	0.509666	0.288967	**0.269639**
c_hi_lo	**0.473802**	0.574349	0.574909
c_hi_hi	0.309931	0.307867	**0.307135**
i_lo_lo	1.940761	1.388678	**1.096692**
i_lo_hi	1.731361	1.394846	**1.234525**
i_hi_lo	**0.725419**	1.494926	1.310443
i_hi_hi	1.170902	1.199991	**1.253376**
s_lo_lo	1.007404	1.456794	**0.918079**
s_lo_hi	0.566386	**0.460995**	0.942587
s_hi_lo	**0.995581**	1.012534	1.396551
s_hi_hi	0.597712	0.835977	**0.461892**

obtained by DE [15], DPSO [3], and HDPSO, respectively. From Tables 1, 2, and 3, the values in bold indicate an optimal value.

Tables 1, 2, and 3 show that the HDPSO algorithm can provide better MS, MFT, and RC in most of the ETC instances and DE, DPSO attain acceptable results only in few ETC instances.

The RPD [3] is used for comparing the results of the HDPSO with DE [15] and DPSO [3]. It is calculated using Eq. (7).

$$\text{RPD} = (\text{AC}_i - P)/P * 100 \tag{7}$$

Table 4 shows that HDPSO is able to give better performance in terms of MS by 21.36 and 25.80 %, MFT by 41.12 and 66.45 %, and RC by 19.68 and 9.82 % compared to DE and DPSO across all ETC instances, respectively. From the results

Table 4 Performance comparison of RPD (%) of HDPSO algorithm with DE and DPSO

ETC instances	MS		MFT		RC	
	DE	DPSO	DE	DPSO	DE	DPSO
c_lo_lo	4.19	31.48	38.39	31.48	4.95	0.038
c_lo_hi	10.99	−14.90	76.10	−14.90	211.40	7.17
c_hi_lo	6.07	13.64	−38.66	13.64	−17.59	-0.10
c_hi_hi	14.46	11.54	2.81	11.54	0.91	0.24
i_lo_lo	34.40	−0.28	74.26	−0.28	76.96	26.62
i_lo_hi	20.16	19.16	43.12	435.46	40.25	12.99
i_hi_lo	−43.70	-30.97	82.36	82.46	−44.64	14.08
i_hi_hi	93.95	78.43	89.53	82.86	−6.58	−4.26
s_lo_lo	80.88	93.23	69.65	67.71	9.73	58.68
s_lo_hi	−67.66	−55.34	−61.07	−55.34	−39.92	−51.09
s_hi_lo	29.52	64.74	44.70	56.75	−28.71	−27.50
s_hi_hi	73.15	98.95	72.30	86.05	29.41	80.99
Average	**21.36**	**25.80**	**41.12**	**66.45**	**19.68**	**9.82**

obtained, the HDPSO is found to be more suitable for high task and high machine heterogeneity.

The simulation results and comparisons show that the HDPSO algorithm performs better than DE and DPSO algorithms.

4 Conclusion

TS is the challenging issue in a HS. This paper presents a new, efficient HDPSO algorithm to the task scheduling problem to find optimal schedules for meta-tasks to minimize the MS, MFT and RC. The proposed HDPSO algorithm adapts the IW based on the hamming distance of the particles from Gbest. Simulation results and comparisons with existing algorithms demonstrated the effectiveness of the HDPSO algorithm.

References

1. Kennedy J, Eberhart R (1995) Particle swarm optimization. IEEE international conference on neural networks, vol 4, pp 1942–1948
2. Kaur K, Tiwari MSP (2012) Grid scheduling using PSO with SPV rule. Int J Adv Res Comput Sci Electron Eng 1:20–26
3. Kang Q, He H (2011) A novel discrete particle swarm optimization algorithm for meta-task assignment in heterogeneous computing systems. In: Microprocessors and microsystems, vol 35. Elsevier, Amsterdam, pp 10–17
4. Dongarra J et al (2007) Bi-objective scheduling algorithms for optimizing makespan and reliability on heterogeneous systems. ACM Parallel Algor Archit 9:280–288
5. Sarathambekai S, Umamaheswari K (2013) Comparison among four modified discrete particle swarm optimization for task scheduling in heterogeneous computing systems. IJSCE 3:371–378
6. Sarathambekai S, Umamaheswari K (2014) Task scheduling in distributed systems using discrete particle swarm optimization. Int J Adv Res Comput Sci Engg 4(2)
7. Kim IY, de Weck OL (2006) Adaptive weighted sum method for multi-objective optimization. Struct Multi Optim 31(2):105–116
8. Shi Y, Eberhart R (2002) A modified particle swarm optimizer. IEEE international conference on EC, pp 69–73. doi:10.1109/ICEC.1998.699146
9. Xin J, Chen G, Hai Y (2009) A particle swarm optimizer with multistage linearly-decreasing inertia weight. IEEE international conference on computational sciences and optimization, vol 1, pp 505–508
10. Bansal JC et al (2011) Inertia weight strategies in particle swarm optimization. IEEE world congress on NaBIC, vol 3, pp 633–640
11. Suresh K et al (2008) Inertia-adaptive particle swarm optimizer for improved global search. IEEE international conference on intelligent systems and design, vol 2, pp 253–258
12. Lindeke R et al (2005) Scheduling of Jobs. Spring: www.d.umn.edu/rlindek1/.../scheduling%20of%20Jobs_Sset11.ppt
13. Ali S, Siegel HJ (2000) Representing task and machine heterogeneities for heterogeneous computing systems. Tamkang J Sci Engg 3(3):195–207

14. Braun TD et al (2001) A comparison of eleven static heuristics for mapping a class of independent tasks onto heterogeneous distributed computing systems. J Parallel Distrib Comput 61:810–837
15. Krömer P et al (2010) Differential evolution for scheduling independent tasks on heterogeneous distributed environments. Adv Intell Soft Comput 67:127–134
16. Luo W et al (2007) Reliability-driven scheduling of periodic tasks in heterogeneous real-time systems. IEEE international conference on advanced information networking and applications, vol 1, pp 1–8
17. Qin X, Jiang H (2001) Dynamic, reliability-driven scheduling of parallel real-time jobs in heterogeneous systems. ACM international conference on parallel processing, vol 2, pp 113–122

Pedagogue: A Model for Improving Core Competency Level in Placement Interviews Through Interactive Android Application

N.M. Dhanya, T. Senthil Kumar, C. Sujithra, S. Prasanth and U.K. Shruthi

Abstract This paper discusses about developing a mobile application running on the cloud server. The Cloud acclaims a new era of computing, where application services are provided through the Internet. Though mobile systems are resource-constrained devices with limited computation power, memory, storage, and energy, the use of cloud computing enhances the capability of mobile systems by offering virtually unlimited dynamic resources for computation and storage. The challenge faced here is that traditional smartphones do not support cloud, these applications require specialized mobile cloud application model. The core innovativeness of the application lies in its delivery structure as an interactive android application centered on emerging technologies like mobile cloud computing–that improves the core competencies of the students by taking up online tests posted by the faculty in the campus. The performance of this application has been presented using scalability, accessibility, portability, security, data consistency, user session migration, and redirection delay.

N.M. Dhanya (✉) · T. Senthil Kumar · C. Sujithra · S. Prasanth · U.K. Shruthi
Department of Computer Science and Engineering, Amrita School of Engineering, Amrita Vishwa Vidhyapeetham, Coimbatore, India
e-mail: nm_dhanya@cb.amrita.edu

T. Senthil Kumar
e-mail: t_senthilkumar@cb.amrita.edu

C. Sujithra
e-mail: sujithrachandramohan@gmail.com

S. Prasanth
e-mail: prasanthsiva93@gmail.com

U.K. Shruthi
e-mail: shruthikrishnan01@gmail.com

L.P. Suresh and B.K. Panigrahi (eds.), *Proceedings of the International Conference on Soft Computing Systems*, Advances in Intelligent Systems and Computing 398, DOI 10.1007/978-81-322-2674-1_7

1 Introduction

Cloud computing generally refers to delivery of resources based on the user needs. Network connection is essential for any cloud-based mobile application. In general, the placement tests are conducted online with the help of web server [1, 2]. This is done by hosting an online web service. Mobile system obtains response from the server based on the request sent by the user. Here, the faculty acts as an administrator and send questions to the data store and the students access these questions as a response to the request sent to the server by clicking on to the test modules. The students can answer the questions posted and submit the tests. The marks will be awarded for each questions and the total would be saved as a result for future use.

Cloud Computing (CC) is recognized as the next generation computing infrastructure, which offers various advantages to the users, by allowing the users to use the infrastructure such as servers, networks, storages, and platform such as middleware services[3–5]. Mobile application can be rapidly released with minimum management effort. The support of CC for mobile users made MCC an emerging technology [6–8]. MCC helps mobile users to take full advantage of CC. MCC is a structure where the data processing and data storage happen outside the mobile device. The processing and storage are moved from mobile devices to powerful and centralized platforms in clouds.

In the existing system, the placement tests are conducted with the help of web servers where data consumption is more and also requires more memory. Hence, cloud offloading improves efficiency and also requires less storage [9, 10]. This allows faculty to maintain a record of the students and also helps students in improving their competency by focusing more on that particular part to improve themselves.

The rest of the paper is organized as follows: Sect. 2 discusses about the related works followed by the cloud framework. The proposed work with the architecture diagram and module details are discussed in Sect. 3. The results and future enhancements are proposed in Sects. 4 and 5, respectively.

2 Related Works

Mobile cloud computing is one among the emerging technologies. More and more researches are working on limited onboard computing capabilities of mobile devices and are hampering their ability to support the increasingly sophisticated applications demanded by users.

Lu and Kumar proposed an analysis on computation offloading [11]. Cloud computing provides the possibility of energy savings as a service. Cloud vendors provide computing cycles to reduce the amount of computation on mobile systems by virtualization. The analysis presented here indicates that cloud computing potentially saves energy in mobile systems. Mobile cloud services vary from cloud services in a way that it must save energy. The analysis indicates that computation offloading saves energy with the help of bandwidth, amount of transmitted data and the computational complexity. There is a fundamental assumption that the data is sent to the server. This implies that the server should not store any data. The client needs to offload the program and the data to the server.

Location-based mobile service is an emerging MCC application. Tamai et al. designed a platform for location-based services leveraging scalable computation and large storage space to answer large queries based on location. These services are context aware. In addition to location information, these mobile services also consider the environment and application context [12, 13]. The environment information can be feasibly obtained.

Suhas Holla, Mahima M. Katti discusses about the architecture of android application. These studies are based on the component framework and the application model of android. This presents the architecture of how the android applications are connected through the HTTP requests and response [14]. In addition to the models, this also gives a detailed study of the security framework and also the security issues related to android platform [15]. The integrity of the android platform is maintained through these services [14].

Qureshi et al. [16] have categorized MCC into two broad categories, viz., General Purpose Mobile Cloud Computing (GPMCC) and Application-Specific Mobile Cloud Computing (ASMCC). In GPMCC, application can be outsourced to the cloud based on the computational complexity. ASMCC is similar to GSMCC but offers more computational power compared to GSMCC.

Lei Yang et al. proposed a framework for partitioning and execution of data stream applications in MCC. This optimizes the partitioning of a data stream application [17] between mobile. In addition to dynamic partitioning, it also supports instances of computation to be shared among multiple users.

3 Proposed Work

This android application is useful for students to take placement tests where the question would be obtained from the cloud server.

3.1 General MCC Architecture

The general architecture of a cloud-based mobile application is depicted in Fig. 1.

Fig. 1 General Mobile Cloud Computing Architecture. This shows the cloud service provider connected to the Internet which helps the personal computers and mobile phones to access the services

3.2 System Architecture

This architecture follows Application Specific Mobile Cloud Computing model. This application is designed to support placements. The detailed architecture of the system is depicted in Fig. 2.

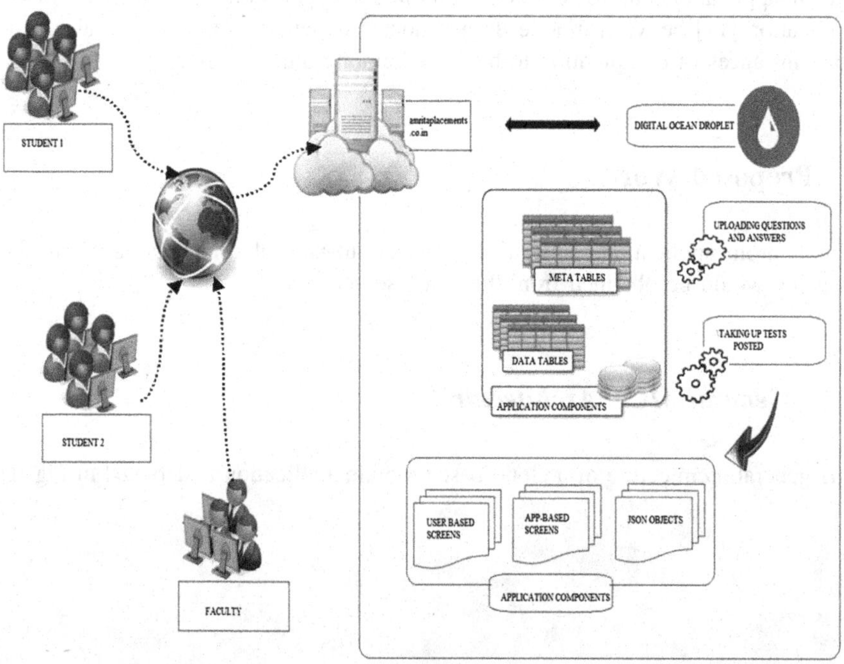

Fig. 2 System architecture

This application is designed to support placement training by implementing a flexible and scalable infrastructure by connecting it to the cloud server. Both the application structure and the application data are stored in the data store. A droplet is created on the digital ocean cloud. The cloud has the LAMP server, Ubuntu 14.04 running on it, where the application API's could be loaded and a domain name is provided. Generate the API token for the application to run on the cloud server, register the application on the server. The amritaplacements.co.in is added to the domain names of the server and it is connected to the IP address of the domain and thereby changing the zone files of the domain with respect to the cloud server. Now, the application is connected to the cloud and with the help of the LAMP server installed on it could run the API's developed for the application. The uploaded questions would be saved as JSON objects in an array which plays an important role in data security. These JSON objects are parsed and are read by the application. With the help of the faculty portal, faculty could add questions to the data store on the cloud and once the student gets logged in he/she would be notified about the tests that are pending for the user. The login credentials are authenticated with the help of database tables and the metadata tables in the cloud server. The student could take up the tests after he/she downloads the questions. The questions could be answered and it gets saved if the students wishes to save the state of his test and continue later; the session variables cloud also let the concurrent users know how many others are taking up tests. These JSON objects are parsed and are read by the application.

3.3 Modules

This applications consists of two parts

1. Administrator
2. Student

Administrator:

1. Faculty Login:
   This module allows the faculty to login with their user name and password that they have registered already on the website with all their personal details. Password can be reset at any instance by the faculty and a reference link would be sent to the mail. For security purpose, there would be various questions that the faculty should answer before resetting the password.
2. Uploading questions:
   This module allows the staff to post questions by clicking on the add questions buttons and also helps them specify the options along with the correct answer. The faculty can post questions based on the subject.
3. Get mark list:
   This module retrieves the marks of all student who have taken the online tests.

Student:

1. Student Registration:
   This module allows the student to register with the details mentioned and then they login with their username and password.
2. Forgot Password:
   This module allows the student to reset his/her password by sending a link to their mail.
3. Selection:
   This allows students to select the tests posted by the staff, i.e., aptitude, verbal, and technical.
4. Display results:
   This displays the marks obtained by the student in that particular test taken by him/her

4 Results

These are the snapshots of the developed application.The Figs. 3 and 4 shows the initial Login screen and the forgot password screen. Figs. 5, 6 and 7 represents how a user takes the test and how choices are made. The Fig. 8 represents the final result displayed to the user. The Fig. 9 is showing how a faculty can add questions to the pool.

Fig. 3 SignUp screen

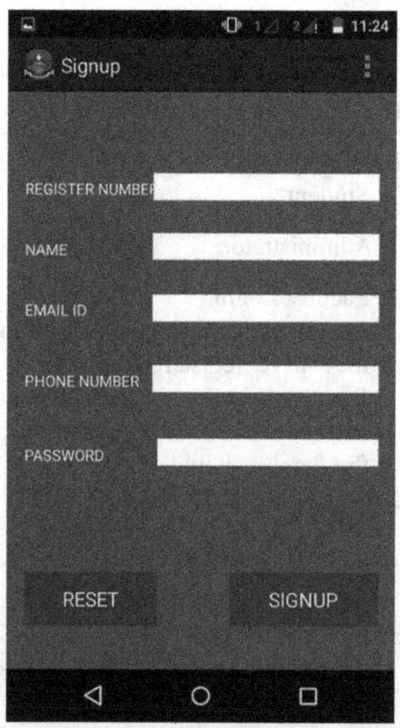

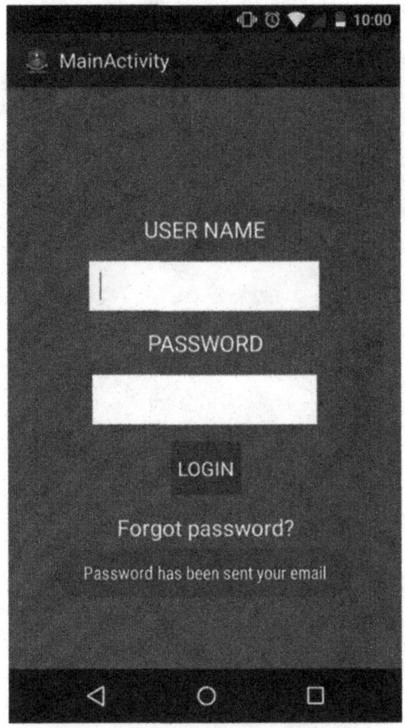

Fig. 4 Login/Forgot password screen

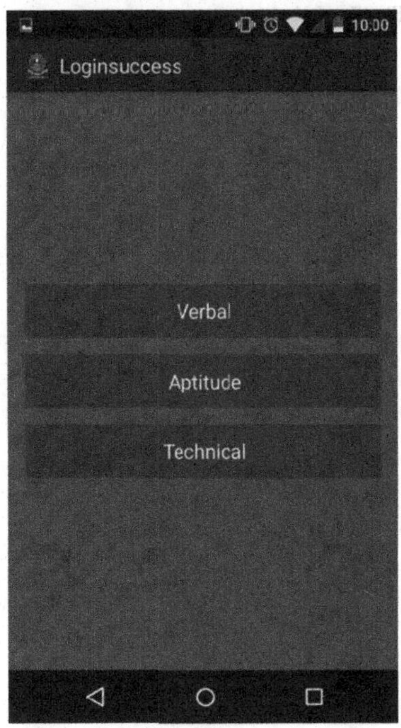

Fig. 5 Selection of test

Fig. 6 Test screen

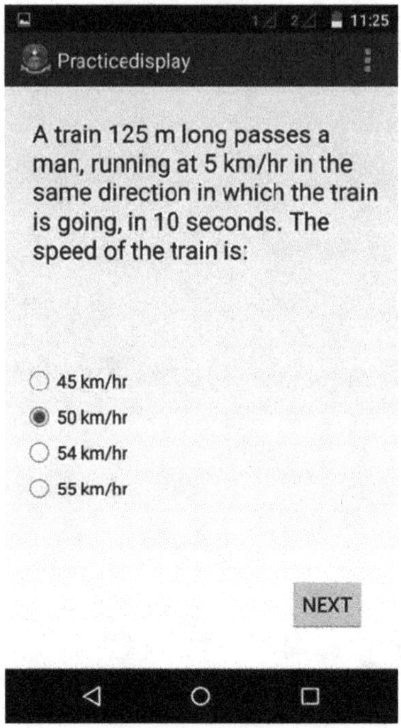

Fig. 7 Verbal selection

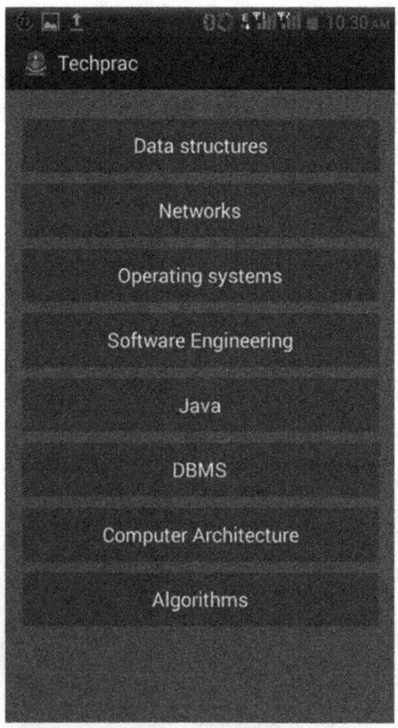

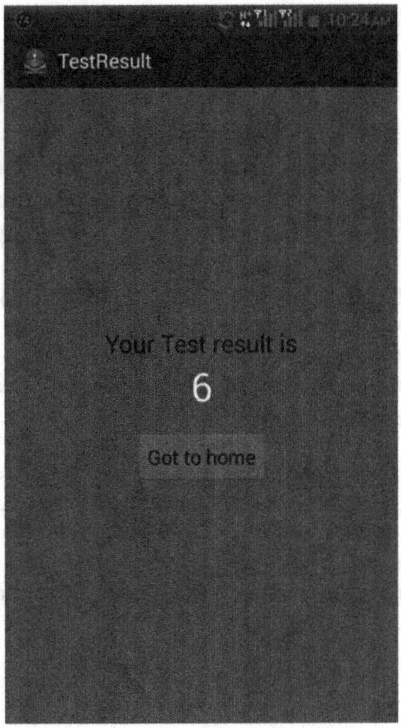

Fig. 8 Test result

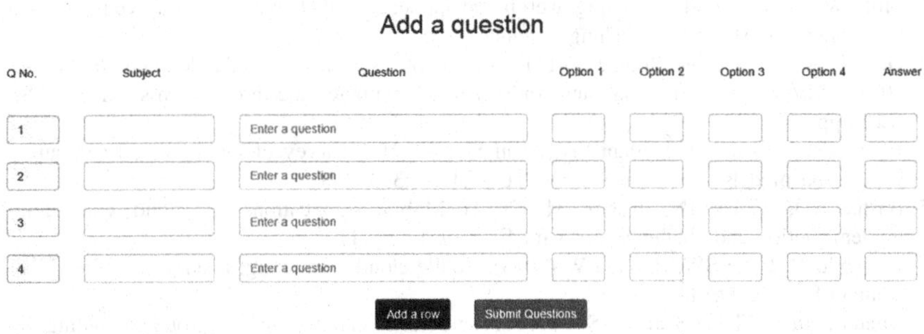

Fig. 9 Add question screen. This helps the staff to enter the questions

5 Future Enchancements

5.1 Score Boards

Application can provide a scoreboard facility which lists the top scorers in the test. Top 30 scorers are displayed in the application with the other student in the batch. This motivates the students to work hard and to get higher scores in the test.

5.2 *Level-Based Questions*

Application can provide a level-based access to questions, i.e., easy, medium, and difficult. He/she can migrate to next level only after they complete the previous levels with the minimum cut-off set. As the level increases, the difficulty of question increases. This helps the students to analyse their competency levels and the areas in which they should work more to improve their core competencies.

Acknowledgments We are highly indebted to Amrita School of Engineering, Ettimadai for their guidance and constant supervision by providing necessary information regarding the project and also for their support in completing the project.

We express our deepest thanks to Dr. Latha Parameshwaran, Head of the Department of Computer Science and Engineering, Amrita University for constant support and valuable guidance.

We express our sincere thanks to Mr. SankaraNarayanan B, Assistant Manager (Aptitude), Learning and Development, Corporate and Industrial Relations Department, Amrita Vishwa Vidhyapeetham, Coimbatore for his constant guidance and support in completing the project.

We express our deepest thanks to CTS lab, Amrita School of Engineering, Amrita Vishwa Vidhyapeetham for supporting our research activities throughout the project.

References

1. Prasad S, Peddoju SK, Ghosh D (2013) AgroMobile: a cloud-based framework for agriculturists on mobile platform. Int J Adv Sci Technol 59:41–52
2. Miller M (2008) Cloud computing: web-based applications that change the way you work and collaborate online. Que Publishing, Indianapolis
3. Guan L, Ke X, Song M, Song J (2011) A survey of research on mobile cloud computing. In: 2011 IEEE/ACIS 10th international conference on computer and information science (ICIS). IEEE, pp 387–392
4. Khan AR, Othman M, Madani SA, Khan SU (2013) A survey of mobile cloud computing application models. Commun Surveys Tutorials 16:393–413
5. Basha AD, Umar IN, Abbas M (2014) Mobile applications as cloud computing: implementation and challenge. Int J Inf Electron Eng 4(1)
6. Fernando N, Loke SW, Rahayu W (2013) Mobile cloud computing: a survey. Future Gener Comput Syst 29(1):84–106
7. Guan L, Xu K, Meina S, Junde S (2011) A survey of research on mobile cloud computing. In: Proceeding of 10th international conference on computer and information science (ICIS), IEEE/ACIS (2011), pp 387–392
8. Dinh HT, Lee C, Niyato D, Wang P (2011) A survey of mobile cloud computing: architecture, applications, and approaches. Wiley online library, Oct 2011, pp 1587–1611
9. Zhao W, Sun Y, Dai L (2010) Improving computer basis teaching through mobile communication and cloud computing technology. In: 3rd international conference on advanced computer theory and engineering (ICACTE), vol 1. IEEE, pp V1–V452
10. Xia Q, Liang W, Xu W (2013) Throughput maximization for online request admissions in mobile cloudlets. IEEE conference on local computer networks, pp 589–596
11. Kumar K, Lu Y-H (2010) Cloud computing for mobile users: can offloading computation save energy. J Comput 43(4):51–56

12. Sanaei Z, Abolfazli S, Gani A, Buyya R (2013) Heterogeneity in mobile cloud computing: taxonomy and open challenges. IEEE Commun Surveys Tutorials 16(1):369–392
13. Verbelen T, Simoens P, Turck FD, Dhoedt B (2012) Cloudlets: bringing the cloud to the mobile user. In: Proceedings of MCS'12, ACM
14. Holla S, Katti MM (2012) Android based mobile application and its security. Int J Comput Trends Technol 3:486
15. Krutz RL, Vines RD (2010) Cloud security: a comprehensive guide to secure cloud computing. Wiley Publishing, Inc, New York
16. Qureshi SS, Ahmad T, Rafique K, Islam SU (2011) Mobile cloud computing as future for mobile applications—Implementation methods and challenging issues. IEEE international conference on cloud computing and intelligence systems (CCIS), 15–17 Sept 2011
17. Yang L, Cao J, Tang S, Li T, Chan A (2012) A framework for partitioning and execution of data stream applications in mobile cloud computing. In: IEEE 5th international conference on cloud computing (June 2012), pp 794–802

Non Functional QoS Criterion Based Web Service Ranking

M. Suchithra and M. Ramakrishnan

Abstract Web Services provides a systematize way to integrate web applications over an internet protocol. As numerous web services exist in the internet, selecting appropriate web service is vital in many web service applications. Quality of Service (QoS) is the predominant parameter which is used for selecting a web service in terms of their quality. Based on user's preference for service quality, services are ranked and best web services are selected. But the difficulty with this approach is that it is strenuous to precisely define QoS property. Hence an enhanced fuzzy multi attribute decision making algorithm for web service selection is presented in this paper. The proposed method periodically collects user's feedback to update the web service QoS ranking. Experimental results show that the proposed method can satisfy service requester's non-functional requirements.

Keywords Web service · Web service discovery · SOAP · WSDL · UDDI · Web service ranking · Non-functional QoS

1 Introduction

Since heterogeneous technologies are available in internet, we need reusable components that can work on different platforms and programming languages. The technologies such as COM, CORBA and RMI can do well to fulfill requirements [1].

M. Suchithra (✉)
Department of Computer Science & Engineering, Sathyabama University, Chennai, Tamil Nadu, India
e-mail: suchithrajai@gmail.com

M. Ramakrishnan
Department of Computer Science, Madurai Kamaraj University, Madurai, Tamil Nadu, India

L.P. Suresh and B.K. Panigrahi (eds.), *Proceedings of the International Conference on Soft Computing Systems*, Advances in Intelligent Systems and Computing 398, DOI 10.1007/978-81-322-2674-1_8

But these components are either language dependent or platform dependent. The solution to this problem is to use web services [2]. Web Services interact with different web applications to exchange data. It may be a piece of software that uses a standard XML messaging system to encode all communications [3]. Web services may be defined as self-contained, distributed, dynamic applications published over the internet to create products and supply chains. Web service may be a collection of open protocols and standards used to exchange data between applications and systems. The components of web service include SOAP, UDDI and Web Services Description Language (WSDL) [4] (Fig. 1).

Simple Object Oriented Protocol (SOAP) is a platform and language independent protocol that allows data transfer between applications. It is based on XML. SOAP also provides a platform to communicate between programs running on different operating systems. Universal Description Discovery and Integration (UDDI) is also a platform independent framework that provides directory service to store information about web services [4]. UDDI can communicate via SOAP, CORBA, Java RMI specifications. WSDL is an XML-based language that allows users to describe web services [5]. It is also used to locate web services. To perform the service, service requester interacts with service registry and suitable service is invoked. Selection of suitable service is based on Quality of Service (QoS) information [5].

QoS is the quality parameter which can be used to select an appropriate web service. QoS calculation is based on several properties such as security, response time, latency, accessibility, availability, etc. These properties represents technical quality of a web service leaving managerial quality. Moreover, QoS property may include several sub properties [6]. The motive of our work is to make the service choice process straightforward by ranking the available services for an equivalent practicality that successively makes the composition of the services more effective and time efficient.

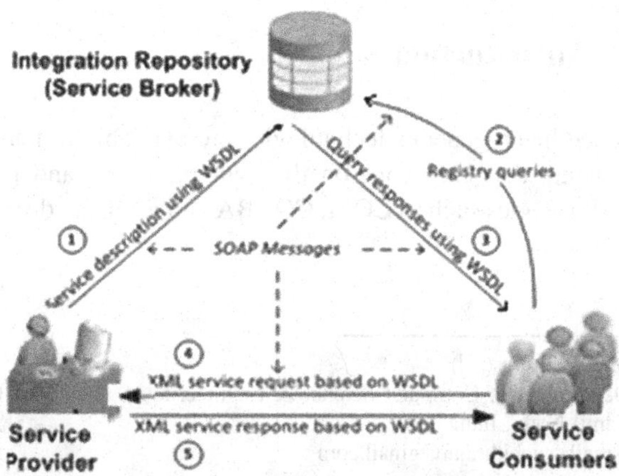

Fig. 1 Web service architecture

2 Related Work

2.1 QoS Criterions

Selecting a web service is based on the value of QoS value. Different QoS values have different value ranges. Hence it is unfair to calculate these values directly. So QoS attributes standardization is needed. The criterions used to calculate the QoS value are: Execution Time, Cost, Service Availability, Reliability and Success Rate.

2.2 Fuzzy Logic

Fuzzy set theory is used to model classes or sets whose frontiers are not defined [7]. Transition of such objects or classes from full membership to full mismatch takes place in a slow manner. Fuzzy logic is a type of problem-solving methodology that can be implemented on any management system regardless of size, complexity of the problem, standalone or networked, multi-channeled or workstation based data acquisition system.

Definition 1 A fuzzy set $\tilde{A}$ is characterized by a member function $\mu A(x)$ that associates with elements x in X to a real number in the range [0, 1]. X is defined as a universe of discourse and $\mu A(x)$ is said to be grade of membership.

Definition 2 A triangular fuzzy number $\tilde{A}$ can be defined by a triplet (a, m, b). Its conceptual schema and mathematical form are shown in the below equation

$$\mu_{\tilde{A}}(x) = \begin{matrix} \frac{x-a}{m-a}, & \text{if } a \leq x \leq m \\ \frac{x-a}{m-a}, & \text{if } m \leq x \leq b \\ 0, & \text{otherwise} \end{matrix}$$

3 Existing System

Existing web service discovery methods fail to include fuzzy based QoS information during web service selection process. A critical mass of researchers is applying fuzzy set theory to include QoS information to solve accurate web service discovery problem [8]. Chen et al. adopts Fuzzy Multiple Criteria deciding (FMCDM) approach to capture. However customers create their analysis of additional services effectively [9]. Huang et al. [10, 11] presents a tempered fuzzy web service discovery approach to model subjective and fuzzy opinions, and to help service customers and suppliers in reaching an accord. To normalize the fuzzy values, various weights are used and entropy weights is one among such methods [12].

Another method called single value decomposition, represents QoS attributes using matrix [13]. Matrix is broken down into three matrices and each matrix represents some set of QoS criterion. The matrix values use 2D space for representation in the form of graph and ranks are allotted as per proximity values. The only problem with this method is that size of the matrix becomes too large for complex attributes [12].

Many QoS aware web service discovery approaches were studied [14–17] as a literature review. Many QoS criterions traditionally exhibit only the non-functional aspects of a web service. The scope of QoS is broad in describing various non-functional properties such as reliability and accessibility [23]. It can also involve properties like response time, cost, availability and reputation. Analysis of QoS problems are based on service provider's perspective and service user's perspective [18]. The list of QoS criterions are accessibility, reaction time, throughput and security.

In [7, 11] provides two templates for the QoS primarily based service composition problem: combinable and graph model. It uses two heuristic algorithms to support applied mathematics for service procedures with a serial flow and a general flow structure. This modifies the choice of QoS directed services. The proposed algorithm is enhanced such that the ranking of the available web services are changed dynamically based on the user experience about the service.

3.1 The Decision Making Process of Services with Qos Constraints

Definition 3 A Quality constraint factor is defined as a triplet (Q, V, and F). Where, Q is the name of the QoS factor, $V \equiv$ Value of quality decisive factor, $F \equiv$ Function that gives the unit of measurement used for each QoS factor.

Definition 4 *QoS Description Model*
Let WS is a set of web services such that WS (f) = $\{S_1, S_2, \ldots, S_n\}$ ($n \geq 2$) that performs a common functionality with set of QoS Constraints $q = \{q_1, q_2, \ldots, q_j\}$. Where '$n$' is the number of candidate services for the specific function f and $_j$ is the no. of quality requirements. For simplicity we denote $K = \{1, 2, 3, \ldots, k\}$ and $N = \{1, 2, 3, \ldots, n\}$. Let WSD represents individual service with default quality criterions set, WSU represents user-defined quality criterions. Therefore a Service Request quality criterions WSR = WSD && WSU. This default quality criterions sets WSD = {Availability, Response Time, Cost, Reliability, Success Rate}.

Definition 5 *QoS description model of candidate service* [18]
If n candidate services are selected based on functional constraints, the service set is represented as: WSC = $\{SC_1, SC_2, \ldots, SC_n\}$, where $SC_i = ((q_{i1}, g_{i1}), (q_{i2}, g_{i2}), \ldots,$

(q_{im}, g_{im})) is the QoS of the *i*th-candidate Web services, $1 \le i \le n$, q_{ij} $(1 \le j \le m)$ is the property name of the first *j*th QoS quality in SC_i, g_{ij} shows the corresponding quality attribute values which is provided by services.

Definition 6 *QoS description model of Service Request* [18]
The set WSR = $((q_1, w_1, g_1), (q_2, w_2, g_2), \ldots, (q_n, w_n, g_n))$ where, q_k $(1 \le k \le n)$ is the *k*th quality criterions in service requests, w_k indicates the weight which requester assigns to the quality criterions, g_k is requester's expectations of the quality attribute.

4 Proposed System

4.1 Service Model with QoS Factor

The decision making algorithm determines which WS (f) with relevant QoS constraints provided by the SP best ensemble the requirement through the following phases:

Step 1: **QoS Matrix**. Suppose candidate service sets by functional match $S = (S_1, S_2, S_3, \ldots, S_n)$, Si has m QoS criterions, $1 \le i \le n$, and S_i $(i \in K)$ then construct an $n \times m$ order decision matrix Q, where each row represents a candidate service corresponding to each QoS property value, each column represents attribute values of all candidate services representing its respective QoS criterion q_j, where $(j \in N)$.

$$Q_n(5) = \begin{bmatrix} q_{11} & q_{12} & q_{13} & q_{14} & q_{15} \\ \vdots & \vdots & \vdots & \vdots & \vdots \\ \vdots & \vdots & \vdots & \vdots & \vdots \\ \vdots & \vdots & \vdots & \vdots & \vdots \\ \vdots & \vdots & \vdots & \vdots & \vdots \end{bmatrix} \quad (1)$$

Let U_n be user feedback of the services for the QoS $U_n = \{u_1, u_2, u_3, u_4, u_5\}$.
Let W be the set of weight vector defined by the SR for each QoS.

$$W = \{w_1, w_2, w_3, w_4, w_5\} \quad \text{such that} \sum_{k=1}^{5} w_k = 1 \text{ and} \quad (2)$$

W_k represent the weight of *K*th quality criterion.

Step 2: **Updating QoS Matrix and Weight of the Services using User Feedback**

Each element in $Q_{n \times 5}$ matrix is updated by using following formula

$$\begin{aligned} q_{ij} &= \frac{(U_n[u_j]*10)+q_{ij}}{2} && \text{if} \quad (q_{ij}-25) < (U_n[u_j]*10) < (q_{ij}+25) \\ q_{ij} &= q_{ij}-3 && \text{if} \quad (q_{ij}-25) > (U_n[u_j]*10) \\ q_{ij} &= q_{ij}+3 && \text{elsewhere} \\ &\text{where} \quad 3 \le j \le 5, \; 1 \le i \le n \end{aligned} \tag{3}$$

Update the weight with user feedback

$$X = \frac{1}{\sum_{k=1}^{5} U_k}. \tag{4}$$

Each element in U_n is multiplied by X such that

$$\sum_{k=1}^{5} U_k = 1 \text{ in all } U_n$$

Now update the set W with U_n

$$W_k = \frac{W_k[w_k] + U_k[u_j]}{2} \quad \text{where} \quad 1 \le j \le 5, \; 1 \le k \le n. \tag{5}$$

Step 3: **QoS Normalization**

Based on the tendency of QoS attribute, it can be either positive or negative criteria.

Every QoS attributes differ each other, hence each attribute need to be normalized. It is an essential step as certain value should be maximized and certain to be minimized to get best results. For negative criteria such as cost and time the QoS value need to be minimized and for the positive criteria such availability The following formula is used for scaling the $Q_{n \times 5}$ matrix. The values of negative criterions are normalized by Eq. 6. And the values of positive criterions are normalized by Eq. 7.

$$q_{ij} = 1 - \frac{q_{ij}}{\max(q_{ij}) + \min(q_{ij})}. \tag{6}$$

$$q_{ij} = \frac{q_{ij}}{\max(q_{ij}) + \min(q_{ij})} \tag{7}$$

Wre max (q_{ij}) is maximum value in jth column and min (q_{ij}) is minimum value in jth column associated with Candidates Service.

Step 4: **Define Quality Vector and Euclidean Distance**
Let g be the quality vector of the positive ideal solution

$$\begin{aligned} g &= (g_1, g_2, g_3, g_4, g_5) \\ &= (\max(q_{j1}), \max(q_{j2}), \max(q_{j3}), \max(q_{j4}), \max(q_{j5})) \\ & 1 \le j \le n \end{aligned}$$

Let b be the quality vector of the negative ideal solution

$$\begin{aligned} b &= (b_1, b_2, b_3, b_4, b_5) \\ &= (\min(q_{j1}), \min(q_{j2}), \min(q_{j3}), \min(q_{j4}), \min(q_{j5})) \\ & 1 \le j \le n \end{aligned}$$

A distance measure is required for grouping the services. To compute the distance between the service vectors, we apply Eqs. 8 and 9.

$$d_{ij} = \sqrt{\sum_{k=1}^{5} W_k(g_j - q_{ij})} \quad 1 \le i \le n \tag{8}$$

$$d_{ib} = \sqrt{\sum_{k=1}^{5} W_k(q_{ij} - b_j)} \quad 1 \le i \le n \tag{9}$$

Step 5: **Calculate Degree of membership**
A Fuzzy approach is applied to find the membership value of each service according to the quality, which the Service Requestor prefers more.

5 Experiments and Results

We have prototyped our approach in Python and Django Web Framework on an Intel® Core ™ Duo CPU, 2.53 GHz, 4 GB RAM and Window 7 Operating System. The interface is in such a way that the required parameters are clearly explained when the page is viewed by the user without any further explanation. Django provides MVC controller that helps to separate business logic from presentation. This eases the design process to handle dynamic changes throughout the web application. At the ranking logic uses dynamic updating of the QoS constraints of the web service. These dynamic changes are based on the user feedback regarding this service.

The Client Interface to enter the weight and Feedback points for each QoS constraints of the candidate Web services is shown in Figs. 2 and 3. The number of candidate web services are eight. The number of QoS constraints is fixed to five (i.e. cost, time (*T*), availability (*A*), reliability (*R*), success rate (*S*). The reference dataset [18] used is given in Table 1.

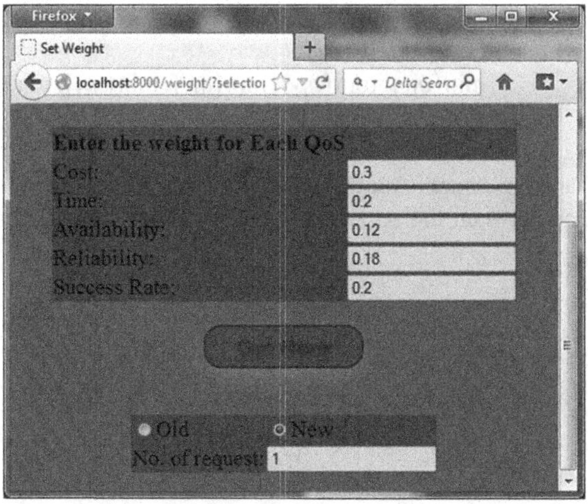

Fig. 2 Client interface showing weights for QoS

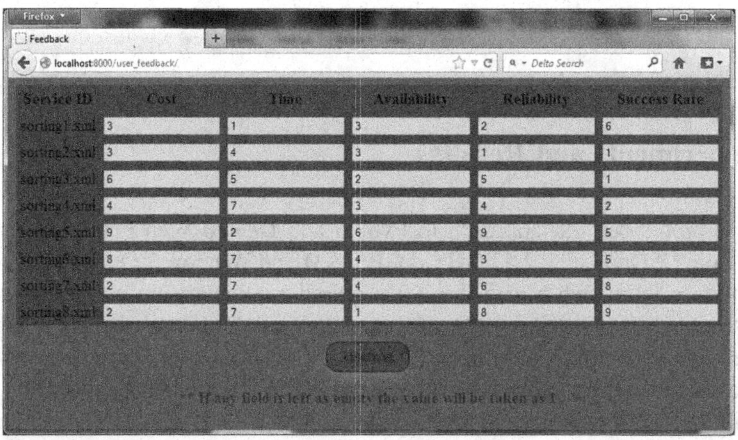

Fig. 3 Client interface for user feedback

Table 1 Decision dataset

WSC tendency	Rank	Cost (Rs) Min.	T (Sec.) Min.	A (%) Max.	R (%) Max.	S (%) Max.
Sorting6.xml	6.859	56	43	90	84	85
Sorting4.xml	6.711	29	37	71	55	32
Sorting7.xml	6.289	12	81	78	24	35
Sorting8.xml	5.441	32	75	98	65	26
Sorting3.xml	4.602	76	28	52	96	70
Sorting1.xml	3.805	49	68	91	91	12
Sorting2.xml	3.757	85	56	62	72	94
Sorting5.xml	0.684	68	95	83	43	25
Weight		0.3	0.2	0.12	0.18	0.2

5.1 Performance Results

The need for our experiment is to compare our model with [18]. The weight vector, $W = (0.3, 0.2, 0.12, 0.18, 0.2)$ the resultant decision table of our approach is given in Table 2.

Analysing the two results, we can find the variant in the order the service is ranked. Consider the service8 which is up in the order is pushed down by three services1,2 and 3. The reason for that though cost is the most preferred QoS by the SR, other QoS also has a role in determining the rank of the service. And in case of service1 all the QoS constraints fall in the average values. It does not excel in one QoS and retard in other QoS. The rank of the services will be updated according to the user feedback in our approach. By comparing chart rank is calculated it is clear, services are ranked exactly how the requestor needs, as it uses the feedback of the user used in real time.

Table 2 Resultant dataset

WSC tendency	Rank	Cost (Rs) Min.	T (Sec.) Min.	A (%) Max.	R (%) Max.	S (%) Max.
Sorting6.xml	8.389	56	43	90	84	85
Sorting4.xml	6.527	29	37	71	55	32
Sorting7.xml	5.618	12	81	78	24	35
Sorting1.xml	5.405	49	68	91	91	12
Sorting3.xml	5.208	76	28	52	96	70
Sorting2.xml	4.847	85	56	62	72	94
Sorting8.xml	3.251	32	75	98	65	26
Sorting5.xml	0.722	68	95	83	43	25
Weight		0.3	0.2	0.12	0.18	0.2

Fig. 4 Rank score of each WS on 10th request

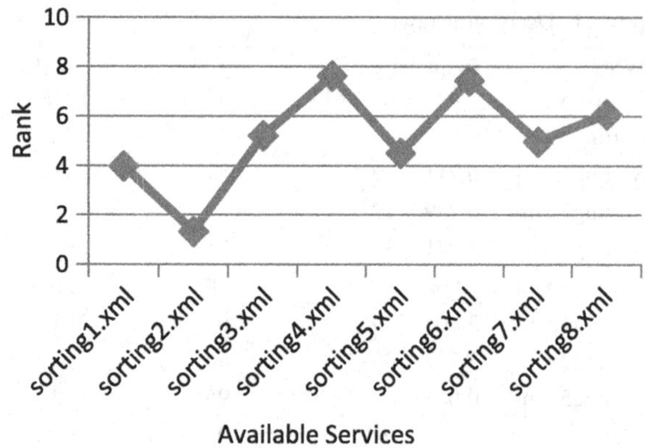

Fig. 5 Rank score of each WS on 40th request

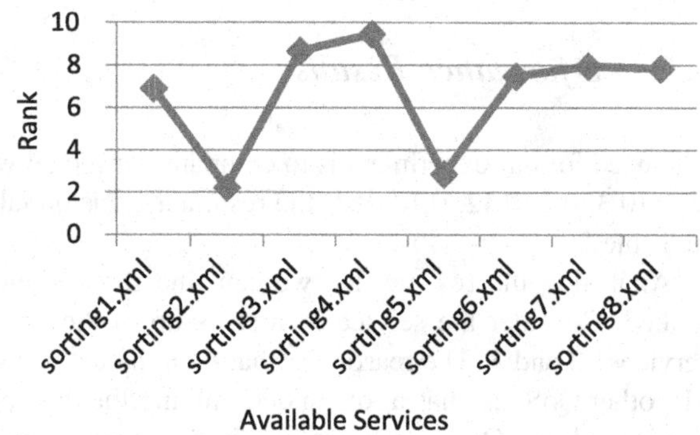

Figures 4 and 5, shows changing service ranking score based on 10th and 40th time of service request. The higher the number of request is generated, higher is the accuracy of the service rank.

6 Conclusion

Web services are the emerging technology in the current computer field. This is language neutral, so it gains more interest from the scholars and attaining new development. But the complexity comes in the selection process of the web services. To overcome this problem a ranking methodology with user feedback is illustrated in our approach. Some of the noteworthy advantages of our proposed approach are as follows.

- Irrespective of user domain knowledge we can prioritize the QoS constraints.
- Dynamic ranking algorithm increases the performance.
- Both functional and non functional properties are given equal importance.
- Simple algorithm which reduces the complexity of the computation.

- Users directly determine the ranking process instead of some third party of the SP itself. This increases the accuracy and satisfies the needs of the SR very efficiently.

As our approach uses a dynamic strategy it is capable to change its nature according to the user speed and also the QoS errors are minimized automatically. In future, the QoS validation process is to be automated which will make the selection process more easily without any human intervention, considering the real dataset on the web.

References

1. Sommerville J (2006) Software engineering, 8th edn. Addison Wesley, Boston
2. Erl T (2005) Service-oriented architecture: concepts, technology, and design. Prentice Hall, Englewood Cliffs
3. Gwo-Hshiung T (2010) Multiple attribute decision making: methods and applications. Multiple Attribute Decision Making: Methods and Applications
4. Wang J, Hwang W-L (2007) A fuzzy set approach for R&D portfolio selection using a real options valuation model. Omega 35(3):247–257
5. David B (2004) Web services architecture. W3C working group note 11, February, W3C technical reports and publications
6. Zeng L (2004) QoS-aware middle for web services composition. IEEE Trans Softw Eng (IEEE Press, NJ USA)
7. Ramakishnan M, Suchithra M (2013)A Review on semantic web service discovery algorithm. Intl J Soft Comput 8(4)
8. Tran VX, Tsuji H (2008) QoS based Ranking for web services: fuzzy approaches. In: 4th international conference on next generation web services practices, Seoul, Republic of Korea, pp 77–82, 20–22 Oct 2008
9. Önüt S, Kara SS, Işik E (2009) Long term supplier selection using a combined fuzzy MCDM approach: a case study for a telecommunication company. Expert Syst Appl 36(2):3887–3895
10. Huang CL, Lo CC, Chao KM, Younas M (2007) Reaching consensus: a moderated fuzzy web services discovery method. Inf Softw Technol 48(6)
11. Huang C-L et al (2005) A consensus-based service discovery. First international workshop on design of service-oriented applications, Amsterdam, 12–14 Dec 2005
12. Xiong P, Fan Y (2007) QoS-aware web service selection by a synthetic weight. In: FSKD proceedings of the fourth international conference on fuzzy systems and knowledge discovery, vol 03
13. Hu R et al (2011) WSRank: a method for web service ranking in cloud environment. In: IEEE ninth international conference on dependable, autonomic and secure computing (DASC), IEEE
14. Zeng L, Benatallah B, Dumas M, Kalagnanam J, Sheng QZ (2003) Quality driven web services composition. In: Proceedings on WWW
15. Canfora G, Penta MD, Esposito R, Villani ML (2005) An approach for QoS-aware service composition based on genetic algorithms. In: Proceedings on GECCO
16. Canfora G, Penta MD, Esposito R, Perfetto F, Villani ML (2006) Service composition (re) binding driven by application-specific QoS. In: Proceedings on ICSOC
17. Nguyen XT, Kowalczyk R, Han J (2006) Using dynamic asynchronous aggregate search for quality guarantees of multiple web services compositions. In: Proceedings on ICSOC
18. Sathya M, Swarnamugi M, Dhavachelvan P, Sureshkumar G (2010) Evaluation of qos based webservice selection techniques for service composition. Int J Software Eng 1(5):73–90

19. Zheng Z et al (2011) Qos-aware web service recommendation by collaborative filtering. IEEE Trans Serv Comput 4(2):140–152
20. O'Sullivan J, Edmond D, Hofstede AHMT (2005) What's in a service? Distrib Parallel Databases
21. ITU-T Rec. E.800 (2008) Terms and definitions related to quality of service and network performance including dependability
22. Wang S, Sun Q, Yang F (2012) Quality of service measure approach of web service for service selection. IET Softw
23. Subashini S, Kavitha V (2011) A survey on security issues in service delivery models of cloud computing. J Netw Comput Appl 34(1):1–11, 5

Experimental Study on Chunking Algorithms of Data Deduplication System on Large Scale Data

T.R. Nisha, S. Abirami and E. Manohar

Abstract Data deduplication also known as data redundancy elimination is a technique for saving storage space. The data deduplication system is highly successful in backup storage environments. Large number of redundancies may exist in a backup storage environment. These redundancies can be eliminated by finding and comparing the fingerprints. This comparison of fingerprints may be done at the file level or splits the files to create chunks and done at the chunk level. The file level deduplication system leads poor results than the chunk level since it considers the entire file for finding hash value and eliminates only duplicate files. This paper focuses on the experimental study on various chunking algorithms since chunking plays a very important role in data redundancy elimination system.

Keywords Data deduplication · Backup storage · Chunking · Storage

1 Introduction

Data deduplication is a storage optimization technique which is used to improve the storage efficiency in backup storages in a large scale. Data deduplication technique consists of various approaches which could be used to reduce the storage space needed to save data or the amount of data to be carried over the network. Normally, these data deduplication approaches [1,2] are used to detect redundancies from large datasets, like archival storage [3], primary storage, and backup storages [4]. Backup storage system is the predominant accomplishment of data deduplication

T.R. Nisha (✉) · S. Abirami · E. Manohar
Department of Information Science and Technology, College of Engineering, Guindy, Anna University, Chennai, India
e-mail: nishaa1210@gmail.com

S. Abirami
e-mail: abirami_mr@yahoo.com

L.P. Suresh and B.K. Panigrahi (eds.), *Proceedings of the International Conference on Soft Computing Systems*, Advances in Intelligent Systems and Computing 398, DOI 10.1007/978-81-322-2674-1_9

technique which compresses a large scale storage requirement to a fraction of the logical backup data size.

Deduplication is generally done by identifying duplicate chunks of data and maintains a single copy of repeated chunks on the disk to save disk space. New occurrences of chunks that were previously stored are replaced by a reference to the already stored data piece. These references are typically of negligible size when compared to the size of the original redundant data. The technique can be applied within a file or object (intra deduplication) or between different files (inter deduplication). Some obvious applications of deduplication are typical backups, which usually contain copies of the same file captured at different instant; virtualized environments storing similar virtual images [5] and even in primary storage, users sharing almost identical data such as common project files.

Deduplication can be performed either offline or in-line. Offline deduplication is performed on the storage data only after the data have been written on the primary storage. In-line deduplication is done as data flows through a deduplication system. In an offline deduplication system, new data are initially stored on the storage device. After storing the data, deduplication system analyzes the data for finding duplicates. An in-line deduplication system tries to find the duplicates in the data as the data enters the device in real time. In-line deduplication systems try hard to reduce the active data flowing across the storage and network infrastructure and hence improve capacity, bandwidth, and cost savings. In-line deduplication systems optimize the size of the data at the initial phases of its life-cycle so that any further deduplication (e.g., during backup) will ends in a smaller data size.

2 Whole File-Based Data deDuplication

In the whole file-based deduplication method, the entire file contents are used by the hashing algorithms such as MD5 or SHA-1 to calculate hash value and making the comparisons with the previously stored whole file hash values. Once if the file is identified as a duplicate file, then the logical references to be made with previously stored file. Otherwise the file is considered as a new data and it will be stored in the server. This whole file-based approach is very simple and comparatively very fast since it neglects the processing over head in other process like chunking, but it can only detect the file which is exactly duplicated. This approach will not consider, if there is any small changes in the file because of viewing the entire file as a whole part. Extreme Binning [6] approach uses the whole file hash method and exploits the file similarity for deduplication to apply to nontraditional backup workloads with low locality.

3 Chunking-Based Data deDuplication

In the chunking-based deduplication method, each file to be backed up is first divided into a sequence of mutually exclusive blocks, called "chunks." Each chunk is a contiguous sequence of bytes from the file which is processed independently. Various approaches are available in the literatures to generate chunks. Out of that, the two significant chunking methodologies are "static chunking" and "content-defined chunking."

3.1 Static Chunking

Instead of using the whole file as smallest unit, the static chunking approach breaks the file into equally sized chunks where the size of the chunk is based on the multiples of a disk sector or the block size of the file system. Because of these reasons, static chunking can also be termed as fixed-sized chunking or fixed block chunking or fixed chunking. Figure 1 shows the working flow of static chunking technique.

Static chunking has been considered as the fastest among the other chunking algorithms to detect duplicate blocks but the performance is not acceptable. But the static chunking algorithm leads to a serious problem called "Boundary Shifting."

3.2 Boundary Shifting Problem

Fixed-size chunking heavily results in boundary shifting problem due to the modifications in the data. When the user inserts or deletes a byte from the data, fixed-size chunking would generate different results for the subsequent chunks even though most

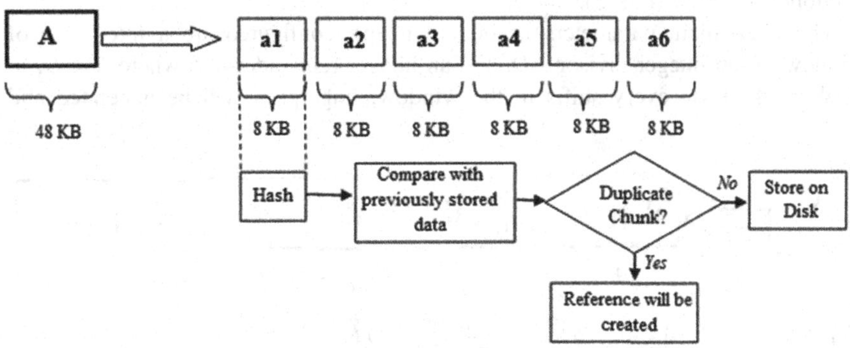

Fig. 1 Fixed-size chunking

of the data in the file are unchanged. This problem is known as the boundary shifting problem. In the second backup, if a new byte is inserted at the beginning of the file, all the remaining chunks may get a different hash value. Hence, everything will be considered as a new data and gets stored in the disk as new ones.

3.3 Content-Defined Chunking

Content-defined chunking is generally preferred in backup systems because it is not subject to the boundary shifting effect, which reduces the amount of duplicate data found by the data deduplication systems. This effect occurs when same data is stored multiple times but slightly shifted, e.g., because new data were inserted into the data stream. In these cases, a system that uses static chunking was unable to detect the redundancies because the chunks are not similar, but in case of content-defined chunking, the chunking realigns with the content and the same chunks as before are created. Instead of setting boundaries at multiples of the limited chunk size, CDC approach defines breakpoints where a specific condition becomes true. This is normally done with a fixed-size overlapping sliding window. Variable-size chunking is also called content-based chunking [7]. In other words, chunking algorithms determine chunk boundaries based on the content of the file. Most of the deduplication systems [8–14] use the content-defined chunking algorithm to achieve good data deduplication throughput. Figure 2 shows the example of variable length chunking.

3.4 Basic Sliding Window

The basic sliding window (BSW) algorithm [8] is the first chunking algorithm that uses the prototype of the hash-breaking (nonoverlap) approach and already been proven [8] to obtain the best performance in balancing capacity scalability and the deduplication ratio (Fig. 3).

The three main parameters needed to be pre-configured are a fixed size of window W, an integer divisor—D, and an integer remainder—R, where $R < D$. In BSW method, on every shifts in the window, fingerprint will be generated and

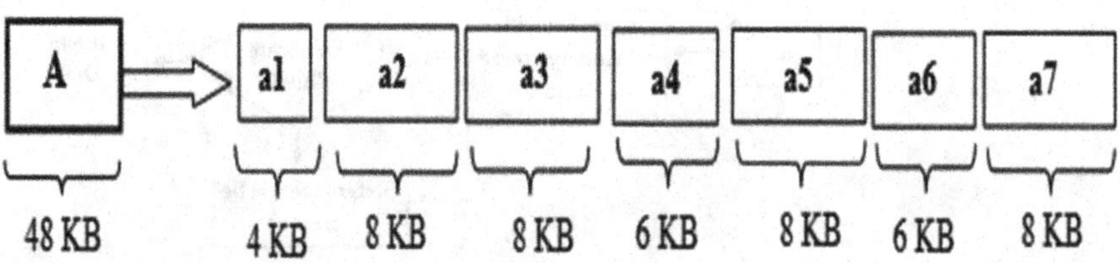

Fig. 2 An example of variable length chunking

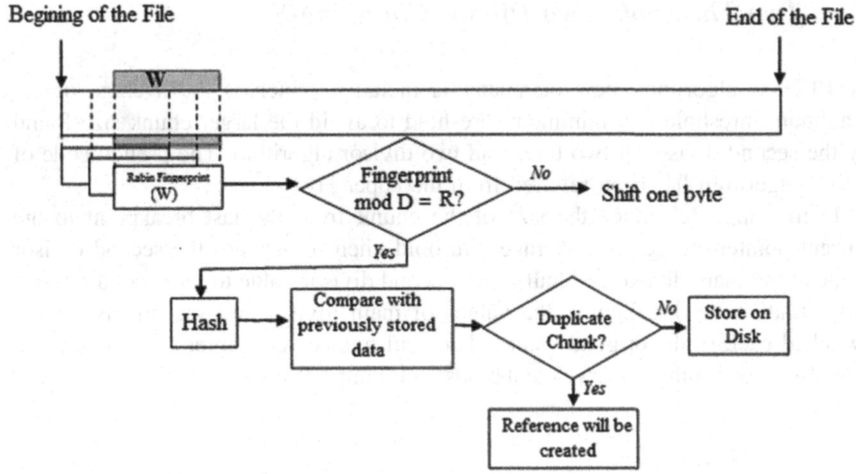

Fig. 3 Basic sliding window chunking

check for the chunk boundary. In the current experiment, this approach involves more calculations in finding the breakup points from the start point to the given threshold even though most of the breaking point found at the maximum threshold.

3.5 Two Threshold Two Divisor (TTTD) Chunking

The two threshold two divisor idea was proposed by HP laboratory [15] at Palo Alto, California. This TTTD algorithm uses same idea as the BSW algorithm does but it uses four parameters, the minimum threshold, the maximum, threshold, the main divisor, and the second divisor in order to avoid the extra processing time taken by the BSW in calculating the breakpoints till the maximum threshold. The pseudo code of TTTD algorithm has been referred from the paper [15].

In this approach, minimum threshold is used to eliminate very large-sized chunks and maximum threshold is used to eliminate the very small-sized chunks. The main divisor works as same as the divisor of the BSW chunking algorithm to determine the breakpoint and the second divisor value is the half of the main divisor. The second divisor in the two threshold two divisor algorithm is used to find the breakup point, in case the algorithm cannot find breakpoint by the main divisor.

TTTD algorithm decides whether to use the backup breakpoint found by the second divisor until it reaches the maximum threshold. And also the breakpoints found by the second divisor are larger and it is very close to the maximum threshold.

3.6 Two Threshold Two Divisor Chunking-S

In TTTD-S algorithm, new parameter is included which is the average of the maximum threshold and minimum threshold to avoid the larger chunk size found by the second divisor of two threshold two divisor algorithm. The pseudo code of TTTD algorithm has been referred from the paper [16].

In this approach, once the size of the chunk from the last breakpoint to the current pointer crosses this average threshold, then it switches the second divisor value to the main divisor and half of the second divisor value to the second divisor. After finding the breakpoint, the values of main divisor and second divisor are switched back to the original values. This will reduce very larger-size chunks and this also avoids unnecessary calculations and comparisons.

4 Results and Discussion

Various chunking algorithms are implemented in an Intel duo core CPU, 4 GB RAM, and 2500 GB HDD in Linux environment with gcc-based C programming language using Eclipse C/C++ development tooling.

Various datasets have been used for experimental study of chunking such as three weekly backups of our university lab backups and Linux-set source code files downloaded from the website [17], consist of 900 versions from 1.1.13 to 2.6.33 of Linux source code files, and represent the characteristics of small files. Linux-set that has a very large number of small files and small chunks. The university lab backups consist of various programs and projects developed by students on weekly basis in many languages, and it contains Word documents, CSS files, C# files, and so on. The details about datasets used in this experimental study are given in Table 1.

The comparisons between the various chunking algorithm in data deduplication is made in terms of the total number of chunks, duplicate detected by the data deduplication approach proposed by [9], disk access in the process detecting duplicates that are already stored in disk and the total storage size after removing the duplicates. Table 2 shows the analysis of fixed-sized, BSW, TTTD, and TTTD-S chunking based on the above-mentioned terms for the Linux version.

Table 1 Test dataset of the experiment

Dataset	Total no. of files	Total input size (GB)
Linux version	139,556	1.3
Lab backup	138,199	1.4

Table 2 Comparison of fixed-sized, BSW, TTTD, and TTTD-S for Linux version

Chunking method	Total no. of chunks	Duplicate chunks	No. of disk access	Total storage size (MB)
Fixed-sized	568,392	433,908	22,985	315.5
BSW	576,080	443,762	23,920	306.0
TTTD	1,461,040	1,290,928	67,419	157.6
TTTD-S	1,494,826	1,320,561	69,178	157.2

From Table 2, it is obvious that the TTTD and TTTD-S algorithm increases the total number of chunks than the fixed-sized and BSW methods. It is also visible that the duplicates found by TTTD and TTTD-S algorithm is more, and it considerably reduces the size of the storage by storing only the unique chunks. Increase in the total number of chunks leads to increased disk access and hence reduced running time.

5 Conclusion

Different chunking algorithms like whole file chunking, fixed-sized chunking, and variable-sized chunking like BSW chunking, two threshold two divisor and two threshold two divisor with switch parameter chunking are explored in this work. Comparisons between the different methodologies of chunking have also been discussed. From these works, it is obvious that still a lot more challenges like reducing the processing time, metadata overhead need to be addressed in the future researches. And creation of the optimized chunking algorithm will increase the throughput of the data deduplication system.

References

1. Kulkarni P, Douglis F, LaVoie J, Tracey J (2004) Redundancy elimination within large collections of files. In: Proceedings of the USENIX annual technical conference, pp 59–72
2. Meyer D, Bolosky W (2011) A study of practical de-duplication. In: Proceedings of the 9th USENIX conference on file and storage technologies
3. Quinlan S, Dorward S (2002) Venti: a new approach to archival storage. In: Proceedings of the first Usenix conference on file and storage technologies, Monterey, California, pp 89–102
4. Wei J, Jiang H, Zhou K, Feng D (2010) MAD2: a scalable high throughput exact de-duplication approach for network backup services. In: 26th IEEE mass storage systems and technologies (MSST), Incline Village, NV, USA, pp 1–14, May 2010
5. Jin K, Miller E (2009) The effectiveness of de-duplication on virtual machine disk images. In: Proceedings of SYSTOR 2009. The Israeli experimental systems conference. ACM, pp 1–12
6. Bhagwat D, Eshghi K, Long D, Lillibridge M (2009) Extreme binning: scalable, parallel de-duplication for Chunk-based file backup. In: Proceedings of IEEE international symposium on modeling, analysis & simulation of computer and telecommunication systems, pp 1–9

7. Geer D (2008) Reducing the storage burden via data deduplication. Computer 41(12):15–17
8. Muthitacharoen A, Chen B, Mazieres D (2001) A low-bandwidth network file system. In: 18th ACM symposium on operating systems principles (SOSP '01), Chateau Lake Louise, Banff, Canada, pp 174–187
9. Xia W, Jiang H, Feng D, Hua Y (2014) Similarity and locality based indexing for high performance data de-duplication. IEEE Trans Comput 1–14
10. Lillibridge M, Eshghi K, Bhagwat D, Deolalikar V, Trezise G, Camble P (2009) Sparse indexing: large scale, inline de-duplication using sampling and locality. In: Proceedings of the 7th conference on file and storage technologies, pp 111–123
11. Debnath B, Sengupta S, Li J (2010) Chunkstash: speeding up inline storage de-duplication using flash memory. In: Proceedings of the 2010 USENIX conference on USENIX annual technical conference. USENIX Association
12. Dong W, Douglis F, Li K, Patterson H, Reddy S, Shilane P (2011) Tradeoffs in scalable data routing for de-duplication clusters. In: Proceedings of the 9th USENIX conference on file and storage technologies. USENIX Association
13. Dubnicki C, Gryz L, Heldt L, Kaczmarczyk M, Kilian W, Strzelczak P, Szczepkowski J, Ungureanu C, Welnicki M (2009) Hydrastor: a scalable secondary storage. In: Proceedings of the 7th conference on File and storage technologies. USENIX Association, pp 197–210
14. Zhu B, Li K, Patterson H (2008) Avoiding the disk bottleneck in the data domain de-duplication file system. In: Proceedings of the 6th USENIX conference on file and storage technologies, vol 18(1–18). USENIX Association Berkeley, p 14
15. Eshghi K, Tang HK (2005) A framework for analyzing and improving content-based chunking algorithms. Hewlett-Packard labs technical report (TR). HPL 2005-30(R.1)
16. Moh TS, Chang B (2009) A running time improvement for two thresholds two divisors algorithm. In: Cunningham HC, Ruth P, Kraft NA (eds) ACM Southeast regional conference. ACM, p 69
17. Linux download. ftp://kernel.org/

A Systematic Review of Security Measures for Web Browser Extension Vulnerabilities

J. Arunagiri, S. Rakhi and K.P. Jevitha

Abstract Web browser is a software application using which we can perform most of the internet-based activities. The commonly used browsers are Mozilla Firefox, Google Chrome, Safari, Opera Mini, and Internet Explorer. Many web applications provide extensions to these browsers to enhance their functionality, while some of the extensions perform malicious activities to get access to the sensitive data without the user's knowledge. This paper presents a review of the research done on the browser extension vulnerabilities. We found that the most of the researches were done for Firefox and Chrome extensions. Static analysis technique was used in most of the solutions proposed by various researchers. There is no ready to use tool for evaluating the vulnerable behavior of an extension. Hence there is need for more research to evaluate and eliminate the vulnerabilities in web browser extensions.

Keywords Web browsers · Extensions · Vulnerability

1 Introduction

Browsers extensions are small JavaScript programs that are used to provide additional features and functionalities to the web browsers. Extensions provided better user interface and convenient operations, which made them an important part of the web browsers.

J. Arunagiri (✉) · S. Rakhi
Department of Computer Science and Engineering, Amrita School of Engineering, Ettimadai, Coimbatore, India
e-mail: jarunagiri5@gmail.com

S. Rakhi
e-mail: rakhi36s@ymail.com

K.P. Jevitha
Department of Computer Science Engineering, Amrita School of Engineering, Ettimadai, Coimbatore, India
e-mail: kp_jevitha@cb.amrita.edu

L.P. Suresh and B.K. Panigrahi (eds.), *Proceedings of the International Conference on Soft Computing Systems*, Advances in Intelligent Systems and Computing 398, DOI 10.1007/978-81-322-2674-1_10

Extensions can be created using HTML, CSS, and JavaScript. They can interact with webpages or servers using content script or cross-origin XMLHTTPRequests. They can also interact with browser features like tabs and bookmarks.

Most of the extensions are developed to provide a specific functionality to the web browsers like email notification, download management, web search and so on. Due to improper secure coding practices and lack of proper understanding by the developers, the most of the extensions become vulnerable to malicious attacks. Some of the malicious activities performed by the vulnerable extensions are remote code execution, data exfiltration, saved password theft, cross-site HTTP Request, cross-site forgeries, and many others.

2 Research Method

This paper presents a systematic analysis and review of research works on web browser extensions vulnerabilities, and attacks. Several papers were found, which were related to browsers, browser extensions and security. Out of this, around 16 papers dealt with problems on browser extension security.

2.1 Research Questions

The research questions addressed in this study are listed below.

Q1: How many academic research papers are available, relevant to browser extension security?
Q2: What are the browser extension vulnerabilities addressed by these papers?
Q3: What are the proposed solutions or techniques to mitigate these problems?
Q4: Which browsers are mainly focused on?

For answering Q1, we referred papers published from year 2008 to 2014, which was the active period of research on browser extension vulnerabilities. With respect to Q2, we wanted to identify the major browser extension vulnerabilities that were addressed in these papers. Regarding Q3, we tried to identify the techniques and solutions proposed to prevent or detect these vulnerabilities. In Q4, we looked at the major browsers that were used in different papers.

2.2 Search Key Words

The following search keywords were used to identify the relevant academic work published in several journals and conferences in this domain.

- Browser Extension Security
- Browser Extension Security Tools
- Browser Security Tools
- Browser Security Measures
- Code Analysis Techniques
- Static Code Analysis Technique

2.3 Categorization and Selection Criteria

The research papers collected using the keywords mentioned in Sect. 2.2 were grouped into several categories based on browsers used, vulnerabilities addressed, and the solution techniques. The papers that focused only on browser extension vulnerabilities were shortlisted for this review. Papers that focused on extension development and issues related to browsers other than extensions were excluded.

3 Comparative Study of Browser Extension Vulnerabilities and Their Solutions

See Table 1.

4 Result

Table 1 summarizes the vulnerabilities addressed and the solutions proposed for browser extensions by various researches.

Table 2 shows the number of papers published per year since 2008 for addressing extension vulnerabilities. It shows that the research in this area is increasing each year. 2014 and 2012, marked the highest amount of research publications done in this domain.

Table 3 depicts the major vulnerabilities that are focused in these papers. From the results, it is clear that majority of researches was focused on addressing high privileges, which could lead to data exfiltration or remote code execution.

Table 4 shows the different code analysis techniques used to mitigate the vulnerabilities identified. Most of the solutions have used static code analysis technique.

Table 5 presents the browsers focused by these research papers. Most of the papers addressed the vulnerabilities in the extensions for Mozilla Firefox followed by Chrome browser.

Table 1 The literature survey of several papers related to our analysis

Reference	Browser	Vulnerabilities addressed	Proposed solution		Implementation approach
			Method	Tool proposed	
1.	Mozilla Firefox	Remote code execution, data exfiltration, saved password theft, and preference modification	Run time policy enforcement for Firefox to give fine-grained control over the actions of JavaScript to the users	SENTINEL–Firefox	Intercepting XPCOM operations from the extensions, and use Policy manager to verify the operation against a local policy database [1]
2.	Google Chrome	Information dispersion (spamming) using internet access privileges and extension updates, information harvesting (password theft), cross-site HTTP request, high-privilege and cross-site forgeries	Micro-privilege management: Differentiating DOM (document object model) elements with Sensitivity levels	PROCTOR (helper extension)	Extension for automatically labeling sensitivity levels of DOM elements, micro-privilege management: Finer permission definitions for each component of extension use new variables to save the permissions for each component (as chrome saves only the host permissions) [2]
3.	Mozilla Firefox	Execution of remote code and API communication across protection domains	Static information flow analysis-taint based analysis	VEX	The tool gives the code path from untrusted sources (*document.popupNode, BookmarksUtils, window.content.document, nsIRDFService, nsILivemarkServices and nsIPrefService*) to executable sinks (*eval, appendChild, innerHTML*) in the extensions' code. This has to be manually examined for exploitability [3]

(continued)

Table 1 (continued)

Reference	Browser	Vulnerabilities addressed	Proposed solution		Implementation approach
			Method	Tool proposed	
4.	Mozilla Firefox	Cross-site scripting, replacing native API, JavaScript capability leaks, mixed content, over privilege	Isolated words, privilege separation and permissions model	Adapted by Google Chrome	**Least Privilege**: instead of extensions running with user's full privileges, they require privileges to be explicitly requested in the extension's manifest file. **Privilege Separation**: Extension to be divided into multiple components such as *Content Script* (JavaScript to interact directly with un trusted web content), *Extension Core* (HTML and JavaScript to control extensions' UI and acts as extension APIs specified in the manifest file), *Native Binary* (execute arbitrary code or access arbitrary files). **Isolated Mechanism**: To isolate extension components from each other, three mechanisms are used: *Origin* (use same-origin web sandbox by running the extension core at an origin designated by a public key), *Process isolation* (extension core and native binaries to execute in their own

(continued)

Table 1 (continued)

Reference	Browser	Vulnerabilities addressed	Proposed solution		Implementation approach
			Method	Tool proposed	
					processes), *Isolated worlds* (content script and untrusted web content to use separate web content heap) [4]
5.	Google Chrome	HTTP network attack, over privilege, silent installation	Isolated worlds, privilege separation and permissions model	Extension checker	**To detect silent installation for Firefox**: Check for fake entry into extensions.sqllite file inside the user profile directory, **To detect silent installation for Chrome**: Checking for fake entry into extensions preferences file (*json*) in the user data directory [5]
6.	Google Chrome	Network attacks on extensions, website metadata attacks on extensions, and vulnerabilities that extensions add to websites	Banning HTTP scripts from core extensions, preventing extension from adding http request to https website, banning inline scripts from extension HTML, banning dynamic code generation (e.g.: using eval and settimeout); Banning HTTP XHR(XML HTTPRequest)—all XHR to use HTTPS Request	Not available	**Black-box testing**, **Source code analysis**: tracking information flow from untrusted source to execution sink, **Holistic testing**: mapping source code to extensions actions, **Privilege separation, Isolated worlds, API Permissions** [6]
7.	Mozilla Firefox	Exploiting DOM using DOM (document object model)-based attacks, exploiting JavaScript	**Limited DOM Access** (prevent access to Https DOM), **Integrity checking** (use hash of	Not available	**Input Grabber**(abusing JavaScript functions like document.getElementBy

(continued)

Table 1 (continued)

Reference	Browser	Vulnerabilities addressed	Proposed solution		Implementation approach
			Method	Tool proposed	
		methods (using *eval, innerHTML and wrapped-JSObject for code injection and privilege escalation attacks*), exploiting API (using XMLHttpRequest for cross-domain request), exploiting XPCOM interface(to interact with network, file system etc.), over privileges	an extension to check its integrity every time the extension is very loaded), **Least Privileges** (limit the privileges that an extension acquires), **Avoiding dangerous method** (use evalInSandbox() instead of eval(), document. createTextNode instead of innerHTML())		TagName, and java script XPCOM interface like nsILocalFile), **Page Modifier** (functions like document. getElementById().innerHTML dynamically modifies web page content), **Overriding wrappers** (wrapper Java Script Object weakens the Firefox wrappers function), **Injecting Brower Exploitation Framework (BeEF)** (malicious codes from browser Exploitation Framework can be used as JavaScript hook into html code when the extension is invoked after the page is loaded), **Bypass Same-Origin Policy (SOP)** (using nsIPrefBranch XPCOM interfaces to change Firefox preferences, and then cross-domain calls made using XMLHTTPRequest method) [7]
8.	Mozilla Firefox	RDF Request from untrusted sources, users are forced to download malicious extensions	Hidden Markov model (HMM) —identifying whether an	Not available	**Model generation** (based on predefined set of rules, and the input conditions, a model will

(continued)

Table 1 (continued)

Reference	Browser	Vulnerabilities addressed	Proposed solution		Implementation approach
			Method	Tool proposed	
			extension is benign, vulnerable or malicious extension		be generated), **Model training** (making the model parameters to fit with the observations), **Model matching/detection** (model that generates maximum probability value corresponding to any of the malicious, vulnerable and benign types will be matched with it, correspondingly) [8]
9.	Mozilla Firefox and Google Chrome	Steal sensitive data using high privilege, browser vulnerabilities	Pointer analysis to detect vulnerability in JavaScript	BEVDT (browser extensions vulnerabilities detection tool)	**Preprocessor module** (extract the JavaScript from the extension), **Rhino module** (extract AST from JavaScript code), **Transformation module** (converts AST into Datalog facts), **bddbddb module** (binary decision diagram based deductive database, detect the vulnerability) and **Output Module** (shows the results and the time taken for its detection) [9]
10.	Mozilla Firefox	**Integrity of the Browser code base** (malicious extension can change the browser actions)	**Protecting code based integrity** (extension code signing), **User data**	Not available	**Protecting Integrity of the Browser code base** (extension code signing by certificate

(continued)

Table 1 (continued)

Reference	Browser	Vulnerabilities addressed	Proposed solution		Implementation approach
			Method	Tool proposed	
		confidentiality and integrity of the user data (malicious extension can lead to data exfiltration), **BROWSERSPY** a proof of concept, malicious extension to demonstrate various attacks like harvesting URL, username and passwords from browser, sensitive user data and browsing history	**confidentiality and integrity** (using run time monitoring with specified policies)		generation for each extension installed, integrity of the extension to be checked when the extension is loaded by calculating the extension hash and verifying it with hash stored in the certificate.) **Confidentiality and integrity of the user data** (Enable the browser to monitor extension code at run time and enforce policy on each extension.) [10]
11.	Mozilla Firefox	Full browser privilege, vulnerabilities due to information flows from sensitive source to executable sinks	Staged information flow (SIF) framework	SIFEX	Parses the extensions' code to find the potentially vulnerable flow from the sensitive sources to executable sinks. Static and flow insensitive analysis is used [11]
12.	Mozilla Firefox, Google Chrome, Internet Explorer, Safari, Opera	Policy authorizations, arbitrary inter leavings of extension code with other untrusted scripts on a web page	Policy language for fine-grained specifications, Semantics and tools support to understand policies, static verification	FINE	**Policy Language for fine-grained specifications** (distinguish data from metadata and ensuring the stability of a security policy and the choice of Datalog and the usage of Datalog facts in the policy), **Semantics and tools support to understand policies**

(continued)

Table 1 (continued)

Reference	Browser	Vulnerabilities addressed	Proposed solution		Implementation approach
			Method	Tool proposed	
					(Calculus for browser extension that represents the model as well as the interaction with DOM, Dynamic semantics for the account of possibility of interleaving between the untrusted, page resident JavaScript and extension code, traditional progress property on reduction relation), **static verification** (using value-indexed types, dependent function types, refinement types, refinements as pre and post conditions, kind language, top-level assumptions, refinement type checking; Refined APIs for extensions.) [12]
13.	Mozilla Firefox	High privileges	**Information flow analysis** (to detect the various suspicious flows) **and Machine learning** (to decide the exploitability of the suspicious flows)	Not available	**AST traversal** and **Taint analysis.** **Information flow type bugs in add-ons** (Taint analysis and points to analysis), **Determining exploitability using Machine learning methods** (Naïve Bayes (NB) and Tree-Augmented

(continued)

Table 1 (continued)

Reference	Browser	Vulnerabilities addressed	Proposed solution		Implementation approach
			Method	Tool proposed	
					Naïve Bayes (TAN) methods are applied to dataset and compared the predicted value with true result obtained) [13]
14.	Mozilla Firefox	XSS attacks on leaking confidential data	**Information flow control** (using reference monitors and protection zones)	BFlow	**Client implementation** (reference monitor implemented as Firefox plugin), **user authentication** (after user authentication HTTP requests are authenticated using authentication cookies), **server implementation** (user gateways), **server storage** (key-value storage system) [14]
15.	Google Chrome	High privilege	Static analysis framework to detect the malicious information flow and violation of the principle of least privilege	Not available	**Analysis of least privilege** (permissions with no corresponding functions are reported as that it violates the principles of least privilege), **Analysis of Malicious information flow** (leaking of sensitive data to remote servers by invoking URL's, extensions downloading malicious files from remote servers to the user's file system) [15]

(continued)

Table 1 (continued)

Reference	Browser	Vulnerabilities addressed	Proposed solution		Implementation approach
			Method	Tool proposed	
16.	Mozilla Firefox	High privilege	Static Taint analysis, code transformation, sensitive information management	Not available	**Static Taint Analysis** (Information flow analysis on Java Script code to identify request to download Java Script files, suspicious source to sink and request for sensitive data.) **Code Transformation** (Replacing evalInSandbox function.) **Sensitive Information Management** (option to create fake profile) [16]

Table 2 Number of papers published per year on browser extension security

Year	Paper
2008	1
2009	1
2010	2
2011	2
2012	4
2013	2
2014	4

Table 3 Summary of vulnerabilities addressed by various research publications

Vulnerabilities mentioned	Number of papers (%)
High privilege	90
Remote code execution	6
Browser problems	1
Data exfiltration	2
Network attacks	1

Table 4 Summary of code analysis technique used

Code analysis technique	References	Count
Static	[2, 3, 5, 10–12, 14–16]	09
Dynamic	[1, 4, 6, 7, 9]	05
Static and dynamic	[8]	01
Other	[13]	01

Table 5 Number of browsers mentioned in papers

Browsers	Papers
Firefox	11
Chrome	4
Combinations of common browsers	2

5 Conclusion

Although extensions provide additional functionality to the web browsers, there is a chance that many of them could be vulnerable due to inadequate focus on security during the development process. This study presents the summary of research works done in the field of web browser extensions. Compared to other web security issues, the number of publications for extension security is low in number. Most of the research works used program analysis techniques. We found that machine learning techniques were hardly used. One of the possible future research directions in this field could be to try different machine learning techniques to identify malicious/vulnerable extensions. Moreover the tools to test the security of the given extension are not readily available for download. So this opens up the scope for more research in this domain.

References

1. Onarlioglu K, Battal M, Robertson W, Kirda E (2013) Securing legacy firefox extensions with SENTINEL. In: Detection of intrusions and malware, and vulnerability assessment. Springer, Heidelberg, pp 122–138
2. Liu L, Zhang X, Yan G, Chen S (2012) Chrome extensions: threat analysis and countermeasures. In: NDSS
3. Bandhakavi S, Tiku N, Pittman W, King ST, Madhusudan P, Winslett M (2011) Vetting browser extensions for security vulnerabilities with VEX. Commun ACM 54(9):91–99
4. Barth A, Felt AP, Saxena P, Boodman A (2010) Protecting browsers from extension vulnerabilities. In: NDSS
5. Rana A, Nagda R (2014) A security analysis of browser extensions. arXiv preprint arXiv: 1403.3235
6. Carlini N, Felt AP, Wagner D (2012) An evaluation of the Google Chrome extension security architecture. In: USENIX Security Symposium, pp 97–111
7. Saini A, Gaur MS, Laxmi V (2013) The darker side of Firefox extension. In: Proceedings of the 6th international conference on security of information and networks, ACM, pp 316–320
8. Shahriar H, Weldemariam K, Zulkernine M, Lutellier T (2014) Effective detection of vulnerable and malicious browser extensions. Comput Secur 47:66–84
9. Wang J, Li X, Yan B, Feng Z (2012) Pointer analysis based vulnerability detection for browser extension. Int J Digit Content Technol Appl 6(1)
10. Ter Louw M, Lim JS, Venkatakrishnan VN (2008) Enhancing web browser security against malware extensions. J Comput Virol 4(3):179–195
11. Agarwal S, Bandhakavi S, Winslett M (2010) SIFEX: tool for static analysis of browser extensions for security vulnerabilities. In: Poster at annual computer security applications conference
12. Guha A, Fredrikson M, Livshits B, Swamy N (2011) Verified security for browser extensions. In: IEEE Symposium on security and privacy (SP), IEEE, pp 115–130
13. (2012) Static information analysis and machine learning methods on automated detection of browser extension vulnerabilities. Accessed from (project not published)
14. Yip A, Narula N, Krohn M, Morris R (2009) Privacy-preserving browser-side scripting with BFlow. In: Proceedings of the 4th ACM European conference on computer systems, ACM, pp 233–246
15. Aravind V, Sethumadhavan M (2014) A framework for analysing the security of chrome extensions. In: Advanced computing, networking and informatics, vol 2. Springer International Publishing, Berlin, pp 267–272
16. Marston J, Weldemariam K, Zulkernine M (2014) On evaluating and securing firefox for Android browser extensions. In: Proceeding

Facial Expression Recognition Using PCA and Texture-Based LDN Descriptor

Roshni C. Rahul and Merin Cherian

Abstract The most expressive way in which humans can display their emotion is through facial expression. The development of automated system that performs this task is very difficult. In this paper, we introduced texture-based LDN descriptor along with PCA which helps to recognize facial expression in efficient manner. In order to obtain directional information, we are using two masks i.e. Kirsch and Robinson mask and we compare their efficiency. Experiments shows that this method is more efficient compared to existing works.

Keywords Facial expression recognition · Local feature · LDN · PCA

1 Introduction

Facial expression recognition plays a major role in human–computer interaction [1, 2] as well as in biometrics that helps to analyse the emotional state of an individual. There exist a large number of systems that perform facial expression recognition, but their efficiency depends on the method that they adopt. The applications include human–computer interaction, multimedia, and security. Facial expression recognition find uses in a host of other domains like behavioural science, telecommunications, automobile safety, video games, psychiatry affect, animations, and sensitive music juke boxes etc.

Features for facial expression recognition can be classified into two namely local (details or parts) and global feature (as whole) refer Fig. 1. Global features focus on the entire image which is less accurate, while local features provide micro information about the face. It helps to identify and verify the persons using unique details in the face image. Hence local feature is more accurate compared to global

R.C. Rahul (✉) · M. Cherian
Computer Science and Engineering, Mahatma Gandhi University, Federal Institute of Science and Technology, Angamaly, Kerala, India
e-mail: roshnicrahul91@gmail.com

L.P. Suresh and B.K. Panigrahi (eds.), *Proceedings of the International Conference on Soft Computing Systems*, Advances in Intelligent Systems and Computing 398, DOI 10.1007/978-81-322-2674-1_11

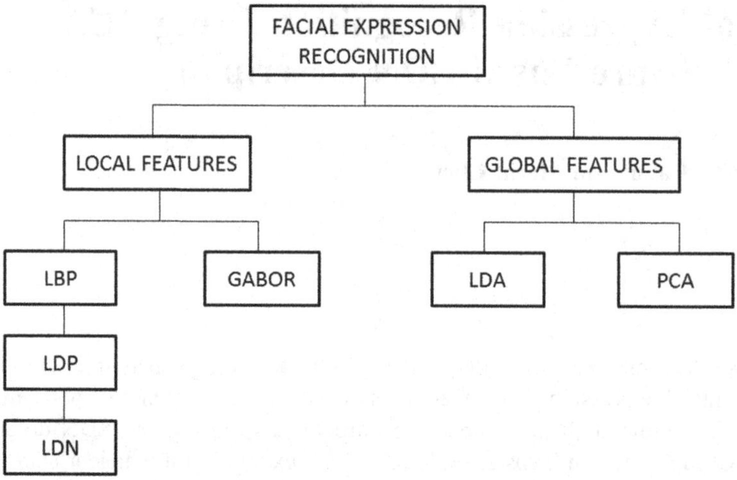

Fig. 1 Generalisation of facial features

feature. The local feature methods compute the descriptor from parts of the face, and then gather the information together to form a global descriptor.

Linear discriminate analysis (LDA) and principle component analysis are the most commonly used global descriptors. LDA makes use of dimensionality reduction technique which search for those vectors in the underlying space that best discriminate among classes. PCA is the most common in which the main objective is to perform feature selection by eliminating irrelevant features.

In the proposed system, local texture-based descriptors are used. Texture is usually defined as the smoothness or roughness of a surface. In computer vision, it is termed as the visual appearance of the uniformity or lack of uniformity of brightness and colour. In our approach, we are using local directional number pattern (LDN) which encodes directional information of face textures. It helps to distinguish among similar structural patterns having different intensity transition. Thus local information regarding face texture obtained using LDN can be further used for analysing facial expression of an individual.

The expression recognition systems consist of four major steps which produce efficient result. First stage is face detection i.e. the region of interest (ROI) is detected from the image so that computation has to be performed only where the required data resides and it increases the computational speed. Along with it, normalization is done which convert data image into a normalized value according to the requirement of application. Next step is feature extraction which extracts the distinct features and irrelevant features are eliminated in feature selection process. Final step of facial expression recognition is classification where the expressions are classified into six basic emotions. In this paper, we mainly focus on feature extraction method which can make the recognition system more real time and robust. There are two common approaches to extract facial features: geometric feature-based [3, 4] and appearance-based methods [5].

There exist several literatures on different local texture-based features. Among these methods are local features analysis [6], graph-based methods [7], Gabor features [8], elastic bunch graph matching [9], and local binary pattern (LBP) [10], [11].

LBP is one of the widely used method in which it is used not only for facial feature extraction, but also for iris recognition, palm vein recognition, hand recognition, etc. Newer methods tried to overcome the shortcomings of LBP, like local ternary pattern (LTP) and local directional pattern (LDP) [12]. Instead of the intensity, LDP encodes the directional information in the neighbourhood. However, some methods such as WLBPH [13] make use of Weber's Law which utilize the space location information and reduce the complexity of recognition which are useful while processing infrared images. Rivera et al. [14] proposed a face descriptor called local directional number pattern (LDN), for robust face recognition that encodes the structural information and the intensity variations of the face's texture by analysing directional information. The main drawback is that some of the irrelevant features are also extracted which reduce the efficiency. To overcome this weakness, we have first preprocessed the input image from CK+ dataset by detecting only the face region with the help of Viola Jones Algorithm and then evaluate the performance while using Kirsch and Robinson mask. Finally, we have reduced the dimensionality of the LDN features using PCA to extract relevant features.

2 Methodology

The directional number pattern is a local feature descriptor for face analysis. It encodes directional information of face textures. Each pixel of the input image is represented as 6 bit binary code. It helps to distinguish among similar structural patterns that are having different intensity transition. Compared to previous work [14], we have preprocessed the input image by cropping the region of interest with the help of Haar feature and also reduced the dimensionality of LDN descriptor using PCA. The main steps that are used to recognize facial expression includes:

- Preprocessing Phase
- LDN code generation Phase
- Histogram generation Phase
- Dimensionality reduction Phase
- Facial expression recognition Phase

Figure 2 shows the flow diagram of our approach. Our aim is to recognize six basic emotions of an individual. We have used CK+ dataset [15] which is publically available. Compared to the previous work [14], we have reduced the dimensions using principal component analysis which increase the efficiency of our system.

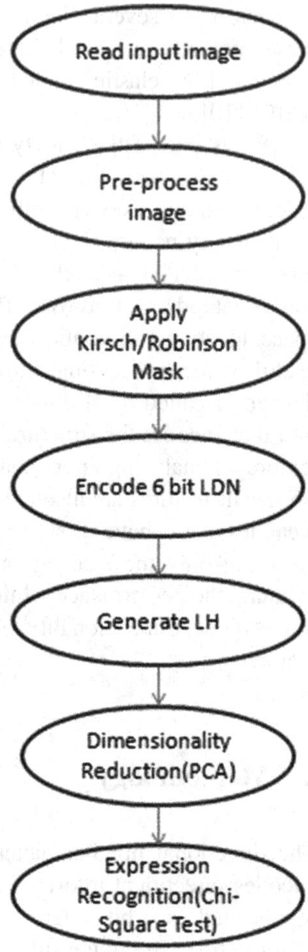

Fig. 2 Flow diagram for expression recognition

2.1 Preprocessing Phase

In our proposed method, we have used face image from CK+ dataset. In order to obtain the region of interest, i.e. only face region and to eliminate background, Viola Jones algorithm is used and then crop the face region. This helps to reduce the computation time. The key steps involved in Viola Jones Face Detection algorithm include:

- Applying Haar Feature
- Integral Image
- AdaBoost classifier
- Cascading

Using the algorithm, we detected ROI and then we extracted it from input image. This extracted image is then used for further processing thus reducing the computation time.

2.2 LDN Code Generation Phase

In order to generate LDN code, we compute edge response of neighbourhood using compass mask. It helps to gather information in the 8 directions. In this paper, we used two compass masks mainly Kirsch and Robinson mask. Kirsch mask (Fig. 3) rotated 45° apart to obtain edge response in 8 different directions. It operates in the gradient space which has more information compared to intensity feature space. Kirsch mask help to reveal the structure of the face and also it is robust against noise.

The edge magnitude is equal to the maximum value found by the convolution of each mask with the image. So the edge direction is defined by the mask that produces the maximum magnitude. Robinson Mask (Fig. 4) is similar to Sobel Mask with mask coefficient as 0, 1, and 2. In this paper, we compare the performance of both Masks to generate LDN descriptor. To generate LDN code, analyse edge response of each mask which represents significant edges in its respective direction. Identify the highest positive and negative values.

Fig. 3 Kirsch mask

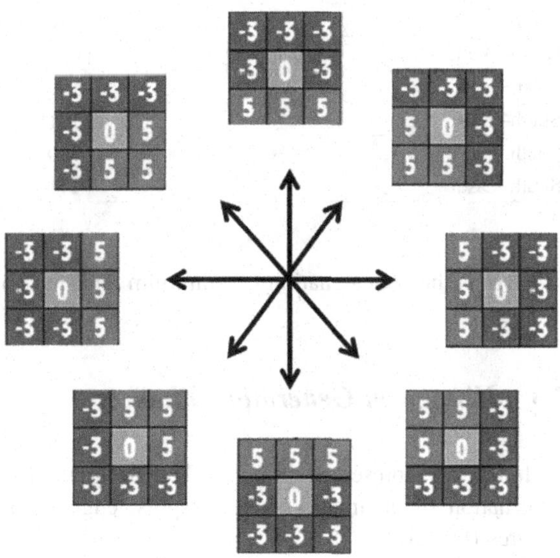

Fig. 4 Robinson mask

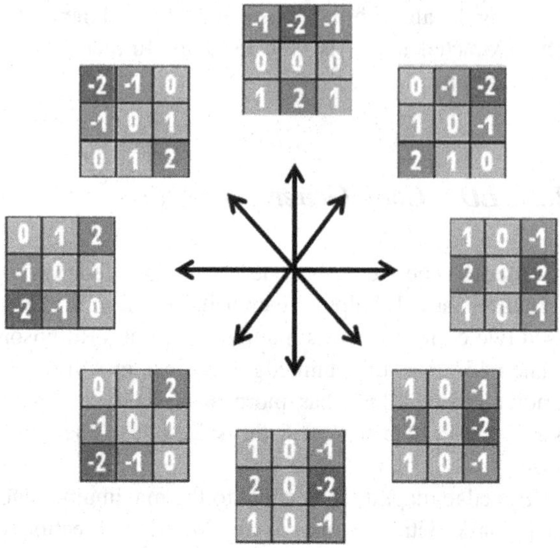

Encode prominent regions i.e. binary codes are the binary numbers of the directions that they represent. e.g.

East:	000
North–East:	001
North:	010
North–West:	011
West:	100
South–West:	101
South:	110
South–East:	111

LDN is the concatenation of minimum and maximum directions.

2.3 *Histogram Generation Phase*

Each face is represented using LDN Histogram (LH). It contains fine to coarse information of an image, such as spots, edges, corners and other local texture features (Fig. 5).

To extract location information, first divide LDN image into small regions $\{R^1, R^2 \ldots R^N\}$. Then extract histogram H^i from each region R^i which represents minute details about local features of the face. We can divide the image into 3×3, 5×5,

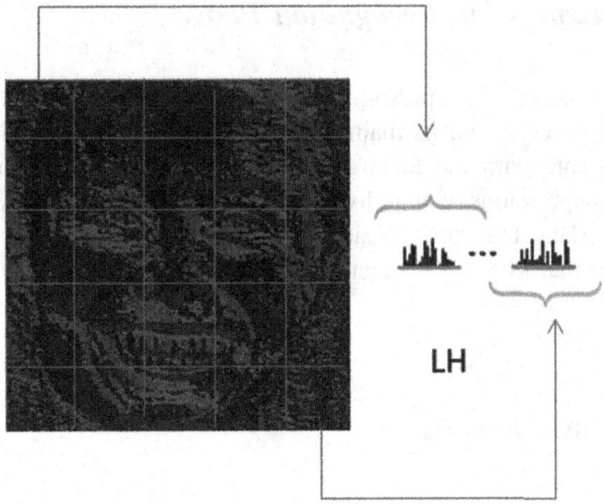

Fig. 5 LDN histogram

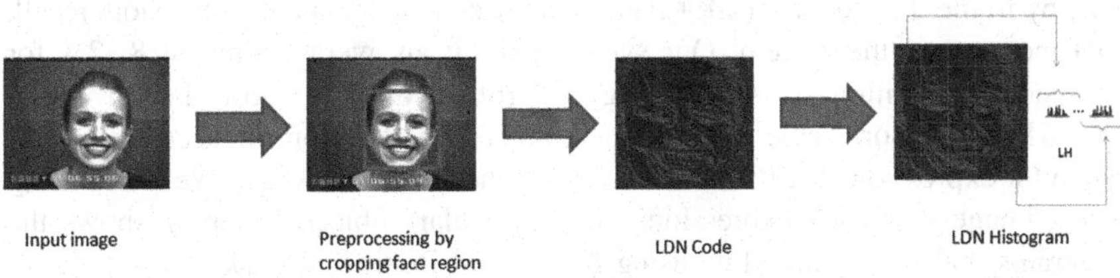

Fig. 6 Computation of LDN code

and 7 × 7. If we are dividing image into 5 × 5 blocks, then $N = 25$. After obtaining histogram from each block, we concatenate these histogram which represent local features to form a global descriptor called LDN Histogram (LH). Figure 6 shows LDN computation.

2.4 *Dimensionality Reduction Phase*

Principal components analysis (PCA) searches for k n-dimensional orthogonal vectors that can best be used to represent the data, where $k \leq n$. The original data is thus projected onto a much smaller space, thus resulting in dimensionality reduction. While using PCA, it is computationally inexpensive. It can be applied to ordered and unordered attributes, and can also handle sparse and skewed data. Data of more than two dimensions (multidimensional data) can be reduced to two dimensions in order to handle the problem. In our approach, we reduced the dimensionality of the LH using PCA, i.e. the number of bins in the LDN histogram is reduced in such a way that most prominent features are extracted.

2.5 *Facial Expression Recognition Phase*

The facial expression of a person can be recognized using LH during the face recognition process, and its main objective is to compare an encoded feature vector of test image with the feature vectors that are already trained. First, we train the basic 6 expressions of an individual and when a test image given as input, it compares with data that are already trained and returns which expression corresponding to that image. This can be done with the help of chi-square dissimilarity test.

3 Experimental Results

In this section, we compared our system with the existing method. We have used CK+ dataset [21] for expression recognition which are available online. They have collected 309 sequences from 90 subjects and 227 sequence of which is labelled frame by frame. Images from CK+ dataset are used to calculate the precision, recall, and f-measure of the system. Our system exhibit an average time of 8.32 s for expression recognition. This technology is profiled on a Windows laptop with a 2.00 GHZ processor. Table 1 shows the comparison study on the accuracy of our system for expression classification for Kirsch and Robinson Mask. We tested using 10 test images for each expression for a particular subject. Figure 7 shows the histograms that we obtain while using Kirsch and Robinson Mask.

For expression classification, using Kirsch Mask is more efficient because some of the distinct features are not accurately extracted while using Robinson Mask. Figure 7a shows our preprocessed image obtained using Viola Jones Algorithm and its corresponding histogram. While analysing the histogram, a large number of features are observed. Our goal is to extract relevant features. To accomplish the task, LDN code generated (Fig. 7b) using Kirsch mask and LDN code generated using Robinson mask (Fig. 7c) are created. From their corresponding histograms, we can conclude that more relevant information are extracted while using Kirsch mask compared to Robinson Mask.

Table 1 Comparison study

Basic six expression	Compass mask used	
	Kirsch (%)	Robinson (%)
Surprise	90	80
Sad	100	100
Disgust	100	90
Anger	90	90
Happiness	90	80
Fear	100	90

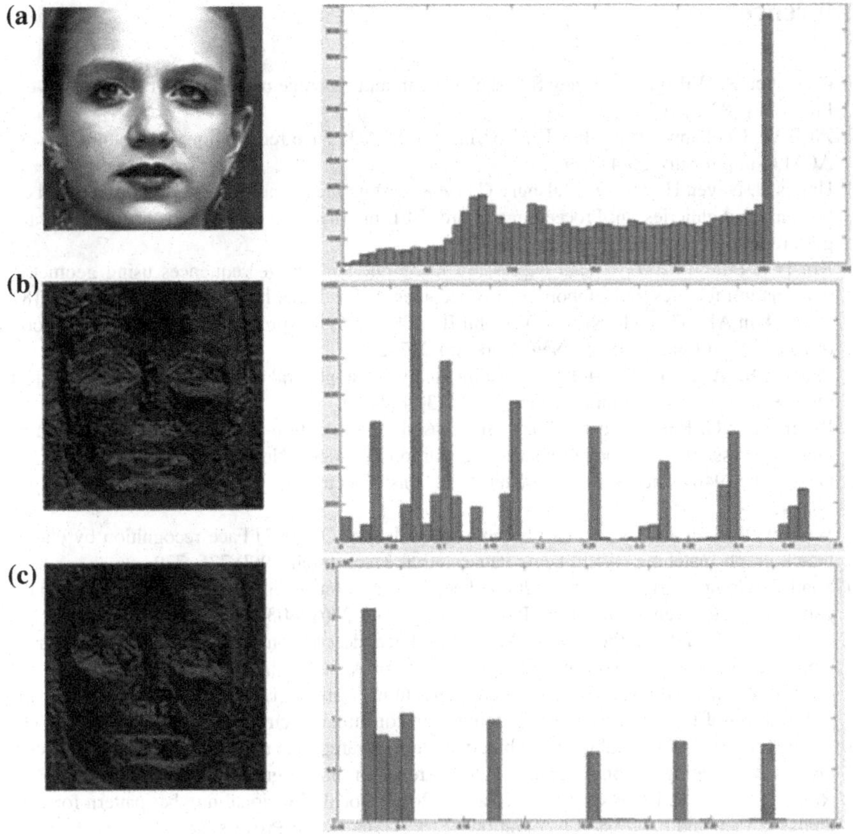

Fig. 7 **a** Preprocessed image and its corresponding histogram. **b** LDN code generated using Kirsch mask and its corresponding histogram. **c** LDN code generated using Robinson mask and its corresponding histogram

4 Conclusion

The objective of this paper is to show the efficiency and consistency of LDN descriptors using two different compass masks. While applying Robinson Mask, it is observed that it occasionally misclassified some of the basic expressions namely surprise and happiness. Therefore accuracy is less compared to Kirsch Mask. PCA is applied to LH descriptor which reduces the dimensionality in such a way that prominent features are extracted as the output. This increases the LDN descriptor efficiency as well as the recognition rate.

Acknowledgments I am thankful to Ms. Merin Cherian, Assistant Professor of Computer Science department, FISAT, Kerala, for her keen interest and useful guidance in my paper.

References

1. Chellappa R, Wilson C, Sirohey S (1995) Human and machine recognition of faces: a survey. Proc IEEE 83(5):705–741
2. Zhao W, Chellappa R, Phillips PJ, Rosenfeld A (2003) Face recognition: a literature survey. ACM Comput Surv 35(4):399–458
3. Hongs H, Neven H, von der Malsburg C (1998) Online facial expression recognition based on personalized galleries. In: Proceedings of 3rd IEEE international conference on automatic face gesture recognition, Apr 1998, pp 354–359
4. Kotsia I, Pitas I (2007) Facial expression recognition in image sequences using geometric deformation features and support vector machines. IEEE Trans Image Process 16(1):172–187
5. Li SZ, Jain AK, Tian YL, Kanade T, Cohn JF (2005) Facial expression analysis. In: Handbook of face recognition. Springer, New York, pp 247–275
6. Penev PS, Atick J (1996) Local feature analysis: a general statistical theory for object representation. Netw Comput Neural Syst 7(3):477–500
7. Bourbakis NG, Kakumanu P (2008) "Skin-based face detection-extraction and recognition of facial expressions. In: Applied pattern recognition. Springer, New York, pp 3–27
8. Gabor D (1946) Theory of communication. JInst Electr Eng III, Radio Commun Eng 93 (26):429–457
9. Wiskott PL, Fellous J-M, Kuiger N, von der Malsburg C (1997) Face recognition by elastic bunch graph matching. IEEE Trans Pattern Anal Mach Intell 19(7):775–779
10. Shan C, Gong S, McOwan PW (2009) Facial expression recognition based on local binary patterns: A comprehensive study. Image Vis Comput 27(6):803–816
11. Ahonen T, Hadid A, Pietikäinen M (2006) Face description with local binary patterns: Application to face recognition. IEEE Trans Pattern Anal Mach Intell 28(12):2037–2041
12. Jabir T, Kabir MH, Chae O (2010) Local directional pattern (LDP) for face recognition. In: Proceedings of IEEE international conference on consumer electronics, Mar 2010, pp 329–330
13. Xie Z, Liu G (2011) Weighted local binary pattern infrared face recognition based on Weber's law. In: Proceedings of 6th international conference on image graph., Aug 2011, pp 429–433
14. Ramirez Rivera A, Rojas Castillo J, Chae O (2013) Local directional number pattern for face analysis: face and expression recognition. IEEE Trans Image Process 22
15. Kanade T, Cohn J, Tian YL (2000) Comprehensive database for facial expression analysis. In: Proceedings of 4th IEEE international conference on automatic face gesture recognition May 2000, pp 46–53 [Online]

Role-Based Access Control for Encrypted Data Using Vector Decomposition

D. Nidhin, I. Praveen and K. Praveen

Abstract In this current era, cloud environment has become essential for large-scale data storage which makes it challenging to control and prevent unauthorized access to the data. At the same time trusting the cloud provider's integrity is also significant. Instead of assuming that the cloud providers are trusted, data owners can store the encrypted data. In 2013, Zhou et al. put forward a role-based encryption scheme (RBE) that integrates encryption with role-based access control (RBAC). In 2008, Okamoto and Takashima introduced a higher dimensional vector decomposition problem (VDP) and a trapdoor bijective function. In this paper, we propose a RBE scheme using VDP.

Keywords Access control · Cloud · Role-based access control · Role-based encryption · Vector decomposition problem

1 Introduction

Along with the growth of computing facilities, the amount of digital information has also increased. Cloud data storage has drawn attention of people as it provides an inexpensive way to manage the unpredictable storage requirements. This lets service providers to focus on other features to improve the user experience,

D. Nidhin (✉) · K. Praveen
TIFAC-CORE in Cyber Security, Amrita Vishwa Vidyapeetham University, Coimbatore, India
e-mail: nidhin.dinesh@gmail.com

K. Praveen
e-mail: ipraveen_k@yahoo.co.in

I. Praveen
Department of Mathematics, Amrita Vishwa Vidyapeetham University, Coimbatore, India
e-mail: praveen.cys@gmail.com

L.P. Suresh and B.K. Panigrahi (eds.), *Proceedings of the International Conference on Soft Computing Systems*, Advances in Intelligent Systems and Computing 398, DOI 10.1007/978-81-322-2674-1_12

reducing the storage headaches and maintenance costs. The commonly used cloud services are software as a service (SAAS), platform as a service (PAAS), and infrastructure as a service (IAAS). The attractive feature of cloud computing is that its services are available as pay-as-you-go. Users are generally concerned about the security issues associated with cloud adoption. There are different types of infrastructures associated with cloud. The cloud which is made accessible to the public is called public cloud. The cloud which is maintained by private organization is called private cloud. Private cloud is considered to be more secure than public cloud. Using a public cloud you are unlikely to even touch one physically. Using a private cloud your data lives behind your firewall and has other security advantages like clarity of ownership, knowledge about to whom the physical access is granted, etc. The vendor might allow physical access to other tenants and your data travels "in the wild" over the internet to your public cloud provider. This calls for suitable access policies to be enforced. These policies must restrict the data access to those intended by the data owners. The cloud has to enforce these policies and the cloud providers have to be trusted.

The existing systems [1, 2] present a design of a secure role-based access control (RBAC)-based cloud storage system where the access control policies are imposed by role-based encryption (RBE) system. In RBE system, the owner encrypts the data in such a manner that decryption is possible only for those users who are members of appropriate roles as per the RBAC policy. In the existing RBE scheme, it is possible to handle role hierarchies, through which roles inherit permissions from other roles. A user could join a role and access the data after the encryption is done by the owner. A user can be revoked at any time which restricts him from any future access and also ensures that other roles and users do not get affected. The part of the decryption computation in the scheme is outsourced to the cloud, in which only public parameters are involved.

Armbrust et al. [3] gave an explanation of public cloud and private cloud. In the proposed scheme, secure cloud data storage architecture using a hybrid cloud infrastructure is used. This hybrid cloud architecture is a combination of private cloud and public cloud. Organization's sensitive information such as the role hierarchy and user membership information is stored in private cloud, whereas the public cloud is used to store the encrypted data. In the proposed scheme, the users who want to access the data interact only with the public cloud. There is no provision for public users to access the private cloud, which greatly reduces the attack surface for the private cloud. It is assured that sensitive information is not leaked and also the public cloud is used for storing large amount of data.

A disadvantage of this scheme is that if a data is to be encrypted for multiple roles, multiple encryptions are to be done which will be stored as different ciphers, that is, same data is stored as different copies for different roles which in turn cause wastage of space. We propose a variant of RBE scheme which makes use of harder mathematical assumption, namely vector decomposition problem (VDP). VDP was initially proposed by Yoshida and further analyzed by Duursma and Kiyavash [4], Galbraith and Verheul [5], and Kwon and Lee [6]. Not only is this system secure under VDP, but it also increases the communication speed and reduces the cost by making use of the homomorphic property of this scheme. Whenever there is a need to establish a new key, it can be derived using the previously established key.

2 Preliminaries

2.1 Role-Based Access Control

A role connects a set of permissions on one side to the set of users on the other. A user in this model could be a computer or a person. A role is a job title or job function having some associated semantics regarding the responsibility and authority conferred on a member of the role in an organization. Permission is the action of officially allowing someone to do a particular thing, consent, or authorization. Role-based access control model [7, 8] with hierarchy ($RBAC_2$) has the following components:

- *R, P, U,* and *S* (Roles, permissions, users, and sessions, respectively),
- PA $\subseteq P \times R$, a many-to-many permission to role relation,
- UA $\subseteq U \times R$, a many-to-many user to role relation,
- *use*: $S \rightarrow U$, a function mapping each session s_i to the single user (s_i) (constant for the session),
- RH $\subseteq R \times R$ is a partial order on *R* called the role hierarchy or role dominance relation, also written as $\geq$ and
- *roles*: $S \rightarrow 2^R$, a function mapping each session s_i to a set of roles *roles* $(s_t) \subseteq roles(s_t) \subseteq \{r|(\exists\ r' \geq r[user(s_t), r') \in \text{UA}]\}$ and session s_t has the permissions $\cup r \in roles(si)\{p|(\exists\ r'' \leq r)[(p, r'') \in \text{PA}]\}$.

2.2 Bilinear Pairing

Let *G*1, *G*2, *GT* be three cyclic groups with prime order *p* and *GT* be a cyclic multiplicative group of prime order $p \cdot g$ and *h* are two random generators where $g \in G1$, $h \in G2$. A bilinear pairing $\hat{e}$: $G1 \times G2 \rightarrow GT$ satisfies the following properties:

- *Bilinear*: for $a,b \in Z_p^*$ we have $\hat{e}(g^a, h^b) = \hat{e}(g,h)^{ab}$.
- *Non-Degenerate*: $\hat{e}(g,h) \neq 1$ unless $g = 1$ or $h = 1$.
- *Computable*: the pairing $\hat{e}(g,h)$ is computable in polynomial time.

2.3 Vector Decomposition Problem

Definition 2.1 *Vector decomposition problem*: Let V be a vector space over the field Fp and $\{b_1, b_2\}$ is a basis for V. Let $Q \in V$. Compute the element $R \in V$ such that $R \in <b_1>$ and $Q - R \in <b_2>$. For a fixed base $\{b_1, b_2\}$ VDP is defined as: given $Q \in V$, find R as above.
The following theorem of Yoshida [9] indicates that VDP is at least as hard as CDHP.

Theorem 2.1 *The VD Problem on V is at least as hard as CDH Problem on $V' \subset V$ if for any $e \in V'$ there are isomorphisms $f_e, ø_e$: $V \rightarrow V$ which satisfy the following conditions.*

- *For any $v \in V$, $ø_e(v)$ and $f_e(v)$ are effectively defined and can be computed in polynomial time*
- *$\{e, ø_e(e)\}$ is an* F- *basis for V*
- *There are $\alpha_1, \alpha_2, \alpha_3 \in F$ with*

$$f_e(e) = \alpha_1 e$$
$$f_e(ø_e(e)) = \alpha_2 e + \alpha_3 ø_e(e),$$
$$\alpha_1, \alpha_2, \alpha_3 \neq 0$$

- *The elements $\alpha_1, \alpha_2, \alpha_3$ and their inverses can be computed in polynomial time.*

Definition 2.2 *Generalized vector decomposition problem*: Let V be a vector space over the field Fp and $\{b_1, b_2, \ldots, b_n\}$ is a basis for V. Let $Q \in V$. Compute the element $R \in V$ such that $R \in <b_1, b_2, \ldots, b_m>$ and $Q - R \in <b_{m+1}, \ldots, b_n>$, $m < n$. For a fixed base $\{b_1, b_2, \ldots, b_n\}$ VDP is defined as: given $Q \in V$, find R as above.

2.4 Trapdoor VDP

Okamoto and Takashima [10] introduced the concept of distortion eigenvector space and based on that they also introduced trapdoor function for solving VDP.

Definition 2.3 Distortion eigenvector space V is an n-dimensional vector space over Fr that satisfies the following conditions:

1. Let $A \leftarrow (a_1, a_2, \ldots, a_n)$ be a basis of F_r—vector space V and F a polynomial time computable automorphism on V. The basis A is called a distortion eigenvector basis with respect to F, if each a_i is an eigenvector of F, their eigenvalues are different from each other, and there exist polynomial time computable

endomorphisms $\o_{i,j}$ of V such that $\o_{i,j}(a_j) = ai$. *We call* $\o_{i,j}$ a distortion map. There exist such A, F and $\{\o_{i,j}\}_{1 \leq i,j \leq n}$.

2. There exists a skew-symmetric non-degenerate bilinear pairing e: $V \times V \rightarrow \mu_r$ where μ_r is a multiplicative cyclic group of order r, i.e., $e(\gamma u, \delta v) = e(u, v)^{\gamma\delta}$ and $e(u, u) = 1$ for all $u, v \in V$ and all $\gamma, \delta \in V$, and if $e(u, v) = 1$ for all $v \in V$, then $u = 0$.
3. There exists a polynomial time computable automorphism ρ on V such that $e(v, \rho(v)) \neq 1$ for any v except for v in a quadratic hypersurface of $V \cong (F_r)^n$.

Lemma 2.1 (Projection Operators): *Let* $A \leftarrow (a_1, a_2, \ldots, a_n)$ *be a distortion eigenvector basis of V, and* a_i *has its eigenvalue* λ_i *of F. A polynomial of F is given by*

$$\mathrm{Pr}_j(v) = \left(\Pi_{i \neq j}(\lambda_j - \lambda_i)\right)^{-1}\left(\Pi_{i \neq j}(F - \lambda_i)\right)(v) \tag{1}$$

i.e., $\mathrm{Pr}_j(a_k) = 0$ *for* $k \neq j$ *and* $\mathrm{Pr}_j(a_j) = a_i$.

Let V be a distortion eigenvector space and $(a_1, a_2, \ldots, a_n)$ be a distortion eigen basis of V. Consider the matrix $X = (x_{ij})$ such that $b_i = \sum_{i=1}^{n} x_{ij}a_j$ and $(b_1, b_2, \ldots, b_n)$ is also a basis for V. Let $v = \sum_{i=1}^{n} y_i p_i$ be the vector in V. Our aim is to decompose v as the sum of two vectors in which one lies in the subspace generated by $(b_1, b_2, \ldots, b_m)$, $m < n$ and the other lies in the subspace generated by $(b_{m+1}, \ldots, b_n)$. If $X^{-1} = (t_{ij})$, the function

$$V\,\mathrm{Deco}\left(v, <b_j>, X, <b_1, b_2, \ldots, b_n>\right) = \sum_{i=1}^{n}\sum_{k=1}^{n} t_{ij}x_{jk}\o_{ki}(\mathrm{Pr}_i(v)) \tag{2}$$

can be used to accomplish this goal.

Lemma 2.2 *If V is a distortion eigenvector space and* $(a_1, a_2, \ldots, a_n)$ *is a distortion eigenvector basis with* $(b_1, b_2, \ldots, b_n)$ *generated by* $X = (x_{ij})$ *such that* $b_i = \sum_{i=1}^{n} x_{ij}a_j$, *then V Deco solves VDP.*

2.5 Encryption Scheme

Using this trapdoor function, a multivariate homomorphic encryption scheme was introduced [10].

Key Generation:

- V: a l_1 dimensional vector space with distortion eigenvector basis A:
- $\{a_0, \ldots, a_{l1-1}\}$

- $X = (x_{i,j})$ with $x_{i,j} \in F_r$, $i, j = 1, \ldots l_1$
- $b_i = \sum_{j=0}^{l_1-1} x_{i,j}, a_j$ and B: $(b_1, \ldots, b_{l1})$
- Secret key—S_k is X and public key P_k is (V, A, B)

Encryption: Enc(P_k,(m_1, ..., m_{l2}))

- message: $(m_1, \ldots, m_{l2}) \in \{1, \ldots \tau - 1\}$
- $r_{l2}, \ldots, r_{l1} \in F_r$
- $c \leftarrow \sum_{i=1}^{l_2} m_i b_i + \sum_{i=m}^{l_1-1} r_i b_i$
- *return* cipher text c

Decryption: Dec(S_k, c)

- $b' \leftarrow V\,\mathrm{Deco}(c, <b_i>, X, B)$
- compute the pairing $e(b_i', b_i) = e_i'$. Let $e_i = e(b_i, b_i)$
- $m_i' \leftarrow D \log_{e_i}(e_i')$ for $i = 1, \ldots, l_1$
- *return* plain text $\left(m_1', \ldots, m_{l_2}'\right)$

3 Proposed System

Our system comprises the following algorithms:

Setup(λ): This algorithm is run by the administrator. A security parameter λ is given as input.

$E[n]$ is a set of n-torsion points for a prime n, and hence it forms a vector space. Let V be a distortion eigenvector space and $(a_1, a_2, \ldots, a_n)$ be a distortion eigen basis of V. Consider the matrix $X = (x_{ij})$ such that $b_i = \sum_{i=1}^{n} x_{ij} a_j$ and $(b_1, b_2, \ldots, b_n)$ is also a basis for V. *Setup*(λ) generates a set of public parameters PP = $(b_1, b_2, b_3, \ldots, b_n)$, a trapdoor X which is the master secret key MSK.

Manage Role: This algorithm is run by the administrator. It creates and maintains for each role R_i, a list of users UL$_i$ who are members of that particular role. It generates a key r_i for each role R_i and sends it to the list of users who are members of R_i through a secure channel.

GenKey: When an owner wants to encrypt a data for a particular role, the owner specifies the public parameters of those particular roles. The private cloud calculates

$$C^0 = \sum_i r_i b_i + \sum_j t_j b_j \tag{3}$$

where $r_i \neq t_j$, r_i is the secret key of roles with public parameter b_i which was specified by the owner. Then the owner applies vector decomposition $V\,\mathrm{Deco}(C^0, <b_i>, X, <b_1, b_2, \ldots, b_n>)$ and obtains r_ib_i for the qualified owners and t_jb_j for the forbidden users.

Now the owner selects n random numbers $k_1, k_2, \ldots, k_n$ and computes

$$C' = \sum_i k_i r_i b_i + \sum_j k_j t_j b_j \tag{4}$$

Owner then calculates

$$K_i = \hat{e}(b_i, b_i)^{k_i} \tag{5}$$

where K_i is used as the encryption key for the data to be encrypted.

AquireKey: User applies vector decomposition:

$$V\,\mathrm{Deco}\left(C', <b_i>, X, (b_1, b_2, \ldots, b_n)\right) = \begin{cases} k_i r_i b_i & \text{for forbidden users} \\ k_j t_j b_j & \text{for forbidden users} \end{cases} \tag{6}$$

User calculates

$$\hat{e}(b_i, b_i)^{\frac{k_i r_i}{r_i}} = \hat{e}(b_i, b_i)^{k_i} = K_i \tag{7}$$

Qualified user extracts the decryption key K_i, whereas for the forbidden users,

$$\hat{e}(b_j, b_j)^{\frac{k_j t_j}{r_j}} \neq K_j \tag{8}$$

4 System Operation

We assume that the *Setup*(λ) algorithm is properly executed by the administrator and all the parameters are defined. Now a user can join the system by giving a request to the administrator. If genuine, the administrator adds the user to the particular role and provides r_i corresponding to the role R_i to which the user got added. This communication is represented by the lines 1 and 2. This communication is through a secure channel as shown in Fig. 1. Administrator updates the list RL_i present in the private cloud denoted by 3. When an owner wants to encrypt a data for a particular role, owner first communicates with the public cloud specifying

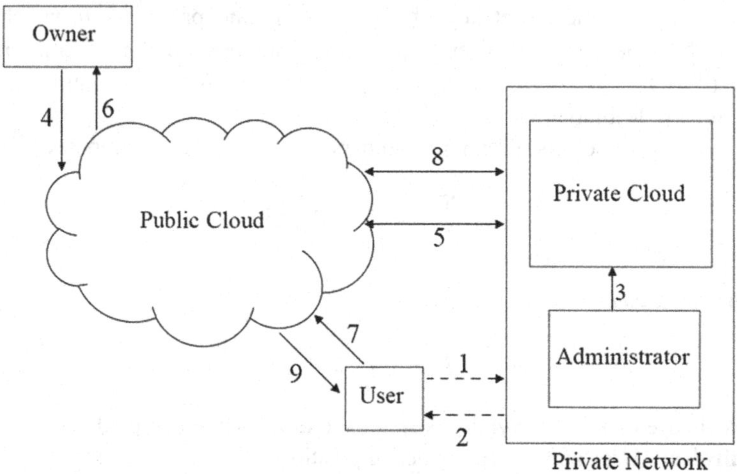

Fig. 1 Architecture

the public parameter of those roles. The public cloud forward this request to the private cloud which returns C^0 to the owner through the public cloud. The owner then calculates C' and K_i which is then used to encrypt the data C' that is stored in the public cloud. This is shown by lines 3–5.

When a data is required, the user requests the public cloud specifying the index of the cipher. The public cloud returns C' to the user along with the cipher. This is shown by lines 7–8. Qualified users can obtain K_i which is used to decrypt the cipher text.

After a key is generated by the algorithm *GenKey,* any symmetric encryption scheme could be used to encrypt the data. The scheme mentioned in Sect. 2.5, which is an asymmetric scheme, can also be used. For encrypting, a particular message for different roles encryption is done as

$$C = mK_1b_1 + mK_2b_2 + \cdots + mK_nb_n \tag{9}$$

For encrypting multiple messages for multiple roles, this scheme could be utilized as

$$C = m_1K_1b_1 + m_2K_2b_2 + \cdots + m_nK_nb_n \tag{10}$$

where K_1, K_2, ..., K_n, etc. are the keys established for roles with parameters b_1, b_2, ..., b_n, respectively, and m_1, m_2, ..., m_n are the corresponding messages. This is more advantageous because of the flexibility in choosing the roles to be specified and also because of the efficient space utilization by incorporating different messages in the same ciphertext.

5 Conclusion

Introduction of RBAC made access control efficient and flexible by assigning permissions to different roles. The present cloud data storage system requires proper access control and a mechanism to counter the mistrust over cloud providers. One way out is to use RBAC policies along with keeping the data encrypted. A solution is proposed in this paper using VDP. With this scheme, multiparty key establishment is possible and multiple plaintexts can be encrypted to a single cipher text. Our key establishment scheme is forward secure. Hence, using the homomorphic property a new key can be established from the existing key if desired. This reduces the communication cost and increases the speed. Use of VDP makes the scheme simple and secure.

References

1. Zhou L, Varadharajan V, Hitchens M (2011) Enforcing role-based access control for secure data storage in the cloud. Comput J, 54(10):1675–1687
2. Zhou L, Varadharajan V, Hitchens M (2013) Achieving secure role-based access control on encrypted data in cloud storage. Inf Forensics Secur IEEE Trans 8(12):1947–1960
3. Armbrust M, Fox A, Griffith R, Joseph AD, Katz R, Konwinski A, Lee G, Patterson D, Rabkin A, Stoica I et al (2010) A view of cloud computing. Commun ACM 53(4):50–58
4. Duursma IM, Kiyavash N (2005) The vector decomposition problem for elliptic and hyperelliptic curves. IACR Cryptology ePrint Arch 31
5. Galbraith SD, Verheul ER (2008) An analysis of the vector decomposition problem. In: Public key cryptography—PKC 2008. Springer, Heidelberg, pp 308–327
6. Kwon S, Lee II-S (2009) Analysis of the strong instance for the vector decomposition problem. Bull Korean Math Soc 46(2):245–253
7. Sandhu RS, Coyne EJ, Feinstein HL, Youman CE (1996) Rolebased access control models. Computer 29(2):38–47
8. Zhang Y, Wu M, Wan Z (2010) Method of information system authority control based on RBAC in web environment. In *2010 2nd international conference on advanced computer control*, vol 2, pp 542–545
9. Yoshida M (2003) Inseparable multiplex transmission using the pairing on elliptic curves and its application to watermarking. In: Proceeding of fifth conference on algebraic geometry, number theory, coding theory and cryptography. University of Tokyo, Tokyo
10. Okamoto T, Takashima K (2008) Homomorphic encryption and signatures from vector decomposition. In: Pairing-based cryptography—pairing 2008, Springer, Heidelberg, pp 57–74

Mobile Controlled Door Locking System with Two-Factor Authentication

S. Priyadharshini, D. Nivetha, T. Anjalikumari and P. Prakash

Abstract In spite of the availability of digital password lockers and advanced door locks, hacking the lock code by an unauthorized person has become a plain-sailing task. Thus, the main goal of this paper is to design a highly advanced and secured home security system using mobile technology, video messaging, and electronic technology with two-factor authentication. The designed system directly communicates to the owner of the house when someone arrives at his door-step by sending a video along with a notification message. The owner sends a one-time password to the visitor's mobile for him to enter in the keypad placed near the door.

Keywords Home security · Video messaging · GSM technology · Door automation

1 Introduction

In today's busy world, wherein everyone is off to work in the day time, it is difficult for a person to receive and greet his guests. So a feel of discomfort occurs for both the guest and the owner. For this sake, we cannot just keep the key near the window sill or give it to our neighbors. An automated door lock system works better here. We would feel secure if we are notified of who visited our house in our absence.

S. Priyadharshini (✉) · D. Nivetha · T. Anjalikumari · P. Prakash
Department of CSE, Amrita Vishwa Vidyapeetham, Coimbatore 641112, Tamil Nadu, India
e-mail: priyaselvaraj1311@gmail.com

D. Nivetha
e-mail: nividurairaj@gmail.com

T. Anjalikumari
e-mail: anjali.arasu44@gmail.com

P. Prakash
e-mail: npprakash@gmail.com

L.P. Suresh and B.K. Panigrahi (eds.), *Proceedings of the International Conference on Soft Computing Systems*, Advances in Intelligent Systems and Computing 398, DOI 10.1007/978-81-322-2674-1_13

The passive infrared (PIR) motion sensor, if it senses any human intervention, the Internet Protocol (IP) camera is activated and the capture of the video is kicked off. The IP camera captures the video for 2 min and it is sent to the owner via the GSM modem. The owner, after viewing the video, decides whether the visitor is to be allowed inside or not. If he is convinced of the fact that the visitor has to be let in, he sends his consent to the system. Simultaneously the system generates a one-time password and sends to the visitor's mobile. If the visitor enters the password correctly, the door is unlocks. The visitor has three chances to enter the password correctly. If the password entered is incorrect for the third time, the siren is set off. Also an alert message is sent to the owner. Thus the guest is greeted and our house is secured in our absence.

2 Related Works

This section deals with the researches that have been done in the area of interest.

In palmtop recognition system [1], the biometric ID of the possible persons who can access and open the door in a database. Once the visitor tries to open the door, the biometric ID of the visitor is checked with the database and if found correct, the door is opened. Here storing the biometric ID of many people in a database becomes a tedious task.

The DTMF technology [2] just enables the opening of door once if the password is typed correct. Anyone can type the correct password or there exists chances where the person who was once our friend/relative may become our enemy due to some personal reasons. We may not change the password periodically and we will not be intimated of who entered the house.

The face recognition system [3] allows the visitor to enter inside if the face of the person matches with that stored in the database. In this case, there is a probability where a robber or an enemy of us can use the photo of the person stored in the database to enter inside. So security may be lost track of.

In password lock security system [4], the 4-digit password is a credential. The visitor who tries to enter in has three chances to type the password correctly. If typed password mismatches the third time, an alert is sent to the owner who has control over the system. The limitation here is only 10 members can be added to this system.

The idea of RF and GSM technology [5] also works more or less similar to the password lock security system wherein the 4 digit password acts as a credential. If the credential is not found to be authentic, the system by itself sends an alert message to the person who gains control over the system. Although the owner of the system is notified that someone tried to open the door, the owner will not have knowledge of who tried to do so.

Knowing about the person who tried to enter our house will be quite safe and it would be better if the owner has hands on control of endorsing who can enter inside. Our project mainly focuses on addressing this scenario.

3 Proposed Methodology

At first, a sensor detects that a person has stepped into the auto-detect sensor area. After sensing, the proposed methodology decides whether to let the person inside the house by opening the door. For this, the methodology follows the steps shown below.

The first step is taking the video when the person comes inside the auto-detect area using a video camera. Next step is video messaging. The recorded video is sent to the user through the Internet. The user can view the video using any web-enabled electronic device and decides whether to unlock the door or not. Then, the user also chooses the mobile number of the visitor stored in his device. When the user prefers the unlock option, a one-time pass code is sent to the mobile of the visitor. The visitor enters the pass code on the keypad placed near the door. The pass code is verified by the system, and the decision is taken whether to open the door (Fig. 1)

If someone tries to break the door, the siren is activated and an alert message is sent to the user's phone and mail id.

4 Experimental Setup

AT89C52 micro controller, PIR motion detector, IP camera, GSM modem, one-time-password (OTP) generation server, a control server for the door lock, 4 × 3 number keypad, and 16 × 2 LCD display are basically used to construct the device. The entire experimental setup block diagram of the proposed method is showed in Fig. 2.

AT89C52 microcontroller is used to control the entire device. It is 8 bit wide, has 8 kb of in-system programmable flash memory, 256 bytes of RAM, and 32 input and output lines. It is a high-performance and low-power microcontroller [6].

PIR motion detector is placed at the auto-detect area which is usually at the entrance of the house. The door handle can also be selected as a motion hotspot. PIR motion detectors are made using an op-amp, Fresnel lens, and pyroelectric

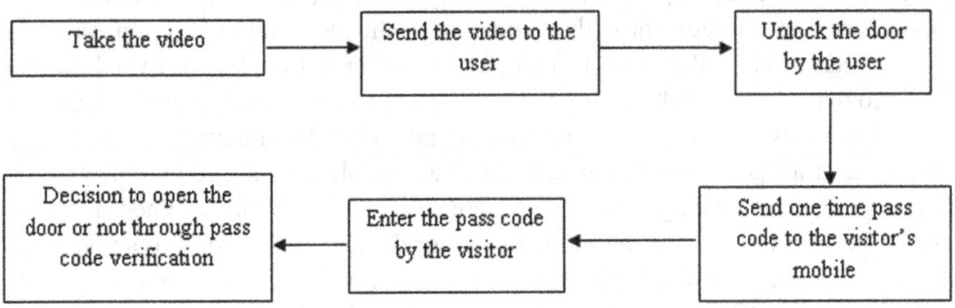

Fig. 1 Steps of the proposed method

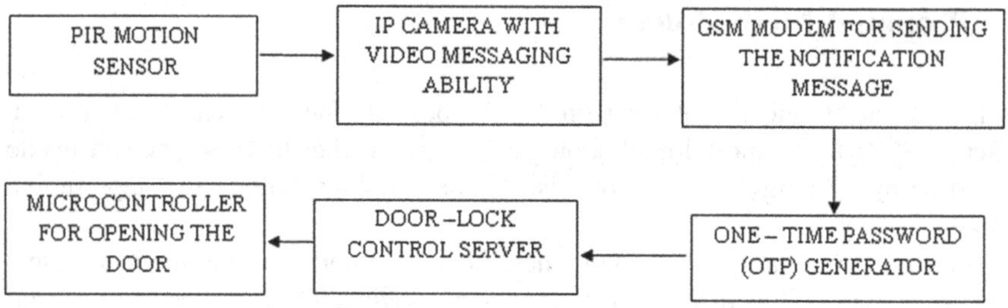

Fig. 2 Block diagram of the proposed system

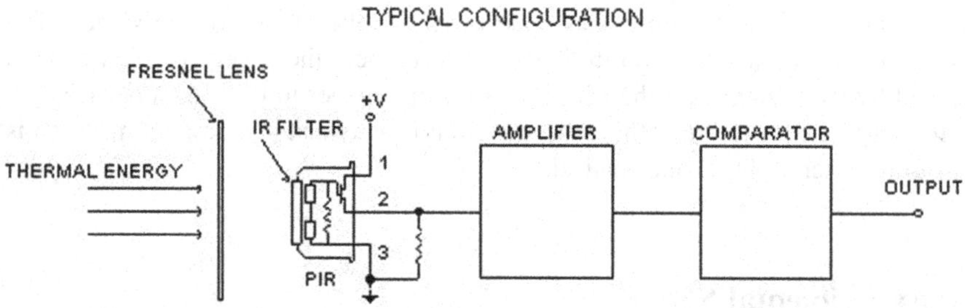

Fig. 3 Typical configuration of PIR motion detector

crystals. The detectors are made to sense the infrared radiation given off by the humans. Radiation of human body is usually at the wavelength of 12 μm [7]. Thus, this device makes use of Fresnel lens that is made up of infrared transmitting material of range 8–14 μm that is most sensitive to human bodies [8]. PIR motion detectors are compact, low-cost, easy-to-use, and a low-power device. Figure 3 shows the typical configuration of the PIR [9], and Fig. 4 explains briefly on the working of a PIR motion detector [9].

Thus, when a human passes through the auto-detect area, the PIR detector generates a sine signal. This signal is detected using a breakout board and a digital signal is produced.

The IP camera is connected to the PIR motion detector. IP camera offers high-definition and high-speed video recording. Moreover, an IP camera provides a well-secured transmission through various authentication and encryption methods such as TKIP, WEP, WPA, and AES. It is connected to a Power-over-Ethernet (PoE) protocol device such that power is supplied to the camera from an Ethernet cable. IP cameras work over a broadband, cellular, or a Wi-Fi network. The camera is installed at a place such that it can cover the whole area at the entrance of the house and also by making sure that the areas are clearly visible. When a person steps in the auto-detect area, the motion detector device sends signals to activate the video camera. The video camera records the video for 1 or 2 min depending on the user's need. The IP camera is records the video directly to a network attached

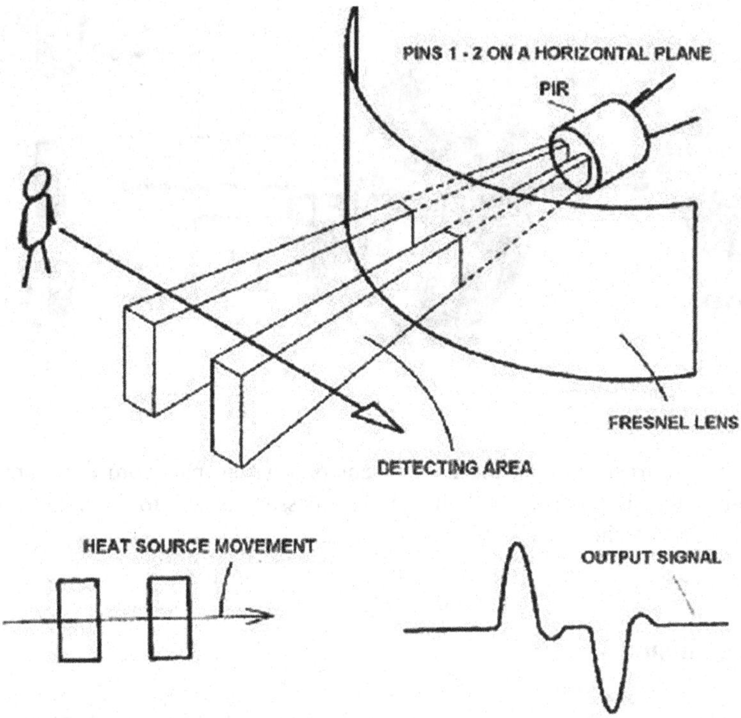

Fig. 4 Principle of PIR motion detector

storage (NAS) device [10]. Using the IP address, the authorized users can view the video recorded by the camera from anywhere using any web-enabled device like laptop, tablet, smart phones, etc. through an Internet connection. Figure 5 shows how the user can access the video taken by the IP camera.

Once the video is taken and sent by the IP camera, a notification message is sent to the user's mobile to view the video using the GSM modem. The user can view the video using the IP address. With the message, an option is provided for the user to select whether the door can be opened or not. If the user chooses the 'unlock the door' option, he is intimated to enter the visitor's mobile number in the pop-up text box or can also select the phone number of the visitor from his contacts list.

The visitor's mobile number is sent to the server. The OTP server generates OTP value and sends it to the control server and also to the visitor's mobile. Then, the value of the OTP is deleted in the OTP generation server. The visitor enters the one-time pass code on the digital keypad.

The control server compares the entered OTP value with the value that was transmitted by the OTP generation server. If the pass code matches with the OTP server generated password, the door is opened. The visitor can take a maximum of three trials to enter the password.

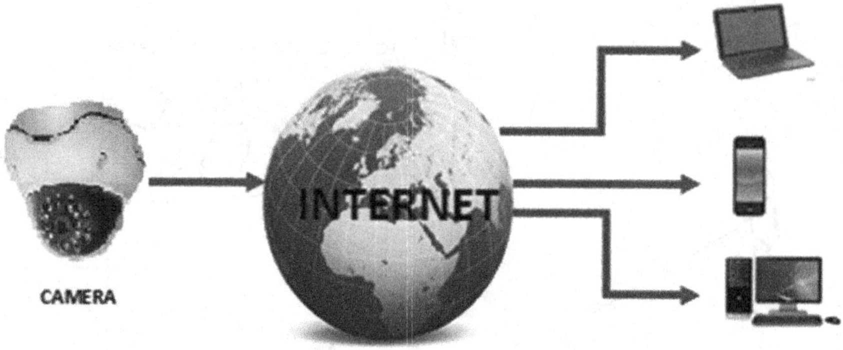

Fig. 5 Video transmissions

If someone tries to break the door or enters a wrong password for more than three times, the siren is activated and an alert message is sent to the user's mobile using the GSM technology.

5 Conclusion

A very well-secured door automation system at a low cost is designed. This system can be used in numerous applications like military, medical equipments, various security systems, etc. This system simplifies the task of managing and extending technology in the home. It is possible to provide high security to places where the confidentiality, privacy, and security are to be preserved. The next stage of this study is to provide more intelligence to the hardware that will increase the response time of the system.

Acknowledgements We extend our deep sense of gratitude to our institution Amrita Vishwa Vidyapeetham for providing this opportunity. We express our sincere thanks to our guide Mr. Prakash, Asst. Professor, Amrita Vishwa Vidyapeetham, for providing us good support and motivation. We also thank our classmates and friends who gave us valuable comments that helped us to improve the proposal.

References

1. Nafi KW, Kar TS, Hoque SA (2012) An advanced door lock security system using palmtop recognition system. Int J Comput Appl 56:18–26
2. Amanullah M (2013) Microcontroller based reprogrammable digital door lock security system by using keypad & GSM/CDMA technology. IOSR J Electr Electron Eng 4(6):38–42
3. Yugashini I, Vidhyasri S, Gayathri Devi K (2013) Design and implementation of automated door accessing system with face recognition. Int J Sci Mod Eng 1(12)

4. Mishra A, Sharma S, Dubey S, Dubey SK (2014) Password based security lock system. Int J Adv Technol Eng Sci 2(5)
5. Supraja E, Goutham KV, Subramanyam N, Dasthagiraiah A, Prasad HKP Dr (2014) Enhanced wireless security system with digital code lock using RF & GSM technology. Int J Comput Eng Res 4(7)
6. 8-bit microcontroller with 8 K Bytes in-system programmable flash. www.atmel.com/images/doc1919.pdf
7. Radiation emitted by human body—thermal radiation. http://www.hko.gov.hk/education/edu02rga/radiation/radiation_02-e.htm
8. How infrared motion detector components work. http://www.glolab.com/pirparts/infrared.html
9. PIR sensor (passive infrared). http://digitalmedia.risd.edu/pbadger/PhysComp/pmwiki/pmwiki.php?n=Devices.PIRsensor
10. Centralized and decentralized IP cameras. http://www.gobeyondsecurity.com/forum/topics/centalized-and-decentralized

Integration of UMTS with WLAN Using Intermediate IMS Network

A. Bagubali, V. Prithiviraj, P.S. Mallick and Kishore V. Krishnan

Abstract Integration of different networks in real-time scenario can provide more benefits related to speed, bandwidth, and connectivity. UMTS and WLAN are two different types of networks with varying specifications. Integrating both the networks will be helpful in making use of all the features simultaneously WLAN networks provide higher data rate but cover smaller areas, while UMTS offers better connectivity with low-speed data rates and also covers larger areas at a time. It means integration of these services can provide higher speed wireless data services with ubiquitous connectivity. This paper focuses on implementing integrated UMTS–WLAN architecture using IP multimedia subsystem (IMS) as an intermediate network. IMS enables the user to switch from one network to another without interrupting a call. The main purpose of using IMS is to minimize the delay incurred during the hand off process and also the authentication, authorization, and accounting (AAA) process servers are inbuilt in its architecture. In this paper, integrating architecture has been implemented by two methods named as loose coupling and tight coupling. To compare the effect of using IMS, different scenarios with and without IMS have been simulated in OPNET Modeler 14. Comparison has been made on the basis of response time and download time for web browsing and FTP applications.

Keywords UMTS · WLAN · AAA · IMS · SGSN · GGSN

A. Bagubali (✉) · K.V. Krishnan
SENSE, VIT University, Vellore 632014, Tamil Nadu, India
e-mail: bagubali@vit.ac.in

V. Prithiviraj
Rajalakshmi Institute of Technology, Chennai, India

P.S. Mallick
SELECT, VIT University, Vellore 632014, Tamil Nadu, India

L.P. Suresh and B.K. Panigrahi (eds.), *Proceedings of the International Conference on Soft Computing Systems*, Advances in Intelligent Systems and Computing 398, DOI 10.1007/978-81-322-2674-1_14

1 Introduction

The trend of using wireless network is increasing day by day as it is more compatible compared to wired networks in terms of mobility and setup cost. Wherever we go, wireless networks provide its services seamlessly with higher speed. But the major drawback is, WLAN does not cover larger areas at a time that means it can provide its services only up to a particular range. This is the reason why integration of networks is more popular among the users. The normal trend today is to use WLANs in hot spot to avail the high bandwidth and data rate, and when the WLAN network coverage is not available or network condition is not sufficient, then switch to UMTS network. In order to achieve this, a handover mechanism should be there which switch over the on-going call to a new network without interfering or dropping the call. But this is not happening now, as the first network before switching, log off the user and the user should sign into the new network as a new user. The WLAN uses access point for the transmission of data packets. The maximum coverage area of WLAN is 820 m and the maximum data rate is about 150 Mbps. The UMTS has been developed in third generation with the improved coverage area of 20 km but data rate is limited to 42 Mbps. In UMTS, the data packets are transferred to the users with the help of gateway GPRS support node (GGSN), serving GPRS support node (SGSN), radio network controller (RNC), and node B [1]. This general procedure will definitely interrupt a call while transferring a call to another network. To overcome this, IP multimedia subsystem (IMS) has been evolved as an advanced methodology, which provides better mechanism to switch from one network to another network. Session initiation protocol (SIP) is used by the IMS networks for session establishment, management, and transformation, due to which different multimedia services can be easily transferred within a single communication with the ability to add and drop services whenever required [2]. The rest of this paper has been designed as follows: next subsections describe about the architecture and different processes involved related to IMS. Section 2 discusses the related work that has been done in past. The model that has been used in this paper is described in Sect. 3. Simulation results have been presented in Sect. 4 and finally the conclusions are made in Sect. 5.

1.1 Integrating Architecture

The main issue involved in achieving the successful integrated network is to select the 3G and WLAN technology architecture. The quality of network integration points depends on the number of factors which includes mobility, handoff latency, security, cost performance, and accounting. The WLAN network can be connected to an UMTS network with various linking points namely SGSN and GGSN. When a WLAN mobile nodes (MNs) wants to exchange information to UMTS network user equipment, it organizes through GGSN Node. The UMTS network (UE) user

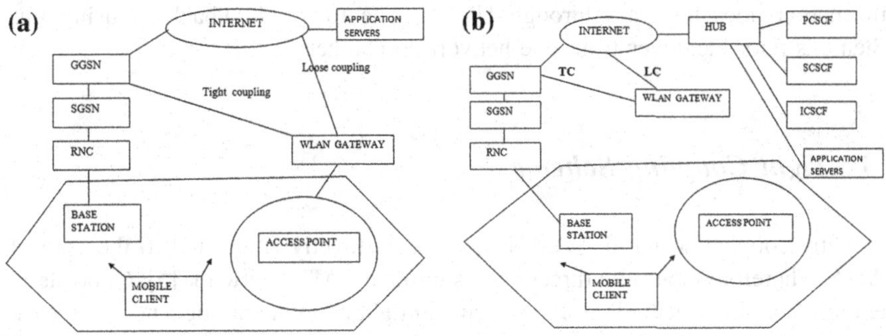

Fig. 1 **a** Integrated UMTS–WLAN block diagram without IMS. **b** Integrated UMTS WLAN block diagram with IMS

equipment first initiates the packet data protocol (PDP). This introduce the operation of UE to its GGSN and also to the WLAN networks using the external data networks.

The main entities of UMTS that play an important role in the integration are GGSN, SGSN, RNC, and node B. The security functions like handling encryption and authentication for the users are the responsibility of the SGSN. In addition, SGSN also have the responsibility of routing packets for the users. The routing function of SGSN is completed by the GGSN as it act as the gateway toward the external network [3].

In 3G network node B also called a base station that converts the radio interface of the signals into a data stream and forwards it to the RNC. In the reverse direction, it formulates incoming data from the RNC for transport over the radio interface. WLAN is integrated with UMTS using two different methods in this paper, i.e., loose coupling and tight coupling.

Both the schemes are defined in later sections of this paper. The diagram given below shows the connection difference while implementing these two schemes. The hexagonal shape shown in Fig. 1 is called as cell which is the area covered by node B of UMTS network. One mobile node is moving in and out of the range of UMTS and switching to WLAN as it moves closer to the WLAN router.

1.2 Loose Coupling Scheme

In loose coupling scheme, the signals are transferred over WLAN to the UMTS network and the traffic flow takes place directly through internet (IP cloud). In this method, WLAN and UMTS are independently connected to each other [4]. In other words, both the networks are connected with each other using a third medium. It uses a common authentication mechanism which setups a link between AAA server of WLAN and home location register (HLR) of UMTS. This method is generally used when any private operator is operating WLAN, because the transmitted information does not pass through UMTS. The major drawback of using this scheme is poor handover from one network to another.

1.3 Tight Coupling Scheme

In tightly coupled scenario, WLAN network is directly connected to the core of UMTS. Therefore the traffic directly goes through UMTS network [5]. Opposite to the loose coupling method, all the systems in tight coupling are dependent on each other. In other words, this scheme does not use any third party medium for connectivity. The integrated networks are not independent. This dependency can be easily justified, because of the direct connection of one network to another. The handover mechanism in this scheme can be considered as cell to cell handover, as the node moves out from one network and directly enters to the range of other network. But the disadvantage of using tight coupling scheme is, due to the more traffic flow through WLAN, overflow situation may arise at the SGSN node of UMTS network [6].

1.4 Integrating Architecture with IMS

As mentioned in earlier sections, the main purpose behind using IMS is to reduce the hand over delay. And also the functioning of AAA server is already inbuilt in IMS, therefore the switching mechanism becomes much easier. The major entities of IMS architecture are proxy-CSCF, serving-CSCF, and interrogating-CSCF. These are the call session control functions (CSCF) having different role in registering and establishing the session between users. All the users need to register to the corresponding IMS terminal of their area. The registration process starts with sending a registration request to PCSCF, which then goes to ICSCF in order to search the location of particular user. Once the location is found, SCSCF carries out the authentication mechanism in order to complete the registration process. Now, if any user wants to make a contact with another user, it needs to send an invite request again through PCSCF. The location of called user will be verified by ICSCF, whether it is related to same network or different network [7]. After this verification SCSCF will be responsible for final establishment of session between calling and called user. This is the basic IMS registration and establishment process, using this IMS architecture, UMTS and WLAN can be integrated in a much easier way with proper hand off. When user moves from UMTS to WLAN, the IMS takes care of all the packet formats according to the switched network. IMS enables an user to continue its session in another network without any disturbance. Figure 1b shown below describes the integration while using IMS. The difference in architecture is only three additional proxy servers which are responsible for the successful session establishment.

2 Literature Survey

Our goal was to implement the two networks as WLAN and 3G cellular network as the integration which will provide us the seamless mobility and best quality of service with authentication, authorization, and accounting (AAA). Many of the previous work related to integration of networks have been surveyed. In [8], Surender et al. have made a comparison between WLAN and UMTS network based on the different performance parameters. In this paper, authors have presented different architectures for the dynamic home agent placement, so that mobile IP can be supported for heterogenous roaming. Results obtained by them shows that UMTS gives poor performance while WLAN provides average outcome. Abu-Amara et al. [9] implemented two different integration schemes, i.e. loose coupling and tight coupling. Along with these integration methods, they have considered mobile IP and mSCTP as the mobility schemes. Observations have been made for the applications HTTP, FTP, and other multimedia. According to the results, it has been concluded that loose coupling method gives the best performance in terms of throughput and delay parameters. In [10], they have surveyed recent advancement in the integration of UMTS and WLAN networks and presented the challenges and issues related to it to achieve seamless integration. They have concluded that IP technologies play a wider role in the integration of different networks. Also, different solution for QoS and the authentication of user has been examined. In [11], Karthika et al., have implemented the integrated networks of WIMAX–WLAN and UMTS–WLAN, and performance have been observed by transmitting voice packets over the integrated networks. The parameters that have been taken into consideration were Jitter, Delay, and MOS value. Conclusions have been made such that the MOS value got increased by the integration of WIMAX WLAN. In [12] authors have proposed an architecture for the integration LTE-WIMAX-WLAN. They have used IMS as an intermediate platform, which enables the user to use any network at any time. They have compared the tight-coupled integration of LTE-WIMAX, LTE-WLAN with tight-coupled integration consisting of UMTS–WLAN and UMTS–WIMAX. Results have shown that successful handoffs can be made between different networks using the proposed architecture with maintained QoS levels. In [6], they have used IMS platform for the integration of UMTS–WLAN networks. They have proposed different measures to avoid the duplication of data transferred. It has been concluded that packet loss can be easily avoided during the handoff while using IMS for the network coupling.

Liu and Zhou [13] have proposed an interworking architecture for the integration of UMTS and WLAN, based on the tight coupling methodology. To achieve the goal of reduced handoff delay, they have adopted a fast handoff algorithm. Based on the results obtained, it has been concluded that tight coupling can bring down the traffic cost and also it provides less burden to the UMTS core network. In [14], Routh proposed an efficient solution for service continuity of mobile users. They have implemented integrated UMTS–WLAN architecture using loose coupling method. Generic routing encapsulation has been used as a intermediate protocol between UMTS and WLAN. In [15] [16], authors have presented a survey on different integrating schemes of various networks. They have defined the vertical handover approach used in heterogeneity of networks.

3 Model Description

3.1 Scenario 1-Loose Coupling Without IMS

The above scenario for integration of UMTS–WLAN has been created in OPNET Modeler. It consists of UMTS network at one side of IP cloud and WLAN network on the other side of it. Since it is a loose coupling scenario shown in Fig. 2a in which the WLAN router is directly connected to IP cloud. Two application servers named as FTP server and HTTP server have been configured to observe the response time. The users in the range of UMTS network has been given mobility from UMTS range to WLAN range. Users can shift to any network wherever the range is more for the particular network. Since the application servers are kept common for both UMTS and WLAN, users can easily handover the call from one network to another. Mobile nodes have been selected from the object pallette in OPNET, so that their trajectory can be provided from one area to another.

3.2 Scenario 2-Tight Coupling Without IMS

In the tight coupling scenario shown in Fig. 2b, all the configurations and attributes have been kept same as loose coupling. The only difference is that the WLAN router is connected directly to the UMTS network instead of connecting through IP cloud. In this model, router is connected directly to **GGSN** node, but it is not necessary to connect it to GGSN, it can be connected to any one of the node of UMTS network.

3.3 Scenario 3-Loose Coupling with IMS

The Fig. 3a shows the implementation of loose coupling scenario using IMS. It consists of three additional proxy servers which are the main entities of IMS architecture. Every user need to register to the corresponding IMS terminal according to the registration process that has been mentioned in earlier section. Main benefit of using IMS is the negligible delay incurred during handoff from one network to another. Here user is moving from UMTS to WLAN network, where IMS is serving as an intermediate. It makes the switching mechanism much easier

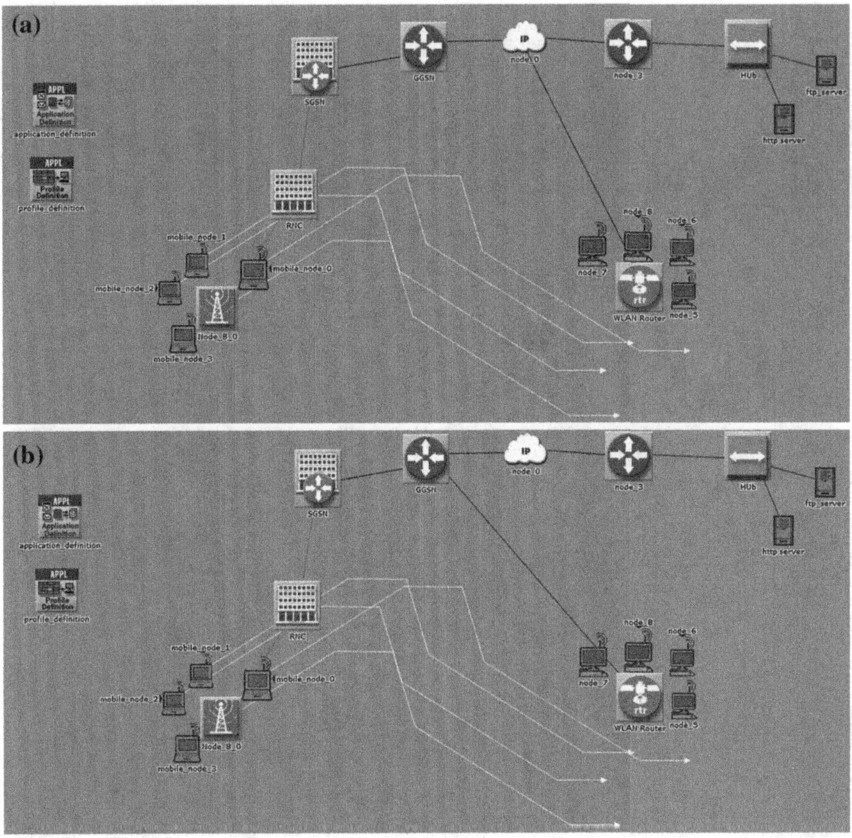

Fig. 2 **a** Loose coupling without IMS scenario in OPNET. **b** Tight coupling without IMS scenario in OPNET

as the AAA process is already inbuilt. Moreover, the response time for any application will be very less as compared to the integration without IMS.

3.4 Scenario 4-Tight Coupling with IMS

Similar to the loose coupling scenario, the Fig. 3b shows the tight coupling scenario with IMS as an intermediate. The functioning of both the networks is completely same, instead of the path used for traffic flow. Here, in tight coupling the traffic flows directly to the UMTS network instead of flowing through the internet. Rest all the process of registration and session establishment is same as the loose coupling scenario.

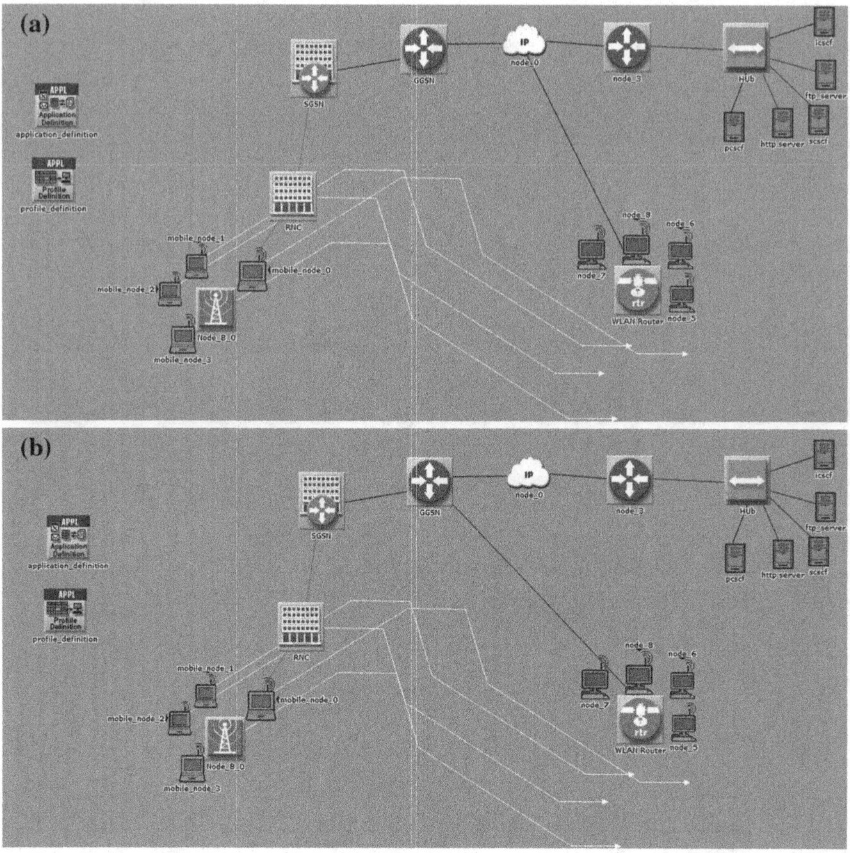

Fig. 3 **a** Loose coupling with IMS scenario in OPNET. **b** Tight coupling with IMS scenario in OPNET

4 Results and Discussions

4.1 Without IMS

This subsection shows the results obtained when IMS is not used for the integration of UMTS and WLAN networks. The simulation results for integrated network without IMS consists of analysis for the following parameters:

(i) **FTP download response time**: Figure 4 shows the download response time for FTP application in loose coupling and tight coupling scenario. As indicated in the scenario, the users are moving from UMTS to WLAN range via the trajectory path that has been given. Since WLAN provides higher data rate than UMTS, therefore as the user moves to WLAN the response time gets

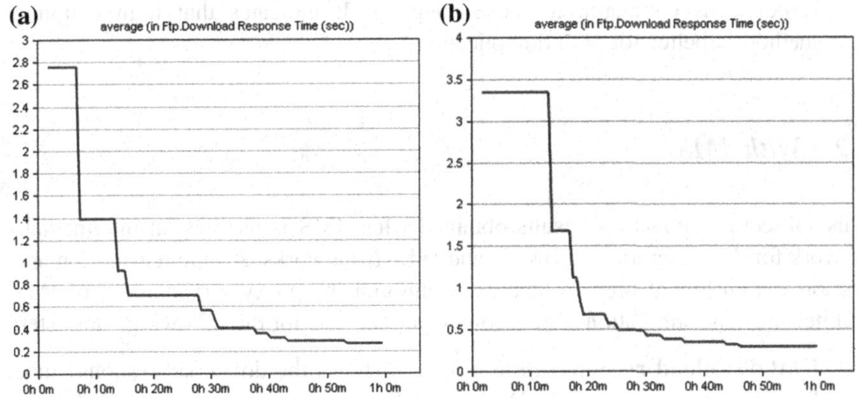

Fig. 4 FTP download response time. **a** Loose coupling. **b** Tight coupling

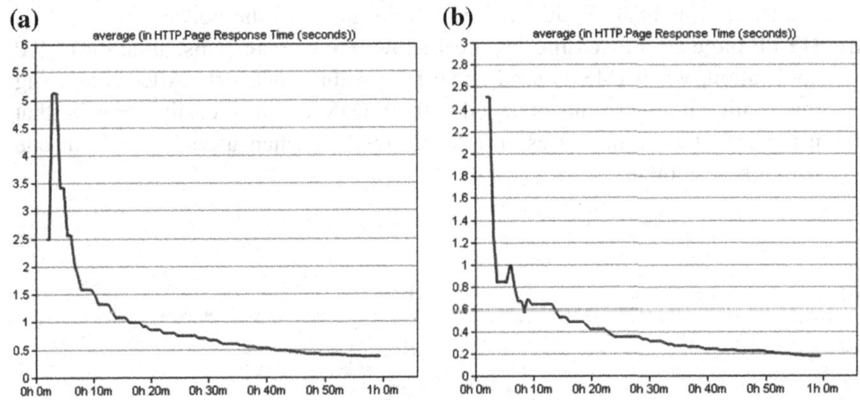

Fig. 5 HTTP page response time. **a** Loose coupling. **b** Tight coupling

decreased. Initially the response time is more as observed from the graph, but as the time increases, the response time gets decreased due to the mobility of nodes. By comparing both the scenarios, it can also be observed that loose coupling method takes less download response time as compared to tight coupling method. It indicates that loose coupling method is better for FTP application.

(ii) **HTTP page response time**: Figure 5 shows the page response time for HTTP application in loose coupling and tight coupling scenario. It indicates the time taken to load a page during the web browsing. Again it is observed that initially the page response time is more, as the time increases it gets decreased due to the shifting of nodes from UMTS to WLAN. But by comparison of both the scenario, it can be seen that tight coupling method takes less page response as compared to loose coupling. It indicates that tight coupling method is better for HTTP application.

4.2 With IMS

This subsection shows the results obtained when IMS is used as an intermediate network for the integration of UMTS and WLAN networks. All the registration and session establishment process take place through the proxy server nodes of IMS architecture. The simulation results consists of analysis for the following parameter.

(i) **FTP download response time**: Figure 6 shows the download response time for FTP application, when IMS is used as an intermediate network for the integration of UMTS–WLAN. As compared to the previous results without IMS, this response time is very less in both loose coupling as well as tight coupling scenario. But as the time increases, the response time still gets decreased, this is again due to the mobility given to the nodes.

(ii) **HTTP page response time**: Figure 7 shows the page response time for HTTP application, when IMS is used as an intermediate network. After comparing this result with the results obtained without IMS, it can be easily observed that using IMS has come out as an efficient method when users move from one network to another.

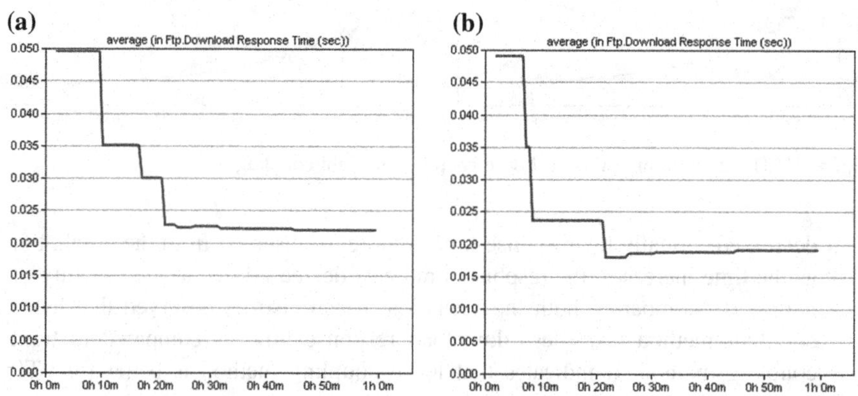

Fig. 6 FTP download response time. **a** Loose coupling. **b** Tight coupling

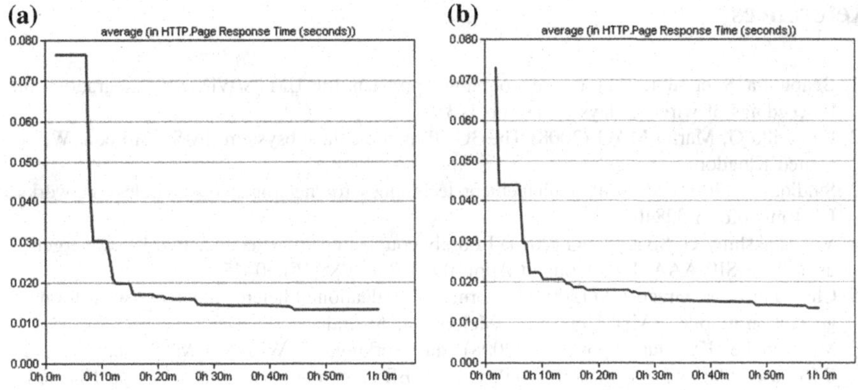

Fig. 7 HTTP page response time. **a** Loose coupling. **b** Tight coupling

5 Conclusions

To achieve the goal of this project, the IMS network has been used in the integration of UMTS and WLAN networks. Results have been taken for two different applications, i.e., FTP and HTTP. The integration scenario has been implemented using two methods which are loose coupling and tight coupling. WLAN has been integrated with UMTS using GGSN node in the UMTS core network. By observing all the graphs obtained after simulating in OPNET modeler, it can be concluded that the response time for any application continuously decreases as the user nodes move from network with low data rate to the network providing higher data rate. In this project, nodes are given mobility from UMTS to WLAN. Another thing that can be concluded is while using IMS in the integration of two networks, the response time is very less as compared to the integration without IMS. It shows the easiness and benefits of using IMS architecture. Therefore, IMS is the best platform to be used in wireless generation, which makes the user the integration has come out as a solution to overcome the situation when any network is having low data rate, lower bandwidth, or less connectivity. The integrated network enables the user to use the features of both the networks according to the requirements. This is the best way to eliminate any drawbacks of the network, because the same drawback can be overcome by other network which has been integrated.

In future, this project can be extended to integrate more than two networks which may include other advanced networks like WIMAX, LTE etc. Integrating more advanced networks with the existing one may provide feasibility to the users in terms of speed, connectivity, and bandwidth availability. The same IMS platform can be used for the integration, as it is able to handle interworking of different networks in a much easier way.

References

1. Benoubira S et al (2011) Loose coupling approach for UMTS/WiMAX integration. In: Proceedings of wireless days (WD), pp 1–3
2. Camarillo G, Martin MAG (2008) The 3G IP multimedia subsystem (IMS), 3rd edn. Wiley, United Kingdom
3. Siddiqui F (2007) Mobility management techniques for heterogeneous wireless networks. UMI microform 3284035
4. Vijayalakshmy G, Sivaradje G (2014) Loosely coupled heterogeneous networks convergence using IMS-SIP-AAA. Int J Comput Appl (0975–8887) 85(16):30–45
5. Chowdhury AS, Gregory M (2009) Performance evaluation of heterogeneous network for next generation mobile. RMIT University, Melbourne, Australia
6. Munasinghe KS, Jamalipour A (2008) Interworking of WLAN–UMTS networks: an IMS-based platform for session mobility. In: IEEE communications magazine, pp 184–191
7. Munir A (2008) Analysis of SIP-based IMS session establishment signaling for WiMax-3G networks. In: Networking and services, 4th international conference, pp 282–287
8. Surender R, Sivaradje G, Dananjayan P (2009) Performance comparison of UMTS/WLAN integrated architectures with dynamic home agent assignments. Int J Commun Netw Inf Secur (IJCNIS) 1(3):70–77
9. Abu-Amara M, Mahmoud A, Sheltami T, Al-Shahrani A, Al-Otaibi K, Rehman SM, Anwar T (2008) Performance of UMTS/WLAN integration at hot-spot locations using OPNET. In: IEEE wireless communication and networking conference (vol 3)
10. Song W, Zhuang W (2007) Interworking of 3G cellular networks and wireless LANs. Int J Wirel Mob Comput 2(4):237–247
11. Karthika AL, Sumithra MG, Shanmugam A (2013) Performance of voice in integrated WiMAX-WLAN and UMTS-WLAN. Int J Inno Res Sci Eng Technol (IJIRSET) 2:1188–1194
12. Hamada RA, Ali HS, Abdalla MI (2014) SIP-based mobility management for LTE-WiMAX-WLAN interworking using IMS architecture. Int J Comput Netw (IJCN) 6 (1):1–14
13. Liu C, Zhou C (2005) An improved interworking architecture for UMTS–WLAN tight coupling. In: IEEE communications society/WCNC, pp 1690–1695
14. Routh J (2013) A study on integration of 802.11 WLAN and UMTS networks. Int J Adv Res Electr Electron Instrum Eng (IJAREEIE) 2(5):2017–2022
15. Atayero AA, Adegoke E, Alatishe AS, Orya MK (2012) Heterogeneous wireless networks: a survey of interworking architectures. Int J Eng Technol (IJET) 2(1):16–21
16. Khattab O, Alani O (2013) A survey on media independent handover (MIH) and IP multimedia subsystem (IMS) in heterogeneous wireless networks. Int J Wirel Inf Netw 20 (3):215–228

Building a Knowledge Vault with Effective Data Processing and Storage

R.R. Sathiya, S. Swathi, S. Nevedha and U. Shanmuga Sruthi

Abstract Till date, any query that is being searched in the web gives us the result in the form of a list. This at times, tends the user to go through a lot of pages to find out what he actually needs! To overcome this issue, our major concern is to categorize the documents that are available in the net. Since there are enormous data available in the net, we first try to classify the documents as abusive or non-abusive. The same algorithm can be extended for different domains and can be used for classification. We have chosen this in particular because the user quite often likes to ignore the abusive documents. So, it is indeed better to hide the abusive documents from the user. This is what we are trying to implement in our paper. So, the idea behind achieving this is to initially crawl all the documents from a sample data set and then store them in a document database. After achieving this, the job is to classify them if they are abusive or non-abusive.

Keywords Data storage · Document database · WUP measure

R.R. Sathiya (✉) · S. Swathi · S. Nevedha · U. Shanmuga Sruthi
Department of Computer Science and Engineering, Amrita School of Engineering, Amrita Vishwa Vidyapeetham University, Coimbatore 641 112, Tamil Nadu, India
e-mail: rr_sathiya@cb.amrita.edu

S. Swathi
e-mail: swathe16@gmail.com

S. Nevedha
e-mail: nevedha248@gmail.com

U. Shanmuga Sruthi
e-mail: ucsabitha@gmail.com

L.P. Suresh and B.K. Panigrahi (eds.), *Proceedings of the International Conference on Soft Computing Systems*, Advances in Intelligent Systems and Computing 398, DOI 10.1007/978-81-322-2674-1_15

1 Introduction

Enormous amount of data are available in the World Wide Web. So, searching the data among them and finding out the accurate information tends to be a tedious task. But nowadays, searching has been achieved efficiently. If we think of a next step of improvisation, it would be categorizing the text document that is available in the net. This is what we are trying to concentrate in this paper. When we think of categorization, too many issues come into existence. The main issues that we consider here are the criteria based on which we categorize them, complexity of the algorithm, and the efficiency of the algorithm. The amount of data that is available is beyond our thinking. So, the categories available for all the domains are obviously inconceivable. So, implementing the algorithm for all the text documents available would be time-consuming. So, for a basic testing of the idea of categorization, we freeze this technique to a single domain. This paper is a step to classify the available documents as abusive or non-abusive documents. The purpose behind this is that, generally when a user searches for something in the net, he will tend to get the documents related to it. Looking at the comments that are abusive to the related query might irritate the user. These kinds of documents should be prevented from the sight of the normal user. So, the basic task that we ought to do now is to check whether the given document is abusive or non-abusive. The data set can be obtained by web-crawling. Following this, we got to classify them into two categories abusive and non-abusive. To achieve this, we divide the entire work into two basic modules namely, Storage of the documents and Categorization of the documents. These are the two main modules that we are to concentrate in this paper.

2 Related Works

Browsing for relevant documents for the user given query is considered as a problem because, sometimes the documents that are irrelevant to the user are also retrieved. The solution to this problem is classifying the documents based on their category to which they belong. There are many approaches that have been proposed to achieve this. Using text analysis method, the terrorism-related articles on the web can be detected. For this analysis, Keselj's context weight is used [1]. Another approach uses keywords for retrieving relevant documents. To get the significant keywords, WUP measurement is used so that significant context words are obtained [5]. There are approaches where WUP uses the WorldNet knowledge base to get the semantic relationship between the words [2]. Some other approaches make use of clustering of the retrieved search result records and query rewriting techniques for categorization [3, 4]. Given a word, all the other words which are allied to it can be extorted from WorldNet dictionary by making use of concept hierarchy [6, 7].

3 Proposed Model

In this paper, we propose a method for categorizing the articles on the web as either abusive or non-abusive.

3.1 Storage of Documents

In this section, we explain how the crawled documents are stored in the database. When a user requests for information, we should now give the non-abusive documents hiding the abusive documents from them. For us to achieve this, we must first store the documents in a database. Normal databases can only store few contents. But the data which we have, as mentioned earlier, is too large to store in a normal database. So, for us to store the documents that we get from crawling the web have to be stored in a document database.

The document database which we have used here is Raven database. This is used as a permanent database. In parallel, there is a temporary database called Redis cache. We use this because document database is efficient only in storage while opening and closing the connection seems to be diminutively slow. To compensate this, we are using a temporary storage which will open and close the connections faster but storage is not efficient. So as and when the crawling module starts, both the databases are allowed to run where the Redis cache starts the connection. Simultaneously, we can process those documents and apply the algorithm (which will be explained in the Sect. 3.2 of this paper) to categorize them. Now, these documents can be stored in the Raven database. The attributes that we store in the database are id, website, URL, hash, contents, crawled at, category, and status code. Now all the documents that are present will be crawled and with the help of the algorithm it will get categorized and stored in the database. As and when the user enters a query, documents that fall under the category non-abusive alone gets displayed to the user.

3.2 Categorization of Documents

This section explains how the documents are categorized. We use two sets of documents. One is training document set which contains documents that are already classified as abusive and non-abusive. And other document set is the test set which contains the documents that are to be classified. The task is to categorize each test document as either abusive or non-abusive. To achieve this, context weight is calculated for the test sets and the document is grouped under the category which yields a higher context weight. To calculate the context weight, context words are to be extracted by using WUP method. So, nouns and verbs are extracted from title and body of each document, because they give the significant meaning to the document. This forms the bag of words. But sometimes, some of them seem to be irrelevant to the document. So, to obtain the most significant words, we go for the WUP measurement. For this calculation, we use the concept hierarchy of the WorldNet dictionary which yields the semantic distance between different pairs of words in the collection. The similarity between the words of title and body is obtained with the help of the following formula:

$$\text{Similarity}_{\text{WUP}} = \frac{2 \times \text{depth (LCS(W1,W2))}}{\text{Depth(W1)} + \text{Depth(W2)}} \tag{1}$$

where

LCS is Lexical Conceptual Structure

N_{mb} indicates the number of matching bigrams between the training and the test document.

Now, we get the similarity value between each word in the title and body. Then, the average similarity value is calculated for each word in the body to all the words in the title. Then, words having the average value greater than 1.0 are considered as context words. After this, bigram frequency is calculated for getting bigram weight. Bigram is the pair of two consecutive words.

The possible number of bigrams in the bag of context words can be calculated using the formula

$$\text{No. of bigrams} = \text{no. of words} + n\text{gram}_{\text{type}} - 1 \tag{2}$$

Here, $n\text{gram}_{\text{type}} = 2$ as we are considering bigrams. Following this, we frame the bigrams for the context words. All these steps are carried out for both the training and test documents.

The bigram set of both the training and the test document are compared. If the bigrams in test document are present in the training bigram set, then the frequency of each matched bigram in the training and the test document is counted. The obtained frequencies are then multiplied.

$$\text{Bigram Weight} = f(\text{Bi}_{\text{tn}}) \times f(\text{Bi}_{\text{ts}}) \tag{3}$$

where

Bi_{tn} is the matched bigram in the trained document set.

Bi_{ts} is the matched bigram in the test document.

Bigram is the pair of two consecutive words.

Following this, context weight is calculated for classifying the test document. Because, the test document may have some matching bigrams with that of the training document if there is a relevancy between them.

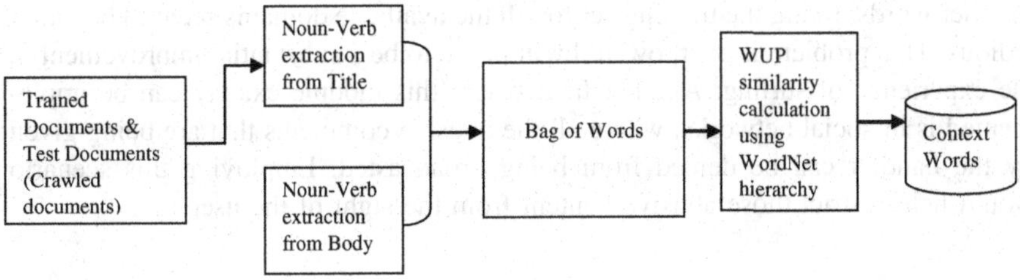

Fig. 1 Extraction of context words from the training and test documents

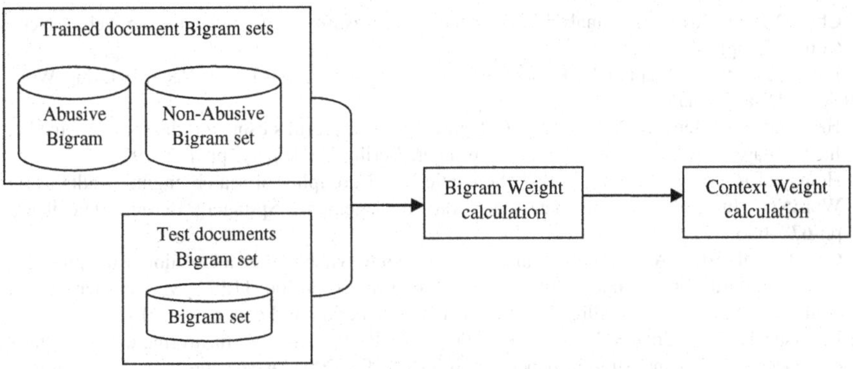

Fig. 2 Classification steps by using bigram similarity

$$CW = \max(\ln(1 + \text{Bigram Weight} \times N_{mb})) \tag{4}$$

where

N_{mb} indicates the number of matching bigrams between the training and the test document.

The test document will belong to the category which yields higher value of context weight. Similarly, all the test documents can be categorized as either abusive or non-abusive based on their context weight (Figs. 1 and 2).

4 Conclusion and Future Work

Now that we have proposed a model for categorization of the documents as abusive or non-abusive, we can apply the same for any domain. Since the results that are retrieved to the user are in the categorized form, it is highly pertinent to the conviction of the user. The only problem which is prevailing in this domain categorization is to restrict the number of subcategories that is there for each domain.

In other words, fixing the training set for all the available domains seem to be rather tedious. This problem if overpowered will prove to be a substantial improvement in the experience of surfing. Another field where this module exactly can be implemented is in social networks, where all the abusive comments that are being given by the handler can be denied from being broadcasted. Employing this scenario would help restrict those abusive content from the sight of the user.

References

1. Choi D et al (2014) Text analysis for detecting terrorism-related articles on the web. J Netw Comput Appl 38:16–21
2. Babisaraswathi R, Shanthi N, Kiruthika SS Categorizing Search Result Records Using Word Sense Disambiguation
3. Hemayati RT, Meng W, Yu C (2012) Categorizing search results using wordnet and Wikipedia. In: Web-age information management. Springer, Berlin, Heidelberg, pp 185–197
4. Hemayati R, Meng W, Yu C (2007) Semantic-based grouping of search engine results using WordNet. In: Advances in data and web management. Springer, Berlin, Heidelberg, pp 678–686
5. Choi D et al (2012) A method for enhancing image retrieval based on annotation using modified WUP similarity in wordnet (vol 12). In: Proceedings of the 11th WSEAS international conference on artificial intelligence, knowledge engineering and data bases, AIKED
6. Pedersen T, Patwardhan S, Michelizzi J (2004) WordNet:: similarity: measuring the relatedness of concepts. In: Demonstration papers at HLT-NAACL 2004. Association for Computational Linguistics
7. Sanderson M, Croft B (1999) Deriving concept hierarchies from text. In: Proceedings of the 22nd annual international ACM SIGIR conference on research and development in information retrieval. ACM

Comparative Performance Analysis of Microstrip Patch Antenna at 2.4 GHz and 18 GHz Using AN-SOF Professional

M. Puthanial and P.C. Kishore Raja

Abstract The main aim of this paper is to simulate and analyze the performance of a rectangular patch antenna at frequencies 2.4 and 18 GHz. The simulation results show that the antenna has achieved a gain of 5 dB at 2.4 GHz and 6 dB at 18 GHz. These two frequencies are used in many wireless applications (2.4 GHz) and Ku-band applications (18 GHz). Here, the patch antenna is designed and simulated in AN-SOF PROFESSIONAL V3.5. The simulated results such as the gain, directivity, radiation pattern, and VSWR at both frequencies are compared and discussed in this paper.

Keywords Microstrip patch antenna · AN-SOF professional v3.5 · Wireless communication · Ku-band · Gain · Directivity · Radiation pattern · VSWR

1 Introduction

Microstrip patch antenna is a directional or omnidirectional antenna that plays a major part in the area of antennas. This antenna is used in many wireless applications from mobile communication to space communication. It has advantages that it is less expensive, low profile, small in size, and ease of fabrication. This antenna can provide both linear and circular polarization. The patch antenna structure has three layers; the patch, dielectric substrate, and the ground plane. The patch is made of copper. The length of the patch is represented by L, width by W, and height by h. Different dielectric substrates are used in patch antenna which determines the antenna size and performance. The patch antenna is also called as the voltage radiator. The gain of patch antenna is >3 dB.

M. Puthanial (✉) · P.C. Kishore Raja
Saveetha School of Engineering, Saveetha University, Chennai, India
e-mail: puthanial@gmail.com

P.C. Kishore Raja
e-mail: pckishoreraja@gmail.com

L.P. Suresh and B.K. Panigrahi (eds.), *Proceedings of the International Conference on Soft Computing Systems*, Advances in Intelligent Systems and Computing 398, DOI 10.1007/978-81-322-2674-1_16

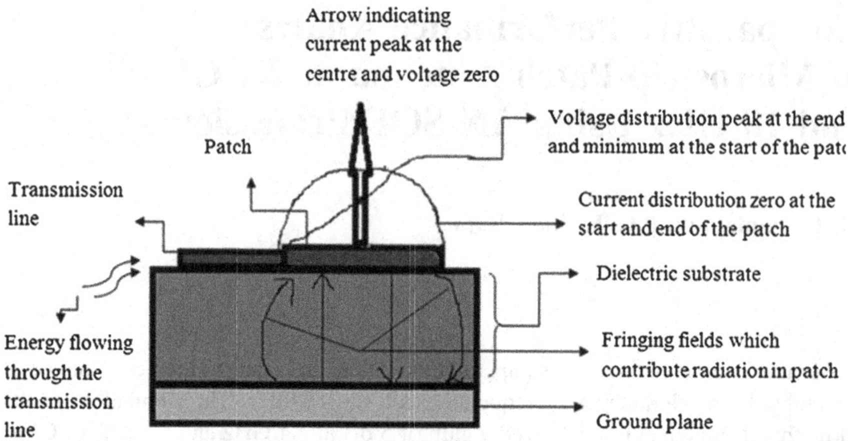

Fig. 1 Side view of patch antenna (the explanation of radiation in patch antenna can be viewed from the above figure)

The patch has an open-circuited transmission line. In that case, whatever the magnitude is the reflection coefficient is equal to 1; the current and voltage distribution is 90° out of phase. In Fig. 1, the current terminating in the short circuit force makes the current distribution look like that and the open circuit force makes the current and voltage out of phase. The current on the patch will add up in phase meanwhile the current in the ground plane will cancel out the radiation which is equal but opposite in direction. Thus, the fringing fields will be going in the same direction with a horizontal component added up in phase, contributing radiation.

2 Antenna Design

The following considerations were taken into count for designing patch antenna.

(i) Calculation of width

$$W = [C_0/2f_r][2/(€_r + 1)]^{-1/2}$$

(ii) Calculation of effective parameters

$$€_{reff} = [(€_r + 1)/2 + (€_r - 1)/2]\ [1 + 12h/W]^{-1/2}$$
$$\Delta L = 0.412[(€_{reff} + 0.3)(W/h + 0.264)/(€_{reff} - 0.258)(W/h\ + 0.8)]$$

Table 1 Input parameters used for designing the patch antenna

Input parameters	2.4 GHz	18 GHz
Substrate used	RT DUROID	RT DUROID
Dielectric constant	2.2	2.2
Length of patch	41.15(mm)	4.460(mm)
Width of patch	49.22(mm)	6.563(mm)
Height of patch	1.58(mm)	1.58 (mm)
Feeding technique	Edge feed	Edge feed

(iii) Calculation of patch length (Table 1)

$$L = [C_0/(2f_r\sqrt{\epsilon_{reff}})] - 2\Delta L$$

where

- C Velocity of light (3 x 10^8)
- ϵ_{reff} Effective dielectric constant
- W Width of the patch
- h Height of the patch
- ϵ_r Dielectric constant
- Δ Extension of length (Table 1)

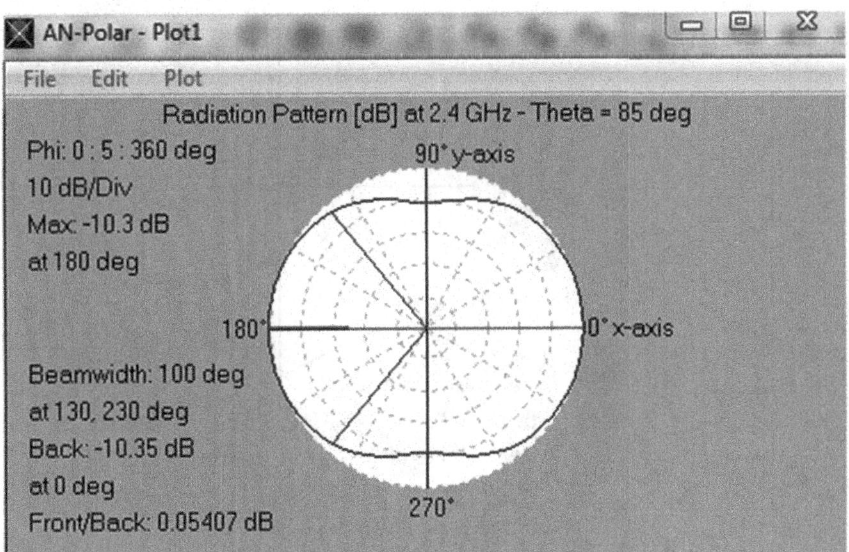

Fig. 2 Radiation pattern at 2.4 GHz in 2D polar plot

3 Simulation Results

The radiation in a patch antenna is maximum only if the transmission line matches with its load. A measure of mismatch in the transmission line is called the voltage standing wave ratio (VSWR). $1 < \text{VSWR} < \infty$, where 1 determines a matched load (Figs. 2, 3, 4, 5, and 6).

The above smith chart represents that the VSWR is 1 at 18 GHz, thus the transmission line is perfectly matched with the antenna. The simulation results at 2.4 GHz shows that the VSWR is slightly varying from the theoretical value required for good antenna performance (Table 2).

The directivity of an antenna is higher than the gain because the directivity takes into count only the direction in which the radiation is maximum, but gain takes into count of radiation in all direction, so that the gain is minimum.

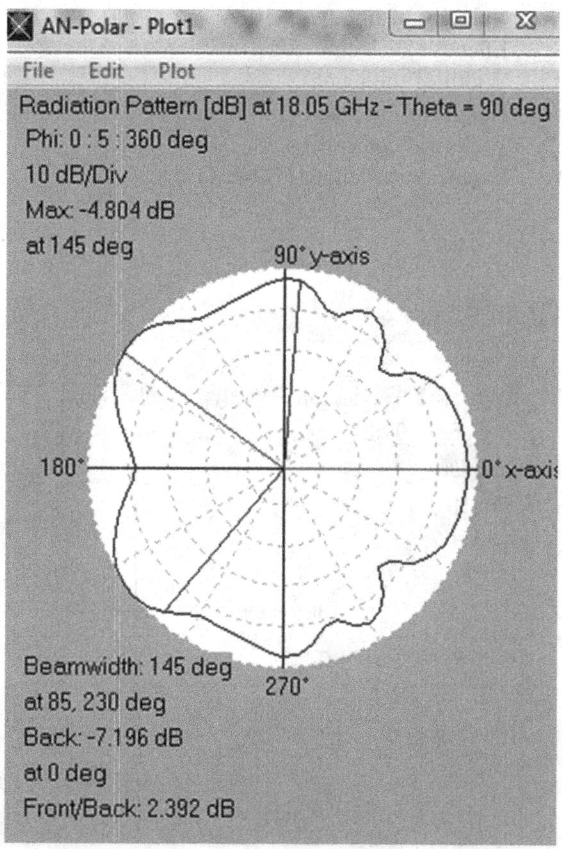

Fig. 3 Radiation pattern at 18 GHz in 2D plot

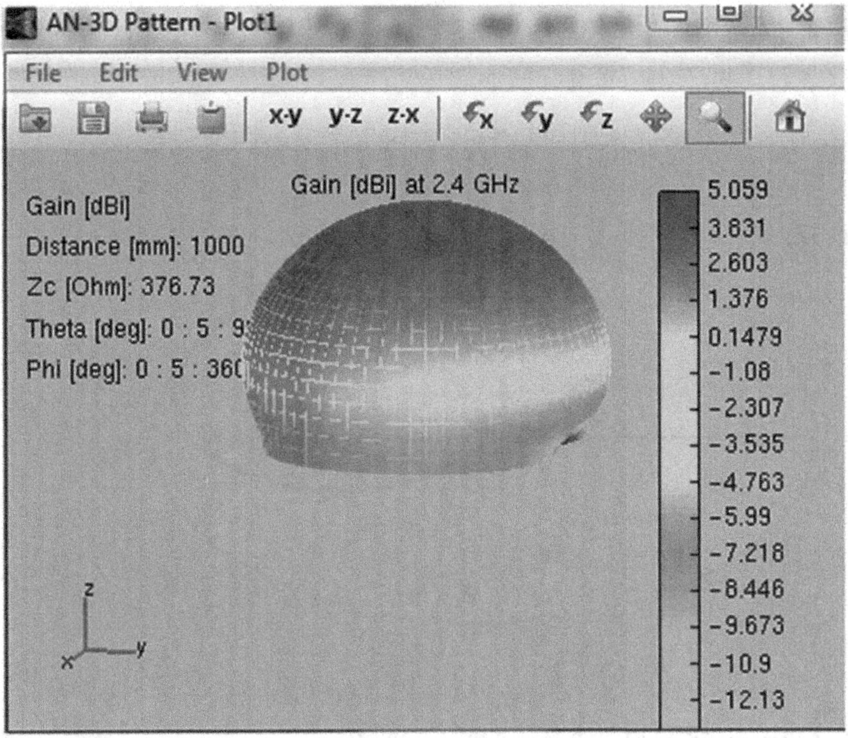

Fig. 4 Representation of gain at 2.4 GHz in 3D plot

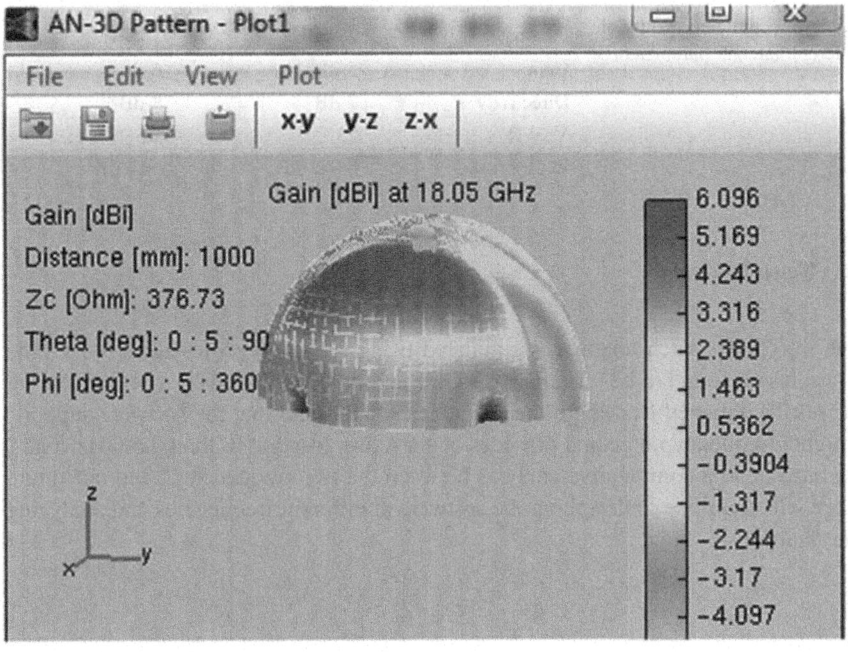

Fig. 5 Representation of gain at 18 GHz in 3D plot

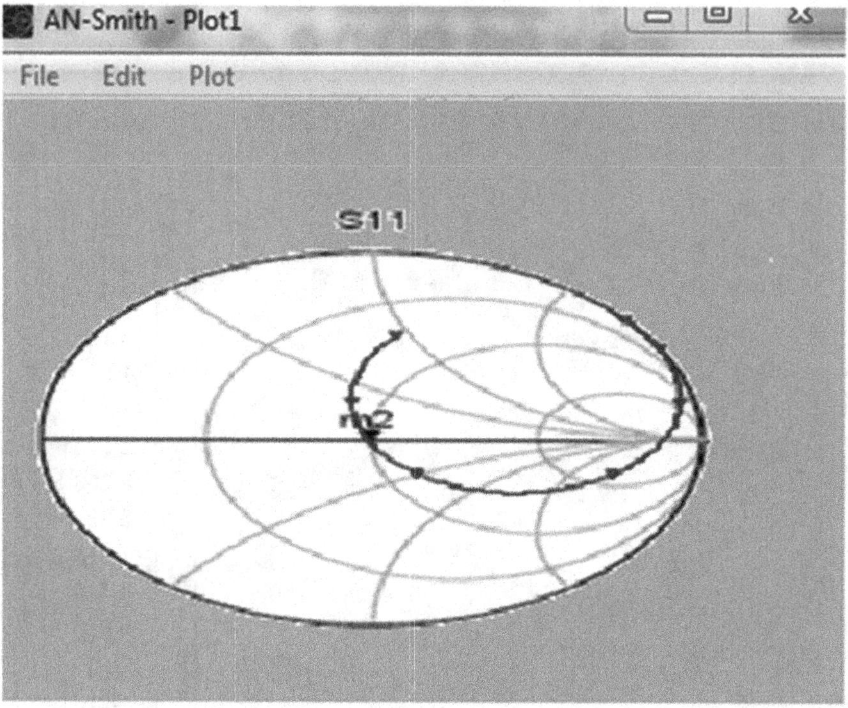

Fig. 6 Representation of VSWR in smith chart at 18 GHz

Table 2 Simulated results of different antenna parameters

Parameters	2.4 GHz	18 GHz
Gain	5 dB	6 dB
Directivity	7 dB	8 dB
VSWR	23	1.5
Return loss	0.087 dB	−14 dB

4 Conclusions

The microstrip patch antenna is simulated using AN-SOF professional V3.5. The return loss is found to be −14 dB at 18 GHz and 0.087 at 2.4 GHz. After comparing the results obtained from both the frequencies, we found that the software supports a higher frequency range and provides better return loss for 18 than 2.4 GHz. Thus, we have done a comparative analysis between the two frequencies. And our future work will comprise of designing the software at different frequencies and analyzing the return loss.

References

1. Sabri H, Atlasbaf Z (2008) Two novel compact triple-band micro strip annular-ring slot antennas for pcs-1900 and WLAN applications. Prog Electromagn Res Lett 5:87–98
2. Abu M, Rahim MKA (2009) Triple band printed dipole TAG antenna for RFID. Prog Electromagn Res C 9(145):153
3. Li RL, Pan B, Wu T, Laskar J, Tentzeris MM (2008) A triple—band low-profile planar antenna for wireless applications, 15 Dec 2008
4. Shambavi K, Alex CZ, Krishna TNP (2009) Design and analysis of high gain millimeter wave Microstrip antenna array for wireless application. J Appl Theor Inf Technol
5. Kraus JD (1988) Antennas, 2nd edn. Mc Graw Hill International, New York
6. Bouhorma M, Benahmed A, Elouaai F, Astito A, Mamouni A (2005) Study of EM interaction between human head and mobile cellular phone. In: Proceedings of information and communication technologies international symposium, Tetuan, Morocco, 3–6 June 2005
7. Balanis CA (1997) Antenna theory, 2nd edn. Wiley, NewYork
8. Jayachitra T, Pandey VK, Singh A (2014) Design of Microstrip patch antenna for WLAN applications. In: International conference on signal processing, embedded system and communication technologies and their applications for sustainable and renewable energy (ICSECSRE'14) (vol 3, special issue 3), April 2014
9. Bancroft R (2006) Microstrip and printed antenna design. Prentice Hall, India
10. Wi SH, Kim JM, Yoo TH, Lee HJ, Park JY, Yook JG, Park HK (2002) Bow-tie shaped meander slot antenna for 5 GHz application. In: Proceedings of IEEE international symposium antenna and propagation, vol 2, pp 456–459
11. https://www.youtube.com/watch?v=4qyb_hslP3A
12. Haneighi M, Toghida S (1988) A design method of circularly polarized rectangular Microstrip antenna by one-point feed. In: Gupta KC, Benalla A (eds) Microstrip antenna design. Artech house, Norwood, MA, pp 313–321
13. Sharma PC, Gupta KC (1983) Analysis and optimized design of single feed circularly polarized Microstrip antennas. IEEE Trans Antennas Propag 29:949–955
14. Rosu J (2010) Small antennas for high frequencies. Yo3dac - Va3iul. http://www.qsl.net/va3iul/
15. Chaturvedi A, Bhomia Y, Yadav D (2010) Truncated tip triangular Microstrip patch antenna. IEEE
16. Howell J Comparison of performance characterization in 2 × 2, 3 × 3 and 4 × 4 array antennas. IJERA 1(4):2091–2095
17. Gupta HK, Singhal PK, Sharma PK, Jadun VS (2012) Slotted circular Microstrip patch antenna designs for multiband application in wireless communication. Int J Eng Technol 3(1):158–167
18. Puthanial M et al (2015) Design of Microstrip smart antenna at discrete versus fitted frequencies at 2.4 GHz Using ADS. Int J Appl Eng Res 10(4):3530–3532. ISSN 0973-4562
19. Kraus JD, Marhefka RJ (2002) Antenna for all applications, 3rd edn. McGraw-Hill, New York
20. Rachmansyah AI, Mutiara AB (2011) Designing and manufacturing Microstrip antenna for wireless communication at 2.4 GHz. Int J Comput Electr Eng 3(5)
21. Pavithra D, Dharani KR (2013) A design of H-shape Microstrip patch antenna for WLAN applications. Int J Eng Sci Invention 2(6)
22. Hasan N (2012) Design of single and 1 × 1 Microstrip rectangular patch antenna array operating at 2.4 GHz using ADS. Int J Eng Res Appl 2(5):2124–2127
23. Agarwal N, Dhubbkarya DC, Mitra R (2011) Designing and testing of rectangular Microstrip antenna operating at 2.4 GHz using IE3D. Global J Res Eng (version 1.0)
24. Albooyeh M, Kamjani N, Shobeyri M (2008) A novel cross-slot geometry to improve impedance bandwidth of Microstrip antennas. Prog Electromagnet Res Lett 4:63–72

25. Kawase D et al (2012) Design of Microstrip antennas fed by four-Microstrip-port waveguide transition with slot radiators. In: International symposium on antennas and propagation (ISAP), pp 54–57
26. Rathod JM (2010) Comparative study of Microstrip patch antenna for wireless communication application. Int J Innov Manage Technol 1(2)
27. Puthanial M et al (2014) Microstrip patch antennas—survey and performance analysis 3(6). ISSN 2277-8179
28. Puthanial M et al (2015) Simulation and analysis of Microstrip patch antenna using AN-SOF professional. IJRASET 3(III). ISSN 2321-9653
29. Jain K, Gupta K (2014) Different substrates use in Microstrip patch antenna-A survey. Int J Sci Res 3(5)
30. Puthanial M, Shubhashini R, Pavithra K, Priyanka Raghu PC, Raja K (2014) Comparative analysis of Microstrip patch antenna using EZNEC and ADS. Int J Eng Trends Technol 16(2):54–57. ISSN 2231-5381

A Study on Security Issues in Cloud Computing

Kaaviyan Kanagasabapathi, S. Deepak and P. Prakash

Abstract Cloud computing focuses on maximizing the effectiveness of shared resources. It is also cost-effective, flexible, quick data storage, and is one of the most successful services in the internet. Cloud services are often outsourced to a third party, increasing the security threats. They are also delivered through the traditional network protocols and so a threat to the protocols is a threat to the cloud as well. In this paper, the major security issues are analyzed and sufficient countermeasures are provided, in order to minimize the security threats concerning cloud computing.

Keywords Cloud computing · Saas · Paas · Iaas · SPI model · Threats · Security · Countermeasures · Vulnerabilities

1 Introduction

Over the decade, cloud computing has got a prominence attention in the computing world as well as the industrial community. Cloud computing enables a shared pool of network and computing resources in a more convenient way.

Cloud computing mainly focuses to provide a quick and convenient computing service delivered over the Internet. It also aims to reduce the cost, by sharing of available resources. Cloud computing is a relatively young field, and there is a lot to experiment. Although the benefits of adopting cloud computing are many, there are

K. Kanagasabapathi (✉) · S. Deepak · P. Prakash
Department of Computer Science and Engineering, Amrita Vishwa Vidyapeetham, 641 112 Coimbatore, Tamil Nadu, India
e-mail: kaaviyan333@live.com

S. Deepak
e-mail: s.dipak99@gmail.com

P. Prakash
e-mail: npprakash@gmail.com

L.P. Suresh and B.K. Panigrahi (eds.), *Proceedings of the International Conference on Soft Computing Systems*, Advances in Intelligent Systems and Computing 398, DOI 10.1007/978-81-322-2674-1_17

a few barriers that make us want to think twice, before we adopt it. Since cloud computing is a relatively new field, there is an uncertainty about the security it offers. Nothing is perfect, so is cloud computing, there are many threats and vulnerabilities that need to be addressed. A threat is defined as an attack, which, when happens would lead to misuse of the information and the security loopholes that makes it happen, is called as a vulnerability. In this paper, we analyzed the different security issues and threats in the SPI model of cloud computing, identifying the major loopholes and vulnerabilities and providing a countermeasure to address the same.

2 Security in the SPI Model

The three types of services provided by cloud model are [1, 2, 3].

- Software as a service (SaaS): The client is able to use the providers' applications which are running on cloud infrastructure. The application can be accessed through a thin client interface such as a web browser. The web-based email can be considered as an instance.
- Platform as a service (PaaS): The client is able deploy his own applications on the cloud infrastructure without installing any kind of tools on his own local machine. PaaS incorporates platform layer resources, including operating system support and software development frameworks that can be used to build higher level services.
- Infrastructure as a service (IaaS): The client is able to use computers both physical and virtual which is also accompanied by other resources.

Among all the three, IaaS offers greater customer control which is followed by PaaS and then the SaaS, because of which the provider will have more control on security in case of SaaS [4].

Let us understand the dependencies and relationship between these cloud service models [5]. PaaS and SaaS are hosted upon IaaS, which means that the security breach in IaaS will directly affect the PaaS and SaaS services. On the other hand, PaaS provides a platform for building and deploying SaaS applications, which shows the higher interdependency. Considering all this it can be said that all the three types of services are interdependent on each other, which means that any attack on one of the service layers will automatically compromise the security of the remaining two. All the service models have their own inherited security flaws, with some flaws common between them. Sometimes the provider of any one of the services may be the customer of the other service, which usually leads to the discrepancies in the security responsibility.

3 Software as a Service (SaaS)

SaaS provides various kinds of application as required by the client for instance email, business applications, etc., [6]. As mentioned earlier, the SaaS users have very limited amount of control on security as compared to the other two. The followings are some of the security concerns in SaaS.

3.1 Multi-Tenancy

The applications provided though the SaaS can be classified into three different maturity models. This classification is based on the characteristics such as scalability, configurability, and multi-tenancy [7]. In the first model, each client has his own customized software. Although this model has some drawbacks, it scores high on better security as compared to the other two. In the next model, the service provider provides different instances of the applications for different customers according to their needs but mostly all the instances have the same application code. The customer is sometimes able to change some configurations options. In the third maturity model, a single instance of the application serves to various customers, which promotes the multi-tenancy [8]. In this model, the data from multiple users is stored in the same database which has the risk of data leakage.

3.2 Data Security

In any kind of service, data security is of prime importance. The SaaS clients depend entirely on the service provider for security [1, 9, 10]. If the data of client is processed in plaintext and stored in the cloud, then the risk increases significantly. Sometimes the cloud service providers can subcontract the services such as backup which also increases the concern. In case of SaaS, the process of compliance is really complex because here the provider has data, which may introduce issues such as data privacy, security, etc., which should be enforced by the provider.

3.3 Issues with Respect to Accessing

Since the cloud service makes the accessibility of the data with any network devices such as public computers and mobile devices, this facility sometimes makes the situation still worse because some devices do not have the required level of security. Because of this increased exposure of the user's data, the security concern increases.

4 Platform as a Service (PaaS)

With PaaS, the client can use the cloud-based application without having the required hardware and software for running the application [5]. PaaS security requires the security of the runtime engine and the security of customer applications which are being deployed on PaaS platform. The PaaS service provider should ensure the runtime engine's security. The followings are the related issues.

4.1 Issues Because of the Involvement of Third-Party

PaaS usually offers the third-party web services. If that particular web service has some security issues, then the PaaS will also inherit security issues [11]. The PaaS users have to depend on both the security of web-hosted development tools and third-party services.

4.2 Involved Infrastructures' Security

If SaaS provides the software to the user, then PaaS provides the development environment tools to create the SaaS applications. The users of PaaS usually have a very little or no access to the underlying layers. The service providers are responsible for ensuring security in the underlying infrastructure as well as application services [12]. The developers who use the PaaS do not have the absolute assurance that the tools provided by PaaS provider are secure. In case of both SaaS and PaaS, the ultimate protection rests only with the provider of the service.

4.3 Application Development Life Cycle

The speed at which the applications will change in the cloud will affect both the system development life cycle (SLDC) and security, which demands the applications to upgraded frequently. Therefore, the developers have to ensure that their applications do not introduce new opportunities for the attackers, for which the security layer should be improved with every upgrade in a way that compatibility with the existing security is not compromised. The virtualized environment is vulnerable to all kinds of attacks for normal infrastructures. In fact, the virtual environments will be more prone to attacks as it has more possible modes of entry and more interconnection complexity [13].

5 Infrastructure as a Service (IaaS)

IaaS, in short for infrastructure as a service, is a model in which computer and networking components (both physical and virtual) are outsourced by an organization. The service provider runs and maintains the equipment and it is rented to a client based on his specific requirements. The client generally pays on a per-use basis. These computers are often referred to as virtual machine (VM). IaaS clouds offer additional resources such as a virtual-machine disk image library, raw block storage, and file or object storage, firewalls, load balancers, IP addresses, virtual local area networks (VLANs), and software bundles [14]. It generally gives the user a full control to the software [15].

A virtual machine monitor (VMM) or hypervisor is a low-level software that controls, monitors, and runs virtual machines. In a computer, there are multiple virtual machines known as host machines, while each of the virtual machines is called as a guest machine.

5.1 Security Issues in IaaS and Their Countermeasures

IaaS, like other services, also has some security issues. Here are few of the main threats and the counter measures to combat each threat.

Resource Sharing. Multiple VMs located on the same server share the same CPU, memory, network, and others, which may decrease the security of each VM. For example, if there is a malicious VM in a pool of VMs sharing memory (or other resources), then using covert channels two VMs can communicate bypassing all the rules defined by the VMM in its security module, without the need of compromising the hypervisor [16, 17].

The most secure way to solve the above problem is to use dedicated servers with a dedicated physical network, instead of using virtual networks to hook the VMs with their respected hosts. Although the above method slightly decreases efficiency and is more expensive, it is more secure [13].

Insecure APIs. Application programming language, or in short API, is a protocol that defines how a third party connects to a service. Cloud APIs are still immature which means that are frequently updated. A fixed bug can introduce another security hole in the application [18]. Also, without proper authentication it poses a direct threat to the cloud server. Improperly configured SSL also poses a threat. If SSL is not properly configured, then there is a chance that an attacker can intercept the data exchanges between two parties (in this case, the server and the client). This is otherwise called as man-in-the-middle attack.

The above vulnerability can be reduced using a strong authentication and encrypting transmission between the server and the client using properly configured SSL or HTTP.

Denial of Service, Zombie attacks. Computers infected by Trojans and used for DDoS (Distributed Denial of Service) are often called as zombies, because they do whatever the attacker commands it to do. A large group of zombie computers is collectively called as a botnet. A DDoS attack is an attack where a cloud service or a computer resource is targeted and is flooded with junk data, such that it even cannot respond to legitimate requests. The CPU is fully loaded such that it is slowed down or in most cases, crashed. When this attack is done by a single computer, it is called as denial of service and when botnet assists it, it is called as distributed denial of service. Several techniques (like the HTTP GET requests and SYN Floods) are used to facilitate this attack [19].

To prevent cloud from such attacks, intrusion detection system (IDS)/intrusion prevention system (IPS) can be used. Computers should be protected with firewalls and anti-viruses such that the Trojans cannot enter into them.

Virtual Machine rollback. In case of any error within the virtual machines, they can have the ability to roll back to a previous state. However, by rolling back, they can be re-exposed to the previously patched vulnerabilities and if left unchecked can pose a serious threat to the server.

Maintain a log noting the vulnerabilities previously fixed and in case if a virtual machine has to rolled back, it is easier to know the unpatched bugs at the time of making the backup [9, 20].

Virtual Machine Monitor (VMM). Since VMM runs the virtual machines, if the hypervisor is compromised then the virtual machine is compromised as well. Due to virtualization, virtual servers can be migrated between physical servers for fault tolerance, load balancing, or maintenance [16, 21]. The network through which the migration takes place is often insecure. Also, if an attacker can compromise the migration module in the VMM, it is possible to intercept the contents of the data [22, 23, 24].

Secure networks, possibly dedicated physical channels, must be used to carry out the migration. Keeping the VMM as simple and small as possible reduces the risk as it will be easier to find and fix any vulnerability.

6 Other General Threats and Issues

The above threats are mostly technology-related threats, but there are also other threats that are common to any organization or individual.

6.1 Insider Attacks

However, secure cloud is maintained, and it is still prone to insider attacks. People continue to be a weak point in information security [25]. Some privileged users have unlimited access to cloud data and if that privilege is used for malicious purposes, then it poses a serious threat to the data and also to the company's integrity.

Proper screening and background checks must be conducted on all employees, in order to reduce this serious threat [21].

6.2 *Phishing and Trojans*

These attacks mostly target the clients than the servers. Phishing is a method of acquiring sensitive information (like passwords, credit card pin, etc.) by misleading the user to enter his credentials in a web service that looks similar to the original. First, the victim is sent a fake copy of website, where he is misled to enter his sensitive details, and then these details are received by the attacker who hosts the fake website. The information thus received are used for malicious purposes. A Trojan horse is a camouflaged application, which in the background collects sensitive data from the victim, without his knowledge [26]. The computers thus affected can be even used as base by attackers to launch their further attacks or zombie attacks.

These types of attacks generally happen due to lack of security education. So proper education related to common security attacks must be given to the people [25].

7 Conclusion

Cloud computing is a growing industry and is receiving attention from both industrial and scientific community. Although the advantages are many, there are also a few security issues surrounding them. We have studied and presented the major security issues in the SPI model of cloud computing. It is to be noted that the issues are not limited to the few presented. Only the major and dangerous issues and threats are presented, while there are also several other threats in the field of cloud computing. The world is forever changing, with new technology comes added security risks. Although a solution is found for a threat, it is only temporary. Malicious attackers soon find a way to bypass all those security measures. The countermeasures and security patches for a vulnerability are only means of slowing down an attacker, not stopping him. The regulatory standards and security measures that are undertaken differ around the globe. Some of the privacy and security requirements must be standardized across the globe. This can unite everybody in working toward a threat-free internet society.

Acknowledgments We thank our computer science faculty Mr. Prakash, for guiding us and supporting us in all possible ways. We also want to thank our University, Amrita University, which encourages us to do these things.

References

1. Subashini S, Kavitha V (2011) A survey on security issues in service delivery models of cloud computing. J Netw CompuT Appl 34(1)
2. Mell P, Grance T (2011) The NIST definition of cloud computing. NIST Special Publication 800–145, Gaithersburg
3. Zhang Q, Cheng L, Boutaba R (2010) Cloud computing: state-of-the-art and research challenges. J Internet Serv Appl 1(1):7–18
4. Mather T, Kumaraswamy S, Latif S (2009) Cloud security and privacy. O'Reilly Media Inc, Sebastopol
5. Cloud Security Alliance "Security guidance for critical areas of focus in cloud computing V3.0." https://cloudsecurityalliance.org/guidance/csaguide.v3.0.pdf
6. Ju J, Wang Y, Fu J, Wu J, Lin Z (2010) Research on key technology in SaaS. In: International conference on intelligent computing and cognitive informatics (ICICCI), Hangzhou, China. IEEE Comput Soc USA, pp 384–387
7. Zhang Y, Liu S, Meng X (2009) Towards high level SaaS maturity model: methods and case study. Services computing conference, APSCC, IEEE Asia-Pacific, pp 273–278, Dec 2009
8. Chong F, Carraro G, Wolter R Multi-tenant data architecture. http://msdn.microsoft.com/en-us/library/aa479086.aspx
9. Rittinghouse JW, Ransome JF (2009) Security in the cloud. In: cloud computing: implementation, management, and security. CRC Press, USA, Aug 2009
10. Viega J (2009) Cloud computing and the common man. Computer 42(8):106–108
11. Xu K, Zhang X, Song M, Song J (2009) Mobile mashup, architecture, challenges and suggestions, International Conference on Management and Service Science, MASS, IEEE Computer Society, Washington, DC, USA, pp 1–4, Sept 2009
12. Chandramouli R, Mell P (2010) State of security readiness, Crossroads—Plugging into the Cloud, vol 16(3), Mar 2010
13. Reuben JS (2007) A survey on virtual machine Security, Seminar on Network Security, Helsinki University of Technology. http://www.tml.tkk.fi/Publications/C/25/papers/Reuben_final.pdf
14. Amies A, Sluiman H, Tong QC, Liu GN (2012) Infrastructure as a service cloud concepts. In: Developing and Hosting Applications on the Cloud, IBM Press, Indianapolis, Jun 2012
15. Dahbur K, Mohammad B, Tarakji AB (2011) A survey of risks, threats and vulnerabilities in cloud computing. In: Proceedings of the 2011 international conference on intelligent semantic web-services and applications
16. Hashizume K, Yoshioka N, Fernandez EB (2013) Three misuse patterns for cloud computing. In: Rosado DG, Mellado D, Fernandez-Medina E, Piattini M (eds) Security engineering for cloud computing: approaches and tools, IGI Global, Pennsylvania, pp 36–53
17. Ranjith P, Chandran P, Kaleeswaran S (2012) On covert channels between virtual machines. J Comput Virol Springer 8(3):85–97
18. Carlin S, Curran K (2011) Cloud computing security. Int J Ambient Comput Intell 3(1):38–46
19. Modi C, Patel D, Borisaniya B, Patel A, Rajarajan M (2013) A survey on security issues and solutions at different layers of cloud computing. J Supercomputing 63(2):561–592
20. Garfinkel T, Rosenblum M (2005) When virtual is harder than real: Security challenges in virtual machine based computing environments. In: Proceedings of the 10th conference on hot topics in operating systems, pp 227–229
21. Cloud Security Alliance "Top threats to cloud computing V1.0." https://cloudsecurityalliance.org/topthreats/csathreats.v1.0.pdf
22. Dawoud W, Takouna I, Meinel C (2010) Infrastructure as a service security: challenges and solutions. The 7th International Conference on Informatics and Systems (INFOS), pp 1–8, Mar 2010
23. Jasti A, Shah P, Nagaraj R, Pendse R (2010) Security in multi-tenancy cloud. IEEE International Carnahan Conference on Security Technology (ICCST), pp 35–41

24. Venkatesha S, Sadhu S, Kintali S "Survey of virtual machine migration techniques," Technical report, Department of Computer Science, University of California, Santa Barbara. http://www.academia.edu/760613/Survey_of_Virtual_Machine_Migration_Techniques
25. Popovic K, Hocenski Z (2010) Cloud computing security issues and challenges. In: Proceedings of the 33rd international convention MIPRO, pp 344–349, May 2010
26. Bhardwaj M, Singh GP (2011) Types of hacking attack and their counter measure. Int J Educ Plann Adm 1(1):43–53

An Analysis of Black-Box Web Application Vulnerability Scanners in SQLi Detection

Shilpa Jose, K. Priyadarshini and K. Abirami

Abstract Web application vulnerabilities enable attackers to perform malicious activities that can cause huge losses to the users. Web application vulnerability scanners are automated Black-Box testing tools that identify the vulnerabilities prevailing in a web application. The scanners have gained popularity with time due to its ability to detect the application architecture weaknesses without accessing the source codes of the target web applications. However, a scanner has its own limitations as well. This paper focuses on analyzing the web application vulnerability scanners' ability to detect SQL injection and therefore we test a set of three open-source scanners against a set of custom-built test samples with various categories of SQL injection.

Keywords SQL injection · Web application vulnerability · Black-Box scanners · Information security

1 Introduction

Over the decades, the World Wide Web has changed the way we communicate and do business around the globe. Web applications have brought data to the people wherever they go and in the cheapest manner. However, in the haste of making applications widely available, they are becoming more exposed and vulnerable to attackers. Today, more than half of the current computer security threats and vulnerabilities found affect the web applications. Attackers can exploit these

S. Jose · K. Priyadarshini · K. Abirami (✉)
Amrita Vishwa Vidyapeetham, Coimbatore, Tamil Nadu, India
e-mail: k_abirami@cb.amrita.edu

S. Jose
e-mail: shilpajose@hotmail.com

K. Priyadarshini
e-mail: k.priyadarshini12@gmail.com

L.P. Suresh and B.K. Panigrahi (eds.), *Proceedings of the International Conference on Soft Computing Systems*, Advances in Intelligent Systems and Computing 398, DOI 10.1007/978-81-322-2674-1_18

vulnerabilities for malicious intents like data breaches, gain user privileges, manipulation of remote files, creation of botnets, etc. This paper concentrates on one such web application vulnerability that has prevailed for many years and yet ranks first as per Open Web Application Security Project (OWASP) TOP 10 for 2013 [1], i.e., SQL injection (SQLi) and the tools used to detect them, i.e., Black-box web application vulnerability scanners.

SQLi is an approach where one attempts to execute malicious SQL statements in the back-end database by injecting it across the web application without the knowledge of the user or the owner(s) of the web application.

Black-Box web application vulnerability scanners are automated tools that discover vulnerabilities by performing attacks and closely observing the corresponding responses generated by the web applications. They categorize the vulnerabilities found as SQLi, cross-site scripting (XSS), information leakage, cross-site request forgery (CSRF), etc. Both commercial and open source web application vulnerability scanners are available for use.

The goal of this paper is to analyze the working and efficiency of web application vulnerability scanners in identifying the known SQLi vulnerabilities. Therefore, our research involves the testing of three well-known open source web application vulnerability scanners against a custom-built test samples for various types of SQLi vulnerability to analyze their behaviors [2, 3, 4]. The scanners selected were:

1. Sqlmap
2. Skipfish
3. w3af

Since the idea is not to compare and rate the scanners, rather to study the behavior of scanners in detecting SQLi vulnerability, we have not presented any comparative data in this paper.

2 Background

2.1 SQL Injection

SQLi is a code injection technique wherein an attacker attempts to pass malicious SQL commands (statements) through a web application for execution by the back-end database [5, 6]. SQLi takes advantage of unsanitized input fields. The high flexibility of SQL gives way to the creation of dynamic SQL queries which makes SQLi possible.

Amir Mohammad et al. informs future researchers and developers of the various possibilities of SQLi attacks [7, 8, 9] as:

Tautologies: This type of attack works by making the "WHERE" clause always true. Considering an example as shown below:

Original Query: SELECT * FROM Users WHERE Username = '$username' AND Password = '$password'

Attack code: 1' or '1' = '1

The result code after injecting the attack code ends up as:

SELECT * FROM Users WHERE Username = '1' OR '1' = '1' AND Password = '1' OR '1' = '1'

The above command is then executed in the backend giving away all the confidential information that is stored in the table Users to the attacker.

Logically Incorrect Queries: These types of attacks force the database to generate an error, giving the attacker or tester some valuable information about the database schema upon which they can refine their injection techniques further. Considering the previous example itself, if an *Attack code*: Abcd''' is injected the web application can throw several database error messages giving clues to the attacker.

UNION query: Here the UNION operator is used in the attack code to join two different queries. Eg:

Original Query: SELECT Name, Phone FROM Users WHERE Id = $id

Attack code: 1 UNION SELECT creditCardNumber,Pinno FROM CreditCardTable

Result after injection: SELECT Name, Phone FROM Users WHERE Id = 1 UNION SELECT creditCardNumber,Pinno FROM CreditCardTable

The above will retrieve all the names and phone numbers from Users table followed by the credit card details present in the CreditCardTable. Here both the queries to be executed should return the same number of columns in order to ensure a successful attack execution.

Piggy Backed Queries: These type of attacks also aim at executing independent queries but unlike the earlier UNION query attack, here the second query is sent to the database under the cover of first query and is executed only after the first query. Eg:

Original Query: SELECT * FROM news WHERE year='.$year.' AND author='.$author.' AND type='.$type.'

Attack code: '; drop table users —

is injected into one of the fields, then,

Result after injection: SELECT * FROM news WHERE year='2013' AND author=''; drop table users — ' AND type='public'

The above would execute the first query till 'author='';' and then execute the second query which will result in deletion of the table users thereby causing tremendous damage.

Alternate Encoding: As many scanners detect based on noted signatures of attack codes, the forms of SQLi query codes are changed using any of the character encoding schemes to deceive the security systems Eg: While the security systems look out for attack codes that look like ' or 1 = 1 –, the attackers can still send the same code in HEX encoding as %31 %20 %4F%52 %20 %31 %3D%31 to accomplish a successful attack without getting noted.

Stored SQLi: Also known as persistent SQLi/second order SQLi, these attacks make use of stored SQLi codes, i.e., SQLi codes are stored in the database (Eg: by passing through a poorly validated User registration form) which is later retrieved and executed (say, like during a User Login).

Inferences: In this type of attack, SQL codes are injected and responses of the web applications are observed to dig further into the web application. This is usually used when the error handling is well performed by the target web application thereby closing the pathway for performing logically incorrect query attacks.

2.2 Vulnerability Assessment

Software vulnerabilities can be decreased by increasing the knowledge on software faults. Some experts think the best way to find software faults is to perform a static code analysis (source code examination), others would like to analyze the systems in terms of the attacker's or victim's perspectives, market trends, etc. Dynamic Security Analysis [10], also known as Black-Box analysis, is one such technique that relies on discovering the vulnerabilities by testing the web application from the attacker's point of view. Tools such as web application vulnerability scanners are used for the same [11, 12].

Threat and vulnerability management majorly consists of two main techniques to rule out the vulnerabilities present in a system:

Vulnerability assessments: Vulnerability assessment is a process of identifying and quantifying the vulnerabilities in an environment.

Penetration Tests (Pen Tests): Pen Tests stimulate actions of an external/internal attacker to breach security.

- White-Box Penetration tests
- Black-Box Penetration tests

White-Box Penetrations Tests are carried out with complete information on the target system, while Black-Box Penetration Tests are conducted in an attacker's perspective with very little knowledge of the target system. Here, it is left to the tester to identify the vulnerabilities of the target system by performing every possible attack combination on it with the intention of compromising the system.

2.3 Web Application Vulnerability Scanners

The scanners operate based on rules and known vulnerabilities recorded in its relative database [12, 10]. The scanner is first configured to the target system as desired, after which it is put to scan. The working of the web application vulnerability scanners consists of the following phases:

1. Crawling: The scanners get an outline structure of the target web application at the end of this phase. Here the scanners crawl the entire web application by following every navigation link.
2. AEP identification: Once the structure of the target web application is fetched, every AEP (attack entry point) where SQLi attack code can be injected is identified.
3. Attack: On the identification of all the attack entry points, it then generates a set of attack codes and starts testing them against the entry points to trigger vulnerability. The responses by the target web application for each attack are noted. The input values for attack codes are usually generated using heuristics or predefined values [13].
4. Analysis: The web applications scanners always test against a set of known vulnerability signatures. The web scanners observe the responses of web application with each attack attempt and then generate appropriate reports citing out the vulnerabilities spotted in the target system.
5. Report generation: At the end, the scanners generate a set of reports with corresponding threat alerts and warnings.

3 Experiments and Analysis

Although all scanners basically follow the stated phases, they still possess varying capabilities. Thus, a set of three widely known open source scanners with SQLi detection proficiency were selected for the study.

Furthermore, since the scanners are more capable of detecting historically recorded vulnerability classes, the test samples were created to contain a vulnerability of each of the known SQLi types documented till date. The test samples were coded in php language and were run on LAMP machines. Each sample consisted of minimum pages just sufficient enough to induce the SQLi vulnerability type alone, while ignoring the other architecture weaknesses and rich user interface design concerns. The test bed was chosen to be a set of custom-built samples rather than one application with several vulnerabilities as this would enable us to induce all the known classes of SQLi as desired and would also make the vulnerabilities prominent to the scanners, thereby eradicating the complexity in crawling and picking out the attack entry points. Additionally, it would also aid in analyzing the scanner behavior better.

The scanners were made to scan each sample to observe their detection competency. The results have not been mentioned against the corresponding scanners as the intention is not to compare them rather to analyze and understand the behavior in general.

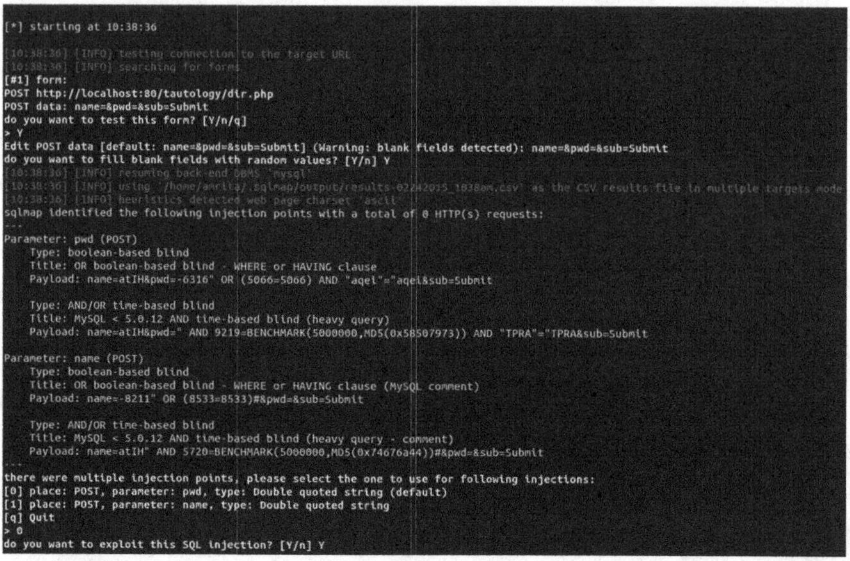

Fig. 1 SQLmap: python sqlmap.py –u http://localhost/tautology/ran.php—forms

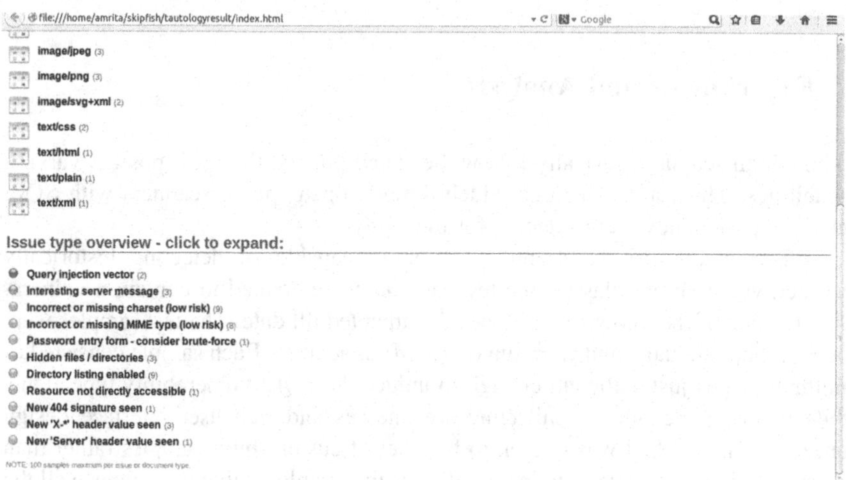

Fig. 2 Skipfish: ./skipfish –o result http://localhost/tautology/ran.php

The results obtained by each of the scanners on scanning the tautology sample are shown below in Figs. 1, 2 and 3.

However, not all the classes of the SQLi present in the samples were identified by all the scanners Union query vulnerability as such was spotted by two of the three scanners. It is observed that in most of the cases where piggybacked queries are injected, a union query can also be injected. Therefore, detection of at least

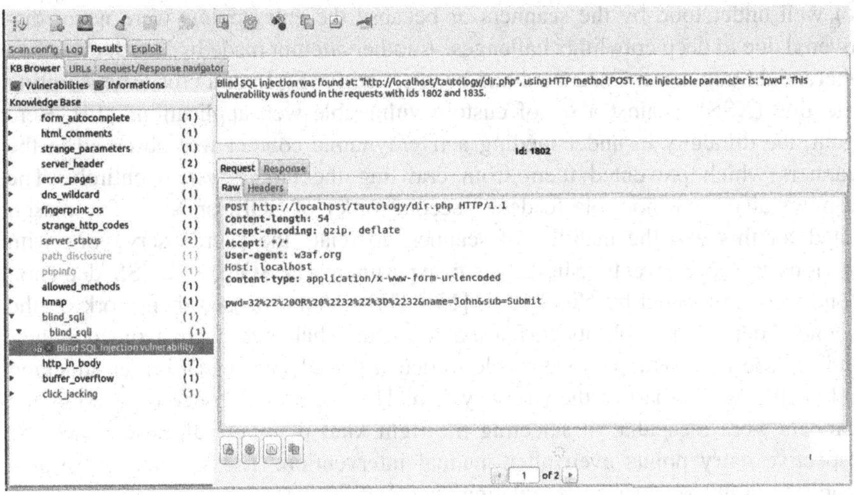

Fig. 3 W3af: ./w3af_gui

union query vulnerability would mostly close the doors for two of the SQLi vulnerability types. Logically, incorrect queries were also not detected by two scanners probably due to its pretentious benign appearance. The fact remains that the seemingly harmless error messages can give away significant information about the backend to the attacker. To detect this, the scanners can perform a comparative analysis of the web application content before and after the attack in terms of the error messages and can then check for keywords in the messages that relate to the database structure. For this, the scanners must possess a mechanism to retrieve the database structure before performing the attack. This mechanism was observed in one of the scanners taken for the study. The scanners equally failed to detect the possibility of second order SQLi because of the lack of ability to logically relate with any historical analysis that it may have performed earlier.

4 Related Work

The working of the web scanners have been analyzed deeply by many authors in an effort to find the causes for their failures in vulnerability detection. Doupe et al. [13] tested 11 scanners (both commercial and open source) against their custom-built test bed known as WackoPicko. The main goal was to identify the tasks that were challenging to scanners. WackoPicko is a realistic web application that contains a total of 16 known vulnerabilities that was induced with the intention of evaluating the scanner's ability to crawl the entire web application and identify the associated vulnerabilities. The authors observed that the scanners were not able to detect whole classes of vulnerabilities. This was either due to the fact that the vulnerabilities were not well understood by the scanners or because the entry points were never discovered due to deep crawling challenges. Another attempt made by Bau J. et al. was where eight commercial scanners had been tested for SQLi and cross-site scripting detection (XSS), against a set of custom vulnerable web applications [14]. Here again, the difficulty in understanding active/dynamic content was observed in the scanners which prevented them from crawling the target system entirely. The scanners also were not able to detect second order of SQLi or XSS. The reason stated for this was the inability of scanners to relate the later observations with previous injection events. Similar such experiments in SQLi and XSS detection alone were conducted by Nidal et al. [15]. The authors focused their work on the scanners' detection capabilities. Hence all crawling challenges were avoided in their study to see if the scanners were able to detect the advanced SQLi after attacking and analyzing that part of the target system. The same was observed to fail as the scanners were incapable of selecting the right kind of input values to attack the respective entry points even after manual intervention. The scanners were also unable to draw conclusions by relating successful events.

5 Conclusion

Though the web application vulnerability scanners have improved with time, many still fail in discovering the known vulnerabilities in spite of containing a vulnerability signature database. The scanners also find it hard to spot logical flaws in the web applications, viz., weak cryptographic functions, etc., and they have a limited understanding in the behavior of dynamic content. Currently, there also involves a considerable amount of manual verification of the results generated by the scanners. Furthermore, the less popularity in vulnerability varieties should not be considered as a reason to treat those vulnerabilities as a lesser threat and therefore all scanners must be finely tuned to detect the same. Though many more reasons for less detection rate like crawling challenges, lack of self-learning capability, etc., have been cited many more are yet to be addressed that would assist the developers in future to improve the overall performance of a scanner.

References

1. Category: OWASP Top Ten Project. http://www.owasp.org/index.php/Top10#OWASP_Top_10_for_2013
2. Sqlmap. Available at https://www.sqlmap.org
3. Skipfish. Available at https://code.google.com/p/skipfish/
4. W3af. Available at https://www.w3af.org
5. Testing for SQL Injection (OTG-INPVAL-005). https://www.owasp.org/index.php/Testing_for_SQL_Injection_(OTG-INPVAL-005)
6. SQL Injection. https://www.owasp.org/index.php/SQL_Injection

7. Sadeghian A, Zamani M, Abdullah SM (2013) A taxonomy of SQL injection attacks, Informatics and Creative Multimedia (ICICM), 2013 international conference on, IEEE, pp 269–273
8. Sadeghian A, Zamani M, Manaf AA (2013) A taxonomy of SQL injection detection and prevention techniques, Informatics and Creative Multimedia (ICICM), 2013 international conference on, IEEE, pp 53–56
9. Sadeghian A, Zamani M, Ibrahim S (2013) SQL injection is still alive: a study on SQL injection signature evasion techniques, Informatics and Creative Multimedia (ICICM), 2013 international conference on, IEEE, pp 265–268
10. Djuric Z (2013) A black-box testing tool for detecting SQL injection vulnerabilities, Informatics and Applications (ICIA), 2013 second international conference on, IEEE, pp 216–221
11. Web Application Security Scanner List. http://projects.webappsec.org/w/page/13246988/Web/%20Application%20Security%20Scanner%20List
12. Web application Security: the role and functions of black box scanners. https://www.acunetix.com/websitesecurity/blackbox-scanners/
13. Doupe A, Cova M, Vigna G (2010) Why Johnny can't pentest: an analysis of black-box web vulnerability scanners, DIMVA'10 Proceedings of the 7th international conference on detection of intrusion and malwares, and vulnerability assessment, Springer, Berlin, Heidelberg, vol 6201, pp 111–131
14. Bau J, Bursztein E, Gupta D, Mitchell J (2010) State of the art: automated black-box web application vulnerability testing, Security and Privacy (SP), 2010 IEEE symposium on, IEEE, pp 332–345
15. Khoury N, Zavarsky P, Lindskog D, Ruhl R (2011) Testing and assessing web vulnerability scanners for persistent SQL injection attacks. In: Seces'11 proceedings of the first international workshop on security and privacy preserving in e-Societies, ACM New York, pp 12–18

Search Optimization in Cloud

U.M. Ramya, M. Anurag, S. Preethi and J. Kundana

Abstract Cloud is the present emerging area to store secured data which may contain sensitive information. This paper provides an optimal search over cloud which allows the user to retrieve only relevant documents with respect to the query. Cloud handles huge set of documents. Thus when a user posts a query to the cloud, it is obvious that the user will receive a mixture of both relevant and nonrelevant documents. After which the user has to post-process the documents and sort out to get the relevant documents. This whole process adds overweight to the user. In this paper, we have come up with architecture and suitable algorithms which serves the user with exact relevant documents. We have used inverted index to index each document and Diffie–hellman algorithm to check for authentication. After the relevant documents are retrieved, we then rank each document based on tf–idf scoring and provide the top K retrievals to the users.

Keywords Diffie–hellman · Inverted index · Tf–idf · Ranking · Cloud

U.M. Ramya (✉) · M. Anurag · S. Preethi · J. Kundana
Department of Computer Science and Engineering, Amrita School of Engineering, Coimbatore, India
e-mail: ramya.u.m@gmail.com

M. Anurag
e-mail: mallemanurag1993@gmail.com

S. Preethi
e-mail: preethi.saravana30@gmail.com

J. Kundana
e-mail: jkssreejani@gmail.com

L.P. Suresh and B.K. Panigrahi (eds.), *Proceedings of the International Conference on Soft Computing Systems*, Advances in Intelligent Systems and Computing 398, DOI 10.1007/978-81-322-2674-1_19

1 Introduction

Memory storage is the primary usage of cloud computing. Cloud provides three services namely, (1) Software as a service (SaaS), (2) Platform as a service (PaaS), and (3) Infrastructure as a service (IaaS). The user uses these services based on their requirement. In this paper, we have used the Infrastructure as a Service (IaaS).

In order to provide an optimal search over cloud which allows the user to retrieve only relevant documents with respect to the query, we need to index all the documents before outsourcing it to the cloud to avoid retrieval of nonrelevant documents [1]. Thus all the documents that have to be stored in the cloud have to be preprocessed (indexed). These indexes are stored in trusted local server. So whenever the user queries the cloud, it first goes to the local server and indexed data is referred [2]. We get the relevant document IDs from the local trusted server which in turn are ranked and then sent to the cloud. The process of getting these relevant document IDs will be explained in the later section. The cloud then sends the relevant documents to the user.

An untrusted cloud server can have multiple data users. The data users are allowed to access the documents only if they are authenticated [3]. Authentication is the process for confirming the identity of the user. So, we have used a three-way Diffie–Hellman algorithm to check for authentication. The user can query the cloud only if he is authenticated [4].

For a given collection of documents, indexing is defined as a word in the collection is called as a term, and corresponding to each term we maintain a list known as posting list, of all the documents in which the particular word appears [5]. Once we get the relevant document IDs from the trusted local server, we rank each document based on tf–idf scoring. Thus, the document with highest tf–idf score will be given higher priority. In later sections, the working of each operation, i.e., inverted index, ranking, and authentication are explained in detail.

2 Literature Survey

Indexing of the documents are done using word-level inverted index [1, 5] in this paper as it provides information about position of each word within a document. There are many cryptographic algorithms available for authentication. Several papers have used RSA, DES, and Triple-DES which are not acceptably efficient. In this paper, the authentication part is done using Diffie–Hellman algorithm because it is an one-way algorithm and the authentication channel provided is more secured [6]. Retrieval of documents for the user is suggested by using Boolean search with the query which is not the effective way of retrieving the documents [7]. In this paper, ranking of retrieved relevant documents is implemented using tf–idf, which is more efficient compared to other scoring algorithm such as vector space model [8], set-oriented model [5]. In tf–idf, *rare keywords* are given more priorities than to frequent occurring words.

3 System Architecture

The three entities in our schema are: Admin, Local Trusted Server, and Untrusted Cloud Server. The admin is the one who stores the data in the cloud and has the authorization to go through the documents. Cloud server is the untrusted one where storage service is provided to the admin's data, where the admin can store their data in the form of plain text. The documents of the admin are indexed and the indexed lists are stored in the local trusted server. The documents are stored in the untrusted cloud server. When the query is made by the user, the local server makes a document request to the untrusted cloud server, where on the ranking basis the retrieval of relevant documents is done and the requested documents are provided to the user by the untrusted cloud server [2] (Figure 1).

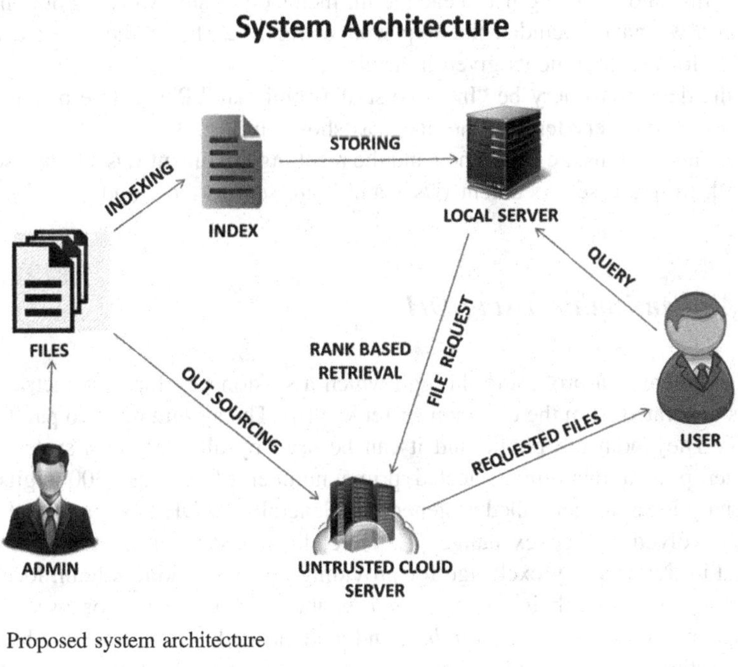

Fig. 1 Proposed system architecture

3.1 Inverted Index

The basic steps carried out in inverted index are-

(1) Collect the documents that are to be indexed.
(2) Tokenize the text, turning each document to a set of tokens.
(3) Normalize the tokens by either converting all the tokens into lower case or upper case, then perform linguistic preprocessing to the tokens.
(4) Index the documents such that for each term occurrence an inverted index is created consisting of a dictionary and posting list.

Each document has a unique serial number known as the document identifier. During index construction, we can assign successive integers to each document when it is first encountered. In this paper, we assume that the first three steps have already been done and the list of normalized tokens for each document has been sent as input for indexing. Now the core step of indexing is to sort the list in alphabetical order. Multiple occurrences of a term in a document are taken as a single entry. Since a term may occur in number of documents, we take only a single instance of that term in the dictionary and a posting list is attached to that term which consists of the entire document ID's in which the term is present. This posting list is implemented using a linked list [1]. Similar operations are performed on all the terms. Thus, finally we will have a dictionary which will have all the unique terms and a posting list to each term. Inclusion of stop words is optional. In this paper, we have excluded the stop words. Figure 2 shows the inverted index generated for the documents given in Table 1.

Let the data user query be "India Australia semifinal." Respective posting lists are retrieved for every term in the query as shown in Fig. 3.

Now, intersections are performed and the resulting document IDs are then sent to cloud [9]. In this case, document IDs 1 and 2 are sent to the cloud.

3.2 Authentication Using DH

Key Agreement is a procedure through which a session key for asymmetric algorithm is generated from the common secret key [6]. The system has two parameters p and g. They both are public and it can be used by all users in a system. The parameter p is a randomly selected prime number of at least 300 digits and parameter g is an integer called as generator. Generally, in DH any number of users can be involved in key exchange [4]. Here, local server, user, and admin are involved in this DH key exchange for providing authentication. Admin, local server, and user generate their private keys *a*, *b*, and *c*, respectively. For a secure key exchange, much larger values of *a*, *b*, *c*; and *p* are needed. *p* is a prime number of at least 300 digits and the private keys *a*, *b*, and *c* can be 100 digit long. Initially, admin uses his private key to compute g^a and sends the computed value to the local

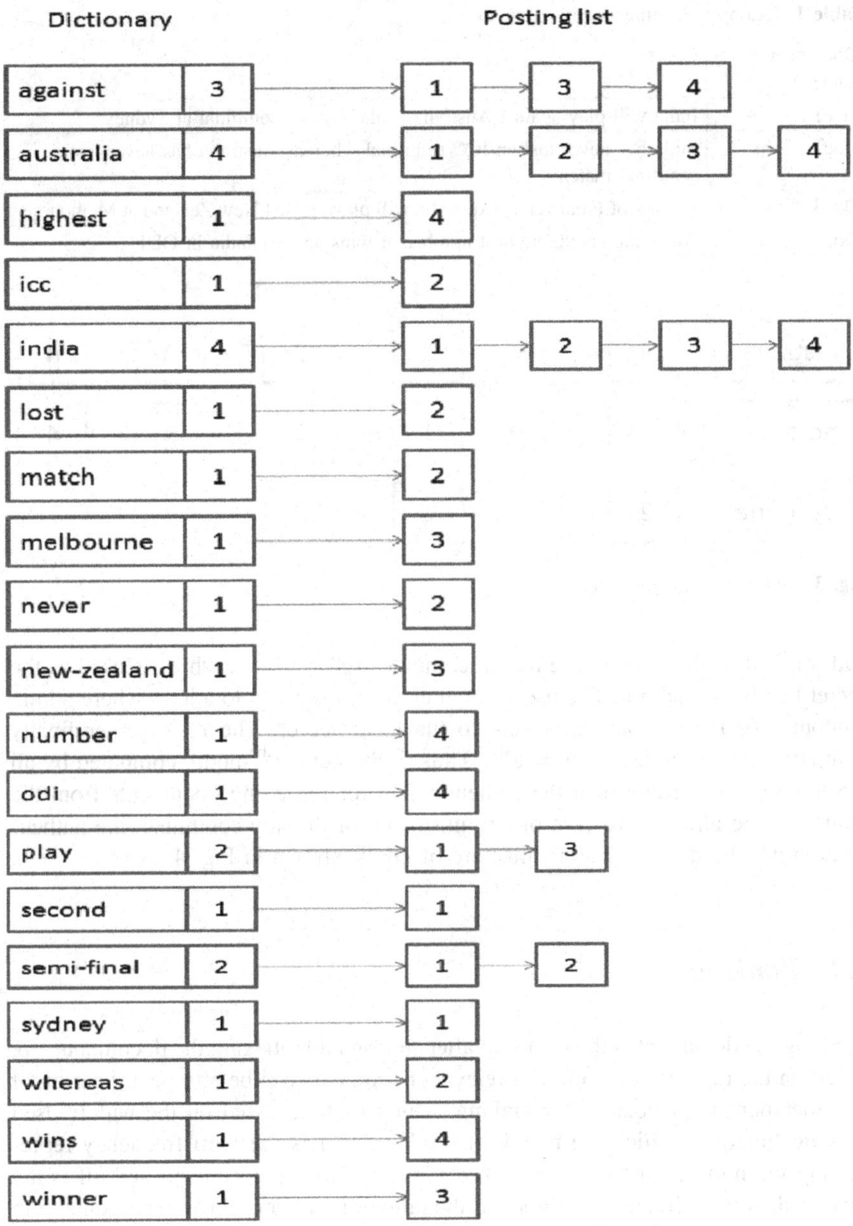

Fig. 2 Inverted index

server where the local server computes $(g^{a})^{b} = g^{ab}$ and it is sent to the user. The user computes $(g^{ab})^{c} = g^{abc}$, which is used as the secret key by the user. Similarly, the local server computes g^{b} and sends it to the user where the user computes $(g^{b})^{c} = g^{bc}$

Table 1 Example documents with contents

Document name	Contents
Doc1	India will play against Australia in the second semifinal at Sydney
Doc2	India has never lost an ICC semifinal whereas Australia has never lost a semifinal match
Doc3	Winner of India versus Australia will play against New Zealand at Melbourne
Doc4	Australia has the highest number of wins against India in ODI

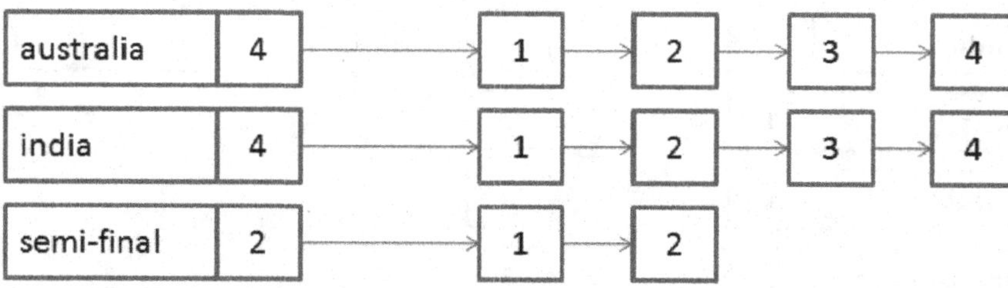

Fig. 3 Inverted index for query

and sends it to the admin. The admin computes $(g^{bc})^a = g^{abc}$, which is used as the secret key by the admin. The user computes g^c and sends it to admin where admin computes $(g^c)^a = g^{ac}$ and it is sent to the local server. The local server finally computes the secret key $(g^{ac})^b = g^{abc}$. Thus, if the keys g^{abc}modp, computed by all the three users is same then the authentication for retrieving documents from the cloud will be allowed. In case of any mismatch of the key computed, the authentication will be denied. The architecture of DH is shown in Fig. 4.

3.3 *Ranking*

Ranking the document is the next step after storing and indexing the documents. So, based on the ranking, the top-most relevant documents can be retrieved. Scoring of the document is mandatory for ranking, which is done based on the widely used ranking functions which are based on tf*idf rule; tf is the term frequency representing the number of times a keyword is present in the document, and idf is the inverse document frequency defined as the ratio of number of documents containing the word to the total number of documents present in the server. Tf–idf is used for ranking because it increases the ratio of the term frequency in the document, but counter balances the frequency of the words so that one can know the fact that general words are more frequent which are mostly stop words [5].

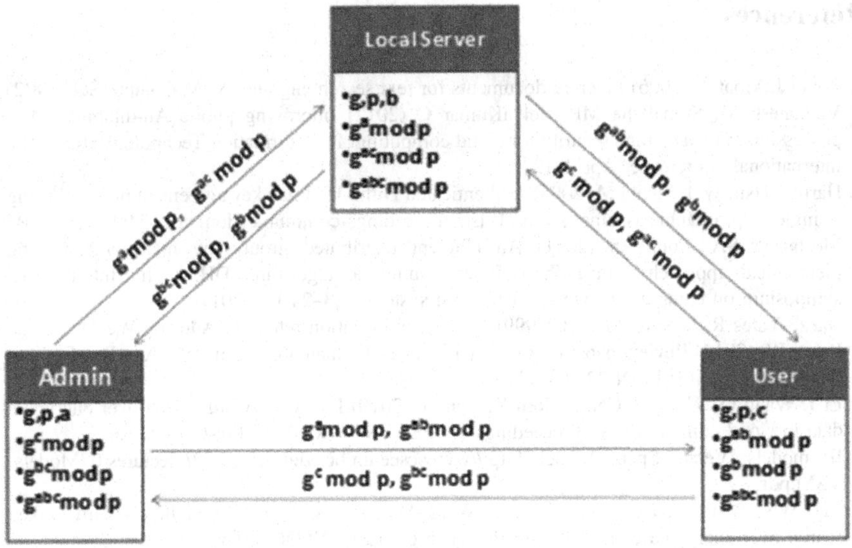

Fig. 4 Diffie–Hellman architecture

Ranking Function:

$\text{Score}\ (W, \text{Fi}) = 1/|\text{Fi}| \cdot (1 + \ln \text{fi,t}) \cdot (1 + N/\text{ft})$

W: Keyword score that is to be calculated
fi,t: Frequency of term in file Fi
|Fi|: Length of the file
N: Total number of files in the collection

4 Conclusion and Future Work

In this paper, we have proposed a design to optimize the search time, i.e., the time taken to retrieve the relevant documents that the user needs, by rank search method, where indexing is done in forehand using IR techniques, thus retrieving relevant documents. We have also provided authentication by using Diffie–Hellman key exchange method. The proposed system can be extended to provide security for documents that are stored in cloud using ECC algorithm and for keyword using RSA algorithm which ensures privacy of cloud owner's documents. The work can also be extended to handle multiple cloud user requests in a single point of time.

References

1. Zobel J, Moat A (2006) Inverted documents for text search engines. ACM Comput Surv 38(2)
2. Venkatesh M, Sumalatha MR, SelvaKumar C (2012) Improving public Au-ditability, data possession in data storage security for cloud computing. In: Information Technology (ICRTIT), international conference, Apr 2012
3. Harn L, Hsin W-J, Mehta M (2005) Authenticated Diffie-Hellman key agreement protocol using a single cryptographic assumption. In: IEEE proceedings-communication, vol 152(4), Aug 2005
4. Mortazavi SA, Pour An (2011) An efficient distributed group key management using hierarchical approach with Diffie-Hellman symmetric algorithm: DHSA. In: International symposium on computer networks distributed systems, 23–24 Feb 2011
5. Baeza-Yates R, Ribeiro-Neto B (1999) Modern information retrieval. Addison Wesley
6. Preeti BS (2014) Review paper on security in Diffie-Hellman algorithm. Int J Adv Res Comput Sci Softw Eng 4(3) ISSN 2277 128X
7. Li J, Wang Q, Wang C, Cao N, Ren K, Lou W (2010) Fuzzy keyword search over encrypted data in cloud computing. In: Proceedings of IEEE INFOCOM10 Mini-Conference
8. IR models: Vector Space Model. http://www.csee.umbc.edu/ ian/irF02/ lectures/07Models-VSM.pdf
9. Cao WN, Wang C, Li M, Ren K, Lou W (2011) Privacy-preserving multi-keyword ranked search over encrypted cloud data. In: Proceedings IEEE INFOCOM 11

Correlated Analysis of Morphological Patterns Between SD-OCT and FFA Imaging for Diabetic Maculopathy Detection: Conformal Mapping-Based Approach

T.R. Swapna, Chandan Chakraborty and K.A. Narayanankutty

Abstract The aim of this work is to identify the morphological patterns associated with macular leaks for diabetic maculopathy from spectral-domain optical coherence tomography (SD-OCT) a noninvasive technique and fundus fluorescein angiogram (FFA) an invasive technique. Here, an attempt has been made to identify the morphological pattern of SD-OCT images, which has association with FFA images affected by diabetic maculopathy based on conformal mapping. The preprocessing step consists of removing the speckle noise using different low-pass filters; we found that wavelet filters are efficient. Out of the 60 eyes, we were able to detect pathologies like micro-cysts in around 52 eyes which resulted in an accuracy of ~ 87 %. The results also showed that when SD-OCT image look normal, the conformal mapping showed angiogram leakages as micro-cysts. This is the first attempt toward correlating the features of two different modalities in retinal imaging from an image processing perspective.

Keywords Diabetic maculopathy · Fluorescein angiogram · Optical coherence tomography · Conformal mapping · Macular leaks

T.R. Swapna
Department of Computer Science and Engineering, Amrita Vishwa Vidyapeetham, Coimbatore 641112, India

T.R. Swapna · C. Chakraborty (✉)
School of Medical Science & Technology, IIT, Kharagpur 721302, India
e-mail: chandanc@smst.iitkgp.ernet.in

K.A. Narayanankutty
Corporate and Industry Relations, Amrita Vishwa Vidyapeetham, Coimbatore 641112, India

L.P. Suresh and B.K. Panigrahi (eds.), *Proceedings of the International Conference on Soft Computing Systems*, Advances in Intelligent Systems and Computing 398, DOI 10.1007/978-81-322-2674-1_20

1 Introduction

In India, diabetes has become the most common disease. At present, there are ~67 million people with confirmed diabetes and another ~30 million in the prediabetes group. The long-term diabetes affects vision leading to a condition called as diabetic retinopathy. It is one of the major reasons of blindness in India often called as the silent killer of vision. Diabetic maculopathy is diabetes affecting the macular region or area of central vision in the eye. Diabetic macular edema (DME) is the major reason for moderate visual loss in people with diabetes. Visual loss from DME is more than that from proliferative diabetic retinopathy (PDR) [1]. Visualization in medicine has been established itself as a research area in the 1980s. The main focus of scientific visualization is exploring the data, testing a hypothesis based on measurements or simulations, and their visualization and the presentation of the results. Visualization has to be considered as the process of understanding the data. It should help the doctors to decide on the pharmacological and surgical interventions [2]. This paper is one such attempt in that direction and would throw more light on the morphological changes that happen during the progress of diabetic retinopathy.

Changes in the macular area are common in diabetes and they are associated with PDR and non-proliferative diabetic retinopathy (NPDR). There are many features in the retinal area associated with PDR and NPDR including microaneurysms, hard exudates, etc. DME occurs due to increase in the permeability of the retinal capillaries [3]. To know more about DME, readers are requested to refer [3]. SD-OCT is a noninvasive procedure useful for ophthalmologists to identify the retinal abnormalities [4]. There are many patterns in OCT associated with clinically significant macular edema (CSME) [5, 6].

2 Dataset Used

The dataset consisting of color fundus, FFA, and OCT images of multiple subjects affected with CSME in the age group of 20–85 years was collected from ophthalmology department of Amrita Institute of Medical Sciences over the period of six months. A set of clinical parameters like age, sex, HBA1C, hemoglobin levels, fasting, and postprandial blood sugars has been considered here while collecting the OCT images. The parameters that are crucial for the evaluation of CSME-like pattern of macular leakage, the presence of PDR, and the presence of pattern traction with respect to FFA images of both eyes were collected. In case of OCT images, the parameters include the central macular thickness and different OCT patterns associated with the edema. The clinical impression on the posterior segment was also noted.

3 Pre-processing

A speckle is an intensity pattern produced by the interference among a set of wave fronts having equal frequency, but phases and amplitudes remain different, which combine together to give an output wave whose amplitude, and therefore intensity, varies randomly. Speckle is a natural property of signals and images and is present in many images like radar, ultrasound, and radio astronomy. Speckle can be considered as a noise and also as a source of information of tissue structure. In case of OCT, the objective of speckle reduction is to suppress the noise and accentuate the information carrying signals [7]. We extracted the green channel of the OCT image as they show a better contrast between the various edemas present in the subject. A preliminary study with different low-pass filters for speckle noise reduction was done. These filters were chosen as they perform well for ultrasound images which also suffer from speckle noise [8]. The filters used are median [9], Fourier, wavelet, and homomorphism filters. The wavelet filtering was found to be the best.

4 Visualization

Correlating the morphological features on spectral-domain optical coherence tomography (SD-OCT) and angiographic leakages in macular edema helps us to understand the pathology and physiology associated with the diagnosis, treatment, and progression of diabetic macular edema (DME). There have been many studies which try to correlate these two important modalities [10–16]. Recent studies have suggested that the angiographic leakages like petaloid, honeycomb, and diffuse are associated with SD-OCT changes such as cystic changes or diffuse pooling. There have also been studies which suggested that both are powerful modalities for the treatment of DME. However, when they are performed alone there is a change for missing the subtle edemas [17]. All these studies are manually done by ophthalmologists after visually examining the angiogram images and OCT images. To the best of our knowledge, nobody has approached this problem from an image processing perspective and tried to analyze the morphological patterns associated with the different forms of maculopathy. We have applied conformal transformation which is popular in various fields of engineering involving natural objects.

5 Conformal Mapping

A conformal mapping is an angle preserving transformation. It can be formally stated as in Eq. 1:

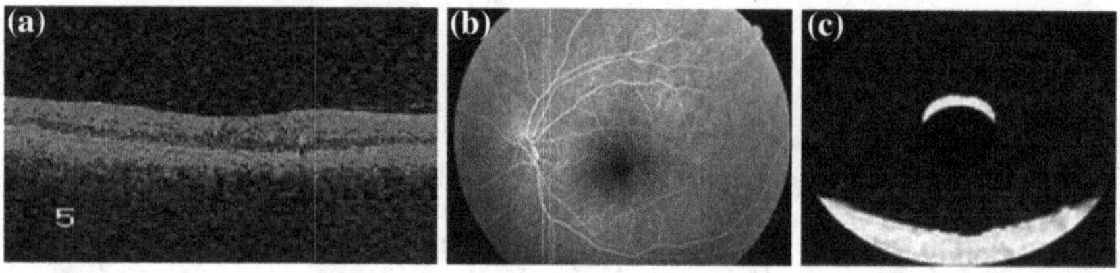

Fig. 1 Normal human retina image: **a** OCT image; **b** FFA image; **c** conformal mapping-based result using (**a**)

$$W = f(z) \tag{1}$$

Conformal maps [18] can be used to represent Euclidean spaces of higher dimension, Riemannian, and semi-Riemannian surfaces. If the geometric structure becomes inconvenient to represent for a physical problem represented as a complex function, then by appropriate transformations it can be represented in a more convenient geometry using conformal mapping.

6 The Conformal Mapping Algorithm

The conformal mapping problem consists of two parts: There should be a mapping function and the texture mapping should be space variant. There are analytical and numerical methods for implementing a conformal mapping. These methods are discussed in detail in [19]. We have used an elementary analytic mapping. The steps of the algorithm are defined below. The conformal analytic function is represented by *f*; we use *f* to distort an image of rectangular shape.

1. The transformation function is represented such as $(u, v) = f(x, y)$ is given by Eq. (1) where $W = u + i \times v$ and $z = x + i \times y$.
2. Defining bounds for mapping the original and transformed images to the input and output complex planes.

An example of conformal mapping on a normal OCT image is given in Fig. 1.

7 Results and Conclusion

We have discussed the commonly used different low-pass filters for noise reduction in retinal OCT images. The wavelet filter of Level-1 decomposition outperformed all other filters. The conformal transformations were applied over both eyes of the subjects affected with diabetic maculopathy. Out of the 70 eyes we collected, ten were discarded as we had incomplete information about the pathologies associated with FFA and OCT. In the remaining 60 eyes, we could detect pathologies like micro-cysts in around 52 eyes which has resulted in an accuracy of 86.66 % that are consistent with the early paper on correlating the morphological patterns of FFA and OCT done by ophthalmologists by visual inspection. Some eyes had resulted in distorted images in the input OCT because of shaking of head by the patients and resulted in a poor visibility of conformably mapped output images. The results for different cases are given in Figs. 1 and 2. The main contribution of this paper is in two aspects. First, this paper has applied different low-pass filters and have removed speckle noise and found that wavelet filters are efficient in removing the speckle noise that is very common in OCT images. The second contribution is in the visualization of the different pathologies affected with diabetic maculopathy using a simple mathematical transformation which to our best of knowledge is not attempted so far in the retinal imaging domain. We can see from Fig. 1 that in case of normal eye the conformal mapping shows no change. When there is an abnormality, we can visualize the presence of different sized cysts which vary according to the different patterns of macular leakages as discussed in the earlier papers [10–17]. In addition, we can also see cloudy regions with variations from mild to dark in the case of different stages of maculopathy. There were cases in which the OCT scans appeared normal but the conformal mapping clearly showed angiogram leakages by the presence of micro-cysts (Please refer Figs. 2d, j and 3j and their conformal mappings 2f, l and 3l). Table 1 summarizes the clinical findings on OCT, FFA, and the proposed algorithm results on the OCT images. In this study, we have emphasized on exploring the hidden features of OCT, which is not identified in a regular A-scans of images. Until now the retinal thickness is a prominent feature which the clinicians have been following for clinical evaluation of macular edema using OCT. One transformation of image enables the visual identification of the pattern of macular leakage which in turn gives a breakthrough in clinical evaluation of macular edema from OCT. This research enhances the importance of OCT being a noninvasive imaging modality through identifying the key clinical features toward the diagnosis. We were able to achieve an accuracy of approx. 87 % in detecting macular leaks. This approach has provided a qualitative comparison of OCT with fundus angiogram imaging technique and a better visualization of the macular leaks than the normal OCT images. We also measured the quality of the resultant conformal mapped images using PSNR as a quality measure. The results are encouraging as the average PSNR values are high and are tabulated in Table 2. All together we conclude that our proposed methodology is a novel and efficient technique for the visualization of clinically significant features from OCT. We believe this research has direct impact on the ophthalmology community and will help the clinician to apply a noninvasive procedure to people who are sensitive to side effects of fluorescein angiogram imaging technique.

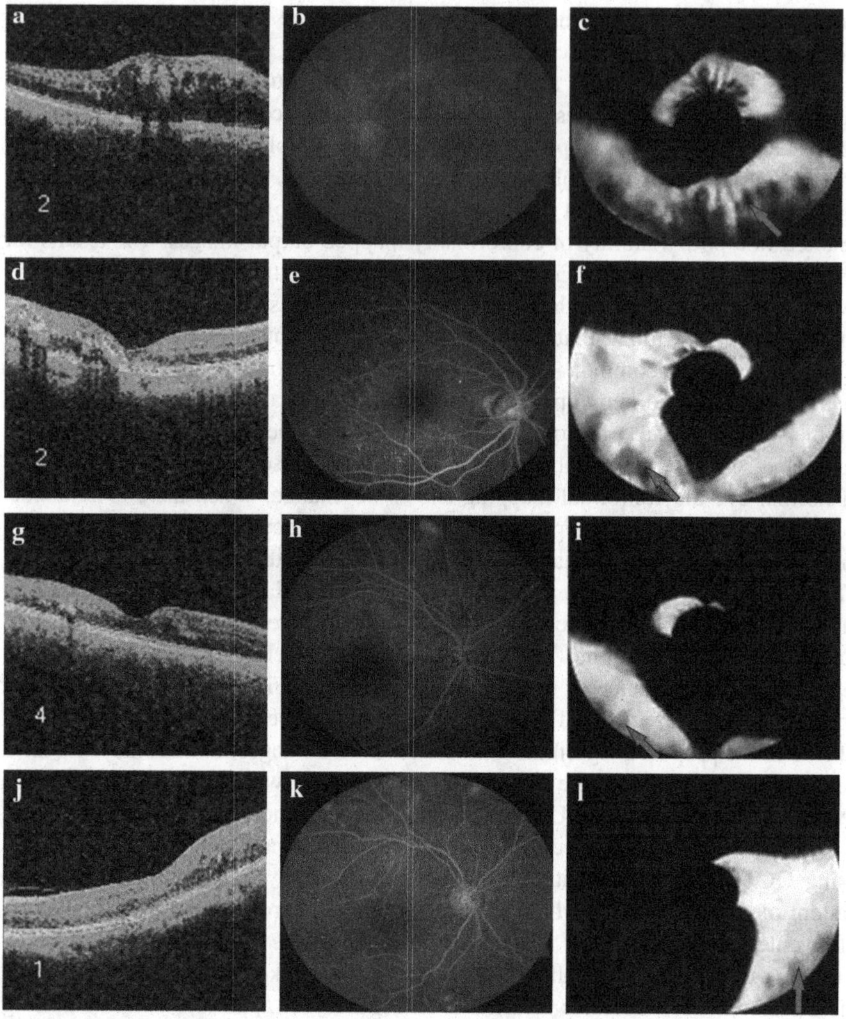

Fig. 2 OCT findings indicate **a** *spongy edema*, **d** and **j** *normal eyes*, **g** *cystoids space*; FFA findings indicate **b** *cystoids leaks*; the posterior segment indicates *severe NPDR with CSME*, **e** *leakage temporal to fovea*; the posterior segment indicates a *lasered macular edema*, **h** *diffuse leaks*; the posterior segments indicate *PDR with CSME*, **k** *diffuse leaks*; the posterior segment indicates *Florid PDR with CSME*; Corresponding conformal mappings show that there are **c** *numerous cysts with cloudy regions*, **f** and **i** show *small cysts with mild cloudy regions*, **l** *small elongated cysts*

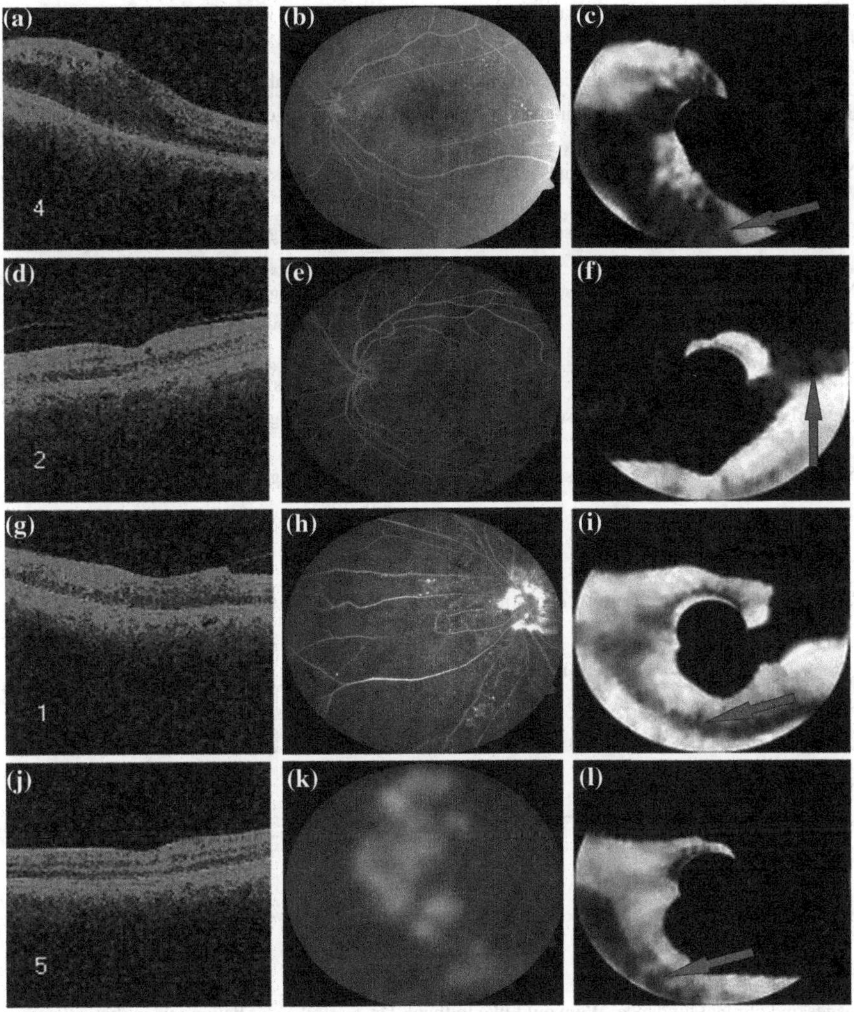

Fig. 3 OCT findings indicate **a** spongy edema with cystoids spaces, **d** and **j** *normal eyes*, **g** *cystoids*; FFA findings indicate **b** *focal leaks*; the posterior segment indicates *NPDR with CSME*, **e** *focal leaks*; the posterior segment indicates *mild NPDR with CSME*, **h** *diffuse leaks*; the posterior segment indicates *PDR CSME*. **k** *diffuse leaks*; the posterior segment indicates *early PDR*; Corresponding conformal mapping shows that there are **c** *small cysts with dark cloudy region*, **f** and **i** *small elongated cysts with cloudy region*, **l** *small broad cysts with cloudy region*

Table 1 Summarizes the clinical findings of OCT, FFA, and contribution of the proposed algorithm

Clinical findings of OCT	Clinical findings of FFA	Proposed algorithm's results on OCT
Spongy edema	Cystoids leaks	Numerous dark cysts with dark cloudy region
Normal eye	Leakage temporal to fovea	Small cysts with mild cloudy region
Cystoid space	Diffuse leaks	Small elongated cysts with mild cloudy region
Normal	Diffuse leaks	Small elongated cysts
Spongy edema with cystoids	Focal leaks	Small cysts with dark cloudy region
Normal eye	Focal leaks	Small cysts with dark cloudy region
Cystoids	Diffuse leaks	Small elongated cysts with mild cloudy region
Normal	Diffuse	Small elongated cysts with cloudy regions

Table 2 The quality measure of conformally mapped images

Image name	PSNR (peak signal-to-noise ratio)	Image name	PSNR (Peak signal-to-noise ratio)
Image 1	30.81	Image 8	38.22
Image 2	30.50	Image 9	32.99
Image 3	34.55	Image 10	31.90
Image 4	35.81	Image 11	33.40
Image 5	36.02	Image 12	32.35
Image 6	35.99	Image 13	46.99
Image 7	36.15	**Average**	**35.056**

Acknowledgments The authors would like to acknowledge Dr. Gopal S Pillai, for providing images and clinical guidance. We would like to thank Dr. Karthika, for helping us in the collection of dataset. Also, the corresponding author (Dr. Chakraborty) acknowledges the Dept. of Biotechnology, Govt. of India for providing partial financial support to conduct this research. The first author thanks Ms. Indu, for compilation of the conformal mapping results.

References

1. Report of a WHO consultation (2006). Prevention of blindness from diabetes mellitus. WHO Press, Geneva, Switzerland
2. Preim B, Bartz D (2007) Visualization in medicine theory, algorithms, and applications, 1st edn. Morgan Kaufmann, USA
3. Khurana AK (2007) Comprehensive ophthalmology, 4th edn. New Age International (P) Ltd., New Delhi

4. Merkene P, Copin H, Yurtserver G, Grietens B (2011) Moving from biomedical to industrial applications OCT enables Hi-Res ND depth analysis
5. Kang SW, Park CY, Ham DI (2004) The correlation between Fluorescein angiographic and optical coherence tomographic features in clinically significant diabetic macular edema. Am J Opthalmol 137(2):313–322
6. Otani T, Kish S, Maryuma Y (1999) Patterns of Diabetic Macular edema with optical coherence tomography. Am J Ophthalmology 127:688–693
7. Schmitt JM, Xiang SH (1999) Yung KM (1999) Speckle in optical coherence tomography. J Biome Opt 4(1):95–105
8. Mateo Juan L, Fernandez-Caballero Antonio (2009) Finding out general tendencies in speckle noise reduction in ultrasound images. Expert Syst Appl 36:7786–7797
9. Chen Y (1996) Broschat San and Flynn, P. Phase insensitive homomorphic image processing for speckle reduction. Ultrasound. Imaging 18(2):122–139
10. Brar Manpreet, Yuson Ritchie, Kozak Igor, Mojana Francesca, Cheng Lingyun, Bartsch Dirk-Uwe, Oster Stephen F, Freeman William R (2010) Correlation between morphological features on spectral domain optical coherence tomography and angiographic leakage patterns in macular edema. Retina 30(3):383–389
11. Otani T, Kishi S (2007) Correlation between optical coherence tomography and Fluorescein angiography findings in diabetic macular edema. Ophthalmology 114(1):104–107
12. Soliman W, Sander B, Hasler PW, Larsen M (2008) Correlation between intraregional changes in diabetic macular oedema seen in Fluorescein angiography and optical coherence tomography. Acta Ophthalmol 86(1):34–39
13. Bolz M, Ritter M, Schneider M et al (2009) A systematic correlation of angiography and high-resolution optical coherence tomography in diabetic macular edema. Ophthalmology 116 (1):66–72
14. Antcliff RJ, Stanford MR, Chauhan DS et al (2000) Comparison between optical coherence tomography and fundus Fluorescein angiography for the detection of cystoid macular edema in patients with uveitis. Ophthalmology 107(3):593–599
15. Ozdek SC, Erdinc MA, Gurelik G et al (2005) Optical coherence homographic assessment of diabetic macular edema: comparison with fluorescein angiographic and clinical findings. Ophthalmologica 219(2):86–92
16. Kang SW, Park CY, Ham DI (2004) The correlation between fluorescein angiographic and optical coherence homographic features in clinically significant diabetic macular edema. Am J Ophthalmol 137(2):313–322
17. Kozak I, Morrison VL, Clark TM et al (2008) Discrepancy between fluorescein angiography and optical coherence tomography in detection of macular disease. Retina 28(4):538–544
18. Ganguli S (2008) Conformal mapping and its applications (http://www.iiserpune.ac.in/~p.subramanian/conformal_mapping1.pdf)
19. Fredrick C, Schwartz EL (1990) Conformal Image warping. IEEE Comput Graphics Appl 10 (2):54–61

Advanced Cluster-Based Attribute Slicing: A New Approach for Privacy Preservation

V. Shyamala Susan and T. Christopher

Abstract Privacy preservation data is an emerging field of research in the data security. Numerous anonymization approaches have been proposed for privacy preservation such as generalization and bucketization. Recent works show that both the techniques are not suitable for high-dimensional data publishing. Several other challenges for data publishing are speed and computational complexity. Slicing is a novel anonymization technique which partitions the data both horizontally and vertically and solves the problem of high-dimensional complexity. To overcome the other challenges of speed and computational complexity, a new advanced clustering algorithm is used with slicing. This advanced clustering technique is used to partition the attributes into vertical columns and tuple grouping algorithm for horizontal partitioning. The experimental results confirm that the advanced clustering algorithm improves the speed of clustering accuracy, reduces the computational complexity, and the outcome of this work is resilient to membership, identity, and attributes disclosure.

Keywords Advanced clustering algorithm clustering · Computational complexity · Clustering accuracy · Dimensionality of the data

V. Shyamala Susan (✉)
PG and Research Department of Computer Science, Government Arts College, Udumalpet, India
e-mail: shyamalasusan@gmail.com

T. Christopher
PG and Research Department of Computer Science, Government Arts College, Coimbatore, India
e-mail: chris.hodcs@gmail.com

L.P. Suresh and B.K. Panigrahi (eds.), *Proceedings of the International Conference on Soft Computing Systems*, Advances in Intelligent Systems and Computing 398, DOI 10.1007/978-81-322-2674-1_21

1 Introduction

Privacy preserving microdata publishing has been studied extensively in recent years. In order to ensure the success of privacy protection, many anonymization techniques have been introduced to anonymize the sensitive data.

The standard methods in the data security are bucketization for ℓ-diversity [1] and the characterization for k-anonymity [2].

Slicing is a novel data anonymization [3] technique carried out by grouping attributes into vertical columns and horizontal columns. To partition the attributes into vertical columns, the correlation between the pair of attributes is evaluated. Two measures which are broadly utilized in the association are classified as Pearson correlation coefficient [4] and mean-square contingency coefficient [4].

After the computation of correlation for each pair of attributes, clustering is used to partition the attributes into columns. The well-known approaches for evaluating non-hierarchical clustering are K-means clustering [5], K-medoid clustering [6], and Partitioning Around Medoids (PAM) [7]. Unfortunately, K-means clustering, K-medoid clustering, and Partitioning Around Medoids are inefficient for large datasets due to the complexity and does not provide high-quality cluster for high-dimensional data. This motivates to propose a new Advanced Clustering Algorithm(ACA) [8] to partition the attributes into columns that would efficiently enhance the speed of clustering and minimize the computational overhead complexity.

The fundamental idea of this algorithm is to create two data structures to hold the labels of clusters and the separation of all the data objects to the closest group amid every object that can be utilized as a part of next cycle. We figure that the separation between the present information object and the new cluster focus, if the registered separation is less than or equivalent to the separation to the old focus, the information item stays in its group that was allotted to in past cycle. In this way, there is no compelling reason to ascertain the separation from this information item to the next $k-1$ grouping focuses, saving the time to the $k-1$ group focuses. Otherwise, we must compute the separation from the present information article to all k clusters and locate the closest group focus. It appoints this point to the closest cluster and afterward independently records the separation to its inside. Since in every cycle some information focuses still stay in the first cluster, it implies that a few sections of the information focuses would not be figured, sparing an aggregate time of ascertaining the separation, accordingly improving the proficiency of the calculation. This can effectively improve the speed of clustering and reduce the computational complexity. Tuple grouping algorithm is used to partition the data horizontally by randomly permuting the values within a bucket.

The author [9] had used the adult dataset that contains the information about the UC Irvine machine learning repository, which has collected data from the US census. This algorithm which is used in novel slicing technique overcomes the limitations such as computational complexity and speed of clustering.

2 Related Work

Although there had been many anonymization techniques, they are not suitable for high-dimensional data due to the lacks in speed and complexity. Here, we have discussed about the existing methods for privacy preserving micro data publishing.

2.1 *Generalization and Bucketization*

These two techniques have segmented the attributes into three classes [10]: (1) Few attributes are used as an detectors that can exactly detect an individual or a person details such as person name or a personal security number; (2) some attributes are quasi-identifiers (QI) that would identify the already well-known antagonist from the common public databases, in which these antagonists have gathered the information about the individuals details such as, birth date, zipcode, and sex; and (3) few attributes are identified as sensitive attributes (SAs), which are strange to the antagonist and are believed as sensitive such as salary and disease.

It has been shown that generalization for *k*-anonymity method has been considered as hazard due to the heavy loss of information, particularly in the high-dimensional data [11, 12]. The bucketization method is better than generalization as it has the better utilization of data, even though it has several drawbacks in it. In the beginning, bucketization approach does not support membership disclosure [13] and it needs a exact detachment between the SAs and OIs. But, in many datasets, it is tough to separate the attributes as QIs and SAs. Third, bucketization breaks the attribute correlations between the QIs and the SAs when the sensitive attributes are distinguished from the QIs attributes.

2.2 *Clustering*

Clustering is a process of grouping objects into patterns and structures that have meaning in the context of particular problem. Clustering is an unsupervised learning technique. Cluster analysis is a challenging task for managing the high-dimensional data sets and their time complexities in data mining. K-means clustering [5], K-medoid [6], and Partitioning Around Medoids (PAM) [7] are well-established methods for analyzing the performance of non-hierarchical clustering.

2.3 *K-means Clustering*

K-means [5] clustering is one of the simplest algorithm to solve the well-known clustering algorithm. It is sensitive to the initial haphazardly chosen cluster centers, noise and didn't work well with global clusters and clusters of different size and different data. Partitioning is done with the mean of each cluster. Euclidean distance is used to measure the similarity between the data. But this method of finding similarity is not suitable for categorical attributes. The computational complexity of the algorithm is given as the $O(nkt)$, where 'n' represents the total number of the objects in the cluster centers, 'k' represents the number of clusters, and finally the 't' the total number of iterations.

2.4 K-Medoid Clustering

K-Medoid [6] is more robust than K-means. In this method, the medoid is chosen as the center along each dimension, instead of the mean, which is used to create the partitioning representative. It is not sensitive to noise and outliers due to less amount of medoids that are determined by outliers or by the extreme mean value. It is highly suitable for large datasets. This algorithm is easy to understand and implement which is fast and convergent in a finite number of steps. But different initial sets of medoids can lead different final clustering. To improve the quality of the cluster, this procedure is repeated several times. The computational complexity for each iteration is $O(k(n-k))$, where n is the total number of the objects and k is the number of clusters.

2.5 Partitional Around Medoids (PAM)

Partitional Around Medoids [7] is known to be more powerful and it is partitioning algorithm. The goal of the algorithm is to minimize the average dissimilarity of objects to their closest selected object. It is more robust against outliers and can deal with any dissimilarity measures. However, PAM also has a limitation that it does not perform efficiently in larger datasets, because of its complexity. It takes longer time, since identifying medoids in a group of points requires calculations of pairwise distances within a group. The computational complexity of the algorithm is $O((1+\beta)$ $\mathrm{K}(\mathrm{T}-\mathrm{K}))$, where n is the total number of the objects, k is the number of clusters.

2.6 Other Clustering Algorithm and Its Time Complexity

Fuzzy based Clustering [14] with the slicing algorithm used to maintain the privacy of micro data. It proposes the fuzzy based data sanitation for partitioning the data into horizontal and vertical. The fuzzy logic is used to categorize the sensitive and non-sensitive attributes so that it increases the clustering accuracy and privacy of micro data and overcomes the data lose and utility problem. In the recent scenario, there is huge necessity for managing large-scale datasets in data mining and in

Table 1 Time complexity of various clustering algorithms

Cluster algorithm	Complexity	Capability of tackling high dimensional data
Fuzzy *c* means	Near $O(N)$	No
Hierarchical clustering	$O(N^2)(\text{time})/O(N^2)(\text{Space})$	No
CLARA	$O\left(K(40+k)^2+K(N-K)\right)^+$ time	No
CLARANS	Quadratic in total performance	No
DBSCAN	$O(N\log N)$time	No
WaveCluster	$O(N)$time	No
DENCLUE	$O(N\log N\beta)$time	Yes
FC	$O(N)$time	Yes
SOM	$O(nkl)$	Yes
HAC	$O(n^3)$	Yes

various fields. Hence, various approaches have been proposed for enhancing the clustering performance. Typical examples consist of ROCK [15], CURE [16], Chameleon [17], and BIRCH [18].

The time complexities for the other clustering algorithms and their capability of tackling high-dimensional data are listed in Table 1.

From Table 1, it is clear that some algorithms fail to provide optimum outcomes, while considering the high-dimensional data in clustering, it develops the complexity and tends to make things more harder when the number of dimensions is added.

So to adequately handle the computational complexity and to improve the speed when there is high-dimensional data, a new algorithm called "Advanced Clustering Algorithm" [8] has been proposed.

3 Slicing Algorithm

Slicing techniques segments the dataset both horizontally and vertically. Vertical segmentation was carried out by grouping attributes into columns, which depends upon the correlations between the attributes. Each column compromises a subset of attributes that are extremely correlated. Horizontal partitioning was performed by grouping tuples into buckets. At last, inside each bucket, points in each column are haphazardly commuted to separate columns. Slicing maintains utility because the grouping attributes are extremely correlated attributes altogether, and maintains the correlations between such attributes. Slicing defends privacy because it breaks the connections between uncorrelated attributes, which are rarely available, but they are detectable. For an example if the dataset compromises quasi-identifiers (QIs) and one sensitive attribute (SA), bucketization has to explore their correlation; slicing, on the other side, can form some quasi-identifiers attributes with the sensitive attributes, maintaining attribute correlations with the sensitive attribute. Slicing algorithm compromises of three stages: attribute segmentation, column generalization and tuple segmentation.

3.1 Attribute Segmentation

To segment the attributes into column the correlation between the pair of attributes are evaluated using two widely used measures of association namely Pearson correlation coefficient [4] and Mean–square contingency coefficient [4]. Having computed the correlations for each pair of attribute, we use clustering to partition the attributes into column. For the shortcomings of the standard clustering algorithm in slicing [3] such as k-means, k-medoid, and PAM, a new advanced clustering algorithm [8] is used to partition the attributes into columns. The essence of this algorithm is to minimize the computation speed and complexity.

3.2 Proposed Advanced Clustering Algorithm for High-Dimensional Data

The procedure of the advanced clustering algorithm for high-dimensional data is stated as follows:

Input: The number of desired clusters R.
Dataset S.
$D = \{n_1, n_2, \ldots n_N\}$ consists of N data objects.
$d_i = \{x_1, x_2, \ldots, x_m\}$//group of attributes of one data point.
Output: Set of K clusters

1. Describe multiple sub-samples $\{S_1, S_2, \ldots, Sj\}$ from the real dataset.
2. Repeat step 3 for $m = 1$ to N.
3. Utilize hybrid approach for sub sample.
4. In each set, assume the middle point as the initial centroid
5. For each data value, find the nearest centroids and allot to closest cluster
6. Select minimal of minimal distance from cluster center based on its condition
7. Now new computation again on dataset S for R clusters
8. Add two nearest clusters into one cluster.
9. Recompute the new cluster center for the combined cluster until the number of clusters reduces into k.

3.3 Tuple Segmentation

In the tuple segmentation process, records are classified into buckets. The important function of the tuple-segmentation algorithm is to assure whether a sliced table meets ℓ-diversity.

Algorithm for Tuple segmentation (T, l)

Q = {T}; SB = Null.
While Q is not Null
Remove the first bucket B from Q;
Split B into two buckets B1 and B2
If DiversityCheck (T, Q U {B1, B2} U SB, l)
Q = Q U (B1, B2) [4]
Else
SB = SB U {B}
Return SB

Algorithm diversity check(T,T_, ℓ)

1. for each tuple t ∈ T, L[t] = ∅.
2. for each bucket B in T_
3. record f(v) for each column value v in bucket B.
4. for each tuple t ∈ T
5. calculate p(t,B) and find D(t,B).
6. L[t] = L[t] ∪ {hp(t,B),D(t,B)i}.
7. for each tuple t ∈ T
8. calculate p(t, s) for each s based on L[t].
9. if p(t, s) ≥ $1/\ell$, return false.
10. return true.

4 Experimental Results

We conduct three experiments. In the first experiment, we evaluate the effectiveness of the advanced clustering algorithm with the standard clustering technique in terms of error rate, execution time, and access time. In the second experiment, we evaluate the various anonymization algorithms with slicing algorithm. In the third experiment, we compare the slicing using ACA with the standard privacy preservation algorithm in terms of total execution time and accuracy.

We use the adult dataset that contains the information about the UC Irvine machine learning repository, which has collected data from the US census. The proposed method developed an advanced clustering algorithm for high-dimensional data to get the initial cluster. The time complexity of the advanced clustering algorithm is given as $O(nk)$. Here, the total time complexity is $O(nk)$. Figure 1 shows the performance comparison based on the execution time. Figure 2 shows the performance comparison in terms of error rate. Figure 3 shows the comparison between various standard anonymous techniques and slicing and Fig. 4 compares the slicing with advanced cluster-based slicing. So the proposed advanced clustering algorithm for high-dimensional data can efficiently increase the speed of clustering and reduce the computational complexity.

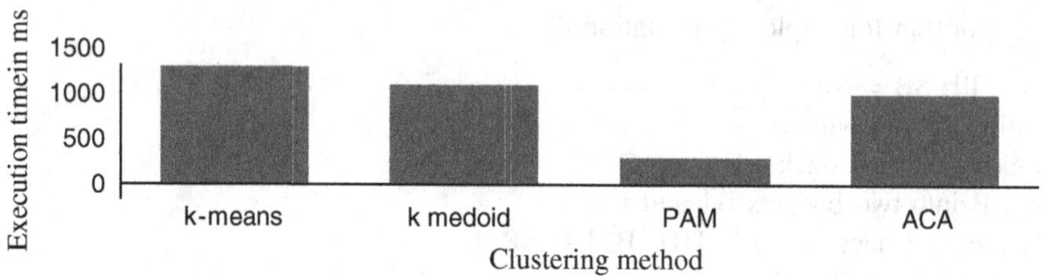

Fig. 1 Performance comparison based on execution time

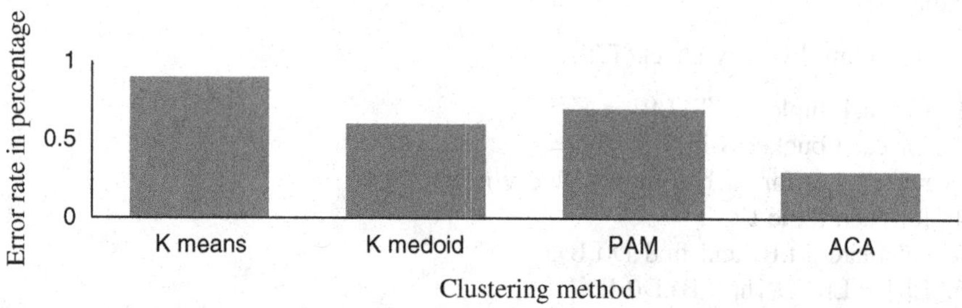

Fig. 2 Performance comparison of clustering algorithm based on error rate

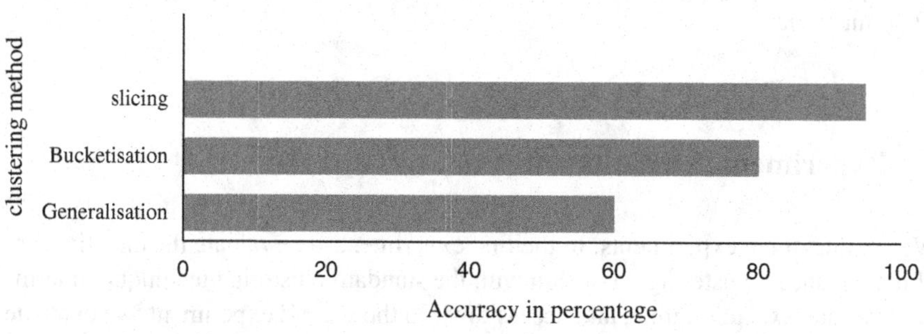

Fig. 3 Performance comparison between various anonymous techniques

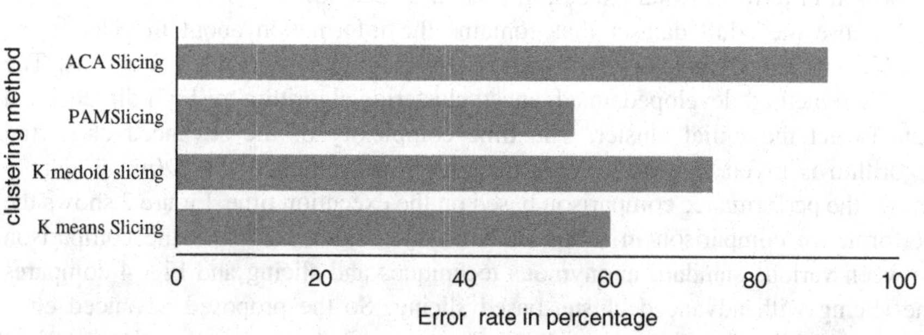

Fig. 4 Performance comparison between standard clustering technique in slicing with advanced cluster-based slicing

5 Conclusion

This proposed algorithm with slicing partitions the data both horizontally and vertically and solves the problem of high dimensionality, speed, and computational complexity. The complexity of this algorithm is *o*(*nk*) time without giving the accuracy of clusters. The outcome of this work is resilient to membership, identity, and attributes disclosure. Experimental results confirm that the advanced clustering algorithm improves the speed of clustering accuracy and reduces the computational complexity.

References

1. Machanavajjhala A, Kifer D, Gehrke J, Venkitasubramaniam ML (2007) Diversity: privacy beyond k-anonymity. ACM Trans Knowl Discov Data 1(1)
2. Sweeney L (2002) Achieving k-anonymity privacy protection using generalization and suppression. Int J Uncertain Fuzziness Knowl Based Syst 10(6):571–588
3. Li T, Li N, Zhang J,(2012) Slicing: a new approach for privacy preserving data publishing. In: Proceedings of IEEE transaction on knowledge and data mining engineering, vol 24(3), pp 561–574
4. Cramt'er H (1948) Mathematical methods of statistics. Princeton
5. MacQueen JB (1987) Some methods for classification and analysis of multivariate observation. In: Proceedings of 5th Berkeley symposium on mathematical statistics and probability, vol 1. University of California Press, Berkeley, pp 281–297
6. Kaufman L, Rousseeuw PJ (1990) Finding groups in data: an introduction to cluster analysis. Wiley, New York
7. Han J, Kamber M (2000) Data mining concepts and techniques. The Morgan Kaufmann Series in Data Management Systems, Morgan Kaufmann
8. Toor A (2014). An advanced clustering algorithm (ACA) for clustering large data set to achieve high dimensionality. Glob J Comput Sci Technol Softw Data Eng 14(2) (Version 1.0)
9. Asuncion A, Newman D (2007) UCI machine learning repository
10. Samarati P (2007) Protecting respondents' identities in microdata release. IEEE Trans Knowl Data Eng 13(6):1010–1027
11. Xiao X, Tao Y (2006) Anatomy: simple and effective privacy preservation. In: Proceedings of the 32nd international conference on very large data bases, ser. VLDB, pp 139–150
12. Kifer D, Gehrke J (2006) Injecting utility into anonymized datasets. In: SIGMOD, pp 217–228
13. Nergiz ME, Atzori M, Clifton C (2007) Hiding the presence of individuals from shared databases. In SIGMOD, pp 665–676
14. Susan VS, Dr. Christopher T (2014) Slicing based privacy preserving of the micro data using fuzzy based clustering. Int J Appl Eng Res 9.
15. ROCK (2000) A robust clustering algorithm for categorical attributes. Inf Syst 255:345–366
16. Guha S, Rastogi R, Shim K (1988) CURE: an efficient clustering algorithm for large databases. In: Proceedings of ACM SIGMOD international conference management of data, pp 73–84
17. Karypis G, Han E, Kumar V (1999) Chameleon: hierarchical clustering using dynamic modeling. IEEE Comput 32(8):68–75
18. Zhang T, Ramakrishnan R, Livny M (1996) BIRCH: an efficient data clustering method for very large databases. In: Proceedings of ACM SIGMOD conference management of data, pp 103–114

Advanced Power Demand Controller and Billing System Using GSM

I. Monah Pradeep and S.G. Rahul

Abstract It is an innovative concept used to eliminate power shutdown to consumers due to increase in power demand. Our proposed project is to distribute the available power proportionally to all consumers thus reducing total power shutdown, where we develop two area concepts, which are two independent power consumers. A GSM system is used where signals are transferred between the user and the distribution station. Whenever power demand increases, the partial shutdown system is implemented and thus power usage to the consumers is limited. This limit is fixed based on the basic power necessity of the consumer. Also, in our project the power used by the consumer is automatically calculated and the corresponding amount for the power used is calculated and sent to the consumer mobile number as a text message.

Keywords Power demand · Billing system · GSM · DTMF · Flow code · Relay · Optocoupler · Automatic billing · Partial shutdown · Total shutdown

1 Introduction

Nowadays, the power demand has been increasing day-by-day. In order to compensate the increasing power demand, we must go for load shutdown process. Due to load shedding, domestic consumers are moving to alternate mode for power, i.e., batteries and inverters. This once again raises the power demand which results in increased time in load shedding patterns. In order to reduce these difficulties, we develop a project which distributes the available power proportionally to all consumers thus avoiding complete power shutdown. And also to wipe the difficulties in manual meter reading, we develop an automatic meter reading system that reads the power usage and calculates the corresponding amount and messages it to the user.

I.M. Pradeep (✉) · S.G. Rahul
School of Electrical Engineering, VIT University, Vellore 632014, India
e-mail: monahprd@gmail.com

L.P. Suresh and B.K. Panigrahi (eds.), *Proceedings of the International Conference on Soft Computing Systems*, Advances in Intelligent Systems and Computing 398, DOI 10.1007/978-81-322-2674-1_22

We incorporated the GSM technique to implement signal transferring between the user and distribution station via SMS intimation.

2 Purpose of Study

As the power demand varies among consumers, some consumers use more power and some use less power. Due to the shortage of power, we go for load shedding. During that time, the power is shut totally for all consumers due to overload and over usage. In order to avoid these problems due, area wise partial or total shutdown is in implementation; so, we can provide uninterrupted power supply to domestic consumers and lighting loads. The billing system is also improved, which is not like the conventional manual calculation. The energy meters are read in every individual house and then they manually calculate the bills, respectively. When the automatic billing system functions, the corresponding bill amounts are sent to the consumers through Short Message Service (SMS).

3 Components Used

- Microcontroller 16F877A
- Transmitter
- Demand controller (Relay, Comparator, and Precision rectifier)
- Power supply unit
- LCD display
- DTMF code convertor
- GSM

4 Block Diagram

4.1 Transmitter Section

See Fig. 1.

4.2 Receiver Section

See Fig. 2.

Fig. 1 Mobile phone

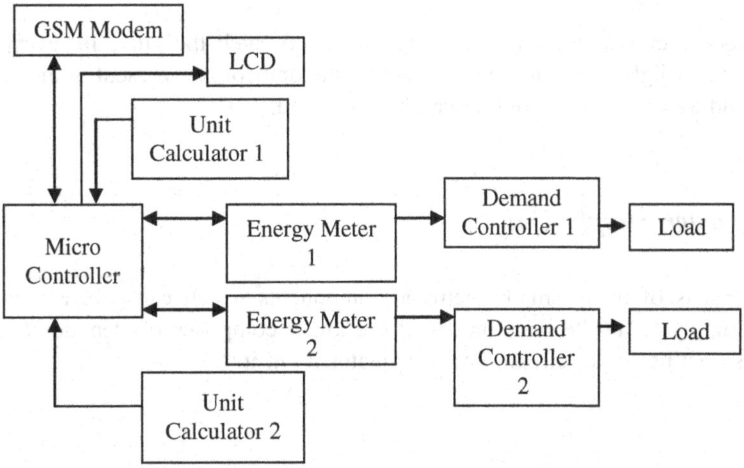

Fig. 2 Receiver side

5 Hardware Descriptions

5.1 Microcontroller

This consists of IC 16F877A microcontroller. It is a 16-bit microcontroller and it has five I/O ports for communication. The flash-based 8-bit microcontroller is upward, compatible with the various PIC devices. This controller is selected because of its wide features such as 200 ns instructions executions, 256 bytes of

EEPROM data memory, two comparators, eight channels of 10-bit analog-to-digital converters, a synchronous serial port that can be configured as either 3-wire SPI or 2-wire I2C bus, self-programming, an ICD, to capture/compare/PWM functions, a USART, and a parallel slave port.

5.2 Relay

This block consists of the 12 V, 300 Ω relay which can act as a switch. The "NO" and "NC" (Normally Open and Normally Closed) contacts of the relay are used to change the direction of current. We use a simple electromagnetic relay which is called as attraction armature type. Here, the relay acts as a limit switch. The limit is varied based on the preset value given for partial shutdown and complete shutdown.

5.3 Opto Coupler

An optocoupler is present in the energy meter to catch the pulse for every unit consumed. A light indication gets sensed by the optocoupler present in the energy meter and accordingly the billing can be processed.

5.4 Power Supply Unit

This consists of many small electronic components which can give a regulated power supply to all the components of the kit. It comprises of step–down transformers, rectifier, filter circuit, and a regulator IC in it.

5.5 Comparator

This consists of IC LM324 which is used to compare two values. It is used to check whether the power consumption in terms of Amps is within the limit. The high–low voltage comparator is designed using the IC LM324. The IC LM324 is a quad comparator, in which two more comparators are used to compare the high–low voltage. The input from voltage rectifier is fed to the inverting pin of two comparators. The reference signal is fed to non-inverting comparator to compare the input voltage signal with reference value. The reference value is set by the 1 k resistor and 820 Ω resistor.

5.6 *DTMF Decoder*

This block contains IC 8870 which is used to convert the DTMF (Dual-Tone Multi-Frequency) into BCD (Binary Coded Decimal). It produces the BCD output for every DTMF signal. The BCD output from the 8870 is then converted into decimal output using IC 4028 which is BCD to decimal decoder. In DTMF to BCD converter circuit, BCD output is latched in each signal receiving, which is not suitable for toggle operation. To eliminate the above problem, AND logic is used. The IC 8870 has a facility to indicate the data acknowledgement signal, which will be taken to one AND input gate. The other input is wired with decimal output of IC 4028. Using this setup for every DTMF signal, we receive a digital signal that is generated.

5.7 *Precision Rectifier*

It is a circuit that acts like an ideal diode. It is designed by placing a diode in the feedback loop of an op-amp. The main disadvantage of ordinary diode is that it cannot rectify voltage below 0.6 (or) 0.7 V. So, therefore, IC LM324 is used for this purpose.

5.8 *GSM*

This consists of GSM modem which communicates between the consumer and the distribution station. And it is also used to send the message regarding the power consumption and the bill amount to the consumers through SMS. The GPS and GSM antenna interfaces uses an OSX type miniature coaxial connector. The antenna is

Fig. 3 GSM modem

connected directly to the GSM35 antenna connectors, or even a short cable can be used. The I/O connections are made using header type terminals (Fig. 3).

5.9 LCD

The LCD display is built in an LSI controller, the controller has two 8-bit registers, one instruction register (IR), and a data register (DR). The IR is used to store the instruction codes, like display clear and cursor shift, address information, etc. The display contains two internal byte-wide registers, one for command and the second for characters to be displayed. There are three control signals called R/W, DI/RS, and En. By making RS/DI signal 0, we will be able to send various commands to be displayed. These commands are used to initialize the LCD, to select the display pattern, to shift cursor or screen, etc. When "1" is pressed in the mobile, then it displays the message "Partial shutdown in Area 1." When "2" is pressed it displays that "The Partial shutdown is reset." When "3" is pressed it displays "Partial shutdown in Area 2." When "4" is pressed it displays that "The Partial shutdown is reseted." When "5" is pressed it displays "Total Shutdown." When "6" is pressed, it displays "Normal Power."

6 Software Implementation

6.1 Flow Code

The software used for the coding purpose has been made easier with less time-consumption. The code is given in the form of a flow chart which can be converted into the controller coding and then it can be loaded into the microcontroller. It is required that we insert flow chart containing the logic of the controller program and on execution of the flow chart, we get the microcontroller coding which is in hexadecimal format. This coding is encoded into the microcontroller using the PIC compiler software and stored in its memory. So, when the kit functions, the program gets loaded and run.

7 Working of the Kit

There are two energy meters each for different areas. Different types of loads are connected to the two meters. In the energy meter, there is an optocoupler based on light signals, which senses the reading of the meter. The power supply unit provides power to all the units. The power supply unit receives its power from the precision rectifier. A call is made from the phone which is received by the GSM modem located at the consumer side. Once the call is detected, then the IC 4028 converts the Dual Tone Multiple Frequency (DTMF) code into Binary Coded Decimal (BCD). Then, we convert the BCD code into decimal code using the IC 4028, so accordingly based on the numbers pressed in the mobile (from 1 to 6) the respective actions and functions will be performed, as discussed earlier. For example, when "1" is pressed in the mobile then it displays the message "Partial shutdown in Area 1." When "2" is pressed it displays that "The Partial shutdown is reset." The optocoupler senses the energy meter readings continuously. After a fixed period of time, an SMS will be sent to the consumers regarding the bill details.

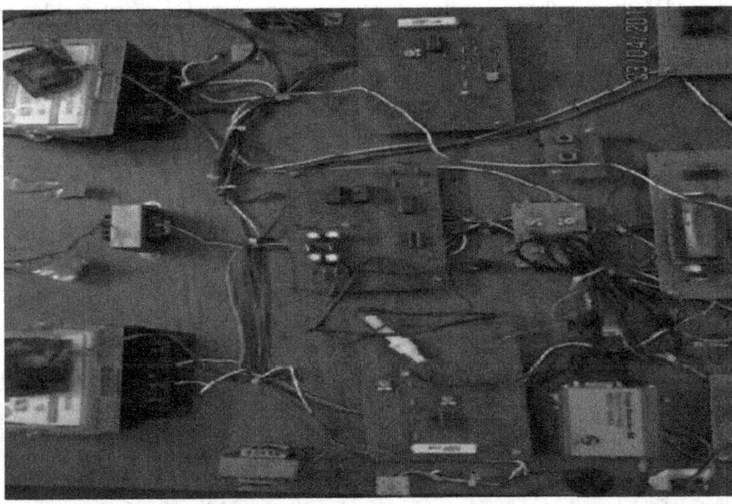

Fig. 4 Overall image of the module containing energy meter, microcontroller, GSM, precision rectifier, comparator, optocoupler, DTMF decoder, and LCD

Fig. 5 Load connected to energy meter

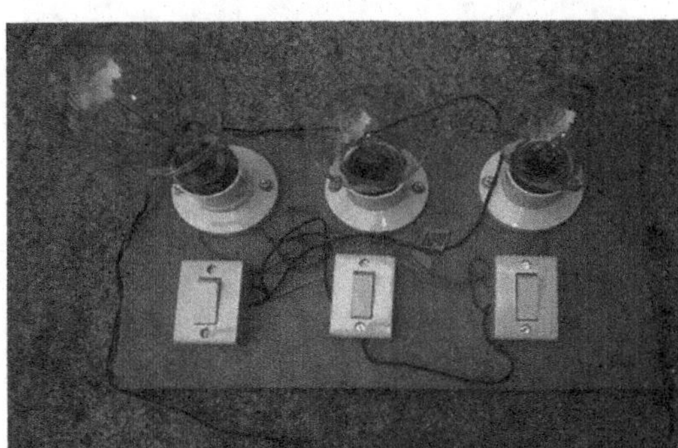

8 Hardware Image

See Figs. 4 and 5.

9 Conclusion

The main motive of our project is to avoid the total shutdown that happens due to increased power demand. Our proposed system helps in distributing the available power proportionally to all domestic consumers with respect to their house size. By installing this system and providing the essential lighting loads to consumers, they do not go for inverters and batteries which are the major cause for increasing power demand. Another major domain in our project is automatic billing system. In our proposed model, the number of units a consumer consumes for every regular period of time can be calculated and the corresponding amount to be paid can also be calculated. The billing details are automatically sent to the consumer mobile through an SMS. This is a great advantage which removes the difficulties in manual meter reading.

Bibliography

1. http://microcontrollershop.com/
2. http://www.microchip.com/
3. Smart meter assisted electric energy management scheme for power distribution and residential consumers. Paper presented at 2014 IEEE conference in Dhaka
4. Yan C (2012) Consumer operational comfort level based power demand management in the smart grid. Paper presented at 2012 IEEE conference in Berlin, Hydropower and Information and Engineering, Huazhong University of Science and Technology, Wuhan, China
5. Application of rural residential hourly load curves in energy modeling, 2003. Paper presented at 2003 IEEE Bologna Power Tech Conference in Italy

A Study on Segmentation-Based Copy-Move Forgery Detection Using DAISY Descriptor

Resmi Sekhar and R.S. Shaji

Abstract Copy-move forgery is one of the most prevalent forms of image forgery and a lot of methods have been developed to detect them. The most important hurdle in the development of such forgery detection methods is that different post-processing operations are done like rotation, scaling, and reflection. JPEG compression, etc., might have been applied on the copied region before it is being pasted, and the method we are using should be invariant to all types of such post-processing. Because of this, most of the methods fail in case of one or the other type of attacks. This paper presents a study on the use of segmentation-based copy-move forgery detection using rotation invariant DAISY descriptors. The paper tries to develop a new method based on three existing methods which have advantages and disadvantages of their own. The expected performance is better than any other existing methods because of the combined effect of three robust methods.

Keywords Digital image forensics · Image forgery · Copy-move forgery · Segmentation · DAISY descriptor · Block-based methods · Key-point-based methods

1 Introduction

Digital image tampering detection is one of the principal branches of digital image forensics. Tampering literally means 'to interfere with something in order to cause damage or make unauthorized alterations' [15]. There are many instances of image

R. Sekhar (✉)
Department of Computer Science, Noorul Islam University, Thuckalay, Tamil Nadu, India
e-mail: sekharresmi@gmail.com

R.S. Shaji
Department of Information Technology, Noorul Islam University, Thuckalay, Tamil Nadu, India
e-mail: shajiswaram@yahoo.com

L.P. Suresh and B.K. Panigrahi (eds.), *Proceedings of the International Conference on Soft Computing Systems*, Advances in Intelligent Systems and Computing 398, DOI 10.1007/978-81-322-2674-1_23

tampering in history even in the nineteenth century. In those days, tampering was a cumbersome work to be performed in the dark rooms of the studio. But today, because of the easily available image processing tools, it is a very common practice to alter the images. Fraudulent images are seen very often in social media, press, and in court rooms. As a result of this, from 2001 onwards, there was a tremendous increase in the methods developed for image forensics and nowadays image forensics has become a routine in physical forensic analysis. Generally, image forensics techniques can be classified into two: active approaches and passive approaches [7]. Active approaches were used traditionally by employing data hiding (watermarking) or digital signatures. Requirement of specialized hardware narrows its field of application. Passive approaches or blind forensic approaches use image statistics or content of the image to verify its genuineness.

Image tampering can be done with a single image or with multiple images. Splicing (combining contents from two or more images into a single image), copy-move forgery, use of image processing operations, false captioning, etc., are treated as different forgery types [7]. Among this, copy-move forgery is the most commonly performed one and the most studied one. It is a type of forgery in which a region from the same image is copied and pasted on the same image in order to hide something or to duplicate something. As it is simple to perform and difficult to identify, a lot of techniques have been developed for copy-move forgery detection since 2004. As the source and the target regions are from the same image, the image features like noise, color, illumination condition, etc., will be same for the both the regions. This is the basis for all copy-move forgery detection algorithms. Some form of post-processing like rotation, scaling, blurring, noise addition, reflection, and compression also are performed on the copied region before it is pasted. This makes the forgery detection more complex. So the important step in such a forgery detection technique would be extraction of features, which are invariant to the abovementioned post-processing operations, from the image. A detection method that is robust to some form of post-processing may not be adequate to detect forgery with another type of post-processing. So, identifying a feature that is invariant to any type of post processing operations is the most crucial step in the development of a copy-move forgery detection technique. Figure 1 shows an example of

Fig. 1 Example for copy-move forgery

copy-move forgery with the copied region given inside the circle. The image has been taken from MICC-F220 database [1]. In Sect. 2, a brief literature review is presented. Section 3 describes the proposed method followed by expected results and conclusion in Sects. 4 and 5, respectively.

2 Related Work

Copy-move forgery detection, generally called CMFD, techniques can be classified into two: block-based approaches and key-point-based approaches. In both the approaches, some form of preprocessing will be there. In block-based methods, the image will be divided into overlapping blocks of specified size and a feature vector will be computed for these blocks. Similar feature vectors are then matched to find the forged regions. In key-point-based methods, feature vectors are computed for regions with high entropy. There is no subdivision into blocks. The feature vectors are matched to find the copied blocks. The common processing steps for copy-move forgery detection are given below [3].

1. Preprocessing (convert the image to gray scale if required)
2. For block-based methods
   (a) Divide the $M \times N$ image into $(M - b + 1)(N - b + 1)$, $b \times b$ sized overlapping blocks.
   (b) Compute feature vector for each block.

   For key-point-based methods
   (a) Extract key-points from the image.
   (b) Compute feature vector for each key-point.
3. Find matching feature vectors by searching its neighborhood. Compute shift vector for each pair of matching feature vector.
4. Remove the matching pairs whose shift vector distance is less than a predefined threshold.
5. Cluster the remaining matches.

If the image contains more than a predefined number of connected pixels, report tampering.

2.1 *Block-based Methods*

Earliest of the block-based methods were developed by Fridrich et al. [4] and Popescu and Farid [14]. Fridrich's method was based on DCT coefficients of blocks and the latter was based on the PCA of blocks. Most of the earlier methods since then are studied extensively in a paper by Christlein et al. [3]. An efficient block-based method was developed by Ryu et al. [16]. The method used Zernike moments to detect copy-rotate-move forgery, but it was found to be weak against scaling and affine transformations. Solorio and Nandi [18] developed a method that extracted color features from the overlapping blocks to detect copy-move forgery with rotation scaling and reflection. The method was claimed to have less computational complexity than the previous methods. Cao et al. [2] developed a method based on circular blocks and a four element feature vector to represent a block. The method was less computationally complex than previous methods. Zhong and Xu [20] used mixed moments from only low-frequency components of the image. The algorithm showed good detection in the case of forgeries with rotation, scaling, and translation and was computationally less complex than the older methods. Li et al. [9] proposed a method using circular blocks and polar sine transform of blocks. The method is claimed to have the same robustness as Ryu et al. [16] with the capability to detect forgeries with scaling, shearing, and perspective transforms. A method using direct block comparison was proposed by Lynch et al. [12] without extracting any features from the block. The blocks are grouped into buckets and a block will be compared with blocks in the same bucket. The method addresses the issue of slight darkening or slight lightening of the copied region before being pasted.

2.2 *Key-Point-Based Methods*

Because of the complexity of block subdivision and searching, many of the recent methods in CMFD are based on key-point extraction and feature matching. These types of methods identify and select the high entropy regions in the image as key-points and use a feature vector to represent them. As the number of feature vector is less, the computational complexity will be less than block-based methods. The popular key-point-based methods till 2011 are evaluated by Christlein et al. [3]. SIFT, SURF, DAISY, MPEG-7, etc., are the feature descriptors found to be used in CMFD till now. Pan and Lyu [13] in 2010 used SIFT to detect key-points and generated a 128-dimensional SIFT descriptor for each key-point. The best-bin-first search method was used to identify the matching key-points. The method could estimate geometric transformations like rotation, scaling, and shearing. Amerini et al. [1] developed another method using SIFT features to handle multiple copy-move forgery. Another SIFT-based method was proposed by Jing and Shao [6]. One of the SURF-based methods was proposed by Shivkumar and Santhosh Baboo [17]. SURF is an improvement of SIFT with the advantage of more computational speed and the same robustness as SIFT. More computational speed is due the four-dimensional SURF feature vector. Euclidean distance between key-points is used to find matching key-points. If the image is having a uniform texture, enough key-points cannot be detected and as a result of this, SIFT-based methods fail. To avoid this difficulty, Guo et al. [5] proposed the use of DAISY descriptors with adaptive non-maximal suppression. DAISY descriptors are supposed to have better performance than SIFT with more invariance to contrast variation, scale change, and low image quality. The use of MPEG-7 features to detect duplicate images were extended to detect duplicate regions in same image by Kakar and Sudha [8]. The extracted features were found to be invariant to rotation, scaling, and translation. Later feature matching is done to identify forged regions.

3 Proposed Method

Christlein et al. [3] conducted a very thorough evaluation of copy-move forgery detection techniques. They proved that key-point-based methods consume very little amount of memory and are faster when compared to block-based methods. This is because of the less number of feature points that need to be matched. Block-based methods consume more resource and time, due to the large number of blocks to be compared. But key-point-based methods cannot give us highly accurate results and they are prone to miss copied regions with little structure. Moreover, there are chances to produce false matches in the case of self-similar structure. Block-based methods can overcome these deficiencies with good precision and reduced number of false matches. So, Christlein et al. [3] proposes a two component copy-move forgery detection system. One should be a key-point-based component, due to its efficiency and low-memory footprint and the other should be a block-based component, for a highly reliable examination of the image. A novel approach to copy-move forgery detection is proposed by Li et al. [10], with the assumption that copy-move forgery might have been carried out with a specific aim. In this paper, the authors segment the image first into nonoverlapping patches and then these patches are matched for forgery detection. We put forward a modified method using [10], and rotation invariant DAISY descriptors [5]. Feature matching step is proposed to be done with the g2NN approach introduced in [1]. The process is outlined in Fig. 2.

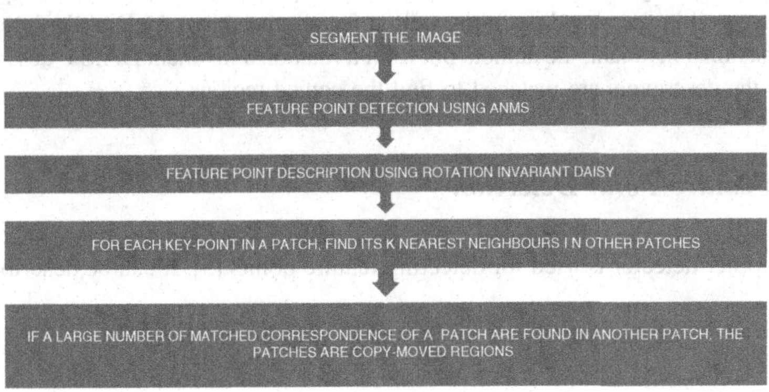

Fig. 2 The proposed method

Fig. 3 Image segmentation

3.1 Image Segmentation

Before performing the patch match, the image has to be segmented into meaningful regions. According to [10], the method used for segmentation has no effect on the CMFD performance. After segmentation, the forged region may be in two or more segments. Coarse segmentation is not done as copied and forged regions are not expected to be in the same patch. Figure 3 is an example for image segmentation, taken from [15]. The two towers in the image are copy-moved regions. After segmentation, the forged region may lie in more than one patch.

3.2 Feature Extraction

A feature descriptor matching algorithm based on Adaptive Non-Maximal Suppression (ANMS) feature-point detection and DAISY descriptor is proposed in [5]. SIFT descriptor, which is considered as the optimal feature descriptor is computationally very complex and requires much time for computation. Moreover, SIFT fails in detecting evenly distributed key-points. So in [5], the authors used ANMS for feature detection. Then for describing the features DAISY descriptor [19] is used. DAISY descriptor is relatively good in speed and performs well in feature matching. As DAISY descriptor is not rotational invariant, the authors put forth a rotational invariant DAISY descriptor. Finally, the descriptors are matched to find the copied regions.

3.3 Feature Point Detection

Harris corner detector is used for detecting feature points [5]. It can be described as

$$E(u,v)|(x,y) = \sum w(x,y)[I(x+u,y+v) - I(x,y)]^2 \quad (1)$$

where $w(x, y)$ is the Gaussian filter and (u, v) is expressed as a minimal distance.

$$w(x,y) = \exp\left[-\frac{1}{2}(u^2+v^2)/\sigma^2\right] \tag{2}$$

where σ is the standard deviation. After Taylor series expansion, the formula can be described as

$$E(u,v)|_{(x,y)} = [uv]v\begin{bmatrix} u \\ v \end{bmatrix} \tag{3}$$

where

$$v = \begin{bmatrix} A & C \\ C & B \end{bmatrix}, \quad A = w \otimes Ix^2, \quad B = w \otimes Iy^2, \quad C = w \otimes IxIy$$

where $\otimes$ denotes the convolution operator. Ix and Iy represent the values of horizontal and vertical directions. V is a 2×2 matrix with two eigenvalues. When these two eigenvalues are large then the point $E(u, v)$ is a key-point. A response function is used to distinguish corner points and edge points as shown below.

$$Z = \det(v) - \alpha \times \mathrm{tr}^2(v) \tag{4}$$

where det(v) denotes the determinant of a matrix and det(v) denotes trace of a matrix.

The number of key-points should be kept as small as possible in order to reduce the cost of matching procedure and at the same time there should be sufficient number of key-points to perform the matching even in regions of uniform texture. So, ANMS is used for selecting key-points from Harris corner detector. Interest points are selected using Eq. (4) around a neighborhood of radius r pixels, where r ranges from zero to infinity, until some desired number of key-points are obtained. ANMS key-points are found to be distributed throughout the entire image.

3.4 Feature Point Description

Now, descriptors are generated for detected feature point. For this DAISY descriptor proposed by Tola et al. [19] is used. DAISY is claimed to have better performance than SIFT and uses circular grids as against SIFT which uses rectangular grids. Because of this structure, DAISY provides better localization properties and is relatively robust to contrast variation, scale change, and low image quality [5]. DAISY is not rotation invariant and so an improved rotation invariant DAISY descriptor is proposed. The working of the traditional DAISY descriptor can be found in [19].

In order to make DAISY rotation invariant, a primary orientation is assigned to each key-point. The local histogram of neighborhood centered at the key-point has to be computed. For this, the gradient magnitude $m(u, v)$ and orientation $\theta(u, v)$ within a circular region of radius R are calculated using the following equations.

$$m(u,v) = \sqrt{(I(u+1,v) - I(u-1,v) + I(u,v+1) - I(u,v-1))^2} \tag{5}$$

$$\theta = \tan^{-1}\left(\frac{I(u,v+1) - I(u,v-1)}{I(u+1,v) - I(u-1,v)}\right) \tag{6}$$

The orientation of each pixel in the circular window is weighted by adding to a histogram to detect canonical orientation. This histogram is quantized into 36 bins. The bins are formulated as below

$$H_i = \sum\nolimits_{(u,v)\in R} m(u,v) \times \frac{1}{\sigma\sqrt{2\pi}} e^{-\frac{(u-u')^2+(v-v')^2}{2\sigma^2}} \tag{7}$$

where (u', v') are the coordinates of the center and σ denotes the standard deviation of the Gaussian weights. Variable H_i denotes the magnitude in each quantized orientation where i is from 1 to 36. After assigning the canonical orientation to the detected key-points, rotation invariant DAISY descriptors are computed at key-point locations as $F = \{f_1, f_2, \ldots, f_M\}$.

3.5 Feature Matching

The next step is to look for a pair of patches that have many feature points in common. A matching operation has to be done in the DAISY space to determine similar key-points. For each key-point in a patch, we have to find K nearest neighbors located in other patches. The nearest neighbor approach based on Euclidean distance between the key-points can be used for finding out matching key-points. The best match for a key-point will be the one with minimum Euclidean distance from it. For efficient key-point matching, the generalized nearest neighbor (g2NN) method proposed in [1] can be used. This is an extension of the 2NN method [11]. In 2NN, the Euclidean distance of a key-point with respect to all other key-points are found out and then sorted in ascending order. If the ratio between the distances of closest neighbor to that of the second closest neighbor is less than a specified threshold, the closest neighbor will be considered as a match for the key-point being considered. This method is not effective in determining multiple copy-move forgeries, i.e., forgeries created by copying and pasting the same region over and over. In the g2NN method, the 2NN test is iterated between every pair of sorted Euclidean distances. Let $D = \{d_1, d_2, \ldots, d_n\}$ be the sorted Euclidean distances of the key-point being considered. Then, the g2NN method will take the ratio

d_i/d_{i+1} until the ratio becomes greater than a predefined threshold. If at a point m the ratio becomes greater than the threshold, then the key-points corresponding to distances $d_1, d_2, \ldots, d_m$ are considered as match for the key-point being considered. Finally, after performing the g2NN test for all the key-points, we obtain the set of matched key-points. As there will be more than a pair of copy-move forged regions, a limit is set on the number of matched key-points as k equal to 10. So, each key-point in a patch corresponds to a maximum of k key-points in the remaining patches. If a large portion of matched correspondences of a patch is found in another patch, then these two patches will contain copy-move forged regions.

4 Expected Results

The authors in [10] used the benchmark database for CMFD evaluation which was constructed by [3] and the MICC-F600 database [1] for evaluating their work. They segmented the images using quick shift image segmentation algorithm and SLIC algorithm in vlFeat software. The performance was evaluated using two parameters: False Negatives (FN) and False Positives (FP).

$$\mathrm{FN} = \frac{|\{\text{Forged images detected as original}\}|}{|\{\text{Forged images}\}|}$$

$$\mathrm{FP} = \frac{|\{\text{Original images detected as forged}\}|}{|\{\text{Original images}\}|}$$

The results showed that this method produced the lowest false negative rate. But the false positive rate was larger. The method was tested for various attacks like JPEG compression, noise addition, rotation, scaling, etc.

Guo et al. [5] tested their DAISY descriptor-based copy-move forgery detection method using the Uncompressed Color Image Database (UCID) dataset. The results were compared with those generated with SIFT and was shown to be better than SIFT-based methods. It was also found that the method outperformed SIFT in case of insufficient key-points. These two methods when combined together are expected to produce results with no false matches and robust to any type of copy-move forgery.

5 Conclusion

In this paper, we have performed a study on a novel approach for copy-move forgery detection. Because of the high computational complexity of block-based methods, key-point-based methods are gaining popularity. But the inefficiency of key-point-based methods to perform well in the case of small forged regions and images with little structure prompted the researchers to think about other better options. One such effort is the work by Li et al. [10]. DAISY descriptor was found to be robust for almost all types of image manipulations and it was proved to be superior to SIFT. Guo et al. [5] used a rotational invariant representation of DAISY descriptor in their work. This paper makes an analytical study on copy-move forgery detection combining the properties of the abovementioned works. The implementation of this paper is taken as the immediate future perspective.

References

1. Amerini I, Ballan L, Caldelli R, Bimbo AD, Sera G (2011) A SIFT based forensic method for copy-move attack detection and transformation recovery. IEEE Trans Inf Forensics Secur 6. doi:10.1109/TIFS.2011.2129512
2. Cao Y, Gao T, Fan L, Yang Q (2012) A robust detection algorithm for copy-move forgery in digital images. Forensic Sci Int 33–43. doi:10.1016/j.forsciint.2011.07.015
3. Christlein V, Riess C, Jordan J, Reiss C, Angelopoulou E (2012) An evaluation of popular copy-move forgery detection approaches. IEEE Trans Inf Forensic Secur 7:1841–1854
4. Fridrich A, Soukal B, Lukas A (2003) Detection of copy-move forgery in digital images. In: Proceedings of digital forensic research workshop
5. Guo J, Liu Y, Wu Z (2013) Duplication forgery detection using improved DAISY descriptor. Expert Syst Appl 40:707–714. doi:10.1016/j.eswa.2012.08.002
6. Jing L, Shao C (2012) Image copy-move forgery detecting based on local invariant feature. J Multimed 7:90–97. doi:10.4304/jmm.7.1.90-97
7. Kakar P (2012) Passive approaches for digital image forgery detection. PhD dissertation, School of Computer Engineering, Nanyang Technological University, Singapore. Retrieved from http://pkakar.droppages.com/publications
8. Kakar P, Sudha N (2012) Exposing postprocessed copy-paste forgeries through transform-invariant features. IEEE Trans Inf Forensics Secur 7:1018–1028. doi:10.1109/TIFS.2012.2188390
9. Li L, Li S, Zhu H, Wu X (2013) Detecting copy-move forgery under affine transforms for image forensics. Comput Electr Eng. doi:/10.1016/j.compeleceng.2013.11.034
10. Li J, Li X, Yang B, Sun X (2015) Segmentation-based image copy-move forgery detection scheme. IEEE Trans Inf Forensics Secur 10(3):507–518
11. Lowe DG (2004) Distinctive image features from scale invariant keypoints. Int J Comput Vision 60(2):91–110
12. Lynch G, Shih FY, Liao HM (2013) An efficient expanding block algorithm for image copy-move forgery detection. Inf Sci 239:253–265. doi:10.1016/j.ins.2013.03.028
13. Pan X, Lyu S (2010) Region duplication detection using image feature matching. IEEE Trans Inf Forensics Secur 5:857–867. doi:10.1109/TIFS.2010.2078506
14. Popescu A, Farid H (2004) Exposing digital forgeries by detecting duplicated image regions. Department of Computer Science, Dartmouth College, Technical Report, TR2004-515
15. Redi J, Taktak W, Dugelay J (2011) Digital image forensics: a booklet for beginners. Multimed Tools Appl 51:133–162. doi:10.1007/s11042-010-0620-1
16. Ryu S, Lee M, Lee H (2010) Detection of copy-rotate-move forgery using zernike moments. In: Proceedings of international workshop on information hiding, Springer, Berlin, pp 51–65
17. Shivakumar BL, Santhosh Baboo S (2011) Detection of region duplication forgery in digital images using SURF. Int J Comput Sci Issues 8:199–205

18. Solorio SB, Nandi AK (2011) Exposing duplicated regions affected by reflection, rotation and scaling. In: International conference on acoustics, speech and signal processing, pp 1880–1883. doi:10.1109/ICASSP.2011.5946873
19. Tola E, Lepetit V, Fua P (2010) DAISY: an efficient dense descriptor applied to wide-baseline stereo. IEEE Trans Pattern Anal Mach Intell 32(5):815–830
20. Zhong L, Xu W (2013) A robust image copy-move forgery detection based on mixed moments. In: IEEE international conference on software engineering and service sciences, pp 381–384. doi:10.1109/ICSESS.2013.6615329

FSM-Based VLSI Architecture for the 3 × 3 Window-Based DBUTMPF Algorithm

K. Vasanth, V. Elanangai, S. Saravanan and G. Nagarajan

Abstract This paper gives a Novel FSM-based architecture for the decision-based unsymmetrical trimmed midpoint algorithm used for high-density salt and pepper noise (SPN) removal in images. The proposed VLSI architecture uses a FSM-based scheduler for the evaluation of unsymmetrical trimmed midpoint in a fixed 3 × 3 window. The proposed scheduler moves between 4 states of the finite state machine for the evaluation of a suitable value to replace the corrupted value in the decision-based algorithm. The proposed architecture consists of sorting network, FSM scheduler and decision unit, which uses 9 values of the current processing window. This setup acts as a sequential architecture. During the simulation the first output of the decision appears after 16 clock cycles. The proposed architecture was targeted for Xc3e5000-5fg900 FPGA and the proposed architecture occupies 857 slices, consumes 298 MW power, and operates at 98.38 MHz frequency.

Keywords FSM-based architecture · Decision-based unsymmetrical trimmed midpoint algorithm · Sequential architecture

1 Introduction

Salt and pepper noise (SPN) gets induced in images due to transmission errors. Over the years it had been found that median filtering is the better choice in removing the SPN by preserving the information content of the image. Fundamental operation in finding median on a computer system is to arrange a given set of data in increasing order. The need for faster sorting algorithms was

K. Vasanth (✉) · V. Elanangai · G. Nagarajan
Department of EEE, Sathyabama University, Chennai, India
e-mail: vasanthecek@gmail.com

S. Saravanan
Department of ECE, Maamallan Institute of Technology, Sunguvarchatram, India
e-mail: elanangai123@gmail.com

L.P. Suresh and B.K. Panigrahi (eds.), *Proceedings of the International Conference on Soft Computing Systems*, Advances in Intelligent Systems and Computing 398, DOI 10.1007/978-81-322-2674-1_24

mandatory owing to its interest and practical importance. Rank ordering architectures are classified into array-based and sorting network-based. The array based architecture has M processing elements of which the kernel is of size M. It was reported that compare and swap operations used in sorting-based operations were relatively faster than the former. Further different algorithms had been formulated over the years for sorting the data. These algorithms are based on word-level and bit-level as illustrated [1]. The word-level algorithms exhibit high throughput capability and these act as a suitable solution for many real-time image/video systems. A fast bubble sorting algorithm for evaluating median [2] using a reduced preprocessing elements that yields a latency of N cycles, where N is the number of input samples. The size of the hardware is complex and proportional to the square of the input samples. Along with the processing elements, extra data buffer are required and hence the overheads increase as the input samples increase. This increases the bandwidth of the memory. To eliminate the flaw of limited bandwidth, message passing algorithms [3] and systolic array architecture [4] were proposed. Even in this case the latency depends on the number of input samples N. This N level complexity does not support real-time operations.

A bitonic sequence-based parallel sorting algorithm was proposed [5] and found to yield execution time of $O(\log_2 N)$ for N input samples using $O(\log_2 N)$ processors. Also a bitonic sorting-based shuffle network [6] and achieved an execution time of $O(\log_2 N)$ using $O(N\log_2 N)$ processors. A digital architecture [7] that has an execution time of $O(\log_2 N)$ time for $O(N\log N)$ processors. In a linearly connected array that runs on $O(\log N)$ processors will require $O(N)$ time for the parallel sorting algorithm proposed by Horiguchi and sheigei [8]. Mesh connected arrays with $N2$ processors were used for the above said purpose and literatures [9, 10] reported that digital architectures extended Bacher's algorithm for $N2$ processors. This architecture obtained the execution time of $O(N)$ for $N2$ data. Vasanth and Karthik [11] proposed VLSI architecture for arranging data in descending order. The architecture used configurable arrays which is not possible to synthesize in FPGA realization. The architecture was bit-level and operated on decomposition and regrouping phase. Vasanth and Karthik [12] proposed a FSM-based architecture for decision-based unsymmetrical trimmed median as estimator of impulse locations. This algorithm used a modified snake-like sorting for arranging 9 pixels in increasing order followed by a usage of scheduler to evaluate the trimmed median value. The main aim of the work is to develop a novel VLSI architecture for faster sorting and subsequent decision-based operation for the removal of SPN. Hence a suitable VLSI architecture is to be proposed for a fixed 3 × 3 decision-based unsymmetrical trimmed midpoint algorithm [13]. Section 2 deals with the decision-based unsymmetrical trimmed midpoint algorithm. Section 3 deals with the proposed architecture for the discussed algorithm. Section 4 gives the simulation results and discussions and Sect. 5 deals with conclusion of the work.

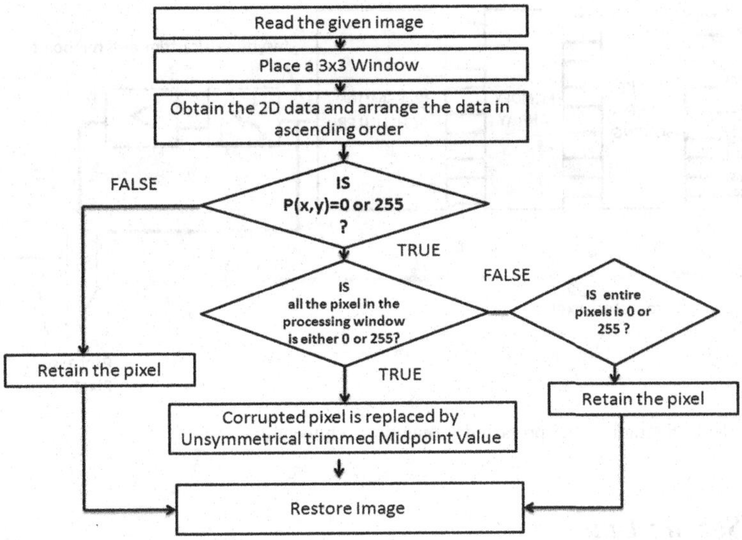

Fig. 1 Flowchart of the proposed algorithm

2 Decision-Based Unsymmetrical Trimmed Midpoint Algorithm

The decision-based unsymmetrical trimmed midpoint filter [13] (DBUTMPF) is an algorithm that removes high density SPN at very high noise densities. The flowchart of the proposed algorithm is given in Fig. 1.

3 Proposed Architecture for Unsymmetrical Trimmed Midpoint Algorithm

The decision-based algorithm uses a 3 × 3 window for removal of SPN for increasing noise densities. The proposed sequential architecture consists of sorting unit, memory unit, FSMD scheduler, and decision-maker unit. The sorting unit arranges the 9 inputs in ascending order using modified snake-like shear sorting.

The sorted data is passed into the memory unit, where it is subsequently used for further processing. FSM scheduler performs unsymmetrical trimmed midpoint operation using the sorted data stored in memory array. After performing the trimming operation two values are passed to decision unit. The fifth element of the input array (referred as processed or centered pixel) and unsymmetrical trimmed midpoint value. The decision unit checks the processed pixel with 0 initially and later with 255. If the above condition was found true, then the output is replaced with unsymmetrical trimmed output, otherwise the processed pixel is left unaltered. The block diagram of the sequential architecture is shown in Fig. 2.

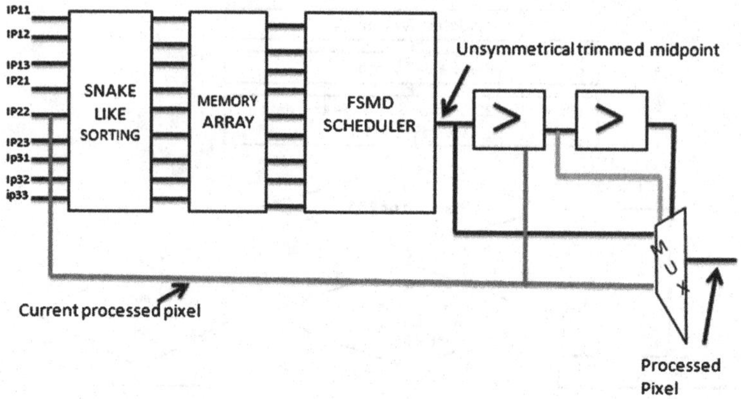

Fig. 2 Block diagram of the proposed sequential architecture

3.1 Sorting Unit

The proposed architecture uses modified snake-like sorting for arranging the data in ascending order [12]. The proposed methodology of the modified snake-like sorting is shown in Fig. 3.

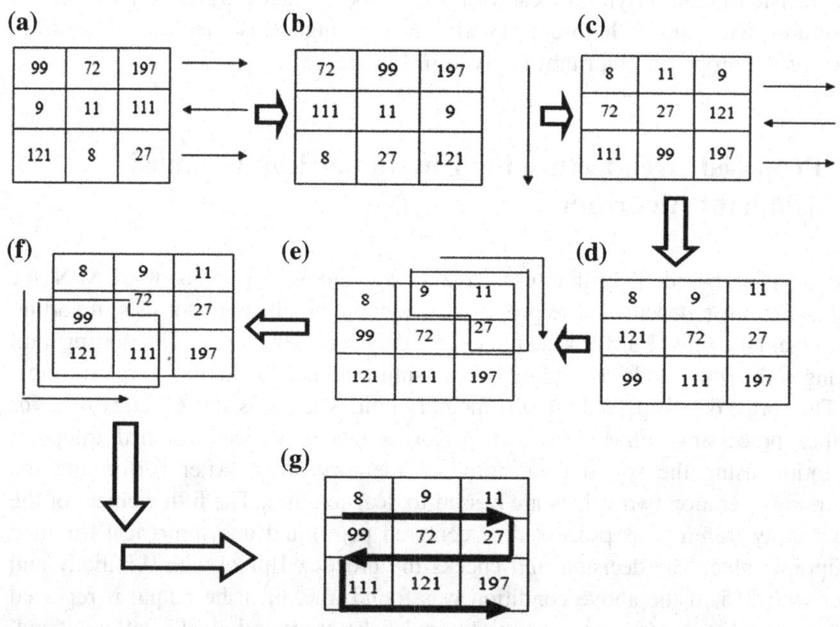

Fig. 3 Modified snake-like sorting [12]

The operation of modified snake like sorting is arranging rows and columns in ascending order and descending order. The 3 × 3 window has only three elements in the row. Hence three elements ought to be arranged in ascending order. Hence the basic processing element in sorter unit is a three-cell sorter. In this work two different three-cell sorters are used. The first three-cell sorter will arrange the data in ascending order and the second three-cell sorter will arrange the data in descending order. The function of the ascending three-cell sorter is to order the data and produce outputs in maximum, middle and minimum values, and the descending three-cell sorter will sort the data in the other way around. Both the three-cell sorter works on the principle of comparing each value with each other using a two-cell comparator. This results in a min and max value (other way for descending three-cell sorter). The third element is compared with the maximum value obtained in previous comparison. This operation results in minimum and the maximum values. The maximum value is the largest value in the sorted array. Now minimum value obtained from second comparison and first comparison is compared. These values are the minimum and median values of the three-element array. The procedure gets reversed for descending three-cell sorters. The block diagram of the ascending and descending three-cell sorter is shown in Figs. 4 and 5 respectively.

The modified snake-like sorting arranges the pixels inside the 3 × 3 window in increasing order. A novel parallel architecture is proposed for the modified snake-like sorting which consists of ascending (indicated in light color in Fig. 6)

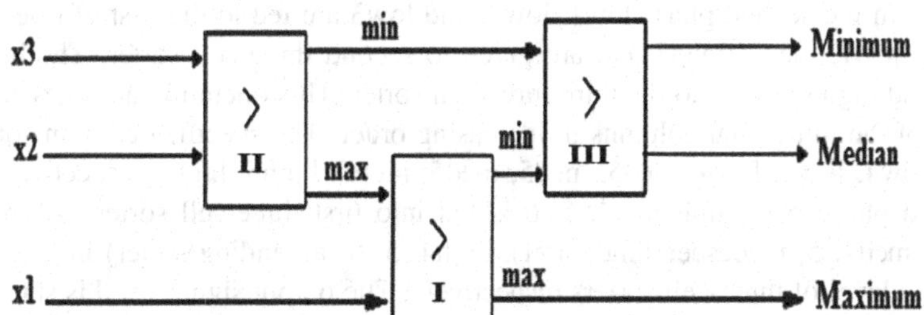

Fig. 4 Basic ascending three-cell sorter

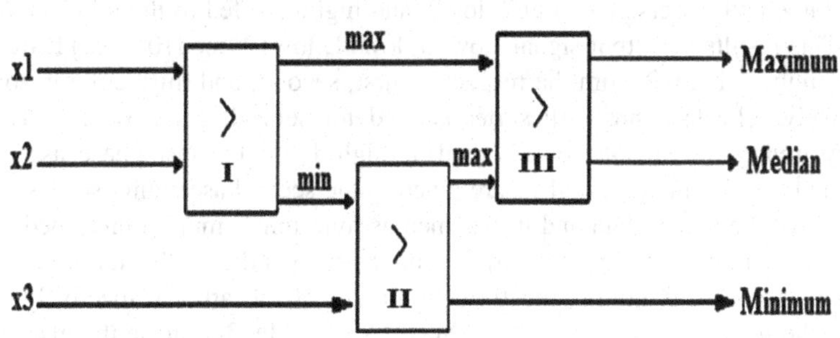

Fig. 5 Basic descending three-cell sorter

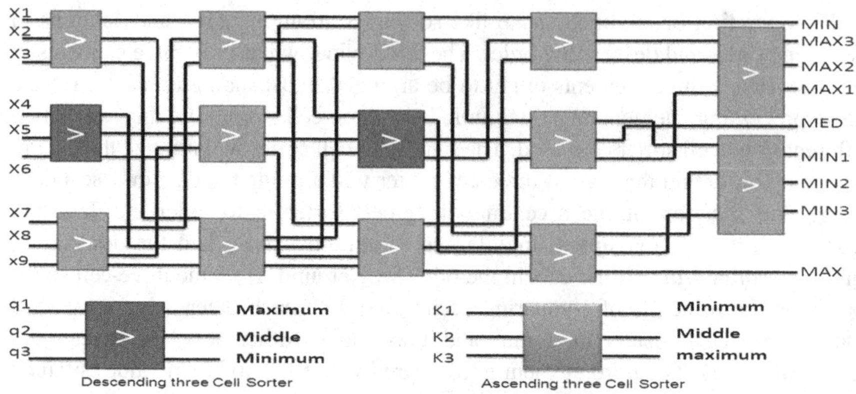

Fig. 6 Proposed parallel sorting technique for the sorting unit

and descending (indicated in dark color in Fig. 6) three-cell sorter. The proposed architecture is illustrated in Fig. 6.

The three pixel elements of the first and third rows are sent inside the first set of parallel three-cell sorter as part of arranging it in ascending order. The second row is arranged in descending order. This results in low1, low2, low3, med3, med2, med1, high1, high2, high3, which is the output of ascending and descending sorters respectively. These output will act as input for the second phase (i.e., column sorting). In the second phase low1, low2, and low3 are fed to the first of three-cell sorters, med1, med2, med3 and are given to second three-cell sorter. The high1, high2 and high3 are fed to the third three-cell sorter. This operation arranges all the values of the individual columns in increasing order. The specified column sorting yields low4, low4, low4, med5, med5, med5, high5, high5, high5 respectively. In the third phase of sorting low4, 5, 6 is fed into first three-cell sorter (ascending sorter), med4, 5, 6 (descending sorter), high4, 5, 6 (ascending sorter) into second and third level of three-cell sorters respectively. The output signals of this stage are marked as low7, low8, low9, med9, med8, med7, high7, high8, high9. Next the column sorting is done again. The column sorting is done by sending low7, high8 and low9 is fed to the first of three-cell sorters, med7, med8, med9 and are given to second three-cell sorters. The high7, low8, and high9 are fed to the third three-cell sorter. This results in output signals low10, low11, low12, med10, med11, med12, high10, high11, high12 from the respective first, second, and third column sorting respectively. The last three-cell sorter is used for semi-diagonal sorting. This is done by giving the signals low11, med11, high 11 to the last phase ascending sorters and med10, high10, high11 are given to the second ascending sorters. After this we have the sorted data and it is named as min, min1, min2, min3, med, max, max1, max2, max3. In this min, min1, min2, min3 indicate the minimum, first minimum, second minimum, and third minimum of the array respectively. Med refers to the median of the array. Max, Max1, Max2, Max3 indicate the maximum, first maximum, second maximum, and third maximum of the array respectively.

This results in a sorted array from the input data. The sorted data is then sent inside a random access memory for storage and for further processing. The RAM implemented in the design is referred as "Memory". The RAM consists of 9 locations and each of the locations will hold a data of size 8 bit. Sorted data is stored in the RAM during the rising edge of the clock.

3.2 *Sorting Complexity Analysis*

In the proposed parallel architecture the operation involves arranging three elements in ascending order. According to our proposed three-cell sorter (which is the processing element) we now need three 8-bit comparators for arranging three elements in ascending order/descending order. The proposed parallel sorting architecture operates in the row and column sorting. Each of the row and column sorting requires three 3-cell sorters. Each three-cell sorter requires three 8 bit comparators. So we have three rows and 3 columns in a 3×3 window. Each row will now require three 3-cell sorter ($3 \times 3 = 9$ eight bit comparator). Similarly for the column sorting we require three more three-cell sorter ($3 \times 3 = 9$ eight bit comparators). The row and column operation is repeated. Hence we requires another 9 (eight bit comparator for row) +9 (eight bit comparator for column) = 18 eight bit comparators. Finally, we require two three-cell sorter for the diagonal sorting. Now we require $2 \times 3 = 6$ eight bit comparators. Total number of eight bit comparator required for sorting the given 9 input data array is given by $9 + 9 + 9 + 9 + 3 + 3 = 42$. The total number of comparators required for sorting 9 input array is 42. Hence the complexity analysis is summarized in terms of eight bit comparator is shown in Table 1. From Table 1 it is clear that the proposed algorithm requires 42 number of comparators for arranging 9 inputs in increasing order.

Table 1 Complexity analysis of different sorting techniques in terms of 8-bit comparator

S. no	Sorting techniques with complexities	Number of comparisons required to sort the entire 3×3 window (worst case)
1	Bubble sorting $o(n^2)$	81
2	Insertion sorting $o(n^2)$	81
3	Selection sort $o(n^2)$	81
4	Merge sort $o(n \log n)$	25
5	Heap sort $o(n \log n)$	28
6	Quick sort $o(n^2)$	81
7	Shear sorting $o(\sqrt{n})(2n))$	54
8	Snake like modified shear sorting $o(4n + 2\sqrt{n})$	42

3.3 FSMD Scheduler Unit

The proposed FSMD scheduler unit consists of 4 states. The states are named as idle, check, Dat1, output. When the system reset is inactive the control of the program is transferred to the next state called idle. The state diagram of the proposed FSM architecture is shown in Fig. 7.

(a) **Idle State**:
The counters such as *F*, *L*, (noise determination counter) and the sum accumulator is initialized to zero and points to check as next state.

(b) **Check State**:
Every element of the 3 × 3 window is checked for 0 or 255. If processed pixel lies between 0 or 255, it is considered as noisy and the scheduler points to Dat1 state else points back to idle.

(c) **Dat1 State**:
Every element of the 3 × 3 is checked for 0 or 255. If 0 is encountered counter *F* is incremented by 1. If 255 are encountered *L* is incremented by 1. The total number of noisy pixel in the array is stored to a variable called "*i*". When the counters *F* and *L* reach the maximum value as 9 or zero respectively, the processed pixel is retained as output else the unsymmetrical trimmed midpoint output becomes the output.

(d) **Output State**:
If the current processing window has all zeros, then forward counter will hold "9" then the processed pixel is considered as texture and retained as output. If the current processing window has all "255" then the reverse counter "*l*" will hold "0" then the processed pixel is considered as texture and output is retained as output. When the above conditions fails the sorted memory array would point to a value "*f*" and "*l*". The midpoint of the trimmed array is the output. The output of the architecture is processed in 16 clock cycles.

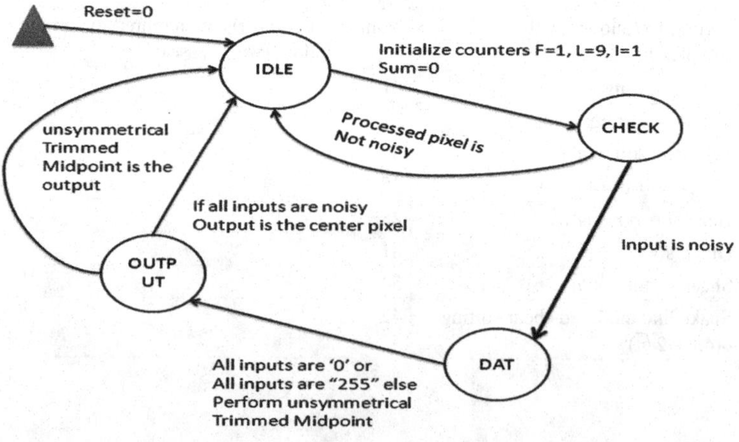

Fig. 7 Proposed FSM scheduler to compute unsymmetrical trimmed midpoint value

Table 2 State encoding assigned by the synthesis XST tool

States	Encoded value
Idle	00
Check	01
Datl	10
O/P	11

4 Results and Discussions

The proposed sequential architecture is implemented for the target device XC3s5000-5fg900 using Xilinx 7.1 compiler tool for synthesis and ModelSim 5.8IIi for simulation as a third party tool using VHDL. The different sorting algorithms such as Bubble sort, Heap sort, insertion sort, selection sort, modified decomposition sort (MDF), Threshold Decomposition Algorithm (TDF) were implemented in VHDL language for the above target device specified. Table 2 illustrates the comparison of other sorting techniques with the proposed logic in terms of area, speed, power.

Figures 8 and 9 illustrates the floor plan and routed FPGA of different sorting algorithm with the proposed parallel architecture. Table 3 gives the area, speed, and power consumption of proposed sequential architecture. Figure 10 gives the simulation result of the proposed sequential logic performed in ModelSim simulator. Figure 11 illustrates the different RTL schematic of the proposed sequential architecture for the decision-based unsymmetrical trimmed midpoint algorithm. Figures 12 and 13 give the Floor plan and routed FPGA for the proposed architecture. The logic for algorithms such as Bubble sort, Heap sort, Insertion Sort, Selection Sort were used from any standard Data Structures book. The MDF

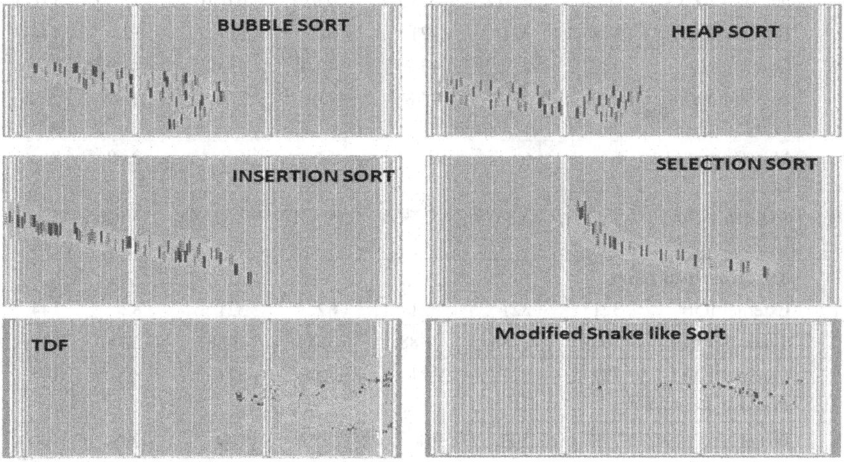

Fig. 8 Floor plan occupied by different sorting techniques for the targeted FPGA

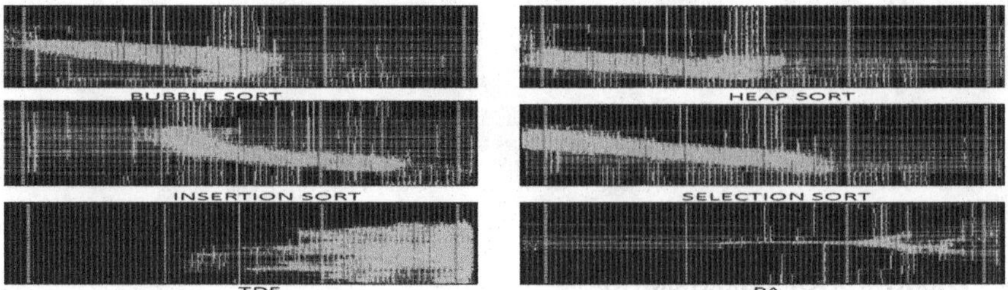

Fig. 9 Routed FPGA occupied by different sorting techniques for the targeted FPGA

algorithm is used from the paper [11]. The values from Table 3 we infer that the proposed parallel architecture for the snake like sorting algorithm requires 709 slices, which is 16 % lesser than other conventional 1D sorting algorithms. The number of 4 input lookup table required by proposed algorithm was found to be 1033. This value is 14.61 % lesser than other conventional 1D sorting algorithm.

Table 3 Comparison of different sorting algorithms for the target FPGA Xc3s5000-5fg900

S.no	Parameters	Buble sort	Heap sort	Insertion sort	MDF	Selection sort	TDF	PA
	After synthesis							
1	No. of slices	4375	3810	4375	4021	4375	4132	709
2	No. of 4 I/P LUT	6080	5312	6080	6854	6080	7066	1033
3	Bonded IOB	328	321	321	82	321	82	144
4	MAX COMB Delay path (ns)	151.715	327.557	151.715	188.933	151.71	190.9	77.3
	After mapping							
5	No. of 4 I/P LUT	6080	5,312	6080	6,922	6,080	7,139	1,033
6	Bonded IOB	328	321	321	82	321	82	144
7	Gate count	43075	37,699	43,075	42,055	43,075	43,927	7,281
8	Avg fan out of LUT	2.78	2.74	2.78	2.6	2.78	2.69	2.72
9	Max fan out of LUT	6	6	6	63	6	2.69	8
10	Avg fan in for LUT	3.33	3.3373	3.3368	3.5717	3.3368	3.6365	3.32
	After place and route							
11	External IOB	321	321	321	82	321	82	144
12	Slices flip flop	3088	2783	3088	3689	3088	3800	517
13	External IOB	321	321	321	82	321	82	144
	Power consumed							
14	Power consumption	298	100	100	298	100	298	100

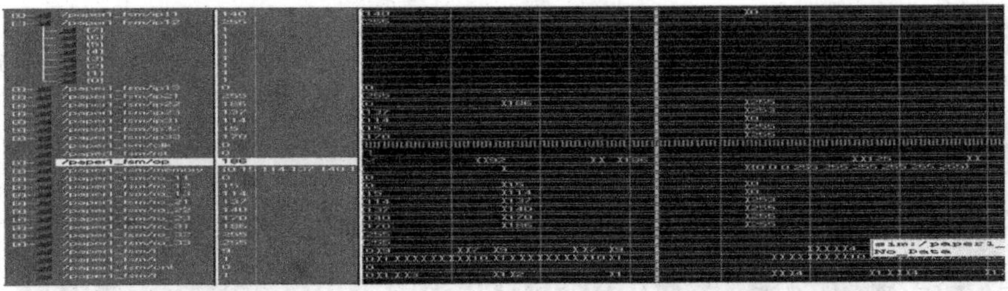

Fig. 10 Simulation of the proposed decision-based unsymmetrical trimmed Midpoint algorithm on ModelSim simulator

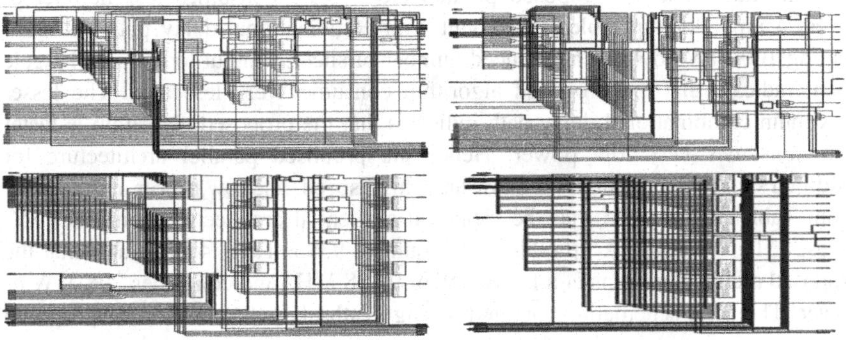

Fig. 11 Different RTL schematic of the proposed decision-based unsymmetrical trimmed midpoint algorithm

Fig. 12 Floor plan of the proposed Decision based unsymmetrical trimmed midpoint algorithm

The maximum combinational delay path required by the proposed parallel sorting algorithm was found as 77.307 ns. This shows the proposed logic has a lesser path delay. This indicates the proposed parallel architecture requires is faster when compared to 1D sorting counterparts. The gate count consumed by the proposed architecture is 7281. This value is 7 times lesser than the other techniques. The power consumed by the proposed algorithm is also found to be very less. The proposed algorithm was found to consume 100 MW. The graphical comparison

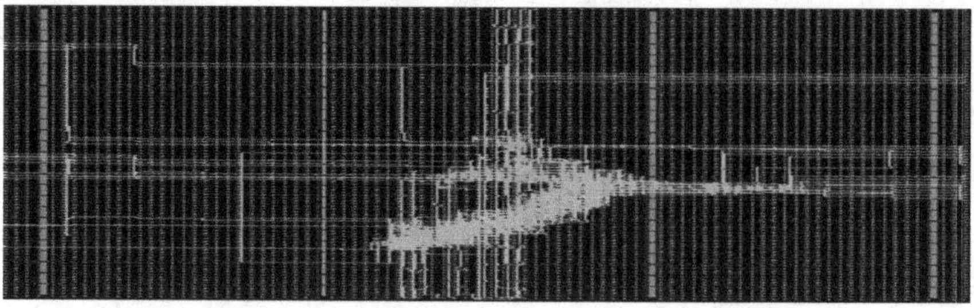

Fig. 13 Routed FPGA of the proposed decision-based unsymmetrical trimmed midpoint algorithm

also illustrates that the proposed parallel architecture consumes less number of slices, 4 input lookup table, gate count with combinational delay path and consuming 100 MW power. The reduced number of slices, 4 input look up table, gate count indicate that the proposed algorithm consumes very less area. The lesser maximum combinational delay path indicates that the proposed algorithm is faster also consumes very low power. Hence the proposed parallel architecture for modified snake-like sorting was a low area, high speed, and low power architecture. From Table 3 it is vivid that the proposed sequential architecture consumes 857 slices, 1311 4 input look up table, gate count of 9393. Table 4 illustrates that the proposed architecture runs at a frequency of 98.38 MHz and consumes 298 MW of power. The RTL schematic generated during synthesis indicates that the proposed

Table 4 Device utilization summary of proposed sequential architecture for the target FPGA Xc3s5000-5fg900

S. no	Parameters		Proposed algorithm
1	Synthesis report	8-bit REG	3
2		8-bit COMP	42
3		No. of slices	857
4		Slices flip flop	48
5		No. of 4 I/P LUT	1311
6		Bonded IOB	82
7		GCLKS	1
8		Min period (ns)	10.16
9		Min I/P arraival time (ns)	79.93
10		Max O/P REQ after CLK (ns)	6.28
11		Max frequency (MHz)	98.38
12	Map report	No. of 4 I/P LUT	1311
13		Bonded IOB	82
14		GCLKS	1
15		Gate count	9393
19	Place and route report	No. of BUFGMUX	1
20		External IOB	82
24		Power consumption	298

architecture is implementable inside the proposed FPGA. The proposed FSM architecture was developed in VHDL and found to have 4 states and has a total of 7 transitions during the operation. The clock transition is assumed to be on the rising edge of the clock. The proposed state machine will have idle state on power up. The clock enable is done using asynchronous reset input with negative logic (i.e., rst = 0). The state encoding is done by the synthesis tool XST. The automatic state encoding for different states of the FSM is given in Fig. 7 as shown below. The states Idle, Check, Dat1, and O/P is encoded with 00,01,10,11 respectively.

5 Conclusion

A FSM-based VLSI architecture for fixed 3 × 3 decision-based unsymmetrical trimmed midpoint algorithm is proposed. The algorithm introduces a parallel architecture for the sorting algorithm which consumes less area, works in high speed, and consumes less power when compared to other sorting algorithms. The unsymmetrical trimmed midpoint is evaluated using an FSM-based scheduler. The proposed architecture is novel and operates at 98.38 MHz.

References

1. Lee CL, Jen CW (1992) Bit-sliced median filter design based on a majority gate. IEE proc-G 139(1):63–71
2. Offen J, Raymond R (1985) VLSI image processing. McGraw-hill, New York
3. Fisher AL (1981) Systolic algorithms for running order statistics. In: Signal and image processing, Department of Computer Science, Carnegie Mellon University, Pittsburgh
4. Kung HT (1982) Why systolic architectures. IEEE Comput 15(1)
5. Batcher KE (1968) Sorting network and their applications. Proc AFIPS Conf 32:307–314
6. Stone HS (1971) Parallel processing with the perfect shuffle. IEEE Trans Comput C-20 (2):153–161
7. Preparata FP (1978) New parallel sorting schemes. IEEE Trans Comput C-27(7):669–673
8. Horiguchi S, Sheigei Y (1986) Parallel sorting algorithm for a linearly connected multiprocessor system. In: Proceedings of international conference on distributed computing systems, pp 111–118
9. Thompson CD, Kung HT (1981) Sorting on a mesh-connected parallel computer. Comm ACM 20:151–161
10. Nassimi D, Sahni S (1979) Bitonics sort on a mesh-connected parallel computer. IEEE Trans Comput C-27:2–7
11. Vasanth K, Karthik S (2010) FPGA implementation of modified decomposition filter. International conference on signal and image processing, pp 526–530
12. Vasanth K, Karthik S (2012) Performance analysis of unsymmetrical trimmed median as detector on image noises and its FPGA implementation. Int J Inf Sci Tech (IJIST) 2(3):19–40
13. Vasanth K, Senthilkumar VJ (2012) A decision based unsymmetrical trimmed mid point algorithm for the removal of high density salt and pepper noise. J Appl Theor Inf Technol 42 (2):553–563

Detection and Segmentation of Cluttered Objects from Texture Cluttered Scene

S. Sreelakshmi, Anupa Vijai and T. Senthilkumar

Abstract The aim of this paper is to segment an object from a texture-cluttered image. Segmentation is achieved by extracting the local information of image and embedding it with active contour model based on region. Images with inhomogenous intensity can be segmented using this model by extracting the local information of image. The level set function [1] can be smoothened by introducing the Gaussian filtering to the current model and the need for resetting the contour for every iteration can be eliminated. Evaluation results showed that the results obtained from the proposed method is similar to the results obtained from LBF [2] (local binary fitting) energy model, but the proposed method is found to be more efficient in terms of computational aspect. Moreover, the method maintains the sub-pixel reliability and boundary fixing properties. The approach is presented with metrics of visual similarity and could be further extended with quantitative metrics.

Keywords Autocorrelation · Active contour model · Segmentation · Level set function

1 Introduction

The area of image segmentation is emerged with a number of perspectives. The pros and cons of each perspective can be decided based on the application. Meantime, in recent years, for segmenting and detecting interesting objects, a well-known

S. Sreelakshmi (✉) · A. Vijai · T. Senthilkumar
Department of CSE, Amrita Vishwa Vidyapeetham (University), Coimbatore, Tamil Nadu, India
e-mail: lakshmi.somadasan@gmail.com

A. Vijai
e-mail: v_anupa@cb.amrita.edu

T. Senthilkumar
e-mail: t_senthilkumar@cb.amrita.edu

L.P. Suresh and B.K. Panigrahi (eds.), *Proceedings of the International Conference on Soft Computing Systems*, Advances in Intelligent Systems and Computing 398, DOI 10.1007/978-81-322-2674-1_25

segmentation method active contour model [3] is used commonly. The model is highly accurate and flexible. The existing ACM methods are classified into two categories: [4] edge-based models and region-based models.

The edge-based model uses a ceasing function to attract the silhouette on to the desired object boundary. This can be achieved by utilizing the intensity change in image as an additional obligation. Additional terms can be added for enlarging the capture range of force. A balloon force term that can control the motion of silhouette is incorporated into the evolution function. Nevertheless, if the selection of balloon force is not proper the results will be unenviable. On counter, region-based models make use of statistical information of images and are more advantageous than edge-based models [4]. This model can deal with segmentation of weak boundary objects since, here the intensity variation is not considered. Also, the model does not depend on the initial placement of contour.

In this paper, segmentation of images with nonuniform intensity can be achieved using the proposed ACM model. This project constructs local image fitting (LIF) [5] energy functional using local image information. LIF energy functional is defined as an impediment of the differences between the original image and the fitting image. In addition, Gaussian Kernel filtering stabilizes the level set function in each iteration. Another advantage of this proposed model is the elimination of re-initialization.

2 Related Work

Based on the methods of curve evolution local statistical information and level set method, the energy function consists of three terms: global term, local term, and regularization term. The images with intensity inhomogeneity are efficiently segmented by embedding local information to this model. Ancestral level set methods uses re-initialization step which is time-consuming. The introduction of penalizing energy to the model avoids the use of this step. For texture segmentation, intensity information is added into the classical structure tensor creating an extended structure tensor (EST). Any texture image can be segmented accurately by fusing the EST with LCV model [2] even though the image is tend to have a nonuniform intensity distribution.

For accurately recovering the object boundary, another local binary fitting energy (LBF) [6] functional is introduced. This method is able to deal with segmentation of images with nonuniform intensity, weak object boundaries, and vessel-like structures. The advantage of this method lies in its accuracy and computational efficiency. Moreover, the proposed method has a favorable application for image denoising. This method is showing robust results on synthetic and real images when applied in different manners.

For segmenting images with inhomogenous intensity, a LIF [5] energy is introduced. This model extracts local information of image for segmentation. In addition, Gaussian filtering is introduced for stabilizing the level set function. It ensures a uniform level set function, and eliminates the requirement of resetting the curve evolution, which is very computationally expensive.

Another effective method for object segmentation is the concept of contour model. The model is hierarchically decomposed into fragments which are grouped to form *part bundles* [7]. Each part bundle is found to have overlapping fragments. An efficient voting method is applied to a given image and a similarity measure based on the local shape is calculated. For optimal grouping a global shape similarity is measured between the part configurations and the model contour.

Boosted Landmark descriptors [8] which are context-dependent can be used for detecting and recognizing objects with cluttered background. It is viewed as a multiclass classification problem, in which a series of trained classifiers are defined. This can be achieved by embedding a criteria based on a forest of optimal tree structures with the error correcting output codes technique.

The related methods are experimented on different manners of cluttering. The type of clutter associated with the image can change the complexity of the background. The proposed method is evaluated on images which has a texture-cluttered background. The complex background resulting from texture cannot differentiate the foreground object from its background. Traditional segmentation method will not be able to segment objects from complex texture background. Hence, a new feature based on autocorrelation is fused with segmentation method. This feature is extracted for exact localization of objects. The same can be extended to work on texture cluttered videos.

Table 1 Comparison of previous work

S. No.	Method	Advantage	Disadvantage
1	Chan–Vese model [4]	• Detect weak boundaries and blurred edges	• Less sensitive to image noises • Does not segment image with in-homogeneities
2	Local binary fitting model [6]	• No re-initialization • Segment images with in-homogeneities and weak object boundaries	• Computational complexity
3	Boosted landmarks of contextual descriptors and forest-ECOC [8]	• Invariant to small variations in scale, translation, global illumination, partial occlusions, and to small affine transformations	• Fails with high deformable objects • Cannot detect from different point of view
4	Contour-based object detection using part bundles [7]	• One-one mapping • Accommodate broken edge fragments and deformations	• Cannot be applicable for objects with distinctive texture

2.1 Comparison of Previous Work

See Table 1.

3 Proposed Method

See Fig. 1.

3.1 Segmentation

Segmentation is achieved by minimizing the energy function which is defined as the difference between original image and fitted image. The fitted image can be developed by evolving a contour using level set function. Corresponding level set function can be initialized as a binary function as follows:

$$\phi(x, t = 0) = \begin{cases} -\rho, & x \in \Omega_0 - \partial\Omega \\ 0, & x \in \partial\Omega_0 \\ \rho, & x \in \Omega - \Omega_0 \end{cases} \tag{3.1.1}$$

where $\rho > 0$ is a constant, Ω_0 is a subset in the image domain Ω, and $\partial\Omega_0$ is the boundary of Ω_0. The generated fitted image formulation is defined as

$$I = m_1 H_\varepsilon(\phi) + m_2(1 - H_\varepsilon(\phi)) \tag{3.1.2}$$

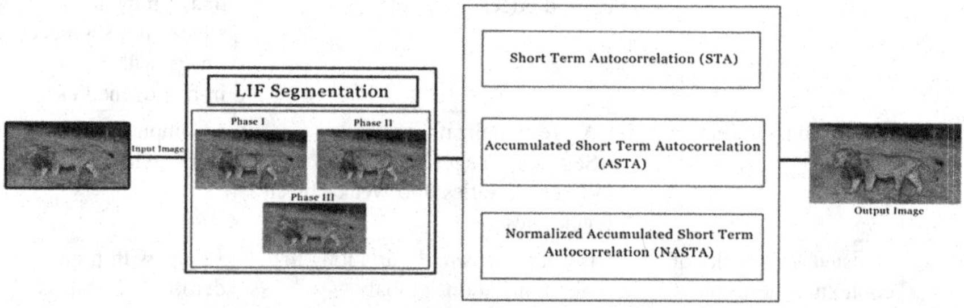

Fig. 1 Architecture diagram

where m_1 and m_2 are defined as follows:

$$\begin{cases} m_1 = \text{mean}(I \in (\{x \in \Omega | \phi(x) < 0\} \cap W_k(x))) \\ \text{mean}(I \in (\{x \in \Omega | \phi(x) > 0\} \cap W_k(x))) \end{cases} \tag{3.1.3}$$

where $W_k(x)$ is a rectangular window function, e.g., a truncated Gaussian window or a constant window. In this experiment, a truncated Gaussian window $K_\sigma(x)$ with standard deviation σ and of size $4k + 1$ by $4k + 1$ is chosen. The constant window can also give a similar segmentation result.

For achieving segmentation the energy function defined as the difference between original and fitted image should be minimized. The energy function is defined as

$$E^{\text{LIF}}(\phi) = \frac{1}{2} \int_{-\Omega}^{\Omega} \left| I(x) - I^{\text{LFI}}(x) \right|^2 \text{d}x, \quad x \in \Omega \tag{3.1.4}$$

Further, contour development can be achieved using calculus of variation, using the calculus of variation and the steepest descent method, to get the corresponding gradient descent flow.

$$\frac{\partial \phi}{\partial t} = \left(I - I^{\text{LFI}}\right)(m_1 - m_2)\delta_\varepsilon(\phi) \tag{3.1.5}$$

After evolving the contour it can be stabilized using the Gaussian kernel, i.e., $\phi = G_\varsigma * \phi$, where ς is the standard deviation. Then the fitted image is checked against the original image whether the curve evolution is stationary. If it is not then the contour is re-evolved by modifying the level set function using Eq. (3.1.5).

This model will be suitable for finding objects from simple background. In case of images with complex background, it will be difficult to detect small objects. Hence a featurebased on autocorrelation function is introduced and fused with LIF model for detecting small objects from complex textured background.

3.2 Feature Extraction

Autocorrelation function gives the similarity measure between original signal *f*(*m*) and a shifted version of itself. The value of autocorrelation determines how similar the signals are. For example, a high autocorrelation value corresponds to highly similar signals. This function can provide useful information on periodic events and hence the same can be used for detecting repeating patterns. Accurate localization of target objects can be achieved by considering windowed autocorrelation and calculating the short-term (time) autocorrelation [9]. Short-term autocorrelation for a real signal is defined as [10]

$$R_l(k) = \sum_m f(m)w(l-m)f(m+k)w(l-k-m) \tag{3.2.1}$$

where $w(l)$ is a small sliding window of length L. Short-term autocorrelation is calculated from $Rl(-L/2)$ to $Rl(L/2)$ in this window $w(l)$ around every index t of the signal $f(t)$. Then, the value in position t is replaced with the sum of short-term autocorrelation. This will create a new signal F_R which is similar to original signal. This new function is known as accumulated short-term autocorrelation [9] and is an appropriate function for target detection from a cluttered signal.

The 2D short-term autocorrelation concept can be used for extending, accumulated short-term autocorrelation to 2D image signals. 2D short-term autocorrelation for the image I is defined at horizontal lag k_1 and vertical lag k_2 as:

$$\begin{aligned} R_{l_1 l_2}(k_1,k_2) = \sum_m \sum_n & I(m,n)w(l_1-m,l_2-n) \\ & \times I(m+k_1,n+k_2)w(l_1-k_1-m,l_2-k_2-n) \end{aligned} \tag{3.2.2}$$

where $w\ (l_1, l_2)$ is a 2D small sliding window with size of $L \times L$. First, 2D short-term autocorrelation is calculated from $R_{l1l2}(-L/2,-L/2)$ to $R_{l1l2}(L/2,L/2)$ in a window defined over each pixel (m, n). Next, the accumulated short-term autocorrelation is normalized in the range of 0–255 and placed in the location index (m, n). A new image F_R^N similar to original image is generated, where each pixel value is replaced with normalized accumulated 2-D short-term autocorrelation [9] value. This new resultant image is known as normalized accumulated short-term autocorrelation [9] that is an effective tool for retrieving objects from cluttered background (Fig. 2).

An image signal comprised of textures and clutters does not provide applicable and significant pixel information. For analyzing such kind of images requires region information instead of pixel information. In addition, the durability of segmentation algorithm can be reduced due to sensitivity of pixel value towards noise. The windowed autocorrelation can thus be a powerful tool for textures and clutters.

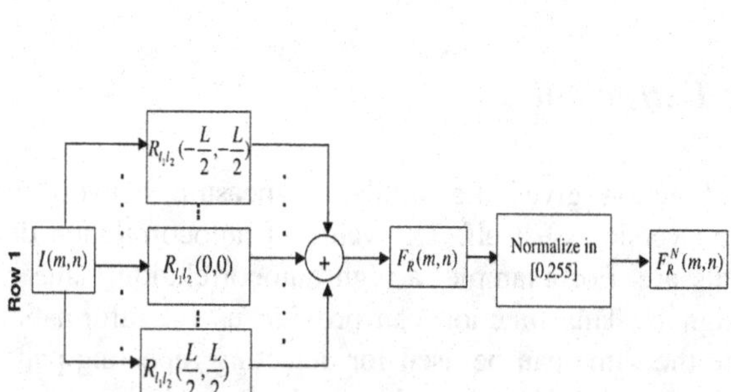

Fig. 2 NASTA [9] calculation for a small window around pixel (m, n) of an image I

4 Experimental Results

The model is evaluated on different real time datasets. (i) natural images with texture cluttered background collected from Caltech datasets (ii) datasets containing synthetic images (iii) datasets containing medical images (iv) images taken in 8 Mb GT-1082 camera. The following results are obtained by applying the segmentation method on the respective datasets. The 2D short-term autocorrelation calculation is considered as a future work for exact localization of object.

The real time datasets shown in Fig. 3 are composed of objects having a pattern similar to its background. The texture-cluttered image when subjected to segmentation, the object is not segmented correctly from the cluttered background. Since the foreground is texture wise similar to its background, a part of the background texture also gets segmented along with the object.

The segmentation is not accurate with simple segmentation, as the texture pattern in the background is also segmenting along with foreground object. The same has been applied on a real time forest image as a proceeding to extend the work to cluttered video. The lion in the image is not segmented correctly meanwhile the

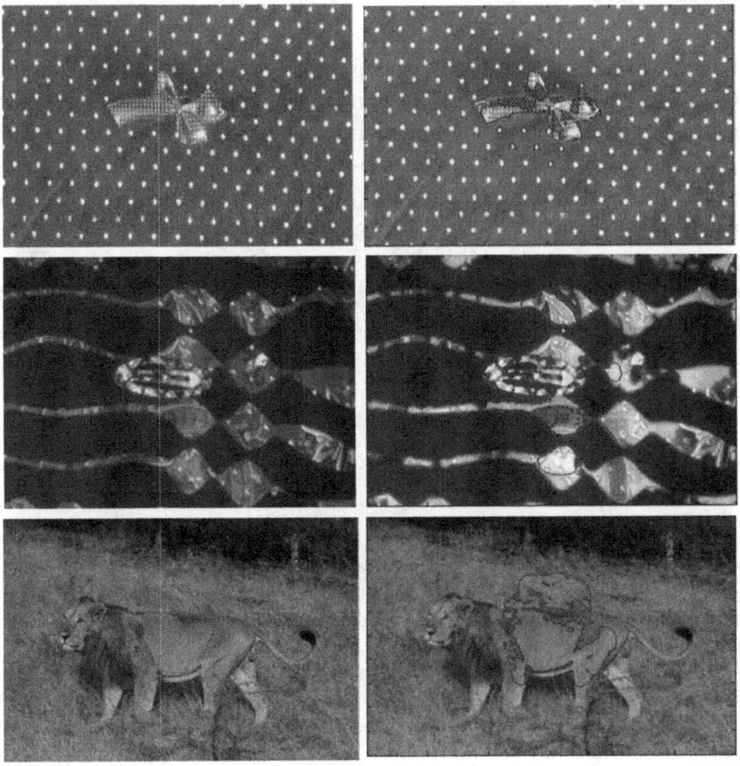

Fig. 3 Column 1: original image. Column 2: LIF segmented image

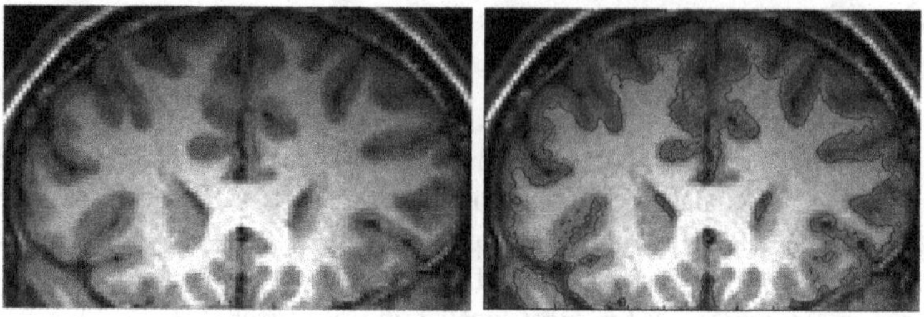

Fig. 4 Result of LIF segmentation on medical image

background against it is also get segmented. While, in Fig. 4 the effect of segmentation when applied on a medical image is shown. The model works efficiently for this dataset where the image background is not complex. The comparison of model is evaluated based on a visual metric and not on quantitative measure. Hence the model works for images with simple background.

5 Future Work

Current work can be extended for extracting the foreground object from a cluttered background using a new feature called normalized short-term autocorrelation [9]. This feature is represented as region information which is obtained from image pixels. The energy function for this region-based model can be defined using the obtained features. Minimizing this energy function makes the algorithm able to detect small objects against textured background. Based on this new feature, the work can be extended for tracking objects in a texture cluttered video. The main application in video includes detecting fast moving in a forest background. Since the application had not been implemented and is assumed as a future work as the results are not predictable.

References

1. Mumford D, Shah J (1989) Optimal approximation by piecewise smooth function and associated variational problems. Commun Pure Appl Math 42:577–685
2. Xia RB, Liu WJ, Zhao JB, Li L (2007) An optimal initialization technique for improving the segmentation performance of Chan–Vese model. In: Proceedings of the IEEE international conference on automation and logistics, pp 411–415
3. Kass M, Witkin A, Terzopoulos D (1988) Snakes: active contour models. Int J Comput Vis 1:321–331
4. Chan T, Vese L (2001) Active contour without edges. IEEE Trans Image Process 10(2): 266–277

5. Zhang K, Song H, Zhang L (2010) Active contours driven by local image fitting energy. Pattern Recogn 43(4):1199–1206
6. Li CM, Kao C, Gore J, Ding Z (2007) Implicit active contours driven by local binary fitting energy. In: IEEE conference on computer vision and pattern recognition
7. Lu C, Adluru N, Ling H, Zhu G, Latecki LJ (2010) Contour based object detection using part bundles. Comput Vis Image Underst
8. Escalera S, Pujol O, Radeva P (2007) Boosted landmarks of contextual descriptors and forest-ECOC: a novel framework to detect and classify objects in cluttered scenes. Pattern Recogn Lett 28:1759–1768
9. Vard A, Jamshidi K, Movahhedinia N (2012) Small object detection in cluttered image using a correlation based active contour model. Pattern Recogn Lett 33:543–553
10. Rabiner L, Schafer R (1978) Digital processing of speech signals. Prentice Hall, New Jersey
11. Chan T, Zhu W (2005) Level set based shape prior segmentation. In: IEEE conference on computer vision and pattern recognition (CVPR), vol II, pp 1164–1170
12. Paragios N, Deriche R (2000) Geodesic active contours and level sets for detection and tracking of moving objects. IEEE Trans Pattern Anal Mach Intell 22:1–15
13. Tsai A, Yezzi A, Willsky AS (2001) Curve evolution implementation of the Mumford-Shah functional for image segmentation, denoising, interpolation, and magnification. IEEE Trans Image Process 10:1169–1186
14. Savelonas MA, Iakovidis DK, Maroulis D (2008) LBP-guided active contours. Pattern Recogn Lett 29(9):1404–1415
15. Vard AR, Moallem P, Nilchi ARN (2009) Texture-based parametric active contour for target detection and tracking. Int J Imaging Syst Technol 19(3):187–198
16. Zhang K, Zhang L, Song H, Zhou W (2010) Active contours with selective local or global segmentation: a new formulation and level set method. Image Vis Comput 28(4):668–676
17. Xu C, Pham D, Prince J (2000) Image segmentation using deformable models. In: Handbook of medical image processing and analysis, vol 2. SPIE Press, Bellingham, pp 129–174
18. Pi L, Shen CM, Li F, Fan JS (2007) A variational formulation for segmenting desired objects in color images. Image Vision Comput 25(9):1414–1421
19. Sum K, Cheung P (2008) Vessel extraction under non-uniform illumination: a level set approach. IEEE Trans Biomed Eng 55(1):358–360
20. Wang L, He L, Mishra A, Li C (2009) Active contours driven by local Gaussian distribution fitting energy. Signal Process 89(12):2435–2447
21. Shi Y, Karl WC (2005) Real-time tracking using level sets. In: IEEE conference on computer vision and pattern recognition, vol 2, pp 34–41
22. Maroulis DE, Savelonas MA, Iakovidis DK, Karkanis SA, Dimitropoulos N (2007) Variable background active contour model for computer-aided delineation of nodules in thyroid ultrasound images. IEEE Trans Inf Technol Biomed 11(5):537–543

Enhanced Automatic Classification of Epilepsy Diagnosis Using ICD9 and SNOMED-CT

G. Nivedhitha and G.S. Anandha Mala

Abstract Epilepsy is a group of neurological disorders characterized by epileptic seizures. The diagnosis of epilepsy is typically made based on the description of the seizures and the underlying cause. It is important that the diagnosis is correct. After the diagnosis process, physicians classify epilepsy according to the International Classification of Diseases (ICD), Ninth Revision (ICD-9). The classification process is time consuming and it demands the realization of integral exams. The existing system proposes an automatic process of classifying epileptic diagnoses based on ICD-9. A text mining approach, using preprocessed medical records is used to classify each instance mapping into the corresponding standard code. The proposed system contributes in enhancing the accuracy in the classification of the diagnosis by identifying the type of epilepsy and mapping it with the standard codes of ICD-9 and Systematized Nomenclature Of Medicine—Clinical Terms (SNOMED-CT). The paper discusses about the related works and proposed system. The experimental results, conclusion, and future work have been discussed in the subsequent topics.

Keywords Text mining · ICD codes · SNOMED-CT · Epilepsy · Seizure · Diagnosis

1 Introduction

Data Mining is becoming popular in healthcare field because there is a need of efficient analytical methodology for detecting unknown and valuable information in health data. Clinical Decision Support Systems (CDSS) [1] are computer systems

G. Nivedhitha (✉) · G.S. Anandha Mala
Easwari Engineering College, Computer Science and Engineering, Chennai, India
e-mail: nivisurjit@gmail.com

G.S. Anandha Mala
e-mail: gs.anandhamala@gmail.com

L.P. Suresh and B.K. Panigrahi (eds.), *Proceedings of the International Conference on Soft Computing Systems*, Advances in Intelligent Systems and Computing 398, DOI 10.1007/978-81-322-2674-1_26

designed to impact clinician decision making about individual patients at the point in time that these decisions are made. These systems may have a great impact in healthcare organizations by reducing the medical error, improving services, and also support better management decisions. They face project management challenges. These challenges need useful project management practices. Such systems do not themselves perform clinical decision making; they provide relevant knowledge and analyses that enable the ultimate decision makers which includes—patients, clinicians, and health care organizations—to develop more informed judgments. CDSS manage information from medical information systems [2].

Epilepsy consists of a number of recurrent and unpredictable seizures that occur through time. A seizure is a manifestation of brain's electrical discharges that will cause symptoms according to the specific location they occur in the brain. Due to these electrical discharges, the brain cannot perform normal tasks causing, e.g., seizures, language disturbances, hallucinations, and absences. But not all seizures are epileptic; an alarm is set only when the seizures occur often (at least two times), not being provoked by alcohol, drogues, poisoning or other diagnosed diseases.

The International Classification of Diseases (ICD) published by the World Health Organization is the standard tool for the diagnosis of epidemiology, clinical, and health management. The ICD is a health care classification system and it monitors the occurrence of diseases and other health problems.

Systematized Nomenclature of Medicine—Clinical Terms (SNOMED-CT) is abbreviated to SNOMED-CT. It is distributed by the International Health Terminology Standards Development Organization (IHTSDO). It is termed as the reference terminology and it is a system for medical language extraction and encoding for different types of diseases. It is a collection of medical terms with codes, terms, synonyms, and definitions which is documented and can be processed by computers.

2 Related Work

Pereiraa et al. [3] proposed an automatic process of classifying epileptic diagnoses based on ICD-9 [4]. In this approach, a text mining [5] approach is put forward, using processed electronic medical records and a K-Nearest Neighbor is applied as a white-box multi classifier approach to classify each instance mapping into the corresponding standard code. Though the results suggest a good performance proposing a diagnosis from electronic medical records, there is a problem of reduced volume of available training data. The proposed approach uses real electronic health records as source, including the preprocessing step that uses Natural Language Processing (NLP), followed by the definition of interpretable models which can be used in real diagnostic scenarios. The results are encouraging with an average measure of 71.05 %. However, with the proposed approach accuracy can be increased only when more medical records and different seizure types are used.

Pereiraa et al. [6] developed a system to support the diagnosis and the classification of epilepsy in children. This work focused in the development of a clinical decision support system to support diagnosis and classification of epilepsy in children. The CDSS are designed to assist physicians or other professionals making more informed clinical decisions, which help in the diagnosis, and analysis. This approach uses text mining techniques namely, text summarization, information retrieval, and clustering for analysing the patient records. A supervised learning approach is then pursued, since there is a classification for each record, i.e., each one has a final diagnosis to build a model to classify future records. For each record classified as probable epilepsy, a new model, previously created also with a supervised learning, suggests the ICD-9 classification. The results on real medical records suggest that the proposed framework shows good performance and clear interpretations, but the issue in this is the reduced volume of available training data. To overcome this hurdle, in this work they have proposed and explore ways of expanding the dataset. They have used the crossover technique for this purpose, as it allows the combination of parts of the existing records, generating new records, when applied to the available medical records.

Roque et al. [7] proposed a general approach for gathering observable characteristics of patients whose characters are different. They demonstrated by extracting the interactive information from the free-text in the structured patient record data, and use it for producing fine-grained patient classification. The proposed approach uses a dictionary based on the International Classification of Disease ontology for text mining. In this approach, they used text and data mining techniques to extract clinical information from the patient records. They performed mining over EPR data and expanded the disease information. The expanded information is then assigned to ICD10 codes. Thus, the system uses structured data, text mining techniques to extract information from text. The complete phenotypic description of patients can be obtained as a result.

Tseng and Lin [8] proposed a scheme for text mining approach. This scheme describes a series of text mining techniques including text segmentation, term association, summary extraction, topic identification, feature selection, cluster generation, and information mapping. They have proposed an approach to verify the usefulness of extracted keywords and phrases, an algorithm for dictionary-free key phrase extraction, an efficient co-word analysis method that produces generic cluster titles. Evaluation of these techniques has shown some success in their proposed framework and the results produces more relevant content words.

Wongthongtham and Chang [9] have analysed the characteristics of software engineering ontology and have defined graphical notations. By using these notations software engineering ontology is modeled. For this, only a part of modeling domain knowledge of software engineering is taken as example. The software engineering ontology is evaluated and deployed to the distributed development. The proposed ontology construction approach is to be applied for the construction of epileptic entities and standard codes.

3 Proposed System

The system is designed to classify the types of epilepsy with improved accuracy. The accuracy is improved by using more types of medical codes that are related to epilepsy. The proposed system classifies the dataset with standard ICD-9 and SNOMED-CT codes. Enhancement of the classification is done by enlarging the dataset with other seizure type classification. Figure 1 shows the overall architecture of automatic classification of epilepsy.

The classification process includes preprocessing of the patient records, which extracts the preprocessed words. From the preprocessed words, medical entities are recognized with which domain ontology will be constructed with the help of rules and relation extraction and mapped to standard codes of ICD9 and SNOMED-CT.

3.1 Preprocessing

In preprocessing, the patient diagnostic records [10] are given as input and the records are preprocessed using different algorithms for each processing. In pre-processing, the documents are first cleaned, to identify and extract information, using a spell checker, replacing acronyms, removing duplicated characters and applying grammar rules. Then tokenizer is used to classify words, sentences, and punctuation marks. Finally, stop word removal algorithm identifies and removes words which are less or no relevance. Figure 2 depicts the preprocessing of the diagnostic records.

Tokenization. Tokenization includes splitting the text into words . In tokenization, first the medical record is loaded and the file is read into buffer and scanned till the end of file is reached. The words are added into an array. Thus, the file is tokenized and added to an array by scanning each line in the text document.

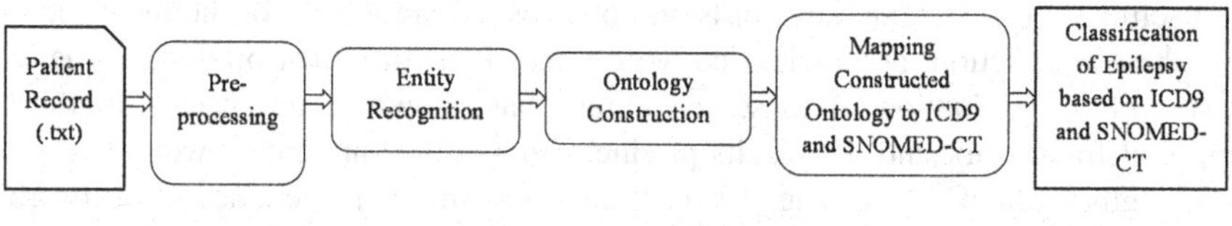

Fig. 1 Overall process of automatic classification of epilepsy

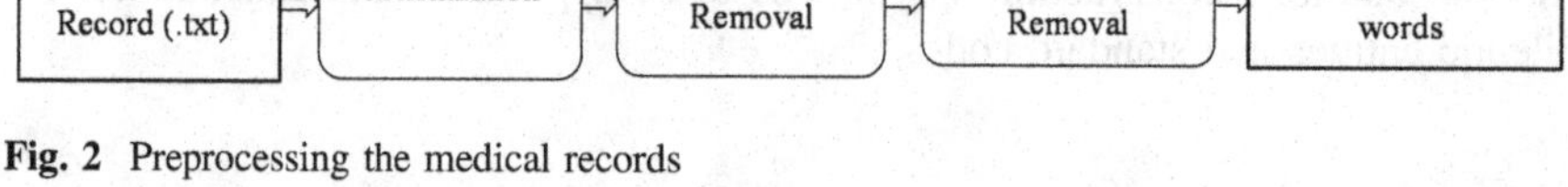

Fig. 2 Preprocessing the medical records

Duplicate Removal. In duplicate removal, words that are duplicated more than once will be replaced or removed. Once words are tokenized, then we have to remove the words which are repeated, in order to remove, our approach will be simple and classic, the tokenized words are in an array. Then the words in the array are converted into hash set which does not allow duplicates. Then, these hashed words are again sent back to the arraylist.

Stop word Removal. Removal of irrelevant words is an important step in preprocessing. Removing such words are done using stop word removal algorithm thereby increasing efficiency in searching, mapping, etc.

3.2 Entity Recognition

The preprocessed words which are obtained from the above preprocessing step will be compared with the existing entities in the database. Entity recognition and ontology tools were required to classify the entity for each word. NLP techniques include approaches such as negation handling and name entity recognition. Name entity recognition is used to classify entities by analysing words, classes, similar abbreviations, and terminology. Thus, the entities related to epilepsy is recognized and stored for retrieval. Figure 3 shows the process of entity recognition where the entities in database are taken for comparison with the preprocessed words.

3.3 Ontology Construction

Ontologies are, generically, a list of concepts organized with classes, subclasses, properties, attributes, and instances that can be useful to retrieve identifiers in documents to describe words and their relations. In ontology construction, the

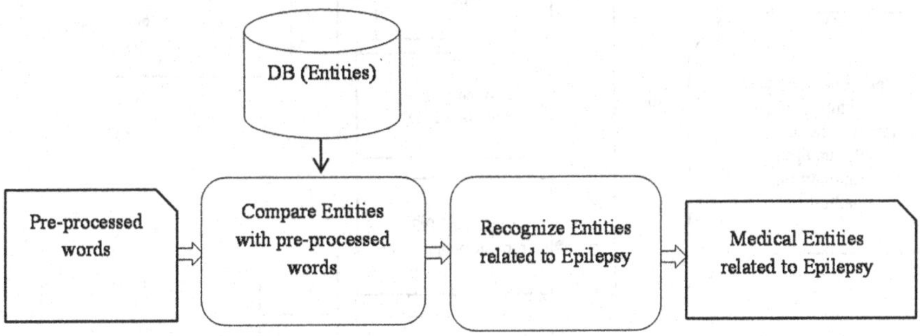

Fig. 3 Entity recognition

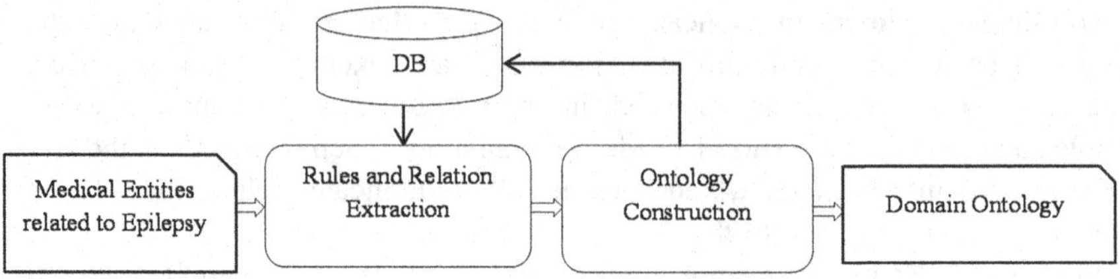

Fig. 4 Ontology construction

entities related to epilepsy is subjected to rules and relation extraction, i.e., family epilepsy history, loss of awareness, etc. Based on these rules, ontology for the epileptic domain will be constructed. Figure 4 shows the ontology construction for a domain.

In ontology construction, the entities related to epilepsy are stored in text file. Then using MapReduce algorithm the entities are divided based on Key–value pairs. Then, with the help of those entities with the same key values, ontology is constructed for each type of epilepsy. Figure 5 depicts the construction of ontology using MapReduce [11].

3.4 Mapping to Standard Codes

The constructed ontology is then mapped with standard classification codes of ICD9 and SNOMED-CT. These standard codes have specific codes for each of the seizure and epilepsy types.

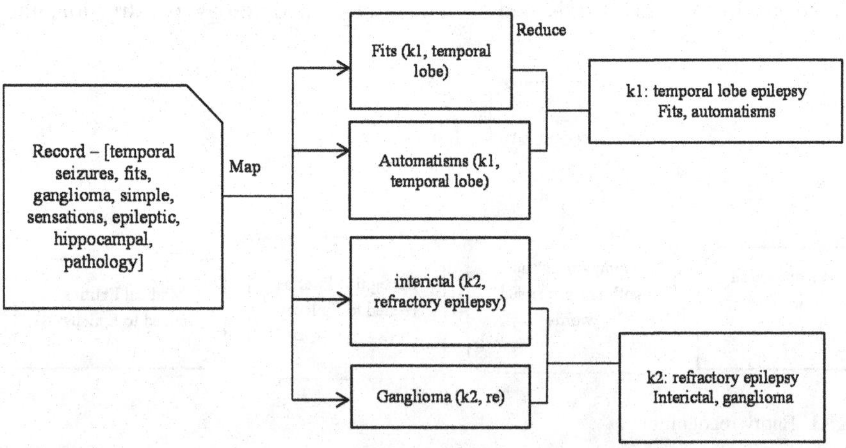

Fig. 5 MapReduce algorithm

4 Experimental Results

In this paper, the patient diagnostic records are handled by preprocessing them. To handle them efficiently, the system takes the text documents containing the patient diagnostic reports in .txt formats. The duplicated words are removed and tokenized. The system then uses stop word removal algorithm for removing the irrelevant words in the text documents. A total of 23 patient records [10] are preprocessed. For entity recognition, the preprocessed words are mapped with the entities stored in the database. A total of 123 entities are stored in the database and more terms are to be stored. Figure 6 charts out the recognized entities with the increase of stored entities in the database. The *x*-axis represents the number of entities that are recognized with the entities stored in the database, i.e., first value shows the number of entities recognized when there are 10 entities stored in the database. The *y*-axis represents the number of entities recognized.

With the help of MapReduce algorithm, the ontology is constructed for each of the epilepsy type. The existing work uses KNN for this classification. The initial result on seizure type classification [3] is shown in Table 1.

Here, FP—False Positive (b), FN—False Negative (c), TP—True Positive (a), TN—True Negative (d). The *F*1 measure [3] is calculated as follows:

$$\mathrm{F1} = 2 * P * R/(P + R). \tag{1}$$

In the above equation precision, P is ($P = a/(a + b)$) and recall, R is ($R = a/(a + c)$). F1 is a measure for text classification.

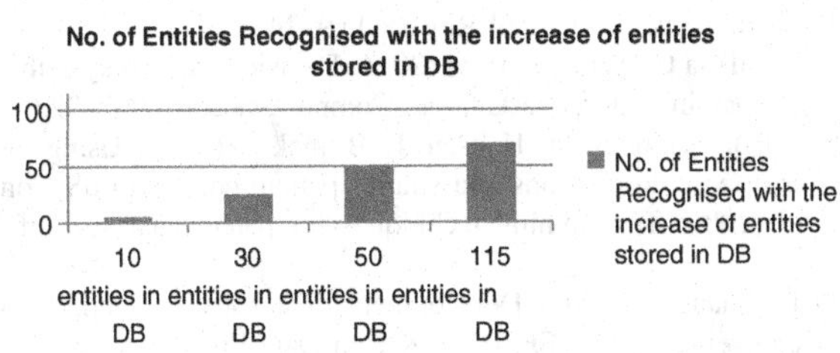

Fig. 6 Result of recognition of entities

Table 1 Initial results on seizure type classification

Seizure type	FP	FN	TP	TN	F1
Complex focal seizure	1	5	9	4	73 %
Generalized convulsive epilepsy	4	10	2	2	62.2 %
Simple focal seizure	3	15	0	1	N/A

5 Conclusion and Future Work

A detailed literature survey has been carried out, which gives us an idea of the recent works been published so far. Eclipse Luna is used as the framework, as it is freely distributed and is an open source. This project had dealt with the automatic classification of epilepsy with corresponding ICD9 and SNOMED-CT codes, thereby enhancing the accuracy. The work consists of preprocessing, entity recognition, ontology construction, and finally mapping the type of epilepsy to the corresponding standard code. To improve the accuracy of epilepsy classification, more types of epilepsy and seizures are taken into consideration.

Future work will be focused on enhancing the accuracy of classification by including diverse set of epileptic and seizure codes and also the dataset has to be expanded for more features.

References

1. Musen MA, Middleton B, Greenes RA (2014) Clinical decision support systems. In: Shortliffe E, Cimino J (eds) Biomedical informatics. Springer, London, pp 643–674
2. Caballero I, Sánchez LE, Freitas A, Fernández-Medina E (2012) HC+: towards a framework for improving processes in health organizations by means of security and data quality management. J Univers Comput Sci 18(12):1703–1720
3. Pereiraa L, Rijoa R, Silvaa C, Agostinhod M (2013) ICD9-based text mining approach to children epilepsy classification. Procedia Technol 9:1351–1360
4. Software A (2013) The international classification of diseases, 9th revision, clinical modification. Available from: http://www.icd9data.com/2013/Volume1/320-389/340-349/345/default.htm 30th May 2013
5. Bukhari AC, Kim Y-G (2011) Ontology-assisted automatic precise information extractor for visually impaired inhabitants. Artif Intell Rev 38(1):9–24
6. Pereiraa L, Rijoa R, Silvaa C, Agostinhod M (2014) Decision support system to diagnosis and classification of epilepsy in children. J Univers Comput Sci 20(6):907–923
7. Roque FS, Jensen PB, Schmock H, Hansen T, Brunak S (2011) Using electronic patient records to discover disease correlations and stratify patient cohort. PLoS Comput Biol 7(8)
8. Tseng YH, Lin CJ (2007) Text mining techniques for patent analysis. Inf Process Manag 43:1216–1247
9. Wongthongtham P, Chang E (2009) Development of a software engineering ontology for multi-site software development. IEEE Trans Knowl Data Eng
10. Zwoliński P, Roszkowski M, Zygierewicz J, Haufe S, Nolte G, Durka PJ (2010) Open database of epileptic EEG with MRI and postoperational assessment of foci-a real world verification for the EEG inverse solutions. Neuroinformatics 8:285–299
11. Zhang H, HU W, Qu Y (2012) VDoc: a virtual document based approach for matching large ontologies using mapreduce. J Zhejiang Univ-Sci C Comput Electron 13(4):257–267

Scalable Casual Data Consistency for Wide Area Storage with Cloud

Kishor Kumar Patel and E. Christal Joy

Abstract Cloud storage administrations have got to be economically main stream because of their mind-boggling points of interest. To give pervasive constantly on access, a cloud administration supplier (CSP) keeps up numerous reproductions for every bit of information on geological servers. A novel key issue utilizes the duplication procedure in the mist is enormously extravagant toward attaining solid constancy on an overall level. This paper, presents a new consistency like an administration (CaaS), which comprise the extensive cloud information and various little review mists. During this CaaS, cloud information is kept up by a CSP, and a gathering of clients that constitute a review cloud can check whether the information cloud gives the assured level of constancy or not. We proposed the two stages examining structural engineering, which just obliges an approximately synchronized check in the review cloud. At that point, we plan calculations to evaluate the seriousness of infringement with two measurements: the shared trait of infringement and the decay of read estimation. At long last, we plan a heuristics auditing strategy (HAS) to uncover, however, lots of infringement as would be prudent. Far reaching trials were performed utilizing a mix of recreations and genuine cloud organizations to accept HAVE.

Keywords Cloud storage · Consistency as a service (CaaS) · Two-stage auditing and heuristic auditing strategy (HAS)

K.K. Patel (✉) · E. Christal Joy
Department of Information Technology, Sathyabama University, Chennai, India
e-mail: patel.kishor17@gmail.com

E. Christal Joy
e-mail: christaljinu@gmail.com

L.P. Suresh and B.K. Panigrahi (eds.), *Proceedings of the International Conference on Soft Computing Systems*, Advances in Intelligent Systems and Computing 398, DOI 10.1007/978-81-322-2674-1_27

1 Introduction

In today's reality, the utilization of mobiles is progressively being used when contrasted with the utilization and reception of Internet. The mechanical enhancements in the later times have made mobile phones extremely helpful and minimal. The prior mobile phones supposed "Hand Phone" was of size equivalent to that of a landline and the main contrast was that it was remote. Presently, we have "Advanced mobile phones" that are in the extent of our palm. Portable computing assumes a crucial part in today's world. Versatile computing has cleared a way for the different present day methods. Versatile computing includes portable correspondence, versatile equipment, and portable programming.

Cloud computing is drawing in enthusiasm through the potential for ease, boundless versatility, and flexibility of expense with burden. The vast assortment of contributions are generally classified as Infrastructure as a Service (IaaS), Platform as a Service (PaaS), and Software as a Service (SaaS). IaaS is exemplified by Amazon Web Services and gives the capability to perform presented projects on a virtual mechanism that is basically the same as a standard box with a standard working framework. The customer has control over the virtual assets. Every PaaS framework offers a dissimilar set of functionalities, as an API permits projects to be composed uniquely to execute in cloud; Google App Engine is a case of this method.

In PaaS frameworks, a persevering and versatile stockpiling stage is a pivotal office. In an IaaS condition, one could.

Basically, introduce a current database motor. For example, MySQL in one virtual machine occurrence, yet the constraints (execution, scale, and adaptation to non-critical failure) of this methodology are well-known, and the customary database frameworks can turn into a restricted access in a cloud phase; consequently, novel stockpiling stages are usually provided inside infrastructure as a service mists as well. These capacity stages work inside the cloud stage, and exploit the scale-out from enormous quantities of shoddy machines; they additionally inside have instruments to endure the shortcomings that are inexorable with such a large number of temperamental machines. Samples incorporate Amazon SimpleDB1, Microsoft Azure Table Storage2, Google App Engine datastore3, and Cassandra4. A term regularly connected to these capacity stages is Not Only SQL (NoSQL). NoSQL record framework is intended to achieve high throughput and high accessibility using surrendering of a few functionalities to conventional database frameworks offer, for example, joins and ACID exchanges. NoSQL information saves might provide weaker consistency property, intended for instance possible consistency. A customer of such a save may watch value that is stale, not from latest compose. This outline gimmick is clarified by CAP hypothesis, which expresses that a parcel understanding appropriated framework can promise the accompanying of two properties: information consistency or else accessibility. Huge numbers of NoSQL database frameworks go for accessibility and parcel resistance as their essential center and accordingly they unwind the information consistency limitations.

Prepared access to a lot of registering force has been a tenacious objective of PC researchers for a considerable length of time. Since 1960s, registered utilities as enveloping as straightforward the phone had ambitious clients and frame work architects. This was perceived in the 1970s that such power could attain to cheaply with accumulations for little gadgets instead of costly single supercomputers. Enthusiasm for plans for overseeing dispersed processors [18, 21, 68] got to be popular to the point stills once a simple contention over importance of statement \distributed.

Previous work made it clear that appropriated processing could attainable; scientist's starts to pay the heed with the aim of dispersed processing could be hard. At the point when messages may be lost, undermined, or postponed, strong calculations must be utilized as a part of request to manufacture an intelligible (not controllable) framework. Such lessons were not lost on these framework architects of the early 1980s. Generation frameworks, for example, Locus and Grapevine grappled with the basic pressure stuck between consistency, accessibility, and execution in appropriated frameworks.

The Condor framework soon turned into a clip of the generation processing environment at College of Wisconsin, somewhat on account of the sympathy toward ensuring entity investments. A generation setting can be both a condemnation and a gift: The Condor task adapted hard teaching as it increased genuine clients. It was easily founded and the burdened mechanism managers would rapidly withdraw from the group and the longstanding prompted a Condor saying: release the holder in direct control, paying little heed to the expense. Axed pattern for speaking to clients and machines was in consistent change thus in the end prompted the improvement of a construction free asset allotment dialect called Classed [58–60]. It has been watched that generally, perplexing frameworks battle through a pre-adulthood of five to seven years and Condor was no exemption occured.

2 Related Works

2.1 Data Consistency Properties in Amazon Simple DB and Amazon S3

The novel services provided by the business cloud storage system has attracted many of the significance nowadays for their capacity to give scalability, availability, and durability at a low price. Another side of these systems generally depends upon their weak consistency properties. Based on the theorem at the rear dispersed systems is the CAP theorem. Considering this work, we analyzed two of these services: Amazon Simple database and Amazon S3 from the consumer side in order to recognize which sort of consistencies they actually provide and if they value what is promised. As a beginning point, we pay attention on the investigation completed by Hiroshi Wada, Alan Fekete, Liang Zhao, Kevin Lee and Anna Liu, described in Mell and Grance [1], trying to replicate their results.

2.2 *Dynamo: Amazon's Highly Available Key-Value Store*

Consistency at a huge scale is one of the major challenges we faced at Amazon.com, one of the main e-commerce operations in the world; still the smallest amount outage has important monetary penalty and their impact client trust. The Amazon.com platform, which gives services for many web sites world wide, is implementing on top of an infrastructure of thousands of servers and network apparatus situated in many datacenters in the region of the world. At this scale, small and large apparatus fail always and the way persistent state is manage in the face of these failure drive the consistency and scalability of the software systems.

2.3 *Data Consistency Properties and the Trade Offs in Commercial Cloud Storages*

A novel class of data storage, called Not Only SQL, has risen to balance the traditional database storage system, with ejection of common ACID interchange as one of its general feature. Dissimilar platforms and indeed dissimilar primitives with in one NoSQL platform can offer similar consistency properties from Eventual Consistency to single-entity ACID. The platform provider, weaker consistency should allow the better capability, lower latency and other advantages. This paper investigates the customers monitor the consistency and performance properties of various contributions. We discover that various platforms seem to carry out to provide more consistency than they would promise; we also discover the similar cases where the platform provide a consumers a choice stuck between stronger and weaker consistency, however, there is no experimental benefit from accepts the weaker consistency property.

2.4 *Quality-of-Data for Consistency Levels in Geo-Replicated Cloud Data Stores*

Providing the individuals and companies with access to use remote computing and storage infrastructures. In order to achieve highly available yet high-performing services, cloud data stores rely on data replication. However, providing replication brings with it the issue of consistency. Given that data are replicated in multiple geographically distributed data centers, and to meet up rasing needs of dispersed application, many cloud data stores adopt eventual consistency and therefore allow running data intensive operations under low latency. This comes at the cost of data staleness. In this paper, we prioritize data replication based on a set of flexible data semantics that can best suit all types of BigData applications, avoiding overloading both network and systems during large periods of disconnection or partitions in the network.

2.5 *Distributed Computing in Practice: The Condor Experience*

Prepared access to huge quantity of compute power has been the determined goal of the computer scientists for the last decades. Vision of computing utilities as enveloping and an easily as the telephone have driven users and system designers. It was documented in such power could be attained reasonably with no of collections of a small devices pretty than classy single super-computers.

2.6 *Client-Centric Benchmarking of Eventual Consistency for Cloud Storage Systems*

Eventually consistent key-value storage systems sacrifice the ACID semantics of conventional databases to achieve superior latency and availability. However, this means that client applications, and hence end-users, can be exposed to stale data. The degree of staleness observed depends on various tuning knobs set by application developers (customers of key-value stores) and system administrators (providers of key-value stores). Both parties must be cognizant of how these tuning knobs affect the consistency observed by client applications in the interest of both providing the best end-user experience and maximizing revenues for storage providers. Quantifying consistency in a meaningful way is a critical step toward both understanding what clients actually observe, and supporting consistency-aware service level agreements (SLAs) in next generation storage systems.

2.7 *Improving Audit Judgment and Decision-Making with Dual Systems Cognitive Model*

The rest of this paper is prepared as follows. We first give details how the heuristics of ease of use, representativeness and anchoring-and-compensating could have non satisfied impact on auditors' judgment and decisions, follow by planning of recency and strength effects. And then bring in the property of the two systems of reasoning put forward by cognitive scientists referred to as the System 1 and System 2 here on in this article. Next to these conditions, we proposed and illustrated the examples of and with no trouble implement dual systems cognitive model. We end our paper with a few directions for future research in a new border of behavioral audit investigation.

2.8 A Loop Correlation Technique to Improve Performance Auditing

Performance auditing is an online optimization plan that empirically calculates the effective of an optimization on an exacting region. It has the possible to very much increase the performance and prevents the degradations due to compiler optimizations. Presentation auditing based on the capability to get the sufficient lots of timings of the code region to make statistically valid conclusion. This work extend the previous state-of-the-art of auditing system using allowing a finer level of granularity for get the timing and thus raising the overall efficiency of the performance of the auditing system. The error is solved by our method is the illustration of the common problem of correlate a program high top performance with its binary commands and thus the uses ahead of the performance of the auditing system. We presented our implementation and evaluation of our technique in a construction Java VM.

2.8.1 Client Interface Design

To interface with server client must give their username and secret word then no one but they can ready to unite the server. On the off chance that the client as of now exits straightforwardly can login into the server else client must enlist their subtle elements, for example, username, secret word and Email id, into the server. Server will make the record for the whole client to keep up transfer and download rate. Name will be set as client id. Logging in is normally used to enter a particular page.

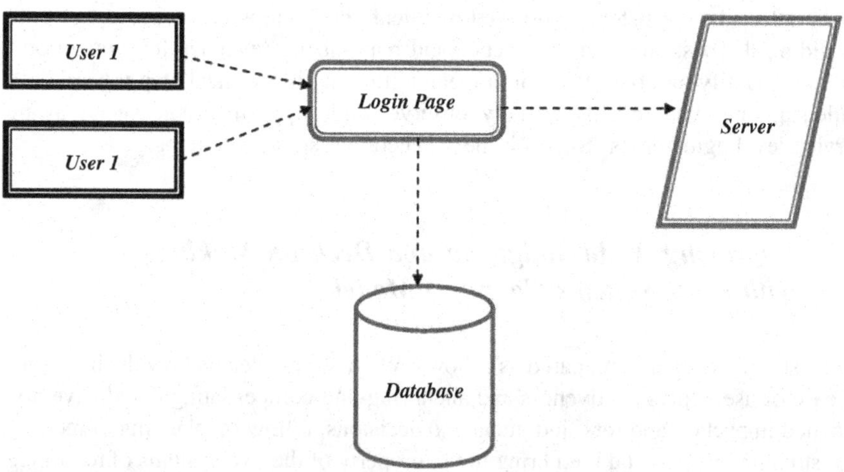

2.8.2 Perused and Write Operations

Every operation is compose W (K, an) or read R (K, a), where W (K, a) methods composed the information quality recognized using key K and R (K, a) methods perusing information distinguished by key K and whose worth is as determined. As we called W (K, an) as R (K, a's) managing compose, and R (K, an) as W (K, a's) directed perused. We expected that the prediction of every new compose. That is accomplished by letting a client append ID and current vectors to predict the composed. Thus, we had the accompanying properties: (1) a read must have a remarkable directing compose. A composes might have zero or else more straight peruses. (2) From the prediction of read, we can identify the legitimate and physical vectors of the managing compose.

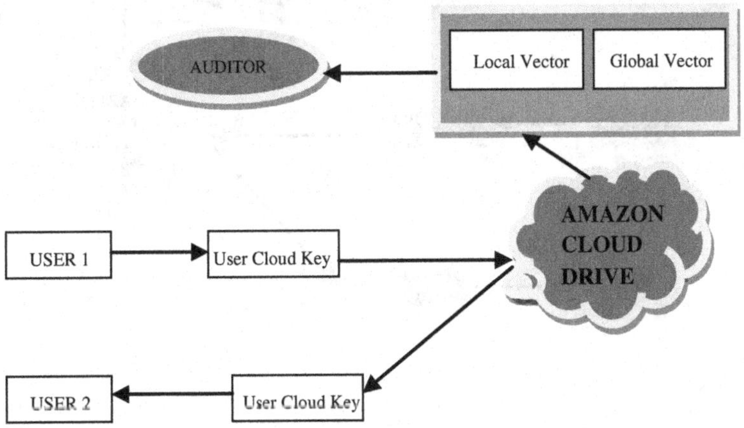

2.8.3 Creating UOT Table

Every client keeps up a UOT for data nearby operation. Every data in UOT is portrayed using three components such as operation, sensible vector and physical vector. During issues in an operation, a client collects the operation records, and also his Current coherent and physical vector, in his UOT. Every process is either a compose W (K, an) or read R (K, a) where W (K, a) methods composing a worth of information. It is distinguished by a key K, and Read (K, a) methods perusing information is recognized by key K and which quality is a determined. As, we called W (K, an) as Read (K, a's) managing compose, and R (K, an) as W (K, a's) directed perused. The estimation of each compose is exceptional. That is accomplished using let a client join his own ID & present vectors to estimate the compose. In this manner, we have the accompanying properties: (1) A read must have an interesting directing compose. A compose might have zero or more straight peruses. (2) and from the read estimation, we can identify the sensible and physical vectors of its compose.

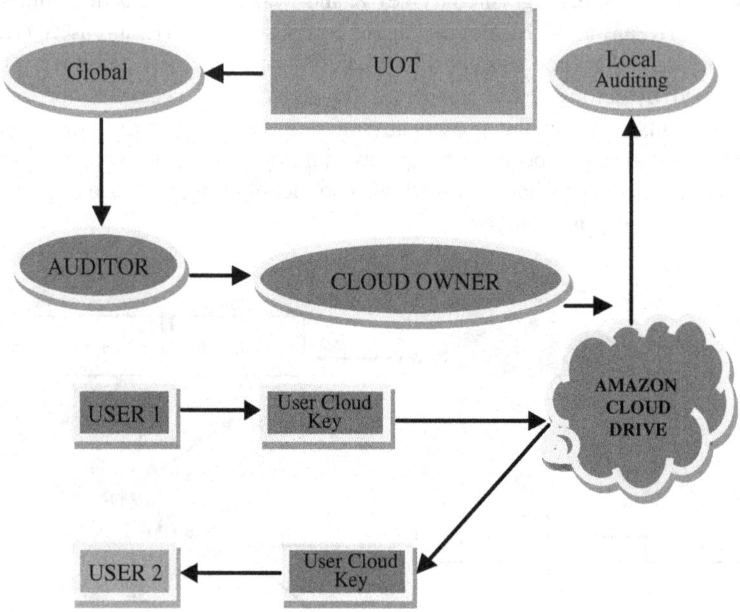

2.8.4 Nearby and Global Auditing

Nearby Auditing: Nearby consistency evaluating is an online calculation every client will record the majority of the user operation in his UOT. During issuing the read operation, the client performed nearby consistency evaluating autonomously. Let R (a) signify the client's present perused who manages the compose is W (a), W (b) denotes the last write in UOT & R (c) mean the last read in UOT who directs the compose in W (c). Perused your compose consistency is damaged if W (a) occurs before W (b), and monotonic read consistency is dis-regarded if W (a) happens before W (c). Note down from the estimation of the read and we can know the legitimate and physical vector of its directing compose. Along these lines, we can request the directing composes by their intelligent vectors.

Worldwide Auditing Worldwide consistency examining is a disconnected from net calculation. Occasionally, a reviewer be chosen from the cloud review to perform the worldwide consistency evaluating. In this condition, every different client randomly sends their UOTs to the concern inspector for acquiring a worldwide hint of operation. In the wake of executing worldwide examining, the inspector will send evaluating results and additionally its vectors to all different clients. Given the inspector's vectors, every client will know other clients' most recent clocks up to worldwide evaluating.

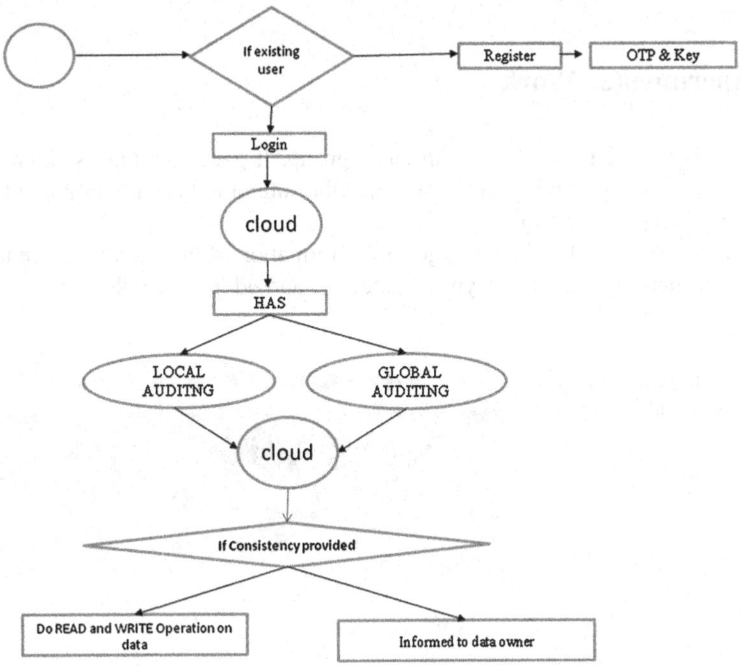

In our proposed system we have been proposed a HAS technique for auditing. Now-a-days cloud is essential for large-scale distributed system where the every piece of data is rde-duplicated on multiple numbers of disseminated servers to attain high capability and high performance. First up all we review the (CaaS) models in scattered system. Since a standard proposed a two stage of consistency models: data-centric and client-centric consistency. Data-centric consistency model finds out the inner state of a storage space system, i.e., how the updates are flowing through the system and what guarantee that the system can offer with value to their updates. However, the customer it actually doesn't matter whether storage or not a storage system internally contain any stale copies. As long as no stale data is received from the client point of view, the customer is satisfied.

2.8.5 Execution Evaluation

We abridge the parameters utilized as a part of the manufactured infringement follows in the Table II. In irregular procedure, the haphazardly pick [1, l] inspecting peruses on every interim, where l is the interim length. To acquire the manufactured infringement follows, physical time is partitioned into 2000 time cuts. We accept that once an information cloud starts the damaged guaranteed consistency, this infringement should proceed only short time cuts, instead of closure instantly. In the reenactment, the span of every infringement d is situated to 3–10 time cuts.

3 Experimental Work

This form is used for user to sign in and sign up. It generates the username and password to submit this form. When we click the submit button, it will move to the sign in page (Figs. 1 and 2).

This page contains details of the general information of the user; we have to list out the card number, amount payment gateway, and address details and register.

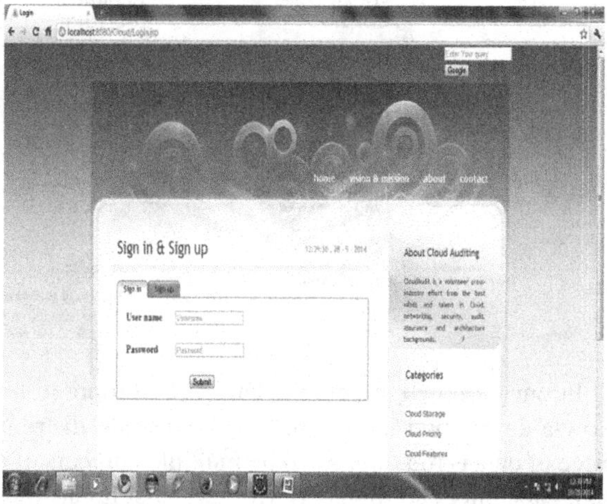

Fig. 1 This form is used for user to sign in and sign up

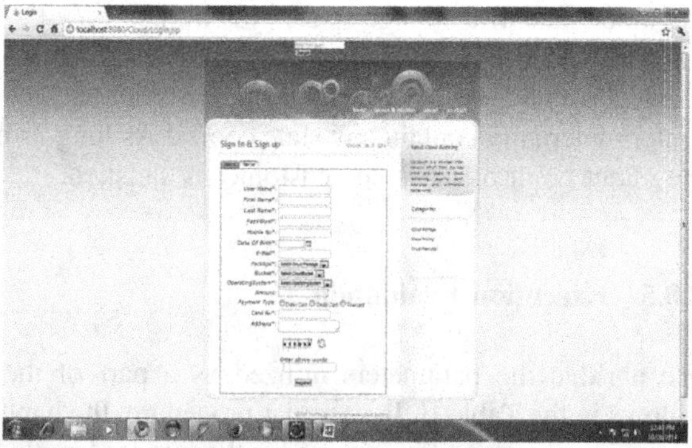

Fig. 2 Amount and payment gateway and address details and Register

In Fig. 3, we can see that the file upload button where we give the key of the cloud is given then the file is uploaded (Fig. 4).

In this figure, we can see public and private file were the public file can be seen by everyone and private file can be seen by only the cloud owner (Fig. 5).

This is a view of data base in the form itself (Fig. 6).

The figure shows the algorithm performance.

Fig. 3 The file upload button

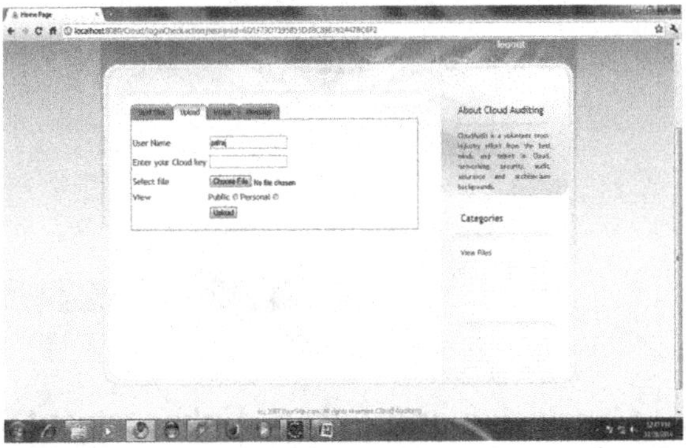

Fig. 4 Public and private file

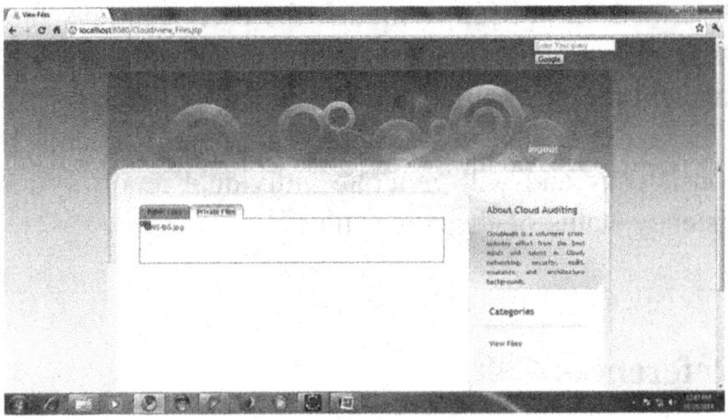

Fig. 5 View of database in the form itself

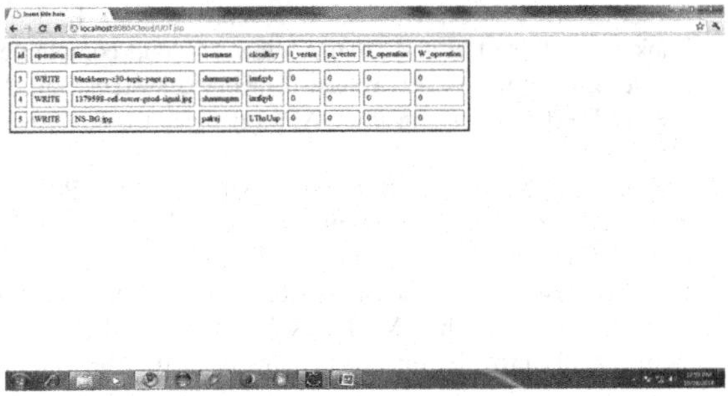

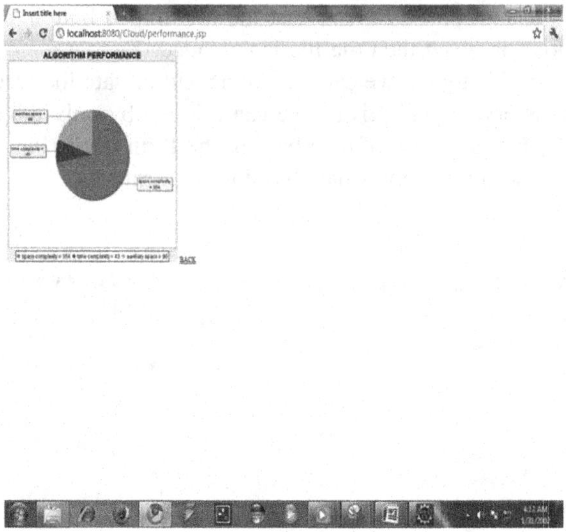

Fig. 6 The algorithm performance

4 Conclusion and Future Enhancement

In future work the detailed design of each component in our framework will be described. In our future model, the users can levy the worth of cloud services and decide a correct CSP amongst different candidates, e.g., the slightest expensive one that still provide sufficient consistency for the users' applications. We will carry out a careful theoretical study of consistency models in cloud and achieve strong consistency and will generate individual report system to user their identify consistency status of their own files.

References

1. Mell P, Grance T (2011) The NIST definition of cloud computing (draft). NIST Special Publication 800–145 (Draft)
2. Armbrust M, Fox A, Griffith R, Joseph A, Katz R, Konwinski A, Lee G, Patterson D, Rabkin A, Stoica I et al (2010) A view of cloud computing. Commun ACM 53(4):50–58
3. Brewer E (2000) Towards robust distributed systems. In: Proceedings of 2000 ACM PODC
4. Brewer E (2012) Pushing the CAP: strategies for consistency and availability. Computer 45 (2):23–29
5. Ahamad M, Neiger G, Burns J, Kohli P, Hutto P (1995) Causal memory: definitions, implementation, and programming. Distrib Comput 9(1):37–49
6. Lloyd W, Freedman M, Kaminsky M, Andersen D (2011) Don't settle for eventual: scalable causal consistency for wide-area storage with COPS. In: Proceedings of 2011 ACM SOSP
7. Anderson E, Li X, Shah M, Tucek J, Wylie J (2010) What consistency does your key-value store actually provide. In: Proceedings of 2010 USENIX HotDep

8. Fidge C (1988) Timestamps in message-passing systems that preserve the partial ordering. In: Proceedings of 1988 ACSC
9. Golab W, Li X, Shah M (2011) Analyzing consistency properties for fun and profit. In: Proceedings of 2011 ACM PODC
10. Tanenbaum A, Van Steen M (2002) Distributed systems: principles and paradigms. Prentice Hall PTR, New Jersey

An Enhanced Tourism Recommendation System with Relevancy Feedback Mechanism and Ontological Specifications

C. Devasanthiya, S. Vigneshwari and J. Vivek

Abstract Data mining is an analytic process used to access data in search of consistent patterns from the database and it is used to get relevant results. This paper describes a recommender system that helps travel agents in recommending tourism options to the customers, especially those who do not know where to go and what to do. This process describes textual messages exchanged between a travel agent and a user through a chat box. Text mining technique analyzes an interesting area in the messages. Then, the system seeks a database and accesses tourist options like attractions and cities. The system provides travel package recommendations to the customers for their choice. Travel agent created travel packages using test Approach. Here, the classification of text queries is done using Rocchio classification algorithm. The final results yielded using ontological specifications (OS) are compared with that without ontological specifications (WOS). OS-based systems' relevancy is comparatively good.

Keywords Data mining · Collaborative filtering · Ontology · Classifications · OS · WOS

C. Devasanthiya (✉) · S. Vigneshwari · J. Vivek
Department of Computer Science and Engineering, Faculty of Computing, Sathyabama University, Chennai, India
e-mail: devasanthiya89@gmail.com

S. Vigneshwari
e-mail: jayam3@rediffmail.com

J. Vivek
e-mail: vivekspy.vivek@gmail.com

L.P. Suresh and B.K. Panigrahi (eds.), *Proceedings of the International Conference on Soft Computing Systems*, Advances in Intelligent Systems and Computing 398, DOI 10.1007/978-81-322-2674-1_28

1 Introduction

Recommender systems are like information filtering system that seeks to decide the preferences that user would give to an item. Tourism is an interesting domain for recommendation research. This process analyzes that a tourism recommendation may indicate cities to go to (destinations, places to visit, number of persons, cost). This system uses the ontology technique to get the accurate data from the database. In this technique, the data are stored in hierarchical structure. Travel recommendation system helps the customer to select the places with the help of the decision support system.

This recommendation system helps the customer to have a chat option directly with the travel agent. Through the chat messages, the travel agent will discover the places the customer aims to visit. Using this chat option, users receive the optimized recommendation from the system. Advantages of this system are that it consumes less time, increased QOS, and provides the optimal solution.

2 Related Work

Poon [1] has specified the tourism structure and data mining methods used in websites. The actual unsolved problem in this paper is that the result out of the data mining process is not an optimized solution for the end user. Sowa [2], in his paper, described that ontology paves the way for the intelligent method in agent's system technology. The ontology may be used in various formats. Applications of ontology are applied in knowledge management systems and recommendation system. Loh et al. [3] represented an approach in discovering texts extracted from the web. In this process, the system uses the text obtained from the customer and uses them as the keywords for mining the data from the database. Using the intelligent system, the travel agent will identify the areas the customer needs to visit.

Liu et al. [4] analyzed the characteristics of the travel packages and developed a tourist-area-season model (TAST) that can be used to extract the details of the tourist and intrinsic features of the places like location, seasons, etc. Blei et al. [5] described the short text classification, which is analyzed classifying short text and systematic summaries of the existing related methods to short text classification using analytical measures. Burke et al. [6] described a recommender system which combines more recommendation techniques together to produce its optimal output. There are a variety of techniques that are problematic especially in hybridization.

Carolis [7] illustrated a recommender system based on mobility especially in the tourism domain. Comparative study is also done on the relevancy based on user's interests. Mary and Jyothi [8] discussed about the feature clustering techniques and applied it on the XML benchmark datasets.

Jyothi et al. [8] analyzed various social media using ontology technique. Here, user feels it difficult to extract relevant data. Bit ontology is used to pick exact data and it works as an effective monitoring tool.

Vigneshwari et al. [9] proposed an ontology merging task to get exact web information from the user profiles and it is based on the mutual information; among the concepts, taxonomy is constructed, and then the relationship among the concept is extracted.

Vigneswari and Aramudhan [10] suggested merging two ontology techniques, which explains polysemy and synonym matching concept using ontology process with a variety of frameworks. Prakash and Rangaswamy [11] explained the concept-based analysis on sentence, document over a web. It is an extracted optimum value with the help of keywords. Vigneshwari [12] drew the mapping model of multidimensional ontology and it was compared with the text-based approaches in HTML documents.

3 Proposed Work

3.1 Problem Description

This paper has given a solution to make the data mining results more feasible and even more accessible. Data mining and text mining play a vital role in analyzing the recommended data. The result after the data mining process is given more as optimal output, so that the user can do decision making on his own with the help of travel agent using chatting module.

3.2 System Architecture

This system provides a feasible solution to the customers. The traditional tourism recommendation system takes more time to create travel package recommendations to the user. However, this system provides related recommendations which are searched by the user. Figure 1 describes the system architecture.

In text mining module, the system identifies the subjects in the messages. Travel agent gets the details from the customer through the text format and those texts are compared in the databases. The themes identified asset as a preference to the customer and they are forwarded to the recommendation module. Here, text mining used Rocchio algorithm (classification) to decrease time consumption.

3.2.1 Classification Algorithm

```
Procedure
Begin
Assign
        x ←a query weight,
        y ←Relevant Document weight.
        z ←Non Relevant Document Weight
        N ←Total Relevant Documetns
        M ←Total irrelevant documetns
        QV ← Query Vector
        RDV={Relvant Document Set}
        NRDV={Non Relevant Document Set}
        A=Sum(RDV)
        B:=Sum(NRDV)
         For Each N
                  For Each M
                      EQV=x. QV +(1/N(A))-(1/M(B))
                 Assign QV ←EQV
         End
       End
```

Recommendation module plays a major role in giving options to the customer. The process of this module automatically starts when it receives the themes from the text module. Henceforth, it searches the databases with the identified themes and recommends the places to the customer.

Chat module is used to communicate between a customer and the travel agent. The customer gets into the system searching for travel information, but he does not know yet where he wants to go in a particular city. Here, collaboration filtering algorithm is used to retrieve specific recommendations among that city from a large

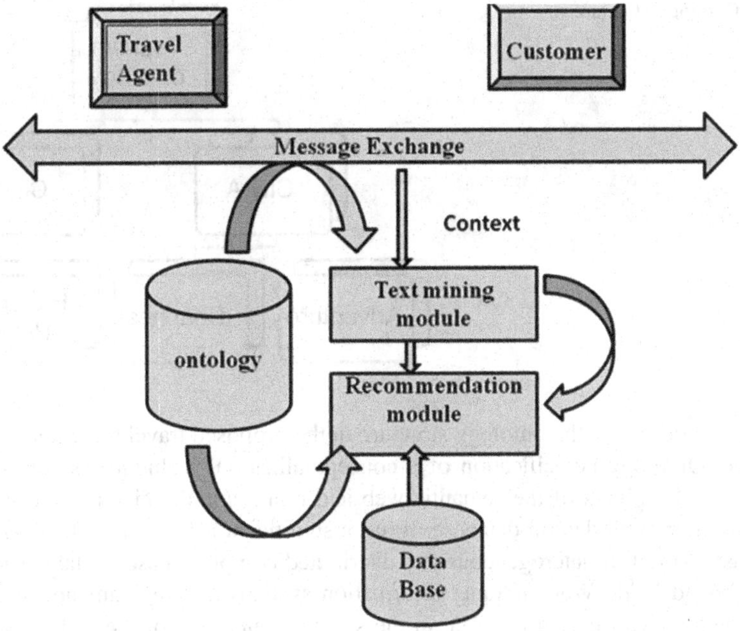

Fig. 1 System architecture

amount of recommendations in database. Then the chat module presents the text mining and tourism ontology techniques. A user can view the list of all user names but is unable to interrupt another conversation; this system provides private chat option between the customers. It is highly secured, e.g., if "X" user wants to know about travel package details personally from "Y" user, then "X" user wants to send a request to the "Y" user. He/she wants to accept that request, so that they can chat with each other in privacy. Figure 2 describes the recommendation module which is part of the chat module.

Customer attraction database is to make recommendations; the system needs a database of options. These options include cities and their tourist attractions. Items in the database are not characterized by attributes. Only a summary of the option, with a few textual lines, is stored.

RECOMMENDATION

AREA	PLACE
Covelong Beach	chennai-beach
Elliot's Beach	chennai-beach
Marina Beach	chennai-beach

Fig. 2 Recommendation module

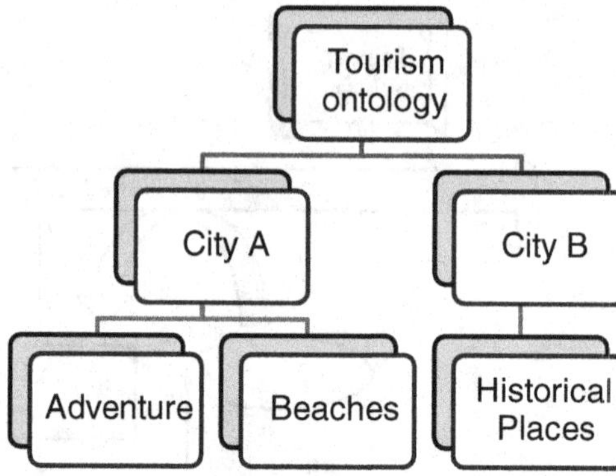

Fig. 3 Ontology structure in database

Figure 3 describes the ontology structure in the proposed travel recommendation database. Ontology (specification of a conceptualization) technique is considered like one of the pillars of the semantic web and represents the hierarchical presentation of the related data in database. It has a significant role in the fields handling with vast amount of heterogeneous and distributed computer-based details, such as www (World Wide Web) intranet information systems and recommendation system. Ontology will play a key role in the second generation of the web, which is

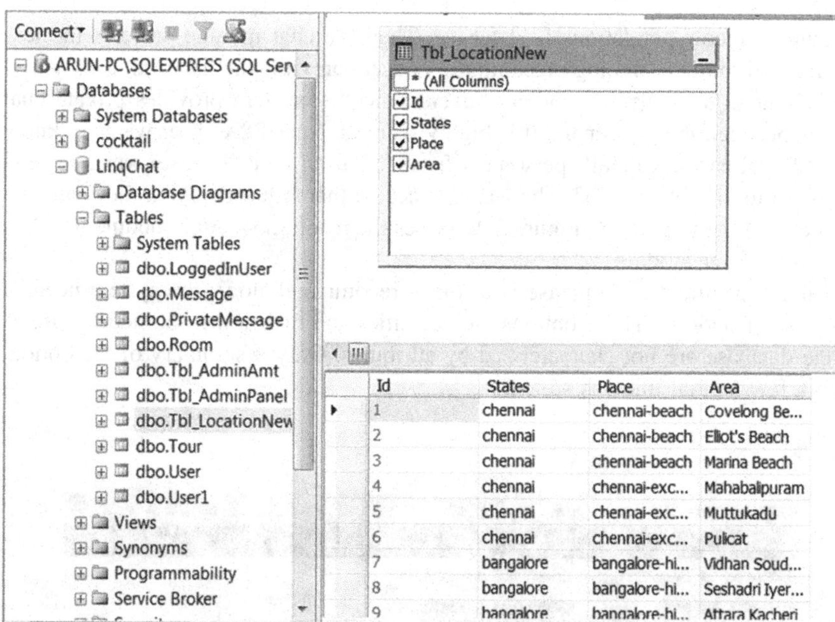

Fig. 4 Tourist places and attractions stored into database by ontology method

called as the "Semantic Web." Here, attractions and places are stored in the database based on ontology method. In this case, user should know the vocabulary, and the exact spelling of places to enter on chat messenger. Then, an exact word and their related words can be retrieved from the database. Figure 4 spots the tourist places and attractions enhanced with ontological specifications.

4 Results and Discussions

This recommender system provides optimal data to the customer and increases time consumption. Traditional systems take more time to create recommendations as customers' expectation and they do not know about user's own interest of the places. However, this system analyzes the recommendations as per user's expectations and is user friendly. Ease to get attractions and places from expected cities by a single chatting message without any distraction. For example, a user wants to go to some historical places in Chennai, but he does not how to ask a travel agent. In this case, when user types Chennai alone, text mining process analyzes the word Chennai and retrieves the related recommendations from the database in ontology manner to user view. Chennai place will retrieve all recommendations from the database and user wants to select specific attraction (historical, beaches, etc.) and it will contain the list of places belonging to the attractions for user selection.

Figure 5 represents the package declaration. Usually, this is maintained by the system administrator. Figure 6 represents the travel package database to be used in the recommendation system.

Fig. 5 Travel package declaration

Start_Palace	End_Place	Amount
chennai	chennai-beache	3000
chennai	chennai-excursion	2000
Bangalore	bangalore-historical	1000
Bangalore	bangalore-religiousplaces	1500
kerala	kerala-beaches	2500
kerala	kerala-hill stations	3100
chennai	bangalore-historical	1000
chennai	bangalore-religiousplaces	1500
chennai	kerala-beaches	800
chennai	kerala-hill stations	1200
bangalore	chennai-beache	1600
bangalore	chennai-excursion	2200
bangalore	kerala-beaches	3000
bangalore	kerala-hill stations	1400

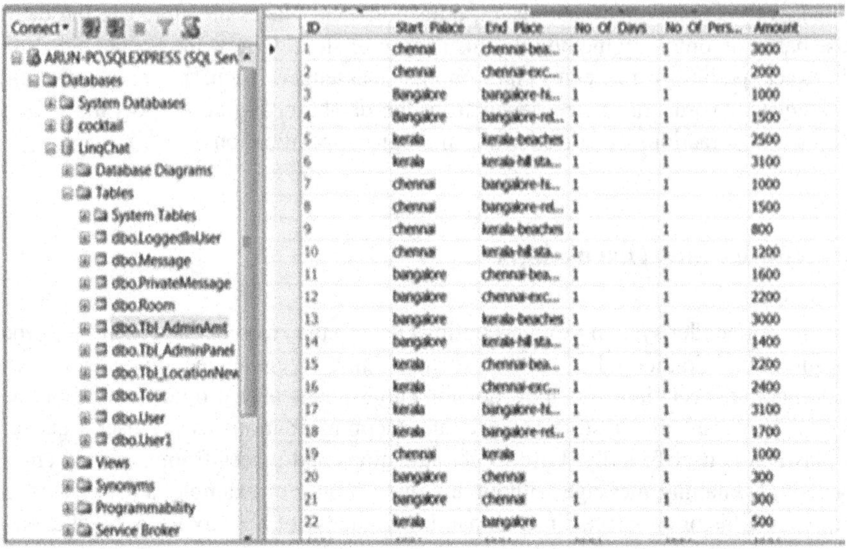

Fig. 6 Travel package storage in database

The recommendation system enhanced with ontological specification (OS) yields more relevant documents when compared to the system without ontological specifications (WOS). Totally, 50 places from the database were considered for recommendation. Of this, more relevant data is retrieved using OS compared to WOS (Table 1).

Table 1 Comparison between OS- and WOS-based systems

Source	Destination	Relevancy using OS: relevant/total	Relevancy using WOS: relevant/total
Chennai	Chennai Beach	0.83	0.52
Chennai	Kerala Beach	1.0	0.54
Chennai	Kerala	0.92	0.51
Bangalore	Chennai	0.89	0.43
Kerala	Chennai	0.90	0.62
Chennai	Kerala Hill station	0.89	0.44
Chennai	Kochi	0.97	0.66
Kerala	Kochi	1.0	0.33
Bangalore	Kerala Beach	0.98	0.56

5 Conclusion

In this proposed work, the customer will be satisfied using this chat for decision making. The architecture section has covered the newly proposed model for collecting data from database and does data mining process based on the customer's need. The Rocchio algorithm has classified the recommended places for customers. It provides feasible recommendation and increasing time consumption to the customers. This system is user friendly to registered user and is a secured process.

References

1. Poon A (1993) Tourism, technology and competitive strategies. 370 p
2. Sowa JF (2014) Building, sharing, and merging ontologies. http://www.jfsowa.com/ontology/ontoshar.htm
3. Loh S et al (2000) Concept-based knowledge discovery in texts extracted from the Web. ACM SIGKDD Explor 2(1):29–39
4. Liu Q, Ge Y, Li Z, Xiong H, Chen E(2011) Personalized travel package recommendation. In: Proceedings of the IEEE 11th international conference data mining (ICDM '11), pp 407–416
5. Blei DM, Andrew YN, Michael IJ (2003) Latent dirichlet allocation. J Mach Learn Res 3:993–1022
6. Burke R (2007) Hybrid web recommender systems. Adapt Web 4321:377–408
7. Carolis BD, Novielli N, Plantamura VL, Gentile E (2009) Generating comparative descriptions of places of interest in the tourism domain. In: Proceedings of the third ACM conference recommender systems (RecSys '09), pp 277–280
8. Mart Posonia A, Jyothi VL (2014) Context-based classification of XML documents in feature clustering. Indian J Sci Technol 7(9):1355–1358
9. Vigneshwari S, Aramudhan M (2014) Web information extraction on multiple ontologies based on concept relationships upon training the user profiles. In: Proceedings of the artificial intelligence and evolutionary algorithms in engineering systems (ICAEES 2014), Vol 2, pp 1–8
10. Vigneshwari S, Aramudhan M (2015) Social information retrieval based on semantic annotation and hashing upon the multiple ontologies. Indian J Sci Technol 8(2):103–107
11. Prakash KB, Rangaswamy MAD (2014) Extracting content in regional web documents with text variation. Int J Inf Sci Comput 8(1):26–31
12. Vigneshwari S, Aramudhan M (2013) Analysis of ontology alignment of different domains: a cross ontological approach. Int J Inf Sci Comput 7(1):66–70

Big Data and Analytics—A Journey Through Basic Concepts to Research Issues

Manjula Ramannavar and Nandini S. Sidnal

Abstract Big data refers to data so large, varied and generated at such an alarming rate that is too challenging for the conventional methods, tools, and technologies to handle it. Generating value out of it through analytics has started gaining paramount importance. Advanced analytics in the form of predictive and prescriptive analytics can scour through big data in real time or near real time to create valuable insights, which facilitate an organization in strategic decision making. The purpose of this paper is to review the emerging areas of big data and analytics, and is organized in two phases. The first phase covers taxonomy for classifying big data analytics (BDA), the big data value chain, and comparison of various platforms for BDA. The second phase discusses scope of research in BDA and some related work followed by a research proposal for developing a contextual model for BDA using advanced analytics.

Keywords Big data · Analytics · Big data analytics · Advanced analytics · Predictive analytics · Prescriptive analytics

1 Introduction

Digital data is ubiquitous and a vital asset for any organization. Today's world is surrounded by ever-growing digital data, which inevitably has become a way of life. Over the years, this data had been managed well using conventional tools,

M. Ramannavar (✉)
Department of CSE, KLS Gogte Institute of Technology, Visvesvaraya Technological University, Udyambag, Belagavi 590008, Karnataka, India
e-mail: manjular@git.edu; manjula.ramannavar@gmail.com

N.S. Sidnal
Department of CSE, KLE Dr. M. S. Sheshgiri College of Engineering and Technology, Visvesvaraya Technological University, Udyambag, Belagavi 590008, Karnataka, India
e-mail: sidnal.nandini@gmail.com

L.P. Suresh and B.K. Panigrahi (eds.), *Proceedings of the International Conference on Soft Computing Systems*, Advances in Intelligent Systems and Computing 398, DOI 10.1007/978-81-322-2674-1_29

Table 1 Examples of embedding advanced analytics in business applications [3]

Business application	Question	Techniques
Price optimization in retail	How should products be priced in order to maximize overall profitability?	Price elasticity models, optimization methods
Markdown optimization ir retail	How should perishable items be marked down, in order to minimize wastage?	Price elasticity models, optimization methods
Portfolio optimization in financial services	How should an investment portfolio be constructed and reviewed?	Optimization methods
Credit scoring	Which applicants/customers for credit are likely to default?	Logistic regression, discriminant analysis, decision trees, neural network, support vector machine
Supply/demand chain forecasting in retail	How to plan the movement of items through the supply chain to maximize availability and minimize inventory levels?	Forecasting models-simple linear regression, regression with smoothing, ARIMA models
Capacity planning, e.g., for sail center management	How to forecast demand and allocate existing resources?	Regression models, neural networks, optimization methods
Customer relationship management [CRM]	Which product/service is each customer likely to purchase next?	Auto-regressive models, maximum-likelihood estimation
Propensity modeling in CRM	Which customers are most likely to respond to a marketing campaign?	Decision trees, logistic regression, neural networks, support vector machine
Customer contact management	How should customers be allocated to a set of marketing campaigns, in order to achieve objectives and satisfy constraints?	Optimization methods
Customer churn/attrition management	Which existing customers are most likely to churn/attrite?	Survival analysis models, auto-regressive models, maximum-likelihood estimation
Lifetime value/duration management	What is the lifetime value of each customer? How long before each customer becomes likely to churn/attrite?	Survival analysis models, proportional hazards models
Marketing mix in retail	How to maximize the income from a product by combining product specification, distribution channel and promotional tactics?	Quadratic optimization with linear constraints
Loan offers optimization	How to optimize and validate financial loan terms?	Constrained optimization
Product life cycle planning and forecasting	How to plan a product life cycle scenario to plan investment returns and avoid risky financial decisions?	Solvers for systems of nonlinear equations, probability distribution functions

methods, and technologies. However, recently, these traditional methods have been challenged by the data deluge and the methods seem to work no longer. According to [1], per minute, Google, the de facto search engine, receives over 4 million requests; Face book users exchange 2.5 million pieces of content; 300,000 tweets are floated by Twitter; and $80,000 of revenue is generated through online sales by Amazon alone. The proliferation of web and social media, sensors, RFID tags, and machine-generated data have all contributed to the big data era that has been evolving over the last decade or two. It will continue to evolve and businesses need to look at it as an opportunity rather than a threat in order to gain an edge. This opportunity maybe realized using the techniques that advanced analytics offer. According to Intel IT [2], "what-if" scenarios and aggregation of data sources in real time can enhance business response to market changes. The authors in [3] suggest several examples of embedding advanced analytics to cater to various business applications. A range of business application scenarios along with the techniques obtainable by advanced analytics in solving diverse business questions has been tabulated in Table 1.

Table 1 illustrates the vast scope for research in creating predictive and prescriptive models to capture a multitude of real-life problems and ideate to engineer pioneering solutions.

The rest of the paper is organized as follows: Section. 2 describes big data, and compares and contrasts it with traditional data. Big data characteristics are specified and taxonomy for classification of big data analytics (BDA) is accentuated. The methodology for BDA through the big data value chain is highlighted and a comparison of various platforms for BDA is also depicted. Section 3 discusses scope of research in BDA through research methodologies and technologies used in existing works and future directions. It also proposes the problem of resume analytics as a use case for unstructured text analytics. Section 4 concludes the paper by providing final thoughts.

2 Big Data Analytics

Traditional data analytics use extract, transform, and load processes [4] that store structured data into data warehouses from enterprise applications such as CRM, ERP, and financial systems. Regular reports run against the stored data created dashboards and limited visualizations that served as analytic results. However, the advent of social networking and media, smart phones, internet transactions, networked devices, and sensors have attributed to the changing nature of data in the form of big data. Hence, there is a need for new methods, tools, and techniques to manage big data; this need has given rise to a new arena called BDA.

2.1 Big Data

Big data refers to massive datasets ranging from hundreds of terabytes to petabytes, exabytes, and beyond, comes from a variety of sources, and hence is heterogeneous (semi-structured or unstructured) [4] and is generated at increasing velocity. Big data is not small data that has bloated to the point that it can no longer fit on a spread sheet, nor is it a database that happens to be very large. There are a number of characteristics that differentiate big data from traditional data which have been tabulated in Table 2.

Thus Table 2 distinguishes big data from traditional data on the basis of the listed characteristics. It is important to comprehend the nature of big data in order to decide on the methods, tools, and technologies to control and manage it effectively.

In [5], big data is characterized by the three dimensions: Volume, Variety, and Velocity. Additional Vs such as Veracity, Value, and Variability further refine the term big data. Big data is mostly of semi-structured and unstructured nature and comes from a variety of sources such as machine-generated data, sensor data, social media data, web data, etc. Data is being generated at an ever-increasing rate and was predicted to double every year reaching 8 zettabytes [6] by 2015. Veracity is concerned with issues of uncertainty of data such as being inaccurate or of poor quality or lacking the trust factor to make decisions. Value is the most important dimension that explores big data using analytics to provide imperative perceptions to businesses thereby allowing them to make decisions for future prospects. Variability [5] deals with data flow inconsistencies for heavy loads; for instance, several concurrent events on the social media platform cause a sharp rise in the load.

2.2 Analytics

Analytics is the process of introspecting data to discover hidden patterns, meaningful relationships, and interesting associations which can be converted into actionable insights. According to [7] in 'Competing on analytics,' analytics has been quoted as "the extensive use of data, statistical and quantitative analysis, exploratory, predictive models, and fact based management to drive decisions and actions." Businesses leverage digitized data, obtained from its internal sources such as ERP data, data warehouses, and data marts by applying analytics in order to describe/summarize, predict, and improve their performance and/or effectiveness. The complexity of analytics may range from a simple exploration, for instance, determining the consumer churn rate over the last couple of months, to complex predictive models that foresee the future, for instance, providing targeted offers to consumers in order to prevent churn.

Table 2 Traditional data versus big data

Characteristic	Traditional data	Big data
Volume	Gigabytes to a few terabytes	Hundreds of terabytes, petabytes, exabytes, and beyond
Nature and location	Typically limited to an organization or an enterprise and intrinsically centralized	Distributed across multiple entities; spread throughout the electronic space such as Internet servers and cloud servers
Structure and content	Highly structured and has a single discipline or may be a sub discipline. This is often in uniform records, such as an ordered spreadsheet	Semi-structured or highly unstructured (free form text, images, motion pictures, sound recordings, etc.). The subject matter may cross multiple disciplines that are not related
Data models	Stable data models	Flat schemas
Goal	Designed to answer very specific questions or serve a particular goal	Designed with a flexible goal to answer protean questions. The goal is vague but is obvious that there really is no way to completely specify what big data resource will contain, how data types are organized, gets connected to other resources, etc.
Data preparation	Data user prepares her own data for her purposes in case of traditional data	Big data comes from many diverse sources and prepared by many people
Longevity	Traditional data projects have a well established time frame and end in finite time	Big data projects typically contain data that must be stored perpetuity. Ideally, data stored in big source will be absorbed into another resource when original resource terminates. Hence, big data projects extend into future and past, accruing data prospectively and retrospectively
Stakes	Being limited to a lab or an institution, can usually recover from occasional small data failures	Being varied in scope, can be irrevocably expensive to recover. Failure may lead to huge losses. Example: the failure of NCI cancer biomedical informatics grid caused $350 million loss for fiscal years 2004–2010
Measurements	Measured using one experimental protocol and the data can be represented using one set of standard units	Delivered in many different electronic formats. Measurements when present may be obtained by many protocols. Verifying quality of big data is a major challenge
Reproducibility	Traditional data projects can be typically repeated	Replication of big data is seldom feasible
Introspection	Individual data points (row and column) can be easily identified	Unless exceptionally organized, it is impossible to locate a data entity

(continued)

Table 2 (continued)

Characteristic	Traditional data	Big data
Analysis	Traditional project itself needs to be analyzed along with data	Big data is analyzed orderly in incremental steps. Data can be extracted, reviewed, reduced, normalized, transformed, visualized, interpreted, and reanalyzed with different methods
Tools and technologies	RDBMS, row-oriented databases, ETL tools, SQL	HDFS, Column-oriented databases, NoSQL, Hadoop, Storm, Spark

2.3 *Big Data Analytics*

BDA is the culmination of big data and analytics, the two key business enablers. BDA exploits massive datasets containing a variety of data types, i.e., big data, to discover unknown correlations, market trends, customer preferences, and other useful business information. The analytical findings can lead to more effective marketing, new revenue opportunities, better customer service, improved operational efficiency, competitive advantages over rival organizations, and other business benefits.

2.4 *Taxonomy for Big Data Analytics*

BDA may be classified along various dimensions. Table 3 presents classification of BDA based on the dimensions of time, techniques, and domain.

2.4.1 Based on Time: Batch/Historical Versus Real time/Streaming

The time over which data is generated discriminates batch analytics from streaming analytics. An enterprise houses its voluminous data over a period of time in its databases or data warehouses or data marts. Such data maybe referred to as batch/historical data, while data that is continually being generated through sensors, RFID tags, social media, etc. maybe referred to as real-time/streaming data. Apache Hadoop [4] is an example of batch analytics requiring time to create information, analyze it, and then create value. For instance, payroll and billing systems employ batch analytics. In contrast, Apache Storm and Apache Spark (for in-memory analytics) are distributed real-time computation engines that analyze data in small periods of time (near real time). Radar systems, ATMs, and customer services are some of the examples.

Table 3 Taxonomy of BDA

Dimension	BDA	Brief description
Time	Historical/batch	Analytics done on data stored over a period of time
	Streaming/real time / near real time	Continuous computations over data flowing into the system and results within deadlines
Techniques	Descriptive	Look into past to draw inferences
	Predictive	Predict future trends
	Prescriptive	Suggest actions and implications for each chosen option
Domain	Human resource analytics	Deals with HR functions such as attrition, recruitment
	Customer analytics	Customer opinions, behavior, and segmentation analysis
	Web analytics	Web structure, content, and usage analysis
	Healthcare analytics	Clinical trials, patient data analysis, and care
	Fraud analytics	Anomalous event detection and management
	Risk analytics	Threat detection and management
	Marketing analytics	Analysis of demand, products, pricing, and sales

2.4.2 Based on Techniques: [8] Descriptive Versus Predictive Versus Prescriptive

Big data techniques refer to computational analyses that lead to business outcomes. The decision of the type of technique(s) to be used is governed by the business problem being solved.

Descriptive analytics is a set of processes and technologies that summarize data to infer what is happening now or what has happened in the past. Standard reporting, dashboards, ad hoc reporting, querying, and drilling down are areas of descriptive analytics. While descriptive analytics look into the past, advanced analytics in the form of predictive and prescriptive analytics provide a forward looking perspective.

Predictive analytics maybe applied in real time (ex-real-time fraud detection) or in batch (ex-predict churn). Predictive modelings, root cause analysis, Monte Carlo simulations, and data mining are some of the categories of predictive analytics. Predictive analytics [9] help businesses to optimize their processes, reduce operational costs, engage deeper with customers, identify trends and emerging markets, anticipate risks, and mitigate them. Author of [10] discusses seven strategic objectives that organizations can achieve by adopting predictive analytics, namely compete, grow, enforce, improve, satisfy, learn, and act.

(a) Compete—Predictive models generated by an organization's data provide competitive edge by offering insights for qualitative differentiation of products.

(b) Grow—Predictive models drive marketing and sales decisions and operations thereby aiding an organization to grow.
(c) Enforce—Predictive analytics maybe extended to information security for detecting intrusions and managing fraud, thus enforcing organizations to maintain business integrity.
(d) Improve—Predictive scores help an organization to realize its core competencies, enhancing which leads to further improvement in the firm's market position.
(e) Satisfy—Targeted marketing, improved business core capacity, improved transaction integrity, and cheaper prices are examples of predictive model outcomes to satisfy the ever-demanding consumers.
(f) Learn—Standard business intelligence and reporting techniques summarize the past, whereas predictive analytics learn from experience to optimize the prediction goal.
(g) Act—Predictive analytics generate conclusive action imperatives guiding the organization to attain its business objectives.

Prescriptive analytics not only enables to look into the future but suggests actions to benefit from the predictions and showing the decision maker the implications of each decision option. It also presents the best course of action to be taken in a timely manner. The author in [11] lists the following as FIVE pillars of prescriptive analytics:

(a) Hybrid Data—essentially 80 % of data would be unstructured.
(b) Integrated Predictions and Prescriptions—both analytics, predictive and prescriptive, have to work synergistically to deliver optimum results
(c) Prescriptions and Side Effects—Operations Research (OR)—the science of data-driven decision making would be needed to generate the prescription without side effects. Optimization and simulation technologies of OR can be used to generate effective prescriptions.
(d) Adaptive Algorithms—in a dynamic environment, algorithms have to be adaptive in order to re-predict and re-prescribe.
(e) Feedback Mechanism—it is important to know whether the prescription was followed or ignored for upcoming predictions and prescriptions.

2.4.3 Based on the Domain

BDA may be applied to a variety of domains [7], such as marketing analytics, customer analytics, web analytics, healthcare analytics, fraud analytics, risk analytics, and human resource (HR) analytics.

Marketing analytics measures and analyzes market performance to optimize return on investment (ROI). Marketing mix modeling and optimization, price, and promotional analyses are instances of marketing analytics.

Customer analytics deals with customer behavior analysis and demographics to better understand customer preferences for customer segmentation, attrition, and lifetime value analyses.

Web analytics, off-site or on-site, monitors and analyzes internet traffic data to comprehend and optimize web usage. It includes web structure, content, and usage analytics as its sub-types.

Healthcare analytics deals with the wellness of people and includes clinical research analytics, ambient assisted living, disease detection and management, customized patient care, etc.

Fraud analytics analyzes several billions of transactions to identify fraudulent ones, those that are anomalous compared to the legitimate ones. Fraud analytics may be applicable to retail, banking, insurance, government agencies, law enforcement industries, and more.

The goal of risk analytics is to understand, quantify, and manage risks associated with an entity. Acquisition modeling, behavioral scoring, and Basel II analytics are some popular risk analyses.

Human resource analytics studies HR data to deal with attrition, training, appraisals, recruitment, and other related functions.

A proper understanding of the taxonomy of BDA is crucial for an organization to undertake a big data project that would aid in effective planning, implementation, and maintenance processes.

2.5 Methodology for BDA

The methodology for BDA is expressed through the big data analysis pipeline [12] or the big data value chain depicted in Fig. 1. It consists of multiple phases shown on top half of the figure.

Acquisition/Recording—big data generated from multiple sources is acquired and recorded into a repository for further phases.

Information Extraction and Cleaning—deals with extracting relevant information from the repository and expressing it in a form suitable for analysis.

Data Integration, Aggregation, and Representation—automated processes that deal with differences in data structure and semantics and computer-understandable representations.

Query Processing, Data modeling, and Analysis—deals with creation of models from data, exploring data to draw inferences, and processing queries.

Interpretation—deals with presentation and visualization of inferences drawn in a comprehensible manner.

The big data value chain thus serves as a blueprint, illustrating the various phases to be followed by an organization in fulfilling its big data objectives. The big data

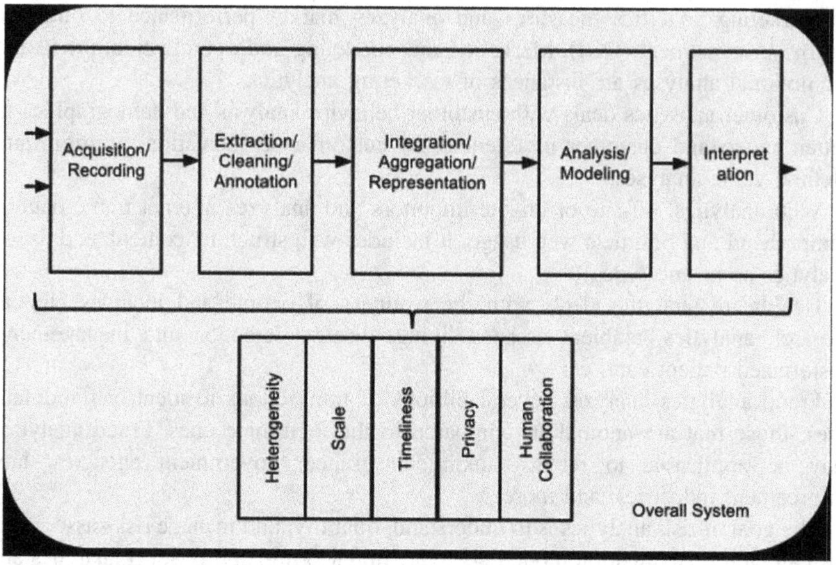

Fig. 1 Big data value chain [12]

value chain is demonstrated through the research proposal for the resume analytics use case, later.

2.6 Comparison of Different Platforms for BDA

The authors in [13] have compared different platforms for BDA along with their communication mechanisms based on various characteristics. Their results are shown in Table 4 where 5 stars denote best possible rating and 1 star denotes lowest possible rating.

Horizontal scaling, also known as "scale out," involves workload distribution across multiple servers or commodity machines. Peer-to-peer networks, Apache Hadoop, and Apache Spark are the most prominent horizontal scale out platforms.

Vertical scaling, also known as "scale up," involves adding more computational power within a single server. High-performance computing clusters (HPC), multicore processors, graphics processing unit (GPU), and field programmable gate arrays (FPGA) are the most popular vertical scale-up paradigms.

The above rating table thus provides a qualitative comparison and brief insight about the general strengths and weaknesses of various platforms with respect to the critical characteristics pertaining to BDA.

Table 4 Comparison of different BDA platforms [13]

Scaling type	Platforms (communication scheme)	System/platform			Application/algorithm		
		Scalability	Data I/O performance	Fault tolerance	Real-time processing	Data size supported	Iterative task support
Horizontal scaling	Peer-to-peer (TCP/IP)	*****	*	*	*	*****	**
	Virtual clusters (map reduce/MPI)	*****	**	*****	**	****	**
	Virtual clusters (Spark)	*****	***	*****	**	****	***
Vertical scaling	HPC clusters (Mapreduce/MPI)	***	****	****	***	****	****
	Multicore (multithreading)	**	****	****	***	**	****
	GPU (CUDA)	**	*****	****	*****	**	****
	FPGA (HDL)	*	*****	****	*****	**	****

*****denotes best possible rating
*denotes lowest possible rating

3 Scope of Research in BDA

This section explores issues and challenges in BDA, reviews some related work done in the area of BDA, and finally puts forth a research proposal for application of unstructured BDA techniques for resume analytics.

3.1 Issues and Challenges

A team of 21 prominent researchers across the US in [12] through collaborative writing have identified the challenges of heterogeneity, lack of structure, error handling, privacy, timeliness, provenance, and visualization along the BDA pipeline shown in Fig. 1. The authors in [14] have discussed issues related to storage and transport, management and processing, dynamic design, and analytics challenges with respect to big data. The "bring code to data" and data triage approaches are proposed to deal with storage and transport issues. The authors summarize that there is no perfect big data management tool yet. Extensive parallel processing and new analytics algorithms will be needed to effectively process big data in order to provide timely and actionable response. The author in [15] states that inconsistency in big data is a huge challenge that can impact BDA. He classifies big data inconsistencies as temporal, spatial, text, and functional dependency inconsistencies. It is important to realize that a proper understanding of these inconsistencies can serve as valuable heuristics to improve performance of BDA. Innovative techniques need to be

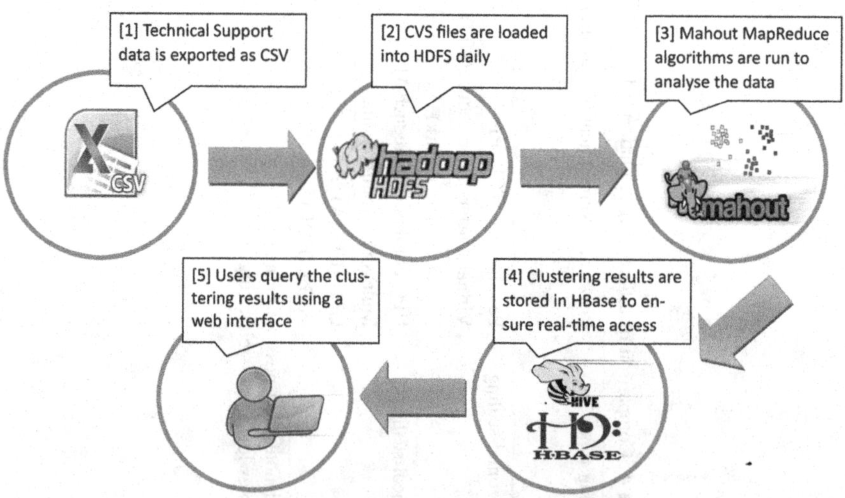

Fig. 2 Proposed open-source end-to-end solution for analyzing large technical support data [16]

devised which allow existing methods to coexist with BDA. BDA should supplement existing techniques rather than replace them until the great transit.

3.2 Related Work

The authors in [16] realize the value of identifying customer patterns, call resolution, and closure rates, generated out of several thousands of customer queries arriving at a technical support call center, which otherwise were discarded. An end-to-end proof-of-concept solution is proposed to categorize similar support requests. Figure 2 shows the proposed architecture and methodology. The solution uses open-source software such as Hadoop, components of the Hadoop extended ecosystem such as HBase and hive and clustering algorithms from the extended Mahout library. Future work may consider alternative advanced methods to transform call center records to vectors and Oozie may be considered for automated orchestration of related support calls.

The authors in [17] describe a real-life case study of a large-scale, predictive analytics system for real-time energy management in a campus micro-grid UCSD (University of California, SanDiego). This work utilizes smart grid data and advanced forecasting techniques to provide a BDA platform in order to analyze and predict building behavior thereby demonstrating energy conservation measures and unwanted energy utilizations. Future work may explore clustering methods, real time, and active learning models employed for predicting peaks and potential outages and parallelization of workflows.

The MapReduce (MR) model is suited for batch data and is not easily expressible for temporal data. In [18], the authors have proposed a framework called TiMR which supplements MR with a time-oriented data processing system and can deal with temporal queries for behavioral targeted (BT) web advertising. Future work may propose new temporal algorithms to improve web advertising with BT.

In the healthcare domain, the authors have devised a proof-of-concept BDA framework [19] for developing risk adjustment model of patient expenditures. Divide and conquer strategy with advanced machine learning is used on health care big data in the framework to improve the risk adjustment model. Their results show that the random forest-based model significantly outperforms the linear regression model to identify the complex relationship in high-dimensional patient big data. More patient risk factors, such as disease history, insurance, income level, and secondary diagnoses, maybe incorporated to enrich the risk adjustment model. The distributed-computing cluster can be expanded to enable learning the model from nationwide dataset (billions of samples), and thus to exploit the power of big data in risk adjustment and other health care service areas.

Law enforcement authorities create a repository that contains web surfing patterns of all users. Such a repository contains valuable forensic information that may help to pinpoint suspected criminals or terrorists, and in certain cases can even preempt an upcoming criminal act. BDA may be employed to perform such detections. In [20], for each user, the proposed detection model derives typical surfing patterns that relate to the topics of interest, frequency of accessing the information, when the information is accessed, etc. Significant deviations from those patterns, particularly when coupled with an event of interest (EOI), such as hit and runs or terrorist attacks, may indicate the subject's active involvement in the event. An outline of the model and the related architecture is proposed which may serve as guidelines for future research.

The above works represent a very small subset of the several kinds of applications of BDA. Nonetheless, they may serve as pointers for future work in the exciting area of BDA.

3.3 A Research Proposal for Developing a Contextual Model for BDA Using Advanced Analytics

3.3.1 The Resume Analytics Problem

Organizations are often faced with the problem of finding the most befitting candidate to fill up a vacancy. They may float call for posts online by submitting job descriptions. Likewise, prospective candidates may upload their resumes anticipating a call for the recruitment process. The problem of finding the most appropriate fit between a job description and a prospective candidate can serve as a use case for applying BDA and is challenging due to the following reasons:

- The unstructured nature of job descriptions and resumes creates scope for research in finding a suitable means for storing them optimally.
- The sheer volume of resumes and job descriptions uploaded by people all over the world calls for a solution to cope up with the scale.

The big data value chain can serve as the outline for the resume analytics use case and may be used as follows:

Acquisition—Resumes are collected from various sources such as emails, job portals, LinkedIn, etc. and recorded for subsequent phases.

Information Extraction—Relevant information pertaining to skills, experience, qualification, etc. is extracted for every resume.

Intermediate Representation—The extracted information is stored in an intermediate representation suitable for subsequent analysis. The representation should be chosen to enable quick and easy access.

Resume Analytics—Every resume in the intermediate representation is analyzed based on the various criteria against the job requirement and assigned a rank.

Interpretation of Results—Based on the ranks, the most appropriate fit between the job description and the candidate(s) serves as the analytic result.

Thus this problem serves as an interesting use case that ignites research in each of the above-mentioned phases of the big data value chain for developing appropriate solutions that would facilitate the HR function of recruitment process.

4 Conclusion

Big data refers to humungous amounts of data that can no longer be managed by traditional approaches. Organizations will, however, need to exploit big data in order to uncover interesting insights that can provide them a competitive advantage. Analytics is the driver that scavenges through big data to discover patterns and hidden relationships to generate meaningful insights in order to create value. Big data and analytics are thus critical enablers for organizations to steer ahead. This paper presented a background of the two topics, big data and analytics and provided taxonomy for classifying BDA. The methodology for BDA in the form of big data value chain was also discussed. It also reviewed some of the existing works done and listed various issues and challenges for research in BDA. The paper also proposed the problem for developing a contextual model for BDA using advanced analytics which shall serve as future work. The conclusive remark based on the survey of related work would be "Advanced analytics in the form of predictive and prescriptive analytics applied to big data would act as the next frontiers of innovation and technology."

References

1. Gunelius S (2014) The data explosion in 2014 minute by minute—infographic, July 12, 2014. http://aci.info/2014/07/12/the-data-explosion-in-2014-minute-by-minute-infographic/
2. Brindle J, Fania M, Yogev I (2011) Roadmap for transforming Intel's business with advanced analytics. Intel IT, IT best practices, business intelligence and IT business transformation, Nov 2011
3. Leventhal B, Langdell S (2013) Embedding advanced analytics into business applications. Barry Analytics Limited and the Numerical Algorithms Group 2013. http://www.nag.com/market/articles/nag-embedding-analytics.pdf
4. Kelly J (2014) Big data: Hadoop, business analytics and beyond, Feb 05, 2014. http://wikibon.org/wiki/v/Big_Data:_Hadoop,_Business_Analytics_and_Beyond
5. Katal A, Wazid M, Goudar RH (2013) Big data: issues, challenges, tools and good practices. In: Sixth international IEEE conference on contemporary computing (IC3), 2013, pp 349–353
6. Intel IT Center (2012) Planning guide: getting started with Hadoop. Steps IT Managers can take to move forward with big data analytics, June 2012
7. Vohra G, Digumarti S, Ohri A, Acharya A (2012) Beginner's guide. Jigsaw Academy Education Private Limited © 2012, Karnataka
8. Lustig I, Dietrich B, Johnson C, Dziekan C (2010) The analytics journey. Analytics Magazine Nov/Dec 2010, pp 11–18
9. Intel IT Center (2013) Predictive analytics 101: next-generation big data intelligence, Mar 2013. http://www.intel.in/content/www/in/en/big-data/big-data-predictive-analytics-overview.html

10. Siegal E (2010) Seven reasons you need predictive analytics today. Prediction Impact Inc., San Francisco, CA (415) 683-1146. www.predictionimpact.com
11. Basu A (2013) Five pillars of prescriptive analytics success. Executive edge, analytics-magazine.org, pp 8–12, Mar/Apr 2013. www.informs.org
12. Agrawal D et al (2012) Challenges and opportunities with big data. A community white paper developed by leading researchers across the United States. http://cra.org/ccc/wp-content/uploads/sites/2/2015/05/bigdatawhitepaper.pdf
13. Singh D, Reddy C (2014) A survey on platforms for big data analytics. J Big Data 1(8). http://www.journalofbigdata.com/content/1/1/8
14. Kaisler S, Armour F, Espinosa JA, Money W (2013) Big data: issues and challenges moving forward. In: 2013 46th Hawaii International Conference on System Sciences
15. Zhang D (2013) Inconsistencies in big data. In: 12th IEEE international conference on cognitive informatics and cognitive computing (ICCI*CC'13), 2013
16. Barrachina AD, O'Driscoll A (2014) A big data methodology for categorising technical support requests using Hadoop and Mahout. J Big Data 1(1):1–11. http://www.journalofbigdata.com/content/1/1/1
17. Balac N, Sipes T, Wolter N, Nunes K, Sinkovits B, Karimabadi H (2013) Large scale predictive analytics for real-time energy management. In: 2013 IEEE international conference on big data, pp 657–664
18. Chandramouli B, Goldstein J, Duan S (2012) Temporal analytics on big data for web advertising. In: 2012 IEEE 28th international conference on data engineering, pp 90–101
19. Li L, Bagheri S, Goote H, Hasan A, Hazard G (2013) Risk adjustment of patient expenditures: a big data analytics approach. In: 2013 IEEE international conference on big data, pp 12–14
20. Kedma G, Guri M, Sela T, Elovici Y (2013) Analyzing users' web surfing patterns to trace terrorists and criminals. In: 2013 IEEE international conference on intelligence and security informatics (ISI), pp 143–145

Efficient Identification of Bots by K-Means Clustering

S. Prayla Shyry

Abstract Botnet has become a major threat to the Internet and has gained a lot of attention from cyber security. Attackers have been increasing on the Internet in order to gain profits by stealing the information of a legitimate user. As per the statistics of Vintcerf, "25 % of internets PCs are part of a botnet". Bots are termed as a collection of compromised computers responsible for various attacks such as phishing attack, DDOS attack, spam e-mails, online fraud, phishing, information exfiltration, etc. In this paper, traditional botminer algorithm is used with k-means clustering. In addition, x-means clustering is used to cluster the traffic and the existing algorithm was executed to compare and validate the results and differences revealed by k-means and x-means clustering algorithms

Keywords Bots · Clustering

1 Introduction

Botnets are becoming one of the most serious threats to Internet security. Many researchers have proposed various methodologies to identify bots. IP's traceback mechanism requires more traffic filtering and more overheads. Multi-layer faceted approach requires changing characteristics of bots without its knowledge. To detect bots in an efficient way, the main objective of this work is as follows:

- To capture the network traffic flow and to fetch parameters such as IPs, port number, time and number of packets transferred.
- To filter the similar communication flows and the similar activities.
- To cluster the filters of communication flows and activities.
- To reduce the clusters to filter the attackers.

S.P. Shyry (✉)
Sathyabama University, Chennai, India
e-mail: suja200165@gmail.com

L.P. Suresh and B.K. Panigrahi (eds.), *Proceedings of the International Conference on Soft Computing Systems*, Advances in Intelligent Systems and Computing 398, DOI 10.1007/978-81-322-2674-1_30

2 Materials and Methods

Chen et al. [1] observed web botnet behaviours and found their abnormal behaviours. They analysed the result in four different network environments and used features such as timeslot, data calculation, mutual authentication and bot clustering.

Arshad et al. [2] proposed a methodology to detect similar NetFlows. They described that bots in the similar network can receive similar type of commands instructed by its masters. They implemented in three phases which include IP mapping, NetFlow generating and alert generating. In each phase the NetFlows are filtered and clustered separately. Clusters are passed into a correlation engine to identify malicious hosts.

Perdiscia et al. [3] proposed clustering of malware based on their behaviours in HTTP networks. They clustered by fine-grained clustering and coarse-grained clustering. Then meta clusters were obtained from the two step clusters. Network signatures are generated automatically from the output of fine-grained clusters for identifying malware variants.

From the literature cited, it is found that the existing work depends on the protocols to detect attackers which are a structure-dependant approach. There may be a chance to diversify the sources of attacks. In the existing work, traffic filtering requires more operational cost. It also requires memory space to maintain the marking fields. Since botnets are normally massive in size, it has been relatively easy to covertly infiltrate a botnet and monitors its transactions. Because of this, botnet monitoring has become a common way to analyse and identify botnets and the destruction they cause. Most research goals in this area have been to identify the command and control of the botnet and shut it down, or to monitor the botnets for statistics without taking actions.

2.1 *Capturing of Network Traffic*

In this module, the traffic in the network is captured using a network protocol analyser. To analyse the traffic of an interface, start capturing the packets on the interface and obtain the necessary parameters such as source IP, destination IP, source port, destination port and protocols. The packets are captured by providing the interface of the network connected to it. The captured packets are stored as pcap files. Capturing the traffic is shown as a use case diagram in Fig. 1.

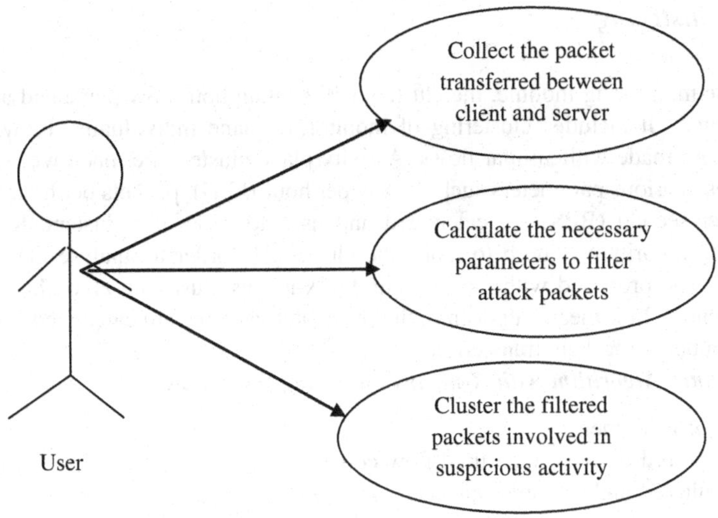

Fig. 1 Use case diagram for packet capture

Pseudocode for capturing packets

```
main(String[] args) throws Exception{
        final String FILENAME = "Filename.pcap";
        final StringBuilder errbuf = new StringBuilder();
// fetch inputs from the pcap file
        final Pcap pcap = Pcap.openOffline(FILENAME, errbuf);
        if (pcap == null) {
                System.err.println(errbuf); // Error is stored in errbuf if any
                return;}
// fetch inputs which are stored in offline
        pcap.loop(10, new JPacketHandler<StringBuilder>() {
                final Tcp tcp = new Tcp();
                final Http http = new Http();
// fetch header information of protocols
```

2.2 Monitoring Planes

The network traffic is monitored using two planes, flow plane and activity plane. The flow plane monitors the communication characteristics of net users. It monitors "who is talking to whom". It maintains the flow log with various parameters such as IP, port number. The activity plane monitors the activity of the net users. It maintains the activity log of the users which includes scanning activity (opened ports), spamming activity and binary downloading (exe downloading). It monitors "who is doing what".

2.3 Clustering

After the monitoring module, the clusters will contain both flow plane and activity plane filters. It includes clustering of monitoring plane individually. Flow plane clusters are made with similar flows. Activity plane clusters are made with similar activities. Various parameters such as flow per hour (FPH), packets per hour (PPF), bytes per second (BPS) are calculated and used to make the clusters. K-means clustering algorithm is used to make the clusters. In order to analyse the performance of the proposed with the existing, the x-means clustering algorithm is also implemented. In k-means algorithm, the mean and variance are calculated from the total number of packets transferred.

Botminer Algorithm with X-means clustering algorithm

Algorithm: Botminer
Input: captured packets, measured flow count,
Output: filtered and clustered packets
Procedure:
1. Fetch parameters such as source IPs, Destination IPs, port number and protocol;
//using packet analyser to capture packets
2. Generate total packet count and flow count;
//calculate the flow measure from capture
3. Check the SYN and ACK activities to identify the opened ports
//using length information determine the connectivity status
4. Calculate rate of flow per hour and bytes per hour
//measure total number of flows and packets per hour
5. if (FPH>certain limit)
Detect it as spam activity
//set the threshold value for the flows
6. ci={fj} j=1….m
fj->is a single TCP/UDP flow
ci->clusters
//cluster the flows f(i)
7. Calculate FPH, PPF, and BPP.
8. Calculate mean and variance.
//compute mean throughout the flows
9. Apply x-means clustering algorithm.
//implement clustering algorithm
10. Filter attack packets.

Packet analyser is used to capture the packets to fetch the parameters such as IPs, port numbers, start time and end time required to implement the algorithm. Flow count and packet count are measured for the captured packets. The length information is analysed to find the connection establishment and termination records. For each flow, the communication characteristics and its activities are captured.

Filters are made with similarities of flow log and activity log. Clusters are formed from the filters using x-means clustering algorithm.

Botminer Algorithm with k-means clustering algorithm
Algorithm: Botminer
Input: captured packets, measured packet count,
Output: filtered and clustered packets
Procedure:
1. Fetch parameters such as source IPs, Destination IPs, port number and protocol;
//using packet analyser to capture packets
2. Generate total packet count and flow count;
//calculate the flow measure from capture
3. Check the SYN and ACK activities to identify the opened ports
//using length information determine the connectivity status
4. Calculate rate of flow per hour and bytes per hour
//measure total number of flows and packets per hour
5. if (FPH > certain limit)
Detect it as spam activity
//set the threshold value for the flows
6. ci={fj} j=1….m
fj-> is a single TCP/UDP flow
ci-> clusters
//cluster the flows f(i)
7. Calculate FPH,PPF,and BPP.
8. Calculate mean and variance.
//compute mean throughout the flows
9. Apply k-means clustering algorithm.
//implement clustering algorithm
10. Filter attack packets.

Packet analyser is used to capture the packets to fetch the parameters such as IPs, port numbers, start time and end time required to implement the algorithm. Flow count and packet count are measured for the captured packets. The length information is analysed to find the connection establishment and termination records. For each flow the communication characteristics and its activities are captured. Filters are made with similarities of flow log and activity log. Clusters are formed from the filters using k-means clustering algorithm in bytes. In x-means algorithm, the mean and variance are calculated from the total number of flows transferred in a count (Fig. 2).

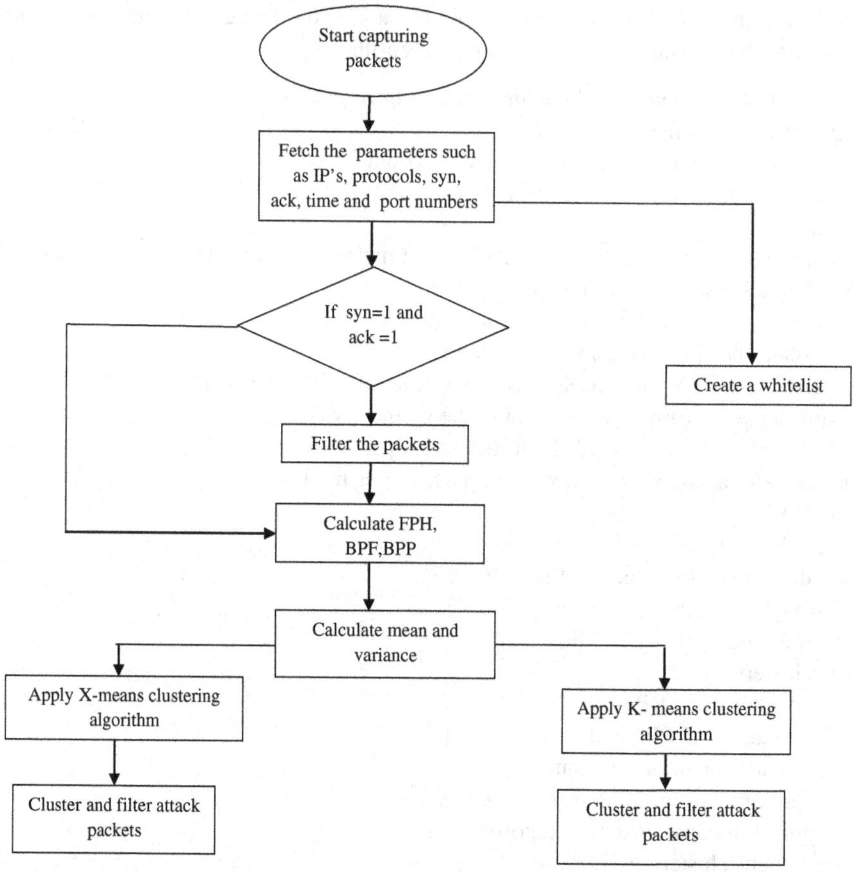

Fig. 2 Data flow diagram of the proposed system

3 Experimental Setup

The experimental setup is done by capturing the network flows over two hours in a Google interface. The network traffics are captured by means of a protocol analyser and the proposed work is implemented over the captured packets to identify the bots. In this section, a thorough discussion about the implementation of each module of the proposed system is provided and the results are discussed briefly.

3.1 Capturing Packets

In order to identify the malicious activity and its attackers, packets transferred between systems are captured using Wireshark. The packets are captured by

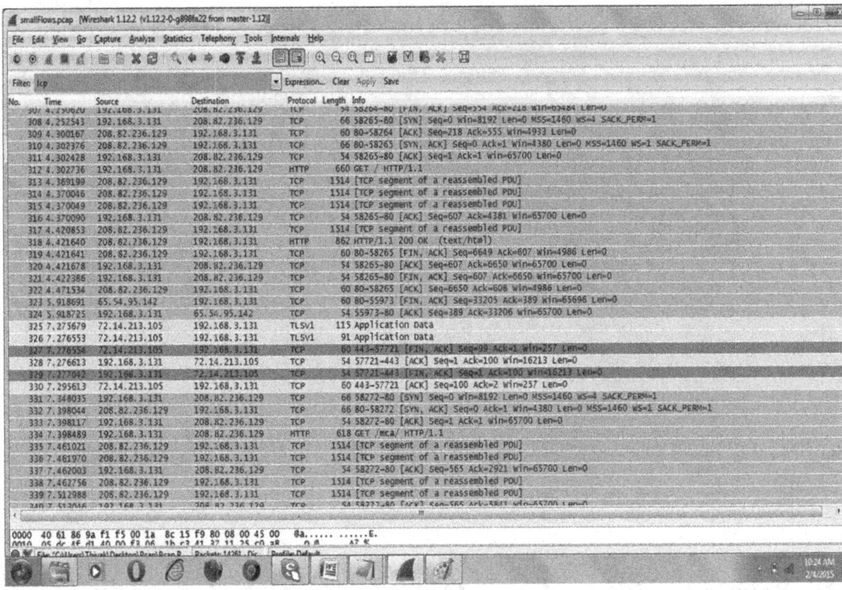

Fig. 3 Structure of captured packets

providing the interface of the network connected to it. The captured packets are stored as pcap files. The captured packets contains various details such as source IP, destination IP, flow number, protocol, time and length information. Figure 3 shows the structure of captured packets. Wireshark consists of various options such as Hexdump, merge and filters to filter out the flows.

Certain interfaces are provided like google.co.in to capture the packets. The input is captured for about 2 h in that interface. It captured about a total of 14261 flows. It shows various parameters such as start time, end time, source IP, destination IP, protocol, length information and number of flows are captured from the pcap files. Then the pcap file is parsed using Java API to fetch the packet information.

Figure 4 shows the graph analysis of the captured packets and the flow graph generated from the feature of Wireshark. This graph is mainly used to extract the IP address of the sending host, and the destination host which is flooded with the packets. The flow graph feature can provide a quick and easy way of checking connections between a client and a server. It can show where there might be issues with a TCP connection, such as timeouts, retransmitted frames, or dropped connections.

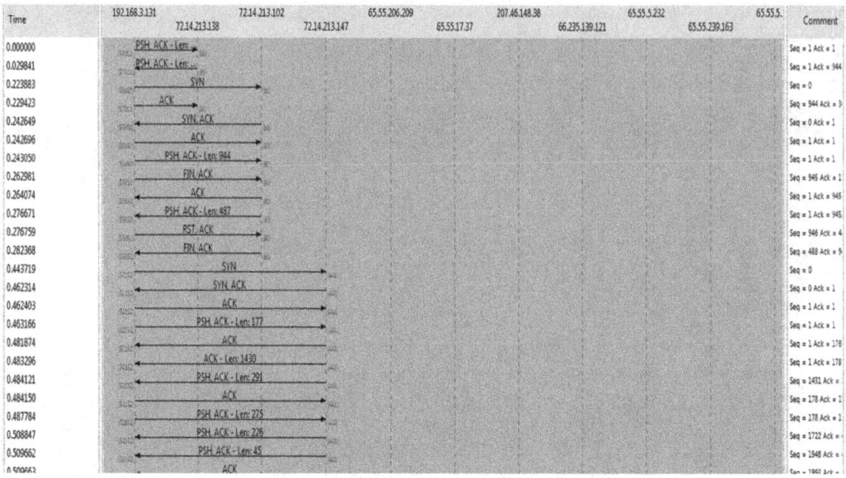

Fig. 4 Flow graph of captured packets

3.2 *Botminer with X-Means Algorithm by Calculating Flow Count*

To obtain the communication characteristics and activities of users the protocol and its header details are fetched with jNetPcap API. It is an open source library provided in Java. Since it is implemented in Windows Os, WinPcap is also included with that API. Various features such as frame size, port numbers, acknowledgements, packet length, headers, and transport layer parameters such as protocol and its headers, network layer parameters such as ethernet details, frame details are retrieved for implementing botminer algorithm. Figure 5 shows the particulars necessary to implement the algorithm.

It also shows the calculated values of packet count and flow count. Packet count is the total number of packets transferred throughout the flows. Flow count is the total number of flows in the captured packets. The given input packets consist of 14196 packet count and 638 flow count.

The packets with similar activities are filtered to identify attackers and the filters are stored separately to make the clusters. Then the filtered packets are clustered using x-means algorithm. X-means algorithm is implemented with parameters such as FPH, PPF, and BPS. FPH is the total number of TCP flows per hour. PPF is the summation of total number of packets in TCP flow. BPS is the total number of bytes transferred in each flow. Figure 6 shows the separated files obtained from the filters.

Figure 6 shows the filters of separated text files. The filters are done with parameter such as IPs which are more in time to filter the communication characteristics, length information such as SYN and ACK to filter scan activities, packet size to filter binary downloading.

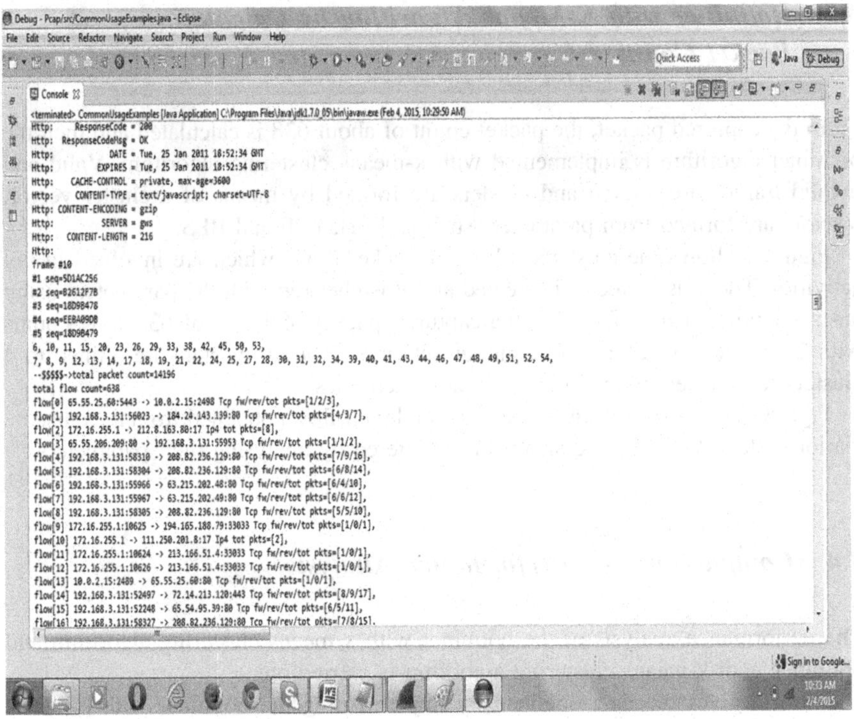

Fig. 5 Fetching captured packets

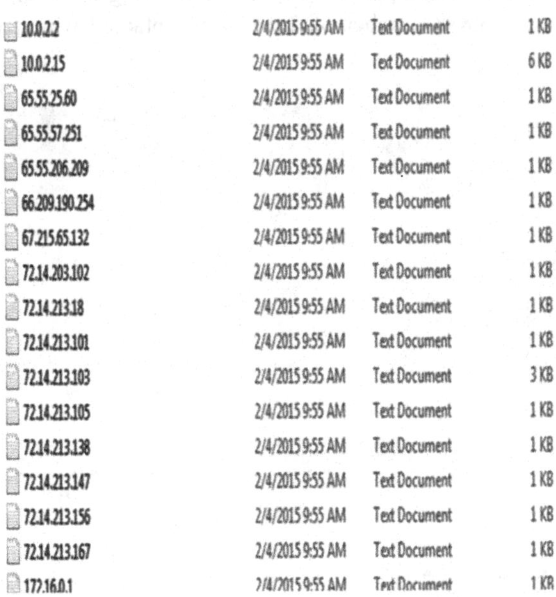

10.0.2.2 2/4/2015 9:55 AM Text Document 1 KB
10.0.2.15 2/4/2015 9:55 AM Text Document 6 KB
65.55.25.60 2/4/2015 9:55 AM Text Document 1 KB
65.55.57.251 2/4/2015 9:55 AM Text Document 1 KB
65.55.206.209 2/4/2015 9:55 AM Text Document 1 KB
66.209.190.254 2/4/2015 9:55 AM Text Document 1 KB
67.215.65.132 2/4/2015 9:55 AM Text Document 1 KB
72.14.203.102 2/4/2015 9:55 AM Text Document 1 KB
72.14.213.18 2/4/2015 9:55 AM Text Document 1 KB
72.14.213.101 2/4/2015 9:55 AM Text Document 1 KB
72.14.213.103 2/4/2015 9:55 AM Text Document 3 KB
72.14.213.105 2/4/2015 9:55 AM Text Document 1 KB
72.14.213.138 2/4/2015 9:55 AM Text Document 1 KB
72.14.213.147 2/4/2015 9:55 AM Text Document 1 KB
72.14.213.156 2/4/2015 9:55 AM Text Document 1 KB
72.14.213.167 2/4/2015 9:55 AM Text Document 1 KB
172.16.0.1 2/4/2015 9:55 AM Text Document 1 KB

Fig. 6 Text files of filtered packets

3.3 Botminer with K-Means Algorithm by Calculating Packet Count

From the captured packet, the packet count of about 638 is calculated and then the botminer algorithm is implemented with k-means clustering algorithm. Valid and invalid traffics are filtered and clusters are formed by means of a cluster vector. Vectors are formed from parameters such as FPH, PPF and BPS.

Figure 7 shows the clustered filters of packet flows which are involved in bot activities. The cluster vector is created and it is checked with the parameters using hash mapping. It describes that the captured packets consists of 638 flow counts with 20 clusters which perform similar activities. The 20 clusters are refined to 4 clusters to attackers with similar malicious activities.

Figure 8 shows the details of attackers and their flows obtained from the clusters. Various attack flows by the attackers are filtered by analysing their parameters.

3.4 Comparison and Performance Analysis

The performance analysis of the botminer with x-means clustering algorithm and botminer with k-means clustering algorithm is carried out.

Table 1 gives the data obtained by implementing the botminer with k-means clustering algorithm. For a capture of 45 min, 126 number of flows were obtained with packet count of 61 showing no attackers in it. For a capture of 90 min, 144 number of flows were obtained with a packet count of 79 showing an attacker in it. For a capture of 120 min, 13,878 number of flows were obtained with a packet

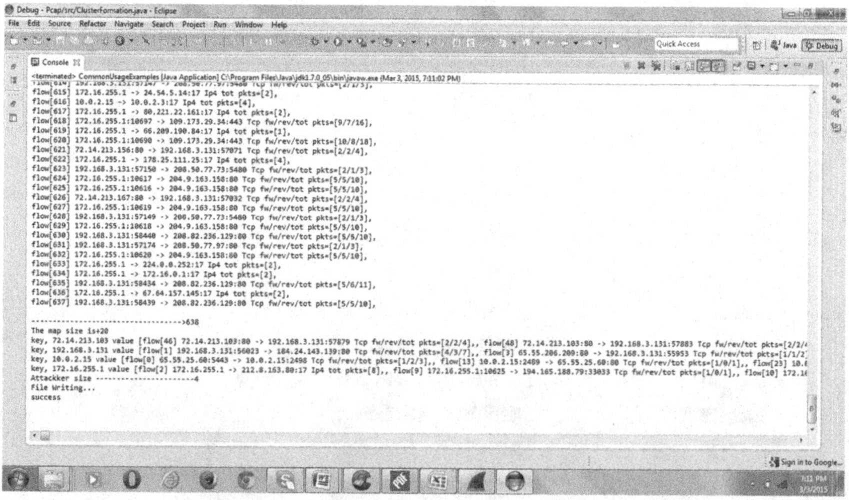

Fig. 7 Clustered filters of k-means clustering algorithm

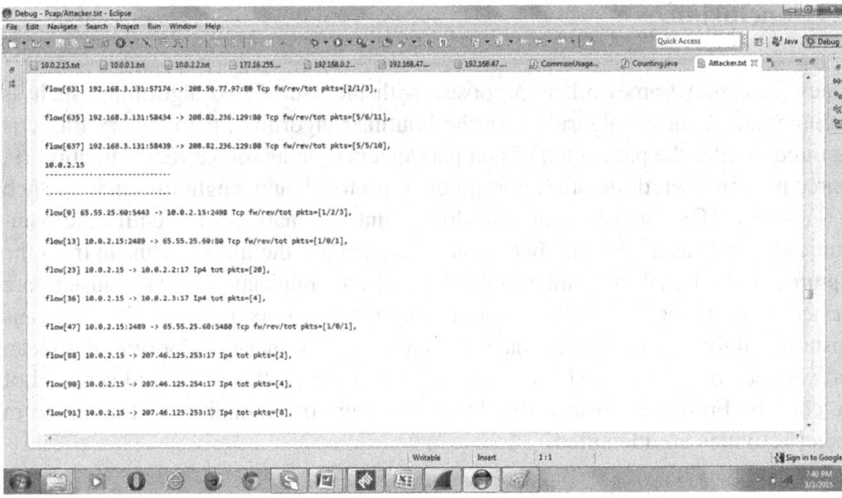

Fig. 8 Flow of attackers

Table 1 Number of packet count using k-means algorithm

Number of flows	Time(min)	Attacker	Packet count
126	45	0	61
144	90	1	79
3878	120	3	3813
3918	135	3	3853

Table 2 Number of flow count using x-means algorithm

Number of flows	Time	Attacker	Flow count	clusters
126	30	0	2	7
144	45	1	4	4
3878	60	3	454	3
3918	75	3	458	18

count of 3813 showing 3 attackers in it. For a capture of 135 min, 3918 number of flows were obtained with a packet count of 3853 showing 3 attackers in it.

Table 2 gives the data obtained by implementing the botminer with x-means clustering algorithm. For a capture of 30 min, 126 number of flows were obtained with a flow count of 2 showing no attackers in it. For a capture of 45 min, 144 number of flows obtained with a flow count of 4 showing an attacker in it. For a capture of 60 min, 3878 number of flows obtained with flow count of 454 showing 3 attackers in it. For a capture of 75 min, 3918 number of flows were obtained with flow count of 458 showing 3 attackers in it. It describes the efficiency of k-means algorithm compared to the existing algorithm.

4 Conclusion

A new detection framework is proposed with the use of two algorithms such as botminer and k-means algorithm. In the botminer algorithm, the network traffic is captured to infer the parameters. Then parameters such as source IP, destination IP, source port number, destination port number, protocol, and length information such as SYN and ACK, packet count and flow count are analysed to identify the communication characteristics and behaviour of host. Then the filters are made from the captured flows based on similar activities and communication flows. Clusters are formed by grouping the similar communication behaviours and activities; k-means clustering algorithm is used to make the clusters. In k-means algorithm, the mean and variance of the captured flows are calculated to find the centre and centroid of the clusters. From the clusters, the flows and their corresponding host involved in botnet activities are identified.

References

1. Chen C-M, Ou Y-H, Tsai Y-C (2010) Web botnet detection based on flow information, || IEEE transactions on computer symposium, pp 381–384
2. Arshad S, Abbaspour M, Kharrazi M (2011) An anomaly-based botnet detection approach for identifying stealthy botnets, IEEE International conference on computer applications and Industrial Electronics, pp 564–569
3. Perdiscia R, Leea W, Feamster N (2011) Behavioral Clustering of HTTP-Based Malware and Signature Generation Using Malicious Network Traces, pp 1–14. www.usenix.org/legacy/event/nsdi10/tech/full_papers/perdisci

Message Authentication and Source Privacy in Wi-Fi

K. Gnana Sambandam and Ethala Kamalanaban

Abstract Wireless fidelity (Wi-Fi) is one of the wireless communication technologies; it will allow transmission of data within a particular range. The critical issues in any wireless transmission will occur due to anonymous users, data loss, corrupted message, and denial of service attack. In a wireless communication, our message can be hacked by anonymous users and they will crack our network and access our data, modify and or delete our data. To avoid these issues we are implementing the message authentication scheme. Message authentication is one of the effective ways to protect our data from unauthorized access and also to handle corrupted messages, while transmitting in wireless networks. Authentication scheme concepts have been developed by generating various key algorithms. There are two approaches—symmetric key and asymmetric key. In asymmetric approach, the public key cryptosystem is implemented using elliptic curve cryptography (ECC) algorithm in wireless sensor networks. It allows us to send bulk messages more than the threshold value. It will solve the threshold and scalability problems. In the proposed system, the symmetric key approach is implemented and it is mainly used in polynomial-based scheme, which will undergo data encryption standard (DES). DES is the best algorithm for symmetric key analysis, so here it is proposed to apply DES algorithm to transfer message from one source to another source through Wi-Fi in a secured manner.

Keywords Network security · Cryptography · Public key · Private key · Triple DES

K. Gnana Sambandam (✉) · E. Kamalanaban
Department of Computer Science and Engineering, Vel Tech University,
Avadi, Chennai 600062, India
e-mail: sambath86@gmail.com

L.P. Suresh and B.K. Panigrahi (eds.), *Proceedings of the International Conference on Soft Computing Systems*, Advances in Intelligent Systems and Computing 398, DOI 10.1007/978-81-322-2674-1_31

1 Introduction

Wireless fidelity (Wi-Fi) [1] is an emerging technology in wireless networking. It has been used for facilitating communication between two devices without using the physical medium. The Wi-Fi communication can happen within limited ranges only. The range of the Wi-Fi is 100–400 m. It is not only used for communication purposes, but can also be used to share the internet with another device. Nowadays Wi-Fi is very popular in mobile devices. Through Wi-Fi data as well as internet can be shared. We can suggest going for an external adapter that can initiate the Wi-Fi connection. The frequency of wireless connectivity is between 2.4 and 5 GHz. The bandwidth can be used based on the quantity of data transferred inside the network. So we need a hotspot zone as an alternative for the infra-oriented counterparts. The Wi-Fi hotspots can be used by the router to access the internet over the wireless local area network (WLAN). Nowadays there are laptops and computers available with inbuilt Wi-Fi. So they have the ability to connect automatically whenever they sense the hotspot. Most of the hotspots available are secure and it requires authentication of some kind. If the inbuilt Wi-Fi is not available we can suggest going for an external adapter that can initiate the Wi-Fi connection. If the user does not need the same, they simply may disconnect the device. The modern gadgets like smartphones, handheld devices, and devices used in military took Wi-Fi technology to the next level, in terms of usage and security. The working style is as similar to the older version techniques like Ethernet cable connectivity. Usually a signal is transmitted through an antenna and requires a router to decode the received signal. It works either way because it supports two-way transmission or traffic.

Cryptography [2] is used to have a secure communication over anonymous users. It has been used for generating original text to cipher text and vice versa. By generating this text, two processes are required. They are encryption and decryption processes. In encryption, plain text will be converted into cipher text. And while decryption the cipher text will be converted into plain text. Once the conversion process is completed the encrypted message will be sent through the network.

Message authentication involves securing a message from anonymous user. The message is attached with the secret key which helps to create a secure signature. There are two keys in network security concept. Symmetric key will use only one key for the encryption and decryption process. Whereas the asymmetric key will be have separate keys for encryption and decryption. Since it has a public key and private key, the public key [3] is used to share with multiple users and private key is used to share with only one person.

The message that accommodates with sender's signature can be processed and verified at the receiver's end by comparing the routing table attached. Once the trustworthiness of the message is confirmed by the receiver, the success acknowledgement will be sent, since the private key attached along with the signature confirms the originality of the message to signer receiver.

2 Background and Related Work

2.1 Background Work

In this part, we discuss the various cryptographic techniques used in the existing and proposed work

2.1.1 Elliptic Curve Cryptography

Elliptic curve cryptography (ECC) [4] is the best and the most accurate algorithm when compared to other algorithms. It is difficult to understand but it obtains possible results. It is the next generation of public key cryptosystems. ECC would work on a structure which is clearly straight forward. It works on mathematical algebraic notations over the set of functions inside a curvy field. It will work on the most common crypto-algorithms.

2.1.2 Data Encryption Standard

Data encryption standard (DES) [5] is a symmetric key analysis which can have same key for encryption and decryption. The electronic data should be encrypted using the key and sent over to an internet. Then in a receiving mode it should be decrypted. It is a simplest algorithm where it uses 64-bit key for cryptography process. To overcome this process, advanced encryption standard has been introduced.

2.1.3 ElGamal Signature Scheme

ElGamal signature [6] is the signature algorithm where it can be used for third party authentication. The digital signature [7] is used to create signature for authentication purposes. First, they generate the key either public or private. Second, they generate the signature and third they will verify the generated algorithm.

2.1.4 MD5

The MD5 [8] is the message digest algorithm. Nowadays the MD5 algorithm is used in most of the cryptographic functions. It has 128-bit hash value which has been uttered in text format of 32 digit hexadecimal number. It will exploit in broad area of cryptographic application and mostly it is used to validate the data integrity. The MD5 processes the unpredictable length message to predictable length message to 128 bits combination.

2.2 Related Literary Work

In this part, we discuss on various existing concepts introduced previously and a detailed study on the system and their efficiency.

In paper [9] li et al. proposed, usage of ECC for sending limitless message that can be sent across wireless network. They use public key for cryptography, so they use different key for encryption and decryption processes. It does not have threshold problem, message can be transmitted more than the threshold value. They have security resilience; suppose if any problem occurs in between it will automatically recover from the problem by itself. It is a computational complexity. The elliptical curve schemes are the basic underlying technique for security. So they found a technique based on the modified ElGamal algorithm and derived a term called source anonymous message authentication (SAMA) [10]. It will authenticate the message and keep the identity of the source secret. If the nodes along with the routing path detect the attack or the corruption in messages, it will simply drop the message. Even though it is flexible it will not compromise on security. There is no such upper limit for message transfer. So this makes it the most stronger than the traditional polynomial algorithms.

In paper [11] Zhu et al. proposed, sensor network mostly meant for the open environment. So this will make this network prone to several attacks that may lead to data duplicity and may result in collapsing of entire network. The most common issue that occurs is draining of resources or nodes. So we require a more reinforced endorsing mechanism that can withstand and never compromise the technique. The proposed technique relies on a stronger solution that is a three-count jump-over leap authentication so that the data duplicity can be curbed at the entry level. It is based on a particular parameter which is system based. So the analysis over performance deals over the correlation between the performance and the stronger security that is the key requirement. It can be implemented over real-time implementations to gain better results.

In paper [12] Albrecht et al. proposed, that the most common problem occur in the regularly used crypto techniques are attacks that ranges from silent to vigorous attacks. Because the underlying techniques are based on mathematical formulation of polynomial-based solution, our proposed solution can perturb the conventional polynomials by adding noise. The initial offerings deal with elasticity over bandwidth without losing any data. But the problem occurs whenever we changed a small portion of the polynomials. Unfortunately the results do not tend to control few attacks likes redistribution, limiting the access schemes and the endorsing the schemes.

In paper [13] Nyberg and Rueppel proposed, there are lot of signature scheme in the cryptography concepts, the discrete algorithm is the first scheme in signature algorithm. It can come under mathematical orientation in the logarithmic approaches. This signature algorithm obtained the message recovery problem. This paper's solution deals with the controlling of message loss and it uses the ElGamal signature. This scheme had the variety of message loss control methods. The ElGamal signature [7] has five types having different procedure in controlling the message loss. The proposed paper is very efficient in message loss controlling by establishing the one time secret keys and ElGamal encryption technique. DES is as efficient as RSA [14] in terms of functionality. It can cope with any larger groups as in ElGamal-type scheme in recovering message. But in RSA the computational efficiency requires in reverse signature verification and generation. Unfortunately RSA is the best ever proven technique for both encrypt and decrypt functions.

3 Design Goals and Pattern

3.1 Attack Models

In this part, we discuss on various possibilities of attack models which may occur inside the network.

3.1.1 Passive Attack

The passive attackers or the inactive attackers are mostly the harmless. They simply intrude into a network and watch the traffic silently. The main motive of the passive attacker is to trace the path of source, destination and the traversing path. Though they seem to be of no harm but the chances of becoming threat to the network is also there. In this attack the intruder eavesdrop the content of data and they will not modify our data. The intruder can access and get our information.

3.1.2 Active Attack

The active attackers are considered as the most dangerous because they do modify the data and sometimes they can collapse the entire system intentionally. The data transferred also become vulnerable in front of active attackers. In this attack the intruders access our information and they can modify the data. The intruder can delete and edit and access the content of our information.

3.2 Terminology Design

In this part, we present our various network terminology designs that may be used in the network.

3.2.1 Polynomial Approach Versus Elliptic Curve Cryptography (ECC)

In existing system, first they apply symmetric approach in polynomial-based algorithm. In this algorithm, they transfer message through wireless sensor networks. In WSN [15], network security is a major problem. Throughout the wireless network lot of intruders will be present and they can access our information. In polynomial-based scheme it will not allow sending bulk amount of messages. Then they apply asymmetric approach, and implement public key cryptosystem using ECC. In this algorithm they transfer message through wireless sensor networks. In this scheme source privacy and message authentication plays a great role. Each and every message can be sent over in the network should be in cipher form and intruders unable to retrieve our information. Then they will allow us to send the bulk messages which can be sent more than the threshold value. By using ECC algorithm the threshold and scalability problems are solved.

3.2.2 Encryption Using DES Over Wi-Fi

DES is the first successive cryptographic algorithm, in which the algorithm first solves the problem associated with the authentication of message in network security. They undergo symmetric key analysis, in which both encryption and decryption will have a same key. By using this algorithm our message will be encrypted and sent over to Wi-Fi networks and in receiver side that message can be decrypted to see the original content of the information. Then the key will be transmitted over the mail for converting the original message and cipher message. The electronic data should be encrypted using the key and sent over to an internet. Then in a receiving mode it should be decrypted. It is a simplest algorithm where its uses 64-bit key for cryptography process. The next version of DES is advanced encryption standard (AES) [16]. This is the modified version of DES, in which it will be having 128-bit keys.

3.2.3 Defensive Mechanism

The major problem faced in the entire process is the active attacks happen while transferring the message. The motives of these attackers are mainly changing the part of the message or even the entire message. Sometimes they intrude along the path that makes the source and destination vulnerable to attacks. The main defensive mechanism used is message authentication. Unless the attackers have the key it is impossible to retrieve the lost message.

3.2.4 Triple DES

It is a triple data encryption algorithm and it is a modified version of DES, where it has three keys for cryptographic process. While encryption, they do two encryptions and one decryption whereas while decryption process they do two decryptions and one encryption process. They have 64-bit block size and they have three keys. Each key will be having 56 bits. First key is having 56 bits and second key is having 112 bits and third key is having 168 bits combination. It is a straight forward method for escalating the size of DES key. Where the DES algorithm will do one-time encryption process the triple DES algorithm will do three times encryption process. For this purpose only they are called as triple DES algorithm.

Algorithm
Let us assume K1, K2, and K3 as the three keys. E denotes the encryption process and D denotes the decryption process.

For Encryption

The plain text should be renovated to cipher text. First they encrypt the plain text with key K1 and then they decrypt the encrypted text with key K2 and at last they once again encrypt the decryption text by using key K3.

$$\text{Ciphertext} = E_{K3}(D_{K2}(E_{K1}(\text{Plaintext}))) \quad (1)$$

For Decryption

In decryption process, the encryption process should be in reverse order. Thus the cipher text should be renovated to the plain text. It has the three decryption process. First the cipher text should be decrypted by using key K3 and then the decrypted text should be once again encrypted by using key K2, at last the encrypted text should be decrypt by using key K1.

$$\text{Plaintext} = D_{K1}(E_{K2}(D_{K3}(\text{Ciphertext}))) \quad (2)$$

4 Performance Evaluation

This section will discuss about the performance evaluation in a real-time environment of the authentication of message and keep the source hidden in Wi-Fi, a performance-oriented technique which allows message transfer through Wi-Fi (Fig. 1) without revealing the source identity. The message sent can be converted into cipher message and transmitted into the Wi-Fi networks and after receiving the

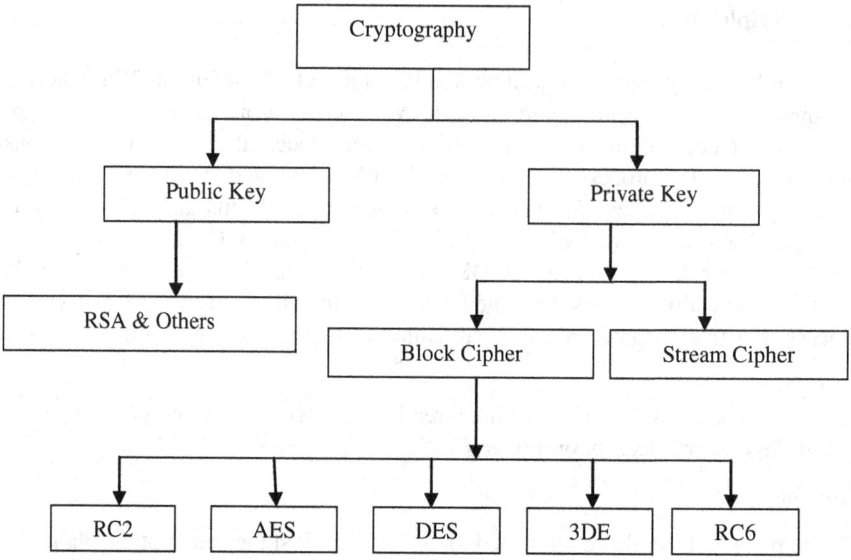

Fig. 1 Various crypto techniques

cipher message it would be converted again to plain text at receiver side. The DES is meant to encrypt-decrypt and considerably performs better and is secure. The previously proposed schemes did not concentrate on message authentication but our paper is first of its kind to provide authentication to both sender and the receiver. Unless the attacker has both the keys it is impossible to affect the network networks (Figs. 2 and 3).

Fig. 2 Effectiveness of triple DES over DES on PDR

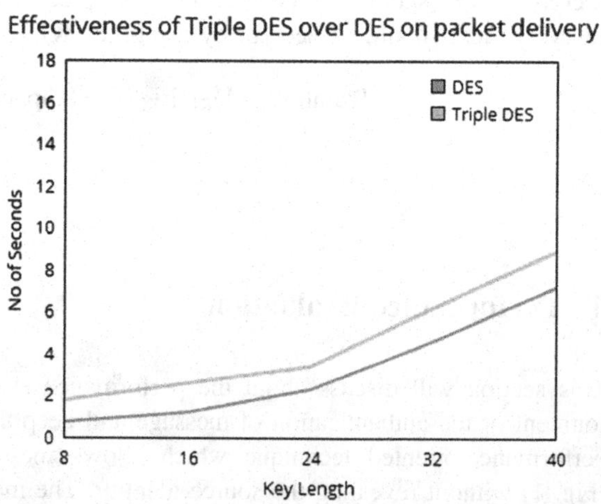

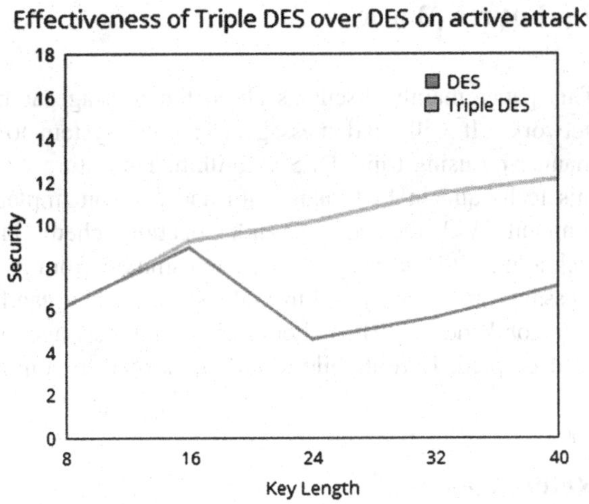

Fig. 3 Effectiveness of triple DES over DES on PDR

5 Comparison Result

The comparison states that the existing system is trying to achieve the near positive results in terms of security or simply tries to safeguard the data whenever the attack happens. Also we can see uses of various available techniques and algorithms across the wireless network that sometimes proved to be efficient but "not to the extent level." The basic concept of the existing techniques is to mainly focus on encryption and decryption using DES, RSA, and other algorithms. Also the terms such as SAMA, ElGamal, perturbation polynomials are at least available to safeguard the sensor network or wireless network. This study tells us the need for a real-secure algorithm that can be a newly written and or a good combination of existing algorithms. So the proposed solution stands above all when compared with existing methods. It follows a simple logic of "unless-until" the intruder have the key the message is safe, which is a greater step in authenticating the message.

6 Conclusion

This paper proposes the message authentication and source privacy in Wi-Fi networks. The original message can be converted into cipher message and transmitted into the Wi-Fi networks and after receiving the cipher message it would be converted again to plain text at receiver side. The symmetric key will be transmitted through the mail and the receiving key is used to decrypt the message. The technique used in this paper is dynamically proved DES. If there is no key, the intruder will be unable to retrieve the information. There is no authentication scheme in Wi-Fi networks. So this is the pilot paper to have the authentication scheme.

7 Future Work

This paper mainly discusses about the message authentication scheme in Wi-Fi networks. It will send message from one system to another system in a secure manner by using triple DES algorithm. The future works lie on implementation of this technique in ECC algorithm and also on implementation in a mobile environment. With the message authentication scheme and source privacy in mobile technology, the message can be transmitted from system to mobile device. The message can be encrypted from the system and it can be sent over a Wi-Fi network to a mobile device. Then after receiving the message into the mobile device it has to be decrypted. Here mobile routing is needed to configure the mobile networks.

References

1. Zhou Z, Wu C, Yang Z, Liu Y (2015) Sensorless sensing with WiFi, vol 20, no. 1. Tsinghua Science and Technology, Feb 2015, pp 1–6. ISSN 1007-0214 01/11
2. Anand D, Khemchandani V, Sharma RK (2013) Identity-based cryptography techniques and applications (a review). 5th international conference on computational intelligence and communication networks (CICN), 27–29 Sept 2013, pp 343–348
3. Rivest RL, Shamir A, Adleman L (1978) A method for obtaining digital signatures and public-key cryptosystems. Commun ACM 21(2):120–126
4. Kapoor V, Abraham VS, Singh R (2008) Elliptic curve cryptography. ACM Ubiquity 9(20)
5. Han S-J, Oh H-S, Park J (1996) The improved data encryption standard (DES) algorithm. IEEE 4th international symposium on spread spectrum techniques and applications proceedings, vol 3, pp 1310–1314
6. ElGamal TA (1985) A public-key cryptosystem and a signature scheme based on discrete logarithms. IEEE Trans Inf Theory IT-31(4):469–472
7. Harn L, Xu Y (1994) Design of generalized ElGamal type digital signature schemes based on discrete logarithm. Electron Lett 30(24):2025–2026
8. Ratna P, Agung A, Dewi Purnamasari P, Shaugi A, Salman M (2013) Analysis and comparison of MD5 and SHA-1 algorithm implementation in Simple-O authentication based security system. International conference on QiR (Quality in Research), 25–28 June 2013, pp 99–104
9. Li J, Li Y, Ren J (2014) Hop-by-hop message authentication and source privacy in wireless sensor networks. IEEE Trans Parallel Distrib Syst 25(5)
10. Zhang W, Subramanian N, Wang G (2008) Lightweight and compromise-resilient message authentication in sensor networks
11. Zhu S, Setia S, Jajodia S, Ning P (2004) An Interleaved hop-by-hop authentication scheme for filtering false data in SensorNetworks. In: Proceedings of IEEE symposium on security and privacy
12. Albrecht M, Gentry C, Halevi S, Katz J (2009) Attacking cryptographic schemes based on 'Perturbation Polynomials'. Report 2009/098. http://eprint.iacr.org/
13. Nyberg K, Rueppel RA (1995) Message recovery for signature schemes based on the discrete logarithm problem. Proc Adv Cryptol (EUROCRYPT) 950:182–193
14. Balasubramanian K (2014) Variants of RSA and their cryptanalysis. International conference on communication and network technologies (ICCNT), 18–19 Dec 2014, pp 145–149

15. Chin E, Chieng D, Teh V, Natkaniec M, Loziak K, Gozdecki J (2013) Wireless link prediction and triggering using modified Ornstein–Uhlenbeck jump diffusion process, 2 July 2013
16. Mestiri H, Benhadjyoussef N, Machhout M, Tourki R (2013) An FPGA implementation of the AES with fault detection countermeasure. International conference on control, decision and information technologies (CoDIT), 6–8 May 2013, pp 264–270

Timely Prediction of Road Traffic Congestion Using Ontology

M. Prathilothamai, S. Marilakshmi, Nilu Majeed and V. Viswanathan

Abstract In developing countries, traffic in a road network is a major issue. In this paper we investigate the tradeoff between speed versus accuracy of predicting the severity of road traffic congestion. The timely prediction of traffic congestion using semantic web technologies that will be helpful in various applications like better road guidance, vehicle navigation system. In the proposed work, ontology is created based on sensor and video data. By using rule inference of ontology on parallel processing of sensor and video data, our system gives the timely prediction of traffic congestion.

Keywords Ontology · Road traffic congestion · Prediction

1 Introduction

Transport facility plays crucial role in economic development in a country. In most of the countries, road network is one of the prominent means of transport. As of now developing countries (India) face challenges of poor rural roads and traffic congestion in major countries. The number of vehicles on road increases day by

M. Prathilothamai (✉) · S. Marilakshmi · N. Majeed
Department of Computer Science and Engineering, Amrita School of Engineering,
Amrita Vishwa Vidyapeetham, Coimbatore, Tamil Nadu, India
e-mail: prathilothamai.m2014@vit.ac.in

S. Marilakshmi
e-mail: marilakshmi123@gmail.com

N. Majeed
e-mail: nilumajeed@gmail.com

V. Viswanathan
School of Computer Science and Engineering, Vellore Institute of Technology,
Chennai, India
e-mail: viswanathan.v@vit.ac.in

L.P. Suresh and B.K. Panigrahi (eds.), *Proceedings of the International Conference on Soft Computing Systems*, Advances in Intelligent Systems and Computing 398, DOI 10.1007/978-81-322-2674-1_32

day. But the infrastructure is not up to that level and it creates congestion on road traffic. There are few external factors that influence the traffic congestion are inclement weather condition, a concert event, road works, accidents, and peak hours. There are two principal categories of causes of congestion, and they are; (a) micro-level factors (e.g., related to traffic on the road) and (b) macro-level factors that relate to overall demand for road purpose. The traffic on the road, more number of peoples when they want to move at the same time, too many vehicles lead to limited road space which is a type of micro-level factor. Hence, it has become a major concern to envisage the severity of road traffic congestion.

Three ways to decrease congestion are: there is need to develop the infrastructure based on road capacity growing rate, promoting public transport in large city which is not convenient always and finally, to determine future road segment which helps in managing the traffic before congestion happens. For example, frequent update in traffic strategy. Historical information is obtained from prediction, or the difficulty in estimating future observations, is an important inference task which is needed for the city traffic managers to know the insight on cities. The traffic control authorities continuously monitor the movement of vehicles on road and take necessary actions to reduce the traffic congestion. Hence, it becomes convenient for people to travel across cities in less time. The drawbacks of the system are the complex statistical calculations. In this paper, a system is developed to monitor the traffic caused on various roads and predict the severity of road traffic congestion. Initially, the raw data are collected using camera and sensors. Once the data is collected, then pre-processing is done by removing noisy data is parking vehicle, bicycles, etc., building an ontology on the top of various data stores like sensor data store and video collected from camera. The ontology will relate sensor data and data which are collected from camera. The processing of sensor data and video data takes longer time. In our system parallel processing of sensor data and camera data has been done. Though the ontology we can timely predict the severity of road traffic congestion.

2 Related Work

To travel smoothly on road during peak hours traffic prediction is required since there will be high traffic congestions. Different methods are used for traffic congestion prediction such as vehicle-to-vehicle communication, spatiotemporal pattern, GSM technologies, etc. Bauza and González [1], Vehicle-to-vehicle communications are used to detect road traffic congestion [2]. In this paper accurately detect and characterize road traffic congestion conditions under different traffic scenarios. They used sensor data. In this proposed system they used continuous exchange of messages between vehicles. Using this method, traffic congestion is more accurate but two wheelers could not consider in this system, also the initial cost is high.

In the paper, Thianniwet et al. [3], they classify different levels of traffic congestion for traffic reports using GPS data. Decision tree algorithm is used for classification. GPS device and webcam are used for data collection. High accuracy is achieved by employing the model on the present traffic report systems. Congestion level is determined using velocity of the vehicle [4]. The consecutive moving average velocities are captured by sliding window techniques. The time interval between consecutive velocities is not taken into account. In the paper, Priyadarshana et al. [5] they proposed a tracking vehicle, motion of vehicle based traffic condition evaluation based on a technique Global Positioning System (GPS) [6]. GPS device is installed in the vehicle. Output of the system is to send an alert SMS to the public in their mobile phone to make aware of the traffic situations, alternative routes, cost, time, and distance to avoid traffic [7]. The main advantage of this proposed system is not only to predict the traffic congestion, suggest alternative path, it does have drawback, in this semi-automated system, only GPS installed device gets the alert message. In the paper, Neha [8] a solution is proposed for congestion control using not only the sensors deployed within the road infrastructure, but the vehicles also collect data. The traffic controlling decision, i.e., the alternative route selection part [9], is provided by the roadside units progress. In this paper, the city is divided into different traffic zones and intelligent wireless traffic lights which are connected to the server are provided to each traffic zone, which helps in taking load balancing and route adjustment decisions. Related to each vehicular node, server maintains the route table which contains three types of information (1) known road segment (2) cost associated with each road segment (3) the segment that a vehicle must follow with the help of road. In this paper cars sense the road infrastructure, pass the information to other vehicles, and to the roadside units because the cars pass each other only in a few seconds cannot totally rely on the vehicle-to-vehicle communication so vehicle-to-infrastructure communication is also suggested in this paper, with the help of routing algorithm it provides alternative route ideas, less congestion, and providing efficient time figures to reach the preferable places.

In the paper, Parag et al. [10], detection of moving vehicle is done by using background subtraction method for a video which is taken from a static camera and it is useful in various essential applications such as video surveillance, human motion analysis, etc., to categorize any new observation as background or foreground, automatically generating and maintaining a representation of the background concept can be used. For modeling a given background scene various statistical approaches have been proposed. However, theoretical framework is not available to choose the features to model different sections of the scene background. A novel framework is introduced to select features for background modeling and subtraction. A Real Boost algorithm [11] is used to select the best combinations of features in each pixel. Kernel Density Estimate (KDE) is used to calculate the probability estimates over a period of time [12, 13], the algorithm chooses the important and valuable ones to differentiate foreground vehicles from the background scene. The outcome explains about the framework which is proposed that selects appropriate features for different parts of the image. In this paper, Banarase et al. [14], it explains about vehicle detection and counting from video sequence method is to tracking the vehicles for high-quality videos [15, 16]. It explains about current approaches of monitoring traffic include manual counting the number of vehicles, or counting number of vehicles using magnetic loops on the road. The main drawback of these approaches are (i) they are expensive (ii) these systems are only for counting. The present image processing method uses temporal differencing method in which full extraction of the shapes of vehicle was the failure, in feature-based tracking method extracted image is not clear, i.e., blur, active counter method is very difficult to implement and in region-based, tracking is very time-consuming, so in this paper adaptive background subtraction method for detection, Andostu's method to overcome these problems [17, 18]. In this result only 96 % of accuracy was there for counting the vehicle [19–21]. 100 % of accuracy count could not be reached because of partial-occlusion. So in our algorithm computational complexity is linear with respect to the video frame size and the total number of vehicles detected [22]. So we have considered traffic on highways, there is no question of shadow of any cast such as trees but occasionally due to illusions such as two vehicles are combined together as a single entity. In this paper, Freddy Lecue et al. [23] predicted congestion using semantic web technology. For prediction they considered all the traffic situations like weather information, road work, accidents, etc. [24, 25]. They correlated recent and historical data, different distance matrices are used for identifying pattern, also future prediction is done using certain rules, this helps for processing sensor data and prediction [26]. There prediction is accurate. Here processing of data is very hectic.

3 Proposed System

In our proposed system, ontology is used to predict the severity of road traffic congestion. Initially, the raw data are collected using camera and ultrasonic sensors and preprocessing (removing noisy data are parking vehicle, bicycles, pedestrian, etc.) is done. Building an ontology related to traffic congestion based on camera and sensor data. Parallel processing of sensor data and video collected from camera are done. By processing sensor data, the data like vehicle count, speed of the vehicle, inbound traffic and outbound traffic are obtained. By processing video data, sensor data, the data like vehicle count, speed of the vehicle, inbound traffic, and outbound traffic are obtained. We may get any of those data at any point of time '*t*' from parallel processing of sensor data and camera data. Our system use ontology to relate and predict traffic congestion at every '*t*' time period based on the data which may be the combination of sensor data and video data.

3.1 Proposed System Architecture

Using two sensors called ultrasonic sensor and passive infrared sensor the data is collected, and then after preprocessing data like vehicle count, speed of the vehicle, inbound traffic, and outbound traffic are obtained. Taking real-time video of road traffic congestion in the city. From this the data like vehicle count, speed of the vehicle, inbound traffic, and outbound traffic are obtained using preprocessing.

In parallel processing (Fig.1) we are using sensor data and video data. By processing sensor data like vehicle count, speed of the vehicle, inbound traffic, and outbound traffic are obtained. By processing video data, the data like vehicle count, speed of the vehicle, inbound traffic, and outbound traffic are obtained. After that noisy data are removed. Noisy data are parking vehicle, bicycles,pedestrian, etc. Ontology is a platform used to define relationship among attributes. It is a structured, hierarchical, controlled term that describes the concept or knowledge regarding particular domain. Ontology help in following processes, share common understanding of the structure of information, allow reuse of domain knowledge, and analyzes domain knowledge. The ontology process includes determining the domain and scope, listing of important terms, defining class and class hierarchies, defining slot and slot restriction.

3.2 Sensor Data Collection and Processing

The data was collecting sensors. Two types of sensors were used. They are (i) Ultrasonic sensors (US) Fig. 2 (ii) Passive infrared sensors (PIR) Fig. 3. Ultrasonic sensors are best for giving an easiest method for measurement of

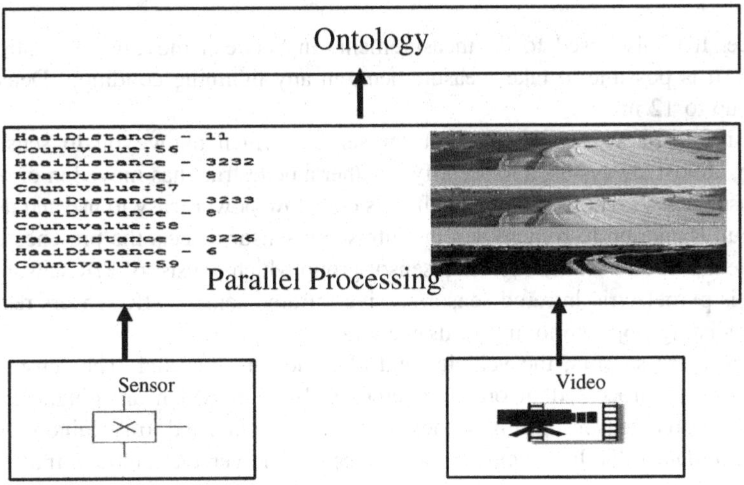

Fig. 1 Architecture of proposed system

Fig. 2 Ultrasonic sensor

Fig. 3 PIR sensor

distance. It is also used to do measurements in between moving and stationary objects. It is possible to take measurements in any lightning condition. Detection range: up to 12 m.

Passive infrared sensors are best for sensors which allow you to sense the motion, almost always used to identify whether a pedestrian has moved in or out of the sensors range. They are very small, less cost, low-power, easy to use and do not wear out. Range up to 6 m. When the ultrasonic sensor detects the presence of an obstacle it will turn on the detection sensor group which consist of PIR sensor. PIR sensor is pyroelectric infrared sensor/passive infrared sensor. PIR sensors respond to IR radiating objects moving in its view range.

Using these sensors, the vehicles and obstacles are detected. The sensor data have to be preprocessed in order to remove obstacles which are obtained from passive infrared sensors. In Fig. 4 these sensors are connected to Arduino board to get sensor data value like vehicle count, speed of the vehicle, inbound traffic, and outbound traffic.

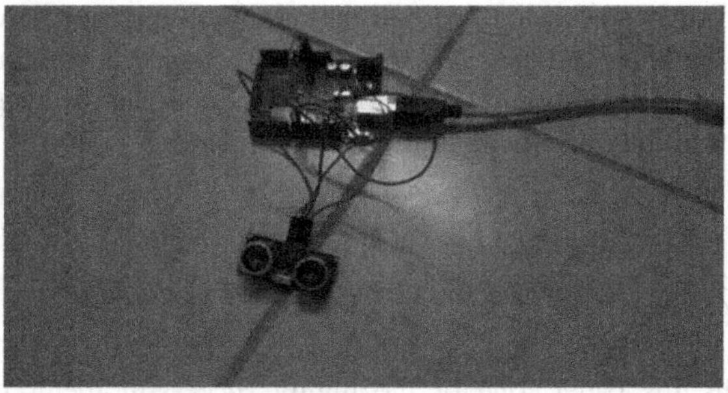

Fig. 4 Arduino connected with ultrasonic sensor

Fig. 5 Video dataset (.avi)

3.3 Video Data Collection and Processing

Real-time video (Fig. 5) of road traffic congestion was taken at different durations using a video camera. While processing this video data is converting into frames Various techniques were considered for detecting and counting total number of vehicles, among which Gaussian mixture model was better as it consumed very less time. In Gaussian mixture model itself was difficult, variations were there in which some variants took 8, 6, and 2 min, etc. Among these variants, the variant that consumed very less time was selected for detecting and counting total number of vehicles in the proposed system. In the selected variant of Gaussian mixture model, initially the foreground subtraction, it identifies the pixels in the converted frame. Foreground subtraction which is used to compare the video frame with the background model. Common approach to detect the foreground is to check the pixel considerably different from the corresponding background and clean foreground subtraction with morphological filter for noise elimination is done, i.e., noisy data are parking vehicle in the road, bicycles,pedestrian, etc. Finally, the vehicle is detected using blob analysis and counting the total number of vehicles is done. Here, blob analysis is blob detection was used to obtain regions of interest for further processing. To find the minimum distance between the two vehicles, using Euclidian distance between all the vehicles are initially calculated and using for loop the minimum one among these Euclidian distance values are considered to be minimum distance between two vehicles.

3.4 Gaussian Mixture Model (GMM)

In this paper, one of the highly successful methods for modeling is Gaussian mixture model. As shown in Fig. 6a–c, that is, why we have chosen this method. The most common way of describing the probability model of the distribution of the background is by Gaussian mixture distribution. This method uses the Gaussian mixture model to describe pixel processes. Depending on the delay and variance of the Gaussian mixture model, we can determine which Gaussian distribution is corresponding to the background colour. The pixels which do not fit the background distribution can be considered as foreground [27, 28]. Gaussian distribution is also called normal distribution. It was developed by the German mathematician Gauss in 1809 [29]

There are two important parameters for Gaussian distribution: the variance, denoted by σ^2 and the mean, denoted by μ. Depending on a pixel value u, $p\ (u)$ is determined. Also, when $\mu = 0$ and $\sigma^2 = 61$, the distribution of Y is a standard normal distribution value. See Eq. 1.

$$P(u, \mu, \sigma) = \frac{1}{\sqrt{2\pi\sigma}} \mathrm{e}^{\frac{(u-\mu)^2}{2\sigma^2}} \partial \tag{1}$$

$p\ (u,\ \mu,\ \sigma)$ means probability density Y—Random variable. μ—mean of the Gaussian distribution, while, σ—variance. If a set of data matches the Gaussian distribution, the data will be concentrated in the center of p within an interval of -2σ to 2σ. Usually, if there are many factors affecting the detection, a Gaussian distribution cannot be used to fully describe the actual background. We need multiple Gaussian models to describe the dynamic background. This means we need to create different Gaussian models for different situations. In the Gaussian

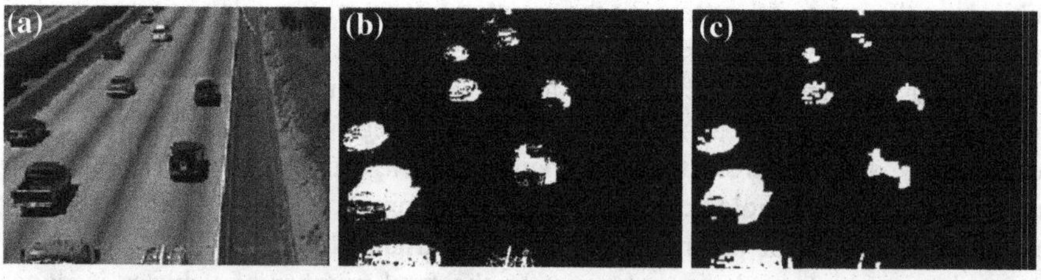

Fig. 6 **a** Original figure. **b** Foreground of the two Frame subtraction. **c** Foreground of the Gaussian model extraction

Model, we used two Gaussian models to represent each pixel. For a point (a, b) in a frame image, the observation value at time t is written as Y_t. At a given point (a, b) a series of observation values are $\{Y_1, Y_2 \ldots Y_t\}$. It can be seen as a statistical random process. We use the number of K Gaussian mixture model to simulate. With the probability distribution for the plotted point (a, b), we can use the equation to estimate the value.

$$P(Y_t) = \sum_{k=1}^{k} V_{k,t}.O(Y_t, \mu_{k,t}, \Sigma_{k,t}) \tag{2}$$

$v_{k,t}$ is the kth Gaussian mixture distribution weight at time t. O is the density function of Gaussian probability. $u_{k,t}$ is the kth Gaussian mixture model mean value. $\sum_{k,t}$ is the variance for the kth Gaussian mixture model. This section will show the basic idea about Gaussian distribution match. For a given point (a, b), we use the value Y_t to match with K Gaussian distributions. K is constant, taking a value from 3 to 5. Set one of the K Gaussian distributions to be O_k. If this Gaussian distribution O_k matches Y_t, then we use the value of Y_t to update the parameters of this O_k. If none of the K Gaussian distributions match Y_t, we use new Gaussian distributions to replace the old one. The definition of match can be shown like this. We arrange Gaussian distributions from big to small by the ratio of weight and variance (the ratio comes from w/σ). Then we choose a Gaussian distribution which has a value similar to X_t and μ_{bt-1} as the matched Gaussian distribution data, as in Eq. 3:

$$N = (|Y_t - \mu_{b,t-1}|^2 < \lambda^2\sigma^2) \tag{3}$$

Λ is a constant always set to 2.5. If Gaussian distribution is not found to match with the recent pixel, then the new Gaussian distribution will replace the smallest Gaussian distribution. In the new Gaussian distribution method, the mean value is the recent pixel value specified. The new Gaussian distribution will have large variance and small weight. The formula for adjusting the weight is shown below:

$$V_{k,t} = (1 - \alpha)V_{k,t-a} + \alpha \tag{4}$$

α is the learning rate. If a Gaussian distribution matches, the value of $N_{k,t}$ is 1. Else, the value of $N_{k,t}$ is 0. For the Gaussian distribution which is not a match, μ and α are of no change. For the Gaussian distribution model which

$$\mu_t = (1 - \beta)\mu_{t-1} + \beta Y_t \tag{5}$$

$$\sigma_t^2 = (1 - \beta)\sigma_{t-1}^2 + \beta(Y_t - \mu_t)^T(Y_t + \mu_t) \tag{6}$$

$$\beta = \alpha O(Y_t|\mu_t, \sigma_k) \tag{7}$$

β is a learning rate which is used to adjust the current Gaussian distribution. If β is large, the degree of matching Gaussian distribution is better. α is also a learning rate, and it reflects the speed of current pixel which merges into the background model. The basic idea of modeling is to extract the foreground from the current video frame. Let us summarize the modeling process of Gaussian mixture model. First, we assign several Gaussian models. Then, we initialized the parameters of the Gaussian model, followed by calculating the new parameters that will be used later. After that, we process each pixel in each video frame to see whether it matches a Gaussian model. If a pixel matches, then this pixel will be included in this Gaussian mixture model, and then we will update the technique under the value of new pixel. If the pixel does not match, then we create a new Gaussian model depending on the values of the pixel and the initialization parameters and replace by most unlikely model in the original model. Finally, select the most likely models as the background model. We establish the Gaussian model for every pixel in the video frame, and then do Gaussian model match with these pixels—this is used for foreground extraction. The Gaussian mixture model uses K—a value ranging from 3 to 5—Gaussian models to represent the characteristics of each pixel in video frame [29]. After getting a new video frame, the Gaussian mixture model will be update. The current video frame with each pixel point matches with the Gaussian mixture model. If successful, this point will be regarded as a background point. Otherwise, it is a foreground subtraction.

3.5 Building an Traffic-Related Ontology

Traffic ontology falls under application ontology. Application ontology describes terms dependent on both of a specific domain and of task solved about it. This is for a particular application which mainly includes the domain and its task. Ontology is defined by classes and their properties. Ontology graphically creates or edit with the help of tool. Protégé is an example for an ontology tool. Protégé helps us for creation and editing of one or more ontologies in single workspace. For interactive navigation of ontology relationship visualization tool is used.

4 Results

Output from sensor and video processing data is shown (Figs. 7, 8 and 9) in the following section:

4.1 Sensor Data

If the vehicle motion is detected in ultrasonic sensor, vehicle count is incremented. Sensor detects movement of the vehicle; it will be helpful to find number of

```
HaaiDistance - 11
Countvalue:56
HaaiDistance - 3232
HaaiDistance - 8
Countvalue:57
HaaiDistance - 3233
HaaiDistance - 8
Countvalue:58
HaaiDistance - 3226
HaaiDistance - 6
Countvalue:59
HaaiDistance - 3230
HaaiDistance - 7
Countvalue:60
HaaiDistance - 3253
HaaiDistance - 6
Countvalue:61
```

```
Motion detected!
7
Motion stopped!
Motion detected!
8
Motion stopped!
Motion detected!
9
Motion stopped!
Motion detected!
10
Motion stopped!
Motion detected!
11
Motion stopped!
```

Fig. 7 Ultrasonic and passive sensor data

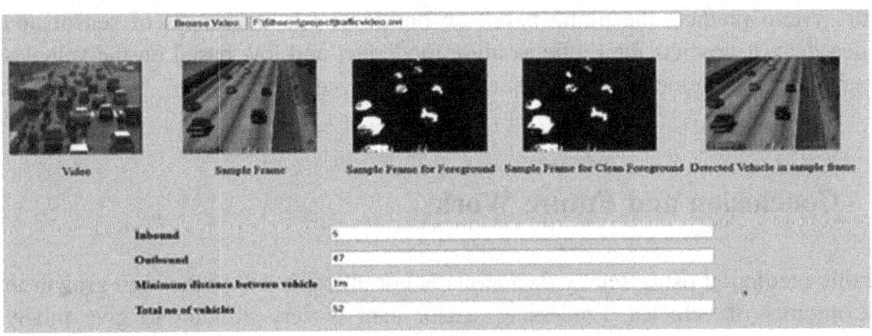

Fig. 8 Video processing result

vehicles based on frequent updates in the count, approximate vehicle speed is also calculated.

4.2 Video Data

In each video frame total number of vehicles are counted and distance between the vehicles is also calculated. Our system will give sum up of all the frames processed data and it will display the total number of inbound and total number of outbound vehicles, minimum distance between the vehicles is calculated.

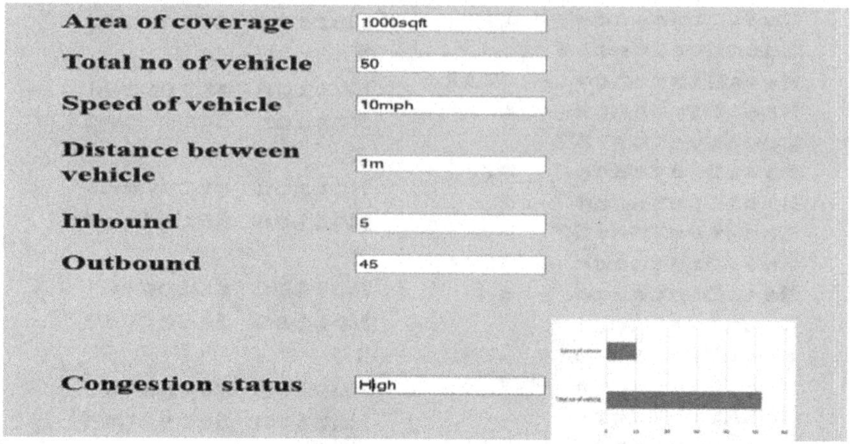

Fig. 9 Traffic prediction

4.3 Our Timely Prediction System

Our system predicts the traffic based on ontology rule inferences of sensor and video data. It predicts the traffic as high, moderate, and low based on the vehicles count, speed of vehicle, and distance between the vehicles.

5 Conclusion and Future Work

Traffic prediction using sensor data alone is not efficient due to the high growth in the number of vehicles. Processing sensor data is very difficult to give timely prediction. Processing video data also have the same problem. In our system, we are doing parallel processing of sensor and video data. If we got partial output from both parallel processing of sensor and video data, then our system infer congestion status through our ontology.

Prediction is done within few seconds. We have used PIR and ultrasonic sensors to collect data and MATLAB is used for video processing. We have created a model to give timely prediction of traffic congestion.

In future, instead of those two sensors, magnetic sensor and instead of MATLAB, TRAIS video processing tool can be used to improve the efficiency of our system which will predict the congestion within few seconds. Congestion is not communicated to public in our system. In future, we have a plan to use location-based SMS services to communicate traffic congestion to public.

References

1. Bauza Ramon, González Javier (2013) Traffic congestion detection in large-scale scenarios using vehicle to-vehicle communications. J Netw Comput Appl 36(5):1295–1307
2. Fukumoto J et al (2007) Analytic method for real-time traffic problems by using contents oriented communications in VANET. In: Proceedings of the 7th international conference on ITS Telecommunications (ITS T). Sophia Antipolis (France), 2007 June. pp 1–6
3. Thianniwet T, Phosaard SC, and Pattara-Atikom W (2009) Classification of road traffic congestion levels from GPS data using a decision tree algorithm and sliding windows. In: Proceedings of the world congress on engineering, vol 1
4. Choocharukul K (2005) Congestion measures in Thailand: state of the practice. In: Proceedings of the10th national convention on civil engineering, 2005 May, pp TRP111–TRP118
5. Priyadarshana YHPP et al (2013) GPS assisted traffic alerting and road congestion reduction mechanism. In: Proceedings of technical sessions, vol 29
6. Tao S, Manolopoulos V, Rodriguez S, Rusu A (2012) Real-time urban traffic state estimation with A-GPS mobile phones as probes. J Transp Technol 22–31
7. Nicolas Lefebvre IVT, Zurich ETH, Balmer M (2007) Fast shortest path computation in time-dependent traffic networks, Conference paper STRC
8. Neha (2014) Implementation and comparison of the improved traffic congestion control scenario. Int J Adv Res Comput Sci Softw Eng 4(4) ISSN 2277 128X
9. Fratila C et al (2012) A transportation control system for urban environments. 2012 Third International Conference on IEEE Emerging Intelligent Data and Web Technologies (EIDWT)
10. Parag T, Elgammal A, Mittal A (2006) A framework for feature selection for background subtraction. 2006 IEEE computer society conference on computer vision and pattern recognition vol 2. IEEE
11. Schapiro Robert E, Singer Yoram (1999) Improved boosting algorithms using confidence-rated predictions. Mach Learn 37(3):297–336
12. Ryan C et al (2008) Real-time foreground segmentation via range and color imaging. IEEE Computer Society conference on computer vision and pattern recognition workshops, 2008. CVPRW'08. IEEE
13. Kim K et al (2005) Real-time foreground–background segmentation using codebook model. Realtime Imaging 11(3):172–185
14. Banarase-Deshmukh RG, Kumar P, Bhosle P Video object tracking based on automatic background segmentation: a survey
15. Gupte S et al (2002) Detection and classification of vehicles. Int Transp Syst, IEEE Trans 3 (1):37–47
16. Waghmare P, Borkar S ((2007) A survey on techniques for motion detection and simulink blocksets for object tracking
17. Alefs B (2006)Embedded vehicle detection by boosting. IEEE intelligent transportation systems conference, 2006. ITSC'06. IEEE
18. Stauffer C, Grimson WEL (1999) Adaptive background mixture models for real-time tracking. IEEE computer society conference on computer vision and pattern recognition 1999, vol 2, IEEE
19. Venugopal KR, Patnaik LM (2008) Moving vehicle identification using background registration technique for traffic surveillance. In: Proceedings of the international multi conference of engineers and computer scientists, vol 1
20. Kamiya K (2012) A framework of vision-based detection-tracking surveillance systems for counting vehicles
21. Raghtate G, Tiwari AK (2014) Moving object counting in video signals. Int J Eng Res Gen Sci 2(3):2091–2730
22. Ambardekar AA (2007) Efficient vehicle tracking and classification for an automated traffic surveillance system. ProQuest

23. Lécué F et al (2014) Predicting severity of road traffic congestion using semantic web technologies. The semantic web: trends and challenges. Springer, Berlon, pp 611–627
24. Lécué F et al () Smart traffic analytics in the semantic web with STAR-CITY: scenarios, system and lessons learned in Dublin City. Web semantics: science, serv agents on the world wide web 27:26–33
25. Lécué F, Schumann A, Sbodio ML (2012) Applying semantic web diagnosing technologies for road traffic congestions. The Semantic Web–ISWC 2012. Springer, Berlin, Heidelberg, pp 114–130
26. Lécué F et al (2014) Star-city: semantic traffic analytics and reasoning for city. In: Proceedings of the 19th international conference on intelligent user interfaces. ACM
27. Power PW, Schoonees JA (2012) Understanding background mixture models for foreground segmentation. In: Proceedings image and vision computing New Zealand vol 2002
28. Gallego J, Pardàs M, Haro G (2009) Bayesian foreground segmentation and tracking using pixel-wise background model and region based foreground model. 16th IEEE international conference on image processing (ICIP), 2009. IEEE
29. Wang Y, Lv W (2011) Indoor video-based smoke detection

Optimal Selection of Security Countermeasures for Effective Information Security

R. Sarala, G. Zayaraz and V. Vijayalakshmi

Abstract Information systems must be protected from misuse by threats that take advantage of the vulnerabilities present in them and cause financial or reputation damage to the organization. The information security official of the organization has to identify suitable controls to mitigate the risks to which the organization is exposed to by considering the risks to be addressed, its impact in terms of revenue and the cost incurred in implementing the security controls. The selection should be made in such a way that (i) the cost incurred by the selected controls should be within the budget constraints and not exceed the losses suffered by the organization, (ii) effectively address a set of vulnerabilities and also minimizes the risks that remain unaddressed. A hybrid approach combining tabu search and genetic algorithm has been proposed to aid in the selection process. The proposed algorithm helps in optimizing the security controls selection process.

Keywords Metaheuristics security controls · Hybrid Tabu-Genetic algorithm *Optimal solution*

1 Introduction

Information systems form part of the critical assets of any organization and securing them from the threats has become an important responsibility for all organizations in this digitized world. The misuse of critical information systems poses serious challenges to organizations leading to loss of productivity, loss of revenue, loss of customers, loss of reputation and sometimes legal issues [1]. The main goal of

R. Sarala (✉) · G. Zayaraz
Department of Computer Science and Engineering, Pondicherry Engineering College, Pondicherry, India
e-mail: sarala@pec.edu

V. Vijayalakshmi
Department of Electronics and Communication Engineering, Pondicherry Engineering College, Pondicherry, India

L.P. Suresh and B.K. Panigrahi (eds.), *Proceedings of the International Conference on Soft Computing Systems*, Advances in Intelligent Systems and Computing 398, DOI 10.1007/978-81-322-2674-1_33

Information security is to protect the information assets of an organization from disclosure, modification, and destruction and ensure their confidentiality, integrity, and availability for the successful functioning of the organization.

1.1 Information Security Risk Management

It is the responsibility of the information security personnel of any organization to identify the security controls required to safeguard the information assets from the threats to them. Information security risk management has an active part in ensuring the security posture of the organization [2]. The information security risk management comprises of activities namely (i) vulnerability identification, (ii) threat identification, (iii) risk assessment, (iv) asset valuation, (v) risk estimation, and (vi) countermeasure suggestion. After assessing the risks to the assets, the risk management team provides a report stating the list of vulnerabilities present in the organization, the threats to those assets and the risk impact if the vulnerabilities are exploited by the threats. The final report also recommends some security controls as countermeasures to mitigate the risks identified.

1.2 Countermeasures

A countermeasure may be an action or a procedure that is employed in reducing the risks due to the vulnerabilities and threats in any organization. The countermeasures that are selected to counteract a given set of vulnerabilities and threats have to focus on maximizing the vulnerability coverage. But, every countermeasure has a cost associated with it. Hence, the vulnerability coverage and cost are the important trade-offs in choosing a specific countermeasure.

1.3 Countermeasure Selection and Metaheuristics

The risk report evaluates the security controls which are already in place in the organization for their ability to mitigate the identified risks. Therefore it requires adding new controls or reconfiguring the existing ones in order to meet the security needs of the organization. This decision on adding up new security controls or modifying the existing ones to tackle the security flaws that were identified is made by considering few important criteriawhich include (i) the severity of the risks identified, (ii) the organizations' spending capacity on security, (iii) the cost of the new security controls, (iv) the loss that the organization has to face if a security breach occurs, and (v) the residual risk that remains even after adding the new security controls or modifying the existing security controls.

This paper handles the problem of selecting the effective security controls by using a decision mechanism that is a hybridization of the metaheuristics genetic algorithm and tabu search. The use of metaheuristics such as tabu search, genetic algorithms, swarm optimization algorithms have been proved to be suitable for solving multiobjective optimization problems in literature [3]. This paper tries to combine a local optimization heuristic which is good at exploitation with a global optimization heuristic which is good at exploration to help in the effective countermeasures selection problem.

2 Related Work

The selection of suitable security controls within the given constraints is a NP-hard problem as the difficulty in identifying the security controls increases with the increase in the number of vulnerabilities and the threats that may exploit these vulnerabilities. Many techniques have been proposed in literature for solving this multiobjective problem, some of which are discussed below.

Tadeusz Sawik modeled the optimal portfolio selection as a single- or bi-objective mixed integer program by calculating the value at risk and conditional var risk measures [4]. Valentina Viduto, Carsten Maple, Wei Huang, and David López-Peréz proposed a multi-objective tabu search based approach to solve the control measure selection problem [5]. The risk assessment and optimization model proposed by them aims at arriving at an optimal countermeasure by employing multiobjective tabu search. The approach is cost effective over other methods in literature, but employing pure tabu search may lead to missing some promising areas of the search space.

Roy et al. [6], proposed attack countermeasure trees for scalable optimal countermeasure selection. Greedy strategies and implicit enumeration techniques are used to compute optimal countermeasure set for various objective functions. The limitation of their approach is that it serves only in the attacker's point of view. Nagata et al. [7] proposed a fuzzy outranking based approach for selecting risk mitigation controls. The countermeasures are compared using a fuzzy relation called outranking. The critical assets are identified using fuzzy decision-making and the degree of risks to the assets is computed using fuzzy inferencing. Checking threats paths is a time consuming process and hence this approach cannot be used for a larger dataset of risk mitigation controls.

A. Ojamaa, E. Tyugu, and J. Kivimaa proposed a hybrid expert system combined with discrete dynamic programming for optimal selection of countermeasures [8]. Bistarelli et al. [9], proposed a qualitative approach for the selection of security countermeasures to protect an IT system from attacks by modeling security scenarios using defense trees. The countermeasure selection is done using conditional preference networks.

Bojanc and Jerman-Blazic [10] proposed the selection of the optimal investment based on the quantification methods for identification of the assets, the threats, and vulnerabilities. Gupta et al. [11], have implemented the countermeasure selection by relating it to minimizing the residual vulnerabilities and matching the information security vulnerabilities to organizational security profiles using a genetic algorithm approach. Rees et al. 12] proposed a decision support system for calculating the uncertain risk faced by an organization and search for the best combination of countermeasures by using a genetic algorithm approach.

3 Proposed Work

A solution set to address the vulnerabilities and threats present in an organization consists of a set of security controls. This paper aims to help in the selection of optimal security controls to form the solution set that helps to enhance the security posture of the organization. Since this problem is identified as a multiobjective problem, the solution to this is a Pareto optimal set of security controls. The proposed work tries to provide a hybridized solution to this multiobjective problem by combining the tabu search algorithm and the genetic algorithm. Genetic algorithms have been used in literature to solve multiobjective optimization problems [13]. This hybridization is inspired by the work of [5] and [11]. But the proposed work differs from [5] in the way that an actual matching between the threats and vulnerabilities, and vulnerabilities and countermeasures is done, whereas the existing work considers only a generic set of vulnerabilities and countermeasures. Also, the idea of combining more than one metaheuristics so as to improve the selection process by combining the advantage of both of them is novel to be included in the risk management approach.

4 Implementation of the Proposed System

The proposed system contains a monitoring module which is used to identify the threats to the organization and the vulnerabilities present in the existing infrastructure. These vulnerabilities could be exploited by the threats to create risks to the assets of the organization.

The sample network architecture for the proposed system is given in Fig. 1.

4.1 Vulnerability Identification

The monitoring activity by using any vulnerability scanning tool such as Nessus identifies the set of vulnerabilities present in the infrastructure. A sample list of vulnerabilities identified is given in Table 1.

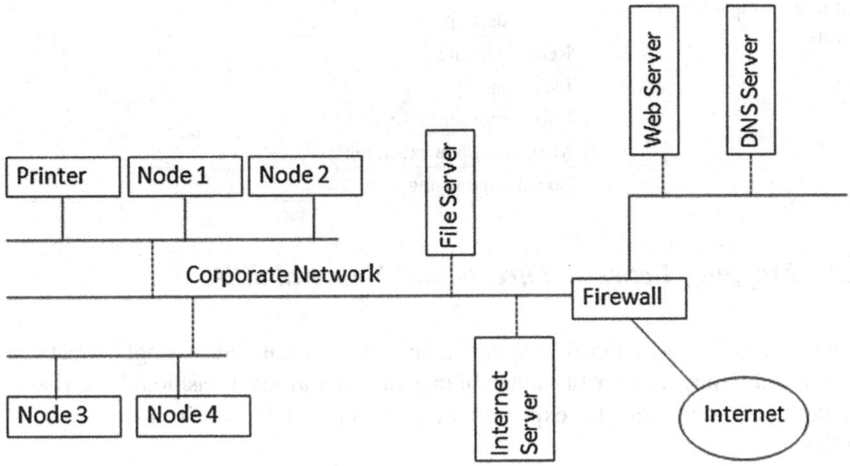

Fig. 1 Sample network infrastructure

Table 1 Sample list of vulnerabilities

CVE-ID	Vulnerability description
CVE-2015-2702	Cross-site scripting vulnerability in the Email Security gateway in Websense appliances allows remote attackers to inject arbitrary web script in an email
CVE-2014-8016	Vulnerability in the Cisco IronPort Email Security appliance allows remote attackers to cause a denial of service
CVE-2011-3579	Vulnerability in IceWarp Mail Server allows remote attackers to read arbitrary files, and cause a denial of service
CVE-2010-4297	Vulnerability in VMware server allows host OS users to gain privileges
CVE-2008-4533	Cross-site scripting vulnerability in Kantan WEB Server 1.8 allows remote attackers to inject arbitrary web script or HTML

4.2 *Threat Analysis*

Threat analysis can be done in order to identify the potential threats to the organization's assets. This could be identified from the organization's historical database or log files. Also attack graphs can be constructed from which we can identify the potential threat sources. This will help in minimizing the time taken for gathering the threats pertaining to an organization rather than going for log analysis. Table 2 gives a sample list of threats identified for the given setup.

Table 2 Sample list of threats

Threat description
Remote spying
Espionage
Information disclosure
Malicious code execution
Social engineering

4.3 Mapping Between Threats and Vulnerabilities

The threats are to be mapped with the vulnerabilities identified. A single threat can be mapped to any number of vulnerabilities in a one-to-many fashion. Likewise, a single vulnerability can be exploited by a number of threats in a one-to-many fashion.

4.4 Mapping Between Countermeasures and Vulnerabilities

A generic list of countermeasures combining technical, administrative, and operational category are taken. A sample list is given in Table 3.

A mapping is done between the countermeasures and vulnerabilities identified in the overall system to identify the number of vulnerabilities that can be addressed by each countermeasure. This helps to choose controls that address the maximum number of vulnerabilities for constructing the solution set. Also, if a control is included in the solution set, it is assigned a cost of one unit. This idea has been taken from [5], as it is difficult to identify the exact cost of any security control. Table 4 gives the number of vulnerabilities addressed by each security control present in Table 3. Table 5 gives the cost associated with the solution set and the number of vulnerabilities addressed by it for a sample set of five solutions.

Table 3 Sample list of countermeasures

Countermeasure	Identifier
Secure network engineering	C1
Penetration tests	C2
Malware defenses	C3
Wireless access control	C4
Data recovery capability	C5

Table 4 Countermeasures and the number of vulnerabilities addressed by them

Countermeasure	Number of vulnerabilities addressed
Installation of antivirus software C1	6
Security policies enforcement C2	66
Implementation of router filtering policies C3	78
Application of vendor patches C4	238
Implementation of firewall filters C5	67

Table 5 Sample solution set with cost and number of vulnerabilities addressed

Solution Set	Cost	Number of vulnerabilities addressed
C2 C3 C4 C5 C6 C7 C8	7	685
C7 C8 C9	3	389
C2 C3 C7	3	456
Cl C3 C4 C5 C6 C7 C8 C9	8	656
C2 C4 C7 C8 C9	5	609

4.5 Hybrid Tabu-Genetic Algorithm for Optimal Selection of Countermeasures

The hybrid Tabu-Genetic algorithm takes an initial solution set from the available list of solution sets composed by combining the security controls as input to identify the Pareto optimal set of solutions. The Pareto optimal set of solutions are those solutions in the search space which are non-dominating with each other with respect to the objective functions.

The first part of the algorithm selects the initial solution set as the best solution. It then generates set of solutions which are called as neighbor solutions. Then, best solution is compared with each solution available in the neighborhood list for the objective functions for the cost and the number of vulnerabilities addressed.

The objective functions are taken as (i) one that is maximizing the number of vulnerabilities addressed by the solution set, and the (ii) one that is minimizing the cost incurred by the solution set. In this comparison, the one with the lesser value for cost and higher value for number of vulnerabilities addressed and non-dominating is taken as the best solution. Those solutions which are used in this comparison are stored in a tabu list and they are not considered again in order to avoid the cycling problem. After running for a predefined number of iterations, a set of Pareto optimal solutions are obtained.

The obtained solutions are then given as the initial population for the genetic algorithm. The genetic algorithm then evaluates the fitness of the solutions and performs the genetic operations of crossover and mutation on the selected parents. The parent selection is done using roulette wheel method.

Step 1: Randomly generate an initial solution.
Step 2: Initialize the tabu list to empty, and assign the initial solution to the best solution.
Step 3: While the termination condition is not met,
- •Choose a candidate solution from the list of neighborhood solutions.
- •Evaluate the candidate solution and the best solution against the objective functions.
- •If the cost (candidate solution) < cost (best solution) && vulnerabilities (candidate solution) > vulnerabilities (best solution)
    - •Best solution = candidate solution.
- •Store Best Solution to Population pool.
- •Update the tabu list.

Step 4: Repeat the step 3 until termination condition is met.
Step 5: Retrieve individuals for the initial population from the population pool.
Step 6: Evaluate the individuals in the population pool by applying the fitness functions.
Step 7: Select the best parents for generating children for the next generation.
Step 8: Apply the cross over operator to generate children.
Step 8: Apply mutation operation to create a diversified population pool.
Step 9: Apply elitism to choose the 10% of fit ones from parents and 90% from children.
Step 10: Repeat steps 6 to 9 until the termination condition is reached.

Fig. 2 Pseudocode for the hybrid genetic Tabu algorithm

The offsprings generated once again evaluated for their fitness. Elitism strategy is applied to choose 10 % of the best parents for the next generation, with 90 % of the fittest children. The proposed algorithm is given in Fig. 2.

The tabu search algorithm works on criteria that among the solutions organized in the Pareto front, the solution vector to be considered for comparison with the neighborhood ones should be the one that has not been already exploited. In that case, only those vectors that have not been taken for evaluation previously can go for comparison in arriving at a best solution. This does not bring in optimality where a Cartesian product combination of vectors is not taken for arriving at best solution. Also tabu search requires evaluating the objective function for every element of the neighborhood of the current solution which would be extremely expensive from a computational point of view.

On the contrary to it, genetic algorithm is specially meant for solving optimization problems. In a genetic algorithm, a pair of individuals is taken for cross over in producing a new offspring thereby forcing the search into previously unexplored areas of the search space. Therefore, a hybrid approach consisting of tabu search and genetic algorithm is employed and it has generated optimal results producing effective solution sets to defend against the vulnerabilities exposed to the organization within the given budget limits.

The effectiveness of the hybrid approach is evaluated based on the number of vulnerabilities addressed and the cost incurred in the security controls selection.

5 Conclusion

Enormous work has been done in literature to support information security decision makers to handle their financial investments in security in an optimized way. However, most of the models described in the existing study are hypothetical and not practical. Tabu search implementation brings in the disadvantage of partial coverage of the Pareto front leading to a nonoptimal solution. A hybrid approach has been followed that exploits the benefits of both the tabu search as well as the genetic algorithmic search techniques in a way that it gives an optimal solution.

References

1. Bojanc R, Jerman-Blazic B (2008) An economic modelling approach to information security risk management. Int J Inf Manage 28:413–422
2. http://csrc.nist.gov/publications/nistpubs/800-137/SP800-137-Final.pdf
3. Zitzler Eckart, Laumanns Marco, Bleuler Stefan (2004) A tutorial on evolutionary multiobjective optimization, metaheuristics for multiobjective optimisation, lecture notes in economics and mathematical systems. Springer, Berlin
4. Sawik Tadeusz (2013) Selection of optimal countermeasure portfolio in IT security planning. elsevier, J Decis Support Syst, pp 156–164
5. Viduto V, Maple C, Huang W, López-Peréz D (2012) A novel risk assessment and optimisation model for a multi-objective network security countermeasure selection problem. J Decis Support Syst, 53, Elsevier, pp 599–611
6. Roy A, Kim DS, Trivedi KS (2012) Scalable optimal countermeasure selection using implicit enumeration on attack countermeasure trees. In: Proceedings of the 42nd annual IEEE/IFIP international conference on dependable systems and networks, USA, pp 1–12
7. Nagata K, Amagasa M, Kigawa Y, Cui D (2009) Method to select effective risk mitigation controls using fuzzy outranking, ninth international conference on intelligent systems design and applications, Italy
8. Ojamaa A, Tyugu E, Kivimaa J (2008) Pareto-optimal situation analysis for selection of security measures. In: Proceedings of IEEE military communications conference, MILCOM 2008, pp 1–7
9. Bistarelli S, Fioravanti F, Peretti P (2007) Using CP-nets as a guide for countermeasure selection. In: Proceedings of the ACM symposium on applied computing, Seoul, Korea, pp 300–304
10. Bojanc R, Jerman-Blazic B (2008) An economic modeling approach to information security risk management. Int J Inf Manage 28:413–422
11. Gupta M, Rees J, Chaturvedi A, Chi J (2006) Matching information security vulnerabilities to organizational security profiles: a genetic algorithm approach. Decis Support Syst 41:592–603
12. Rees LP, Deane JK, Rakes TR, Baker WH (2011) Decision support for cyber security risk planning. Decis Support Syst 51:493–505
13. Abdulla K, Coit DW, Smith AE (2006) Multi-objective optimization using genetic algorithms: a tutorial. J Reliab Eng Syst Saf, Elsevier

Stabilization of Load in Content Delivery Networks

P. Asha

Abstract The proposal is to determine an effective solution to the problem of load balancing in content delivery networks. This system is proposed based on a thorough analysis of a content delivery network system and taking into consideration the packet size and also the individual node size. The obtained result is then used to maximize the performance of the system and incorporate time to live (TTL), piggybacking, and pre-fetching mechanisms for consistency. Every data item is authorized with a corresponding TTL value within which the corresponding data item must reach the destination node. And upon expiration, a call is sent to the data source to revive them. The entire system is validated by means of simulations and upon implementing such a technique, it is seen that the traffic per node is drastically reduced as the packet size increases. The results clearly narrates that, compared with other conventional routing protocols, this mechanism reduces the path discovery overhead, delay and the traffic during transmission by a great margin.

Keywords Content delivery networks · Piggybacking · Pre-fetching · Load balancing

1 Introduction

A content delivery network (CDN) [1–3] is a group of distributed servers (network) that delivers WebPages and other Web related data to a user, based on the topological location of the user by means of a distributed collection of servers. It is an effective approach to support Web Applications and is widely used to minimize congestion issues that occur owing to the immense request rates from the clients, in turn downsizing inactivity and simultaneously augmenting the availability of the content.

P. Asha (✉)
Department of Computer Science and Engineering, Sathyabama University, Chennai, India
e-mail: ashapandian255@gmail.com

L.P. Suresh and B.K. Panigrahi (eds.), *Proceedings of the International Conference on Soft Computing Systems*, Advances in Intelligent Systems and Computing 398, DOI 10.1007/978-81-322-2674-1_34

- In this proposal, the important issue is the implementation and definition of an efficient solution for the problem of load imbalance in "Content Delivery Networks"
- Features like piggybacking [4, 5], time to live (TTL), and pre-fetching are incorporated to enhance the effectiveness of the proposed system.
- The technique suggested is then validated by means of simulations.

The concept of piggybacking [6, 7] is integrated in the network layer. It is a dual-direction data transportation technique that helps the sender in making sure that the data frame sent by the sender was well received by the intended receiver (ACK acknowledged). This concept of piggybacking is implemented in this proposal so as to enhance the overall efficiency of the system and to avoid any loss of data packets during transmission.

TTL is a factor that is obtained by the difference in time among the query duration of the item and its previous time of updation. TTL [8] is the amount of hops that a packet is permitted to navigate, prior to being discarded by a router. A packet refers to the fundamental unit of information transport computer networks and is being implemented in other communications networks as well. And if a particular packet of data is alive in the network for too long without reaching a destination node, that is, if it is subject to an infinite redirection loop, it causes the succeeding packets to wait and thus resulting in an eventual network crash.

Pre-fetching [9] refers to the concept of transferring (data) from main memory to temporary storage so that it is readily available for later use. In this context the temporary storage refers to as the cache. When a sender node intends to send data to the receiver, it initially sends it via intermediate nodes. These intermediate nodes store the destination node information temporarily and during the next transaction this temporary storage (cache node) [10] provides the destination information to the source, thus facilitating quick processing of data.

1.1 Disadvantages of Existing System

- The local server's queues tend to overflow due to unawareness of the total number of requests that could possibly be received by the system at a particular instance.
- Irregular ordering of distributing the client requests lead to packet losses at the receivers end.
- There is a possibility that there could be loss of packets due to the overflow of the local server queues.

2 Proposed System

The primary step is to model a relevant load-balancing rule, which promises stability of queues in the balanced CDN.

- A new mechanism for the even distribution (Fig. 1) of the approaching requests from the clients to the server that is most suitable, hence stabilizing the load of the total system requests is proposed. This is made possible by frequent interaction between the system nodes.
- The client requests will be gathered into the proxy server and it will distribute to n number of nodes and again it has been uniquely distributed to another proxy.
- We try to avoid the overflow of data in the queue, avoiding the number of redirection mechanisms and increasing the computation speed via a "Relative weight dividend algorithm."
- The concept of Piggybacking and Pre-fetching are included in the proposed system design, so as to make node-to-node interactions even more efficient.

2.1 *Modules*

Various modules involved in this work are explained below.

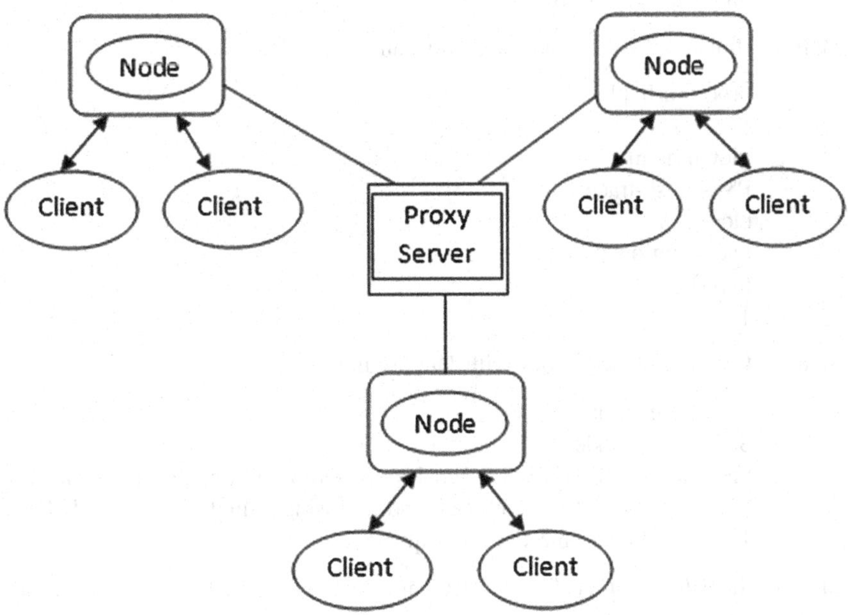

Fig. 1 Architecture diagram

1. Node creation
2. Request Handling
3. Network Monitoring
   - TTL Adaptation
   - Piggybacking
4. Traffic Maintenance
   - Pre-fetching
5. Performance Analysis

2.2 Node Creation

This module concerned with creation of nodes with the help of TCL script using NS2. The steps to create node are as follow:

Step 1: Creating simulator_object;

```
set ns [new Simulator]
```

Step 2: Opening a file in write mode (nam trace)

```
set nf [open out.nam w]
$ns namtrace-all $nf
```

Step 3: 'finish' - Closes trace file, start nam.

```
proc finish {}
{
global ns nf
P$ns flush-trace
close $nf
exec nam out.nam&
exit 0
}
```

Step 4: Creation of two nodes with Duplex link

```
Set n0 [$ns node]
Set n1 [$ns node]
```

$ns duplex-link $n0 $n1 1 Mb 100 ms Drop Tail line tells the simulator object to connect the nodes n0 and n1, using a duplex link, 1 MB BW, 10 ms delay and the Drop tail queue.

Step 5: Save file, initiate the script using 'ns project.tcl'. Now nam boots up automatically.

2.3 Request Handling

This module describes the basic operation of client request to the server and how the request has been handled in the network. It also shows the basic functionalities of the requesting node (RN) and the server.

2.4 Network Monitoring

This module deals with nodes present in the network. It shows how the network adapts itself in order to handle the disconnected. It also deals with the responsibility for maintaining and replacing of the cached data in the acceptor node (AN).

2.4.1 TTL Adaptation

TTL is a factor that is obtained by the difference in time among the query duration of the item and its previous update time. TTL is the amount of hops that a packet is permitted to navigate, prior to being discarded by a router. A packet refers to the fundamental unit of information transport computer networks and is being implemented in other communications networks as well. And if a particular packet of data is alive in the network for too long without reaching a destination node, that is, if it is subject to an infinite redirection loop, it causes the succeeding packets to wait and thus resulting in an eventual network crash.

2.4.2 Piggybacking

The concept of piggybacking is integrated in the network layer. It is a dual-direction data transportation technique that helps the sender in making sure that the data frame sent by the sender was well received by the intended receiver (ACK acknowledged). This concept of piggybacking is implemented in this proposal so as to enhance the overall efficiency of the system and to avoid any loss of data packets during transmission.

2.5 Traffic Maintenance

This module keeps track of the update rate and the request rate of a particular data in the AN and the server. Based on that the caching of data is done which reduces the traffic in the network.

2.5.1 Pre-fetching

Pre-fetching refers to the concept of transferring (data) from main memory to temporary storage so that it is readily available for later use. In this context the temporary storage refers to as the cache. When a sender node intends to send data to the receiver, it initially sends it via intermediate nodes (Fig. 2). These intermediate nodes, store the destination node information temporarily and during the next transaction this temporary storage (cache node) provides the destination information to the source, thus facilitating quick processing of data.

3 Results and Discussions

The performance of the network has been evaluated by varying the number of nodes, varying the query request rate and mobility of nodes in the network (Table 1).

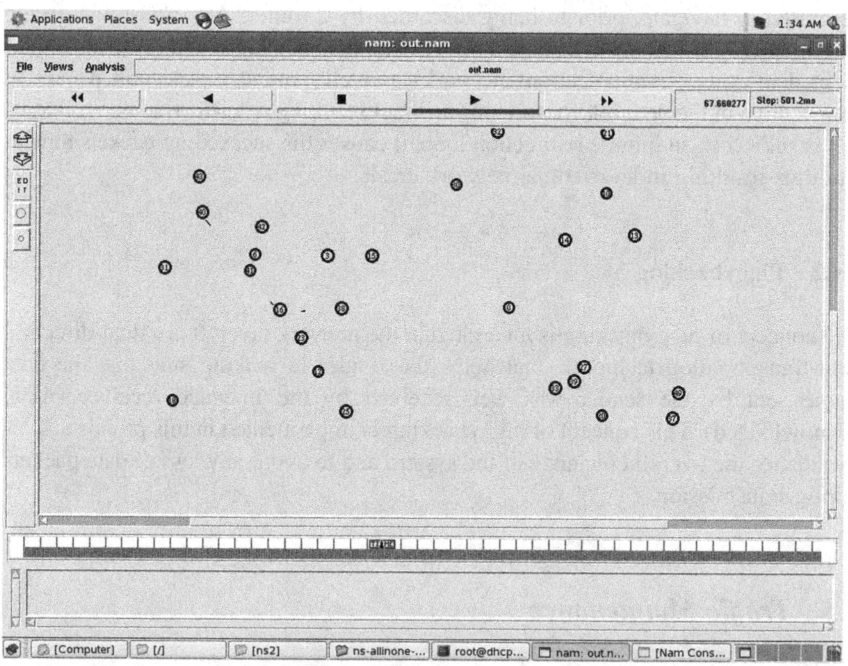

Fig. 2 Sending request to cache node

Table 1 Comparison of existing and proposed system

Packet size ($\times 10^3$)	Traffic per node (pkts/s)	Traffic per node (pkts/s)
0	0	0
0.1	20	19
0.2	34	27
0.3	50	40
0.4	100	90
0.5	150	140
0.6	200	190
0.7	219	200
0.8	232	215
0.9	250	220
1	265	240
1.1	278	259
1.2	290	280

3.1 Relative Weight Dividend Algorithm

Step 1: Initialize all the nodes of the network.
Step 2: Configure the directional nodes and the destination node.
Step 3: Define the packet size and the length of the data packets that are to be transmitted across the network.
Step 4: Configure the starting and ending point of node transactions.
Step 5: Check the weight (load) of node and compare the corresponding node size and packet size,

i.e., Node Size > Packet Size.

Step 6: Increment the packet count upon successful transmission of packet across the nodes.
Step 7: Calculate the distance between one node and the other node for communication.
Step 8: Buffer the packets that were lost during the transmission (Piggybacking).
Step 9: End transmission when all packets have reached the destination node.

After that, while computation is on progress, sometimes load imbalance occurs. To balance the group workload, the system takes up a load-balancing procedure, which is stated below.

3.2 Load Balancing

Step 1: Calculate the workload of every individual node in a Group (Cluster): wccn (workload of cluster compute node)

wccn = processing time * communication time

Step 2: Calculate the workload of a Cluster: chn (cluster head node)

chn = sum of wccn.

Step 3: Let Avg represents the average workload,

Avg = chn/No. of Nodes.

Step 4: Then for every Node, check

- If load value exceeds the average then term it as overloaded.
- If load value is less than the average then term it as underloaded.
- Otherwise assume that the load is balanced.

Now, the work can be migrated from overloaded node to the underloaded node.

3.3 XGraph Representation

Here the graph is obtained for the throughput where it shows the activity of nodes and their data transmission (Table 1) throughout the process. The graph indicates how the traffic rate of the proposed system (represented in green) is relatively lesser than that of the existing system (represented in red). The difference is highly evident when the packet size increases.

As shown in the graph (Fig. 3), the difference in traffic in the network between the proposed and the existing system is greater when the packet size is higher and marginal when the packet size is small.

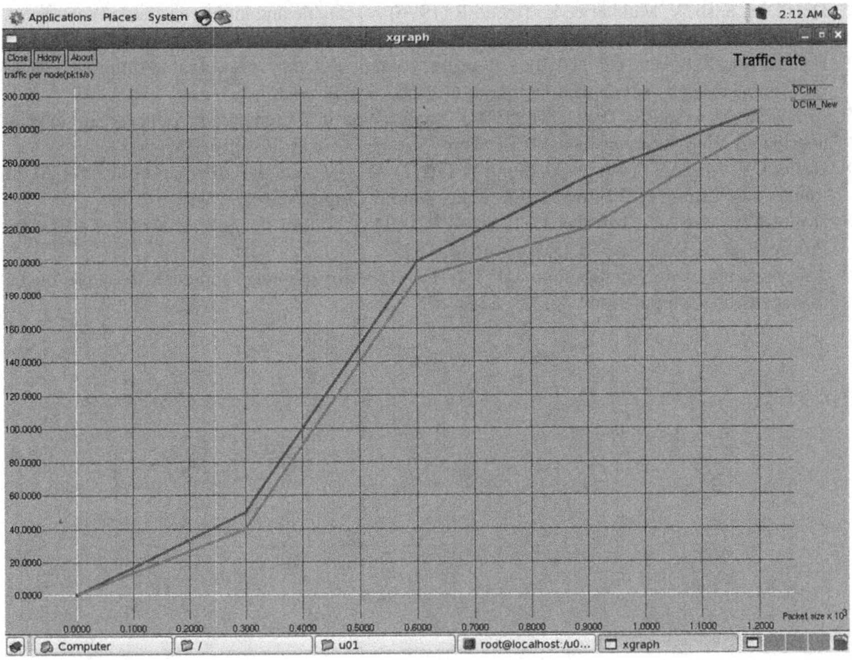

Fig. 3 Performance variations

4 Conclusion

Hence the proposed technique aims at obtaining load balancing in the network, by getting rid of the local queue vulnerability circumstances. This happens by redistribution of possible surplus traffic to its set of neighbors of their overcrowded server. It also provides ways to retransmit lost data packets back to the intended destination.

References

1. Manfredi S, Oliviero F, Romano S (2013) A distributed control law for load balancing in content delivery networks. IEEE Netw 21(1):55–68
2. Merazka F (2013) Packet loss concealment using piggybacking for speech over IP Network Services. 2013 IEEE 7th international conference on intelligent data acquisition and advanced computing systems (IDAACS), vol 1, Berlin, Germany, pp 509–512
3. Manfredi S, Oliviero F, Romano S (2010) Distributed management for load balancing in Content Delivery Networks. In: Proceedings of IEEE GLOBECOM workshop, Miami, FL, Dec 2010, pp 579–583
4. Yin H, Liu X, Min G, Lin C (2010) Content delivery networks: a bridge between emerging applications and future IP networks. IEEE Netw 24(4):52–56

5. Dias DM, Kish W, Mukherjee R, Tewari R (1996) A scalable and highly available Web server. In: Proceedings of IEEE computing conference, pp 85–92
6. Pineda JD, Salvador CP (2010) On using content delivery networks to improve MOG performance. Int J Adv Media Commun 4(2):182–201
7. Colajanni M, Yu PS, Dias DM (1998) Analysis of task assignment policies in scalable distributed Web-server systems. IEEE Trans Parallel Distrib Syst 9(6):585–600
8. Hollot CV, Misra V, Towsley D, Gong W (2002) Analysis and design of controllers for AQM routers supporting TCP Flows. IEEE Trans Autom Control 47(6):945–959
9. Hollot CV, Misra V, Towsley D, Gong W (2001) A control theoretic analysis of RED. In: Proceedings of IEEE INFOCOM, pp 1510–1519
10. Aweya J, Ouellette M, Montuno DY (2001) A control theoretic approach to active queue management. Comput Netw 36:203–235

Eigen-Based Indexing and Retrieval of Similar Videos

R. Sunitha and A. Mary Posonia

Abstract Proficient identification of similar videos is a significant and a consequential issue in content-based video retrieval. Video search is performed on the basis of keywords and mapped to the tags associated with each video, which does not produce anticipated results. This gave rise to the concept content-based video retrieval (CBVR), which utilizes the visual features of video to store and retrieves a video effectively. Proficient retrieval of similar videos aims at creating a video by video retrieval to achieve an ontology-based retrieval of videos.

Keywords Video preprocessing · Key frame generation · Pencil shaded · Pattern extractor · Eigen-based indexing

1 Introduction

In recent days, videos occupy a large part of data in the online server. Video retrieval methods can be categorized as two types such as text based or content based [1]. In text based, search using keywords mapped to the tags associated with each video, which produces inefficient results. We have many Web sites dedicated especially for browsing and viewing videos. Texture, shape, color, object, motion, spatial, and temporal are the most common visual features [2, 3, 4]. It was used in the visual similarity match. In all the video-oriented Web sites are based on the keywords which we type in, which produced an irrelevant output. Each video will have many tags associated with it. This leads to association of multiple tags to a same video. It makes video retrieval inefficient, time-consuming, and inappropriate

R. Sunitha (✉) · A. Mary Posonia
Department of Computer Science and Engineering, Sathyabama University, Chennai, Tamil Nadu, India
e-mail: rsunitha77@gmail.com

A. Mary Posonia
e-mail: maryposonia.cse@sathyabamauniversity.ac.in

L.P. Suresh and B.K. Panigrahi (eds.), *Proceedings of the International Conference on Soft Computing Systems*, Advances in Intelligent Systems and Computing 398, DOI 10.1007/978-81-322-2674-1_35

process [5]. Keywords are typed in to retrieve videos, but not much importance is given to retrieve videos in search engines [1]. This gave rise to the concept content-based video retrieval (CBVR) which utilizes the visual features of video to store and retrieves a video effectively. The existing system exhibits the following flaws: (1) Takes more time for browsing. (2) Results are irrelevant. (3) There is no filtration for redundant videos. Hence, we propose CBVR system to filter redundant problem and produce efficient output of video.

2 Related Work

CBVR has been one of the important research areas in the field of video and image processing. There are many related literature reviews, but here we only take the ones closely related to this work.

Lai and Yang [1] proposed a video retrieval system, one of the existing systems, which promotes CBVR system. The visual features are focused on color model (CIEL*a*b*color space), trajectories utilizes low-level features of object motion, and object appearances utilizes SURF (Speeded Up Robust Feature) and bag-of-words methods. SURF was based on the texture and bag-of-words was based on hierarchically clustered by k-means algorithm. Indexing and searching method used three different databases (path, object appearances, and background databases). Visual features are much prone to errors due to variation caused in color models (CIEL*a*b*color space) and some noises that affected the texture. K-means algorithm does not predict *k*-value and there is no similarity. It needs more memory space because of three different databases.

Gitte et al. [2] proposed a CBVR System, which includes key frame selection, feature extraction, shot detection, and hierarchical indexing scheme. Videos are segmented into shots by step-by-step process of video segmentation [6]. Key frames of shot are selected based on euclidean distance. Feature extraction [7] contains spatial (color, shape, edge) and temporal features (motion, audio) of key frames. Video contents are classified based on SVM and clustered based on k-means algorithm [1]. Indexing is performed by hierarchical indexing algorithm. The color features are extracted based on an average RGB algorithm which does not work well when the color model varies. K-means algorithm does not predict *k*-value and there is no similarity.

Hero, A, O [8] proposed multimodal indexing with direct information which utilizes the visual features, audio features [7], as well as the textual channels of a video. Videos are categorized and indexed based on directed information obtained by the processing of these features. Directed information involves predicting the future event based on the past event which is employed to relate the existing videos in this system. The processing of visual and audio features acts as the baseline for attaining the directed information required for video indexing. Whereas the processing of visual features is much prone to errors due to variation caused in color models, and the processing of audio features is also still incomplete due to the natural language-processing vulnerabilities.

Hamdy and Sahar [9] proposed a video retrieval system that used video segments [6] and key frame selection among both the low-level and high-level semantic object annotation. Low-level feature texture is obtained by entropy mechanism; shape is obtained by SOBEL mechanism. High-level feature object annotations are obtained using supervised learning techniques. Even though the proposed method produces high precision results than color-based retrieval systems, designing of SOBEL operation for obtaining a shape of the object is a bit expensive and the efficiency of learning techniques is questionable with large datasets.

Patel and Meshram [3] proposed CBVR system which utilizes low-level features such as object motion, texture, loudness, power spectrum, pitch, and bandwidth. The retrieval system involves similarity measurement and multidimensional indexing. Similarity measurement was used to find the most similar objects. Multidimensional indexing was used to accelerate query performance in the search process. Texture features were obtained by entropy, object features were obtained by SIFT [10] and audio features were extracted based on CHROMOGRAM. The processing of audio to extract features like pitch and bandwidth is highly expensive. These low-level audio features can never produce ontological outcomes.

Rajendran and Shanmugam [11] proposed CBVR system using query clip. This work utilizes low-level features such as object motion, texture and tries to retrieve results based on video clips. Motion of objects is detected using Fast Fourier transforms and L2-norms. Color features are obtained based on RGB model and then converted into HSV histograms. All the extracted features are stored in the mat file format of MATLAB. Based on the input query clip, similar videos are retrieved. Kullback–Leibler distance acts as the similarity matrix. Fast Fourier transforms employed are highly sensitive to noise and conversion of RGB to HSV is computationally expensive.

Sudhamani and Venugopal [12] proposed a color-based indexing and efficient retrieval images. The color features of the images are obtained using mean shift algorithm. Indexing is performed based on spatiality avoiding the traditional histograms. Even though the spatial indexing proposed enormously reduced the storage, it failed due to the existence and nonstandardization of various color models.

Narayanan and Geetha [13] proposed CBVR which includes shot segmentation, feature extraction, key frame extraction, clustering, indexing similarity search, refinement, and relevance feedback. Color model was applied, which does not work well when color varies. K-means [1, 2] and ISODATA are typical interactive clustering algorithms. K-means does not predict a k-value and ISODATA was poor.

Zhang and Nunamaker [4] proposed a natural language approach which includes audio and visual features. Visual features contain texture, shape, and color histograms. The demerits of a visual feature are that content-based image retrieval has not reached a semantic level and natural language is unpredictable. No standardization of various color models.

Sebastine et al. [14] proposed a semantic Web for CBVR which includes automatic features, extraction, storage, indexing, and retrieval of videos. Video is clustered based on k-means algorithm. Feature extraction contains text, face, audio, color, shape, and texture [3]. RDF technique is also applied. K-means does not predict *k*-value and color, shape, and texture supported MPEG-7 and CEDD descriptor. Color model does not work standardized.

Chen and Tjondronegoro [15] proposed CBVR which includes key-segment extraction, MPEG-7 content descriptions, and video retrieval. Demerit of the MPEG-7 is that it does not address how to extract the files. Feature extraction contains takes time consumption of video clips, segments, and sound. The feature extraction has been automated, but required, separate software has been limited.

Le et al. [16] proposed a sub-trajectory-based video retrieval which includes numeric trajectories, symbolic trajectories, and segmentation. The previous problem was representing a trajectory and had not detected similar model of the trajectories.

Dao et al. [17] proposed a topological information using sub-trajectories for spatial and temporal. Shift- and scale-invariant techniques have been applied. It can be handled on texture. Some noises affect the texture and color models are also applied.

Colombo et al. [18] proposed visual information retrieval using semantic. Motion features have been used for retrieval which includes motion trajectories and motion objects are used to describe spatial–temporal relationship. The spatial and temporal relationships were based on texture. It was more complex dynamic motion and many colors were applied.

3 Proposed Work

We propose an efficient content-based video retrieval system for indexing and retrieving videos from the database. Here, the input video is partitioned into frames using preprocessing techniques. We extract the background key frame and object key frame. Then we remove the shadow for both background and object using image subtraction algorithm, which is in turn converted into matrices and their sum of eigenvalues is computed. The sum of the eigenvalues of background and object key frames is used as a metadata for indexing and retrieving the videos. Advantages of proposed system: (1) The computational space for storing the visual feature has been comparatively reduced since the matrices are used for representation. (2) Eigen-based indexing is expected to achieve high desirable outcomes with accuracy which is of low computational cost and low time consumption. There are five tasks (as shown in Fig. 1), namely video preprocessing techniques, key frame generation, shadow removal, matrix conversion by Eigen-based indexing, and storing video in a database and retrieval of similar videos.

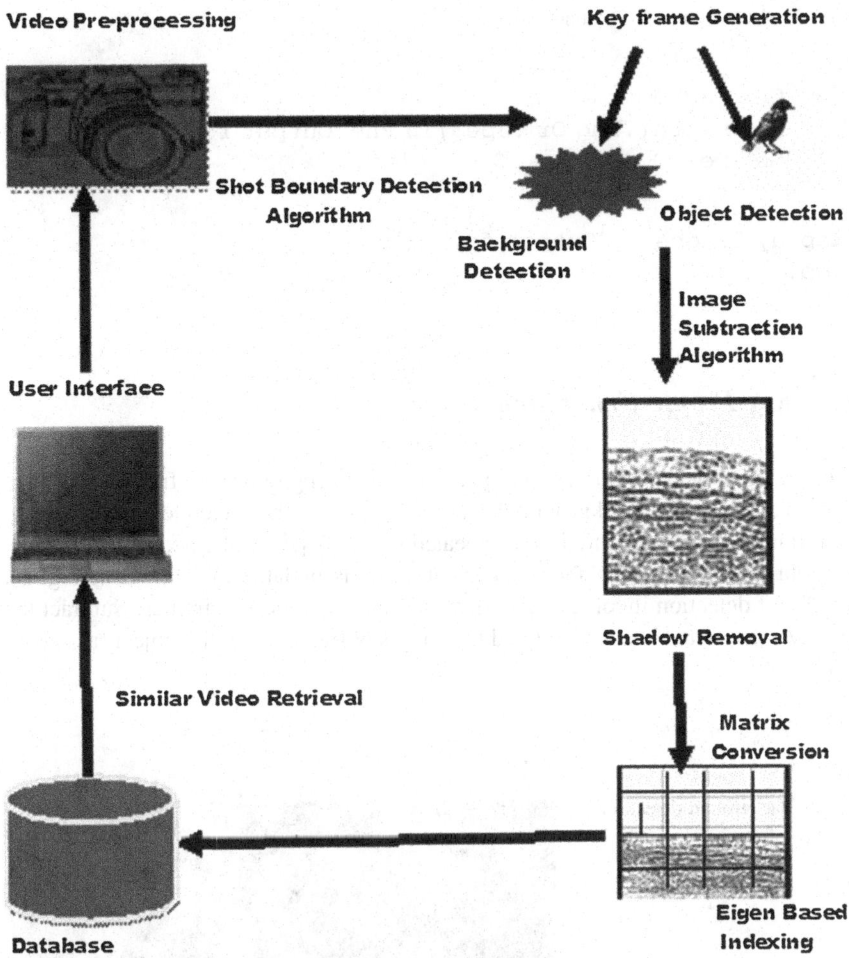

Fig. 1 System architecture

3.1 Video Preprocessing Techniques

Most of the videos are unstructured data. So applying video preprocessing techniques, the video clips are partitioned into frames or images using shot boundary detection algorithm.

Shot boundary detection algorithm

```
Start
Read Video File [AVI Format]
Open Output Folder 'Output Folder'
Calculate Number of Frames
```

```
For 1: Number of Frames

      Get each Frame
      Save with(.Png Or .Jpeg)in the 'Output folder'
      Get Next Frame

End
Display Images in 'Output folder'
Stop.
```

3.2 Key Frame Generation

The key frame generation has two types such as background key frame and object identification. (i) The background detection is based on the principle of temporality. Each frame $F_i[k]$ is compared to its repeated frame $F_{i+1}[k]$. If the pixel value is same for both the frames, then the consequent value is updated (as shown in Fig. 2). (ii) Object detection involves extracting the objects in the key frames. Subtract the already constructed background and take any key frame to get the object (as shown in Fig. 3).

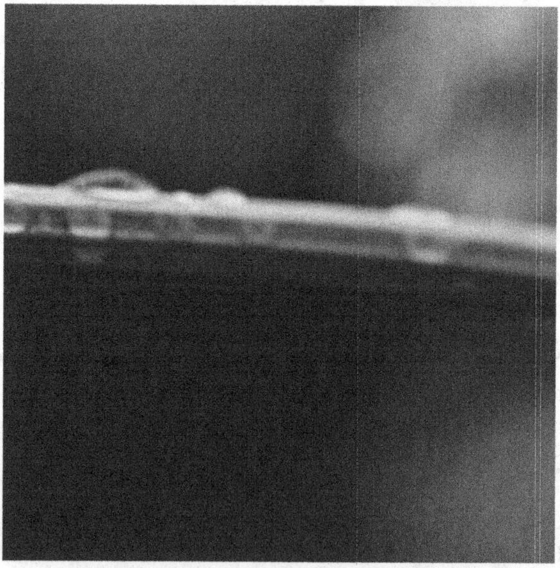

Fig. 2 Background detection

Fig. 3 Object detection

3.3 *Shadow Removal*

The shadow removal was based on a simple image subtraction algorithm (as shown in Fig. 4).

Image subtraction algorithm

```
Start
Read Detection of Background and Object Key frame
It computes Four Adjacent Pixels
If all Four pixels have Same Color

     It is marked as an Interior Pixel

Else

     It is marked as Border Pixel

Stop
```

Fig. 4 Shadow removal

3.4 Matrix Conversion by Eigen-Based Indexing

The shadow removal images of background and object key frames are converted into a matrix (*Vk* and *Ok*). The morphological shapes contain grid values such as 0s and 1s. 0 covers outside of the shape boundary and 1 covers inside of the shape boundary (as shown in Fig. 5).

Let us take a grid matrix *R*, background key frame *Vk* and object key frame *Ok* in order 3 × 3 where the pixel shapes are represented as 0s and 1s.

$$Vk = \begin{matrix} 1 & 0 & 0 \\ 0 & 1 & 0 \\ 0 & 0 & 1 \end{matrix} \quad (1)$$

$$R = \begin{matrix} 1 & 1 & 1 \\ 1 & 1 & 1 \\ 1 & 1 & 1 \end{matrix} \quad (2)$$

Then calculate the matrix values for both *EVk* and *EOk* and then calculate eigen-values for both background and object key frames.

$$EVk = \begin{matrix} R11 * V11 & R12 * V12 & R13 * V13 \\ R21 * V21 & R22 * V22 & R23 * V23 \\ R31 * V31 & R32 * V32 & R33 * V33 \end{matrix} \quad (3)$$

(or)

$$EOk = \begin{matrix} R11 * O11 & R12 * O12 & R13 * O13 \\ R21 * O21 & R22 * O22 & R23 * O23 \\ R31 * O31 & R32 * O32 & R33 * O33 \end{matrix} \quad (4)$$

$$\mathrm{Det}(EVk) - \lambda I = 0 \quad (5)$$

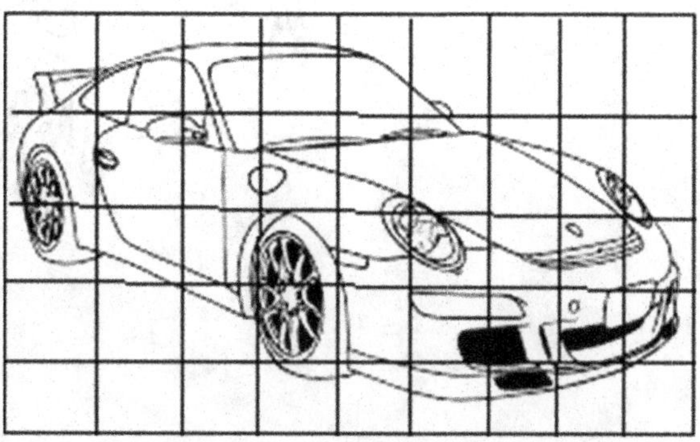

Fig. 5 Matrix conversion by Eigen-based indexing

$$\mathrm{Det}(EOk) - \lambda I = 0 \tag{6}$$

Then calculate sum of eigenvalues for both background and object key frames, such as SAb and SAo. These values act as a metadata.

$$\mathrm{SAb} = \sum (\lambda 0 + \cdots + \lambda n) \tag{7}$$

$$\mathrm{SAo} = \sum (\lambda 0 + \cdots + \lambda n) \tag{8}$$

3.5 Storing Videos in the Database

There are two tables created such as (i) Eigen table, (ii) Path table. The sum of eigenvalues (SAb and SAo) is stored in eigen table and their path stored in path table. Eigen table is maintained for all the eigenvalues.

3.6 Retrieval of Similar Videos

Retrieval of similar videos is more efficient and effective, but one of several key issues is to measure the similarity between videos. Thus, searching the videos from large databases is computationally expensive. To overcome this problem, we detected the background key frame and object key frame, and then found the morphological shape of the images, and the images are converted into a matrix. Eigenvalues are calculated for both background and object key frame and these values summed up to SAb and SAo and then their values are stored in the databases. Finally similar video is retrieved from a database. Video retrieval is simple and effective process.

4 Results

We take 3–5 min of video clip as input. Mostly the videos are unstructured format. So we are applying video preprocessing techniques. The video is sliced into several frames as output using shot boundary detection algorithm (as shown in Figs. 6 and 7).

We compared each and every frame to detect background and image and then go with shadow removal because color model is not standardized. So we are applying image subtraction algorithm. This algorithm computes four adjacent pixels. If four pixels have the same color, they are marked as an interior pixel. If any pixel has a different color, it is marked as a border pixel and shaded (convert into black or gray) (as shown in Fig. 8).

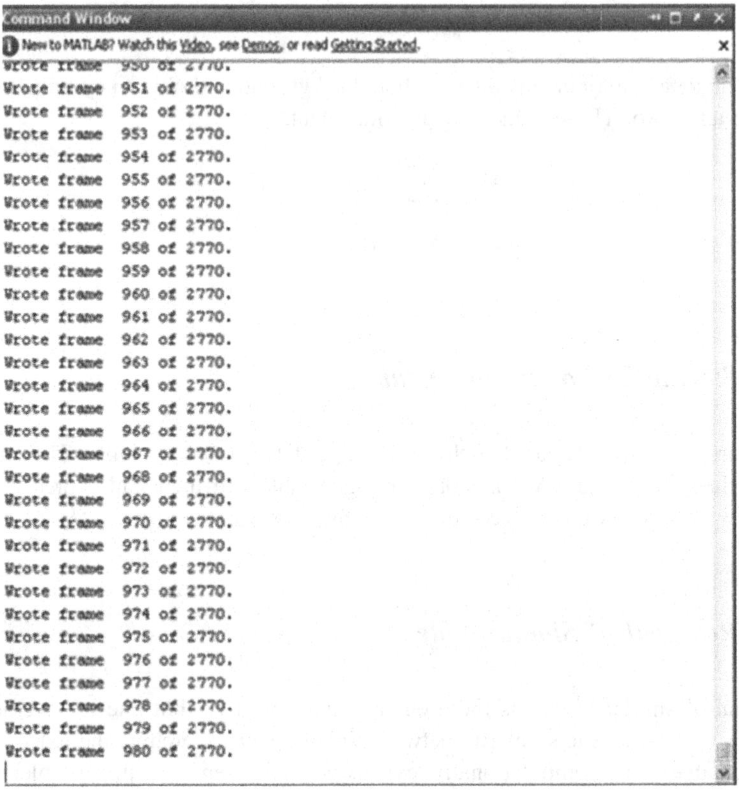

Fig. 6 List of frames

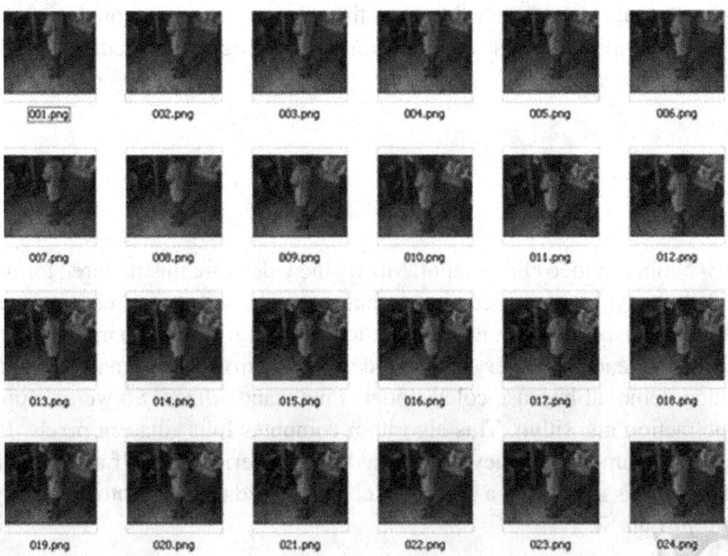

Fig. 7 List of images

Fig. 8 Detection of background and object with shadow removal

5 Conclusion

Proficient retrieval of similar videos aims at creating a video by video retrieval to achieve an ontology-based retrieval of videos. The advantage of the proposed system is the computational space for storing the visual feature that has been comparatively reduced since the matrices are used for representation. Eigen-based indexing is expected to achieve high desirable outcomes with an accuracy which is of low computational cost and low time consumption. In further works, we try to develop a more valuable algorithm to get the background key frame and object key frame to retrieve similar videos.

References

1. Lai Y-H, Yang C-K (2014) Video object retrieval by trajectory and appearance. IEEE Trans Circuits Syst Video Technol PP(99). doi:10.1109/TCSVT.2014. ISSN:1051-8215
2. Gitte M, Bawaskar H, Sethi S, Shinde A (2014) Content based video retrieval system. Int J Res Eng Technol 3(6)
3. Patel BV, Meshram BB (2012) Content based video retrieval system. Int J UbiComp (IJU) 3(2)
4. Zhang D, Nunamaker JF (2004) A natural language approach to CBVI and retrieval for interactive e-learning. IEEE Trans Multimedia 6(3)
5. Packialatha A, Chandra Sekar A (2014) Adept identification of similar videos for web-based video search. IJERSS 1(4)
6. Babhale B, Ibitkar K, Sonawane P, Puntambekar R (2015) CB lecture video retrieval using video text information. Int J Adv Found Res Comput (IJAFRC) 2(NCRTIT 2015). ISSN 2348 - 4853
7. Kumar AS, Nirmala A (2015) A survey of multimodal techniques in visual CBVR. Int J Adv Res CS Softw Eng 5(1)
8. Chen X, Hero AO, Sararese S (2012) Multimodal video indexing and retrieval using directed information. IEEE Multimedia Trans 14(1). ISSN 1520-9210
9. Elminir HK, Sabbeh SF (2012) Multifeature CBVR using high level semantic concept. Int J CS Issues 9(4)
10. Patel DH (2015) CBVR. Int J CA (0975 – 8887) 109(13)
11. Rajendran P, Shanmugam TN (2009) An enhanced CBVR system based on query clip. Int J Res Rev AS 1(3)
12. Sudhamani MV, Venugopal CR (2007) Grouping and indexing color features for efficient image retrieval. Int J Comput Inf Syst Control Eng 1(3)
13. Narayanan V, Geetha P (2008) A survey of CBVR. J Comput Sci 4(6):474–486. ISSN-1549-3636
14. Sebastine SC, Thurai-Singham B., Prabhakaran B (2009) Semantic web for content based video retrieval. IEEE international conference on semantic computing
15. Chen Y-PP, Tjondronegoro D (2002) CBIR using MPEG-7 and X-Query in video data management systems. WWW: Internet and Web Information Systems
16. Le T-L, Boucher A, Thonnat M (2007) Sub-trajectory based VIR (Video Indexing and Retrieval), pp 418–427. Springer, Berlin
17. Dao M-S., Sharma IN, Babaguchi N (2009) Preserving topological information in sub-trajectories based representation for spatial-temporal trajectories. ACM international conference on multimedia, New York, USA. ISBN 978-1-60558- 608-3
18. Colombo C, Del Bimbo A, Pala P (1999) Semantics in visual information retrieval. IEEE Multimedia 6(3). ISSN 1070-986X

FoetusCare: An Android App for Pregnancy Care

M. Bharathi and V. Nirmalrani

Abstract FoetusCare is an android application for pregnant mothers. It calculates the pregnancy due date, shows the pictures of foetus growth, and explains the symptoms and usual discomforts of the mother every week; it tells the needed weight gain of the mother every week; it provides information about daily calorie requirement. This application alerts the pregnant mother for the due dates of doctor consultation, scan, and other tests. And this app allows the mother to share her weight information, new symptoms and minor discomforts with doctors. The mother can store her basic pregnancy data and her weight every week. User data will be stored in the server hosted in the cloud, in order to support multiple devices.

Keywords Android · Application · Women · Pregnancy care · Foetus care

1 Introduction

The next generation of medical applications will provide a big change in performance and sophistication, transforming the way health care professionals and doctors interacts with patients. According to the recent development in the medical field, innovative medical applications have been developed to be used in tablets and smart phones. This trend forces the computer professionals to develop such kind of medical applications. Recently, developing user-friendly medical applications is becoming an integrated part of medical solutions. In the modern world everyone is using smart phones. So it is necessary to develop mobile-based applications to provide convenient access and smart usage.

In the recent world, there are many applications available to guide and help the pregnant women. By downloading and keeping the applications in their smart phone, they can use it anywhere and anytime. These applications guide about

M. Bharathi · V. Nirmalrani (✉)
Department of IT, Sathyabama University, Chennai, Tamil Nadu, India
e-mail: nirmalv76@gmail.com

L.P. Suresh and B.K. Panigrahi (eds.), *Proceedings of the International Conference on Soft Computing Systems*, Advances in Intelligent Systems and Computing 398, DOI 10.1007/978-81-322-2674-1_36

monitoring of pregnancy, health promotion in pregnancy, preparing for parenthood, or caring for the baby at home. When the pregnancy test comes back positive, the mother has begun a life-altering journey. As the baby grows and changes through each stage of pregnancy, she goes through changes, too, in her body, emotions, and lifestyle. She needs lots of information to answer her questions and help her make good decisions for a healthy baby and her healthy.

My “FoetusCare” application is one of such android applications which will give lots of useful information to a pregnant mother. For every woman her pregnancy period is a spring season of her life. Along with her baby she carries lots and lots of memories, expectations, dreams, doubts and fears, etc. It helps the mother to know about the growth of the foetus and explains all kinds of symptoms that the mother faces during her pregnancy. Each and every time while facing new symptoms, the mother may get scared. With the help of this application she can send queries easily to the doctor regarding her discomforts. She can clarify her doubts and can get immediate help or guidance from her gynecologist without visiting her in person.

2 Related Work

Some of the similar mobile applications are:

- “My Pregnancy Today” by baby center—shows the images of fetal development and can watch what’s happening inside the womb with 3D animations [1].
- “Pregnancy disc” by HipoApps—an application for a week-by by-week baby tracking [2].
- “ZBaby” by Sinthanayothin C., Wongwaen N., Bholsithi W., and Xuto P.—an android application to calculate pregnancy due date and to view fetus development simulation and weight gain during pregnancy [3].
- “I’m Expecting” by MedHelp, Inc—Top Health Apps—gives tips and weekly updates of baby’s growth, can view weekly pregnancy videos.

There are so many applications available on the net to guide the pregnant mothers [4]. My application is based on “ZBaby: Android application for pregnancy due date, fetus development, simulation and weight gain during pregnancy.” Features of this application are:

- Calculates the due date.
- Shows fetus growth in 3D pictures.
- Gives the description of the fetus growth pictures.
- Tells the weight gain and calories needed for the mother.
- It finds the nearest hospital through GPS.

3 Proposed System and Architecture

In addition to the existing system features, following features are added to the proposed system:

- Usual symptoms and discomforts that mother feels each week are explained.
- Allows the user to share her weight information and her discomforts to doctors.
- Alerts the user to doctor consultation, scan, and other test due dates.
- Allows the user to store her weight each week.
- Preserves the user data on the server hosted in the cloud for supporting multiple devices (Fig. 1).

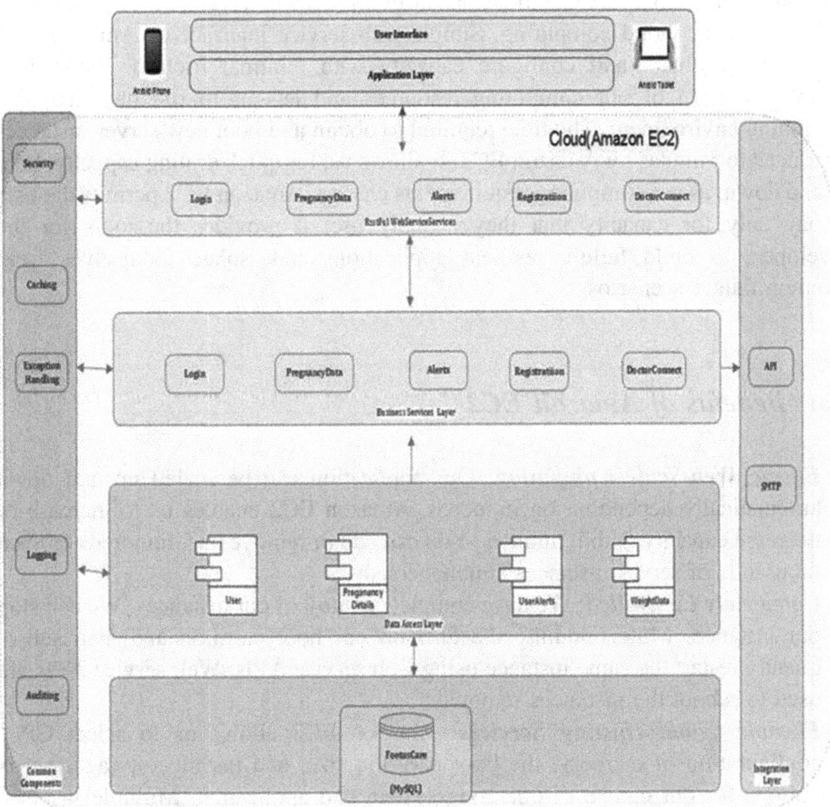

Fig. 1 FoetusCare architecture

4 Proposed System Implementation

4.1 Technologies Used

The front end of this application is developed with the Android SDK [5, 6]. Web services and database processing activities are done in Java. A database is created in MySQL. The tables are created and maintained in Amazon EC2 Instance (type: t2. micro) [7].

4.2 Amazon EC2

Amazon Elastic Compute Cloud (Amazon EC2) is a web service. It is designed to make web-scale cloud computing. Simple web service interface of Amazon EC2 allows us to obtain and configure capacity with minimal friction. It provides complete control of our computing resources and lets us to run on Amazon's computing environment. The time required to obtain and boot new server instances is reduced to minutes by Amazon EC2. It allows us, for quick scaling capacity, both up and down, as per computing requirements change. Amazon EC2 permits the user to pay only for capacity that they actually use. It provides the tools for the developers to build failure resilient applications and isolate themselves from common failure scenarios.

4.3 Benefits of Amazon EC2

1. ***Elastic Web-Scale Computing***: Our application can be scaled up and down automatically depending on its needs. Amazon EC2 enables us to increase or decrease capacity within minutes. You can add or remove one, hundreds or even thousands of server instances simultaneously.
2. ***Completely Controlled***: We have complete control of our instances. We can stop our instance, while retaining the data on our boot partition and then subsequently restart the same instance using web service APIs. Web service APIs are used to reboot the instances remotely.
3. ***Flexible Cloud Hosting Services***: Amazon EC2 allows us to select CPU, configuration of memory, the boot partition size, and instance storage that is optimal for our choice of operating system and application. Multiple instance types, operating systems, and software packages can be chosen by us with the help of flexible cloud hosting services.
4. ***Designed for use with other Amazon Web Services***: Amazon EC2 is designed to work in conjunction with Amazon Simple Storage Service (Amazon S3), Amazon SimpleDB, the Amazon Relational Database Service (Amazon RDS),

and Amazon Simple Queue Service (Amazon SQS) which offers the solution for query processing, computing, and storage across a wide range of applications.

5. ***Reliable***: Amazon EC2 provides a highly reliable environment where replacement instances can be rapidly and predictably commissioned. The service runs within Amazon's data centers and proven network infrastructure. Amazon EC2 ensures 99.95 % availability for their service-offering regions.
6. ***Secure***: Amazon EC2 works in conjunction with Amazon VPC to provide security and robust networking functionality for computer resources.
7. ***Inexpensive***: Amazon EC2 allows us to pay a very low rate for the compute capacity we actually consume. On-demand instances let us pay for compute capacity by hourly with no long-term commitments. This frees us from the costs and complexities of purchasing and maintaining. On-demand instances also remove the need to buy "safety net" capacity to handle periodic traffic spikes.
8. ***Easy to Start***: We can set up the instance quickly and can start the instance whenever we need.

4.4 System Implementation

1. ***Login***: This login module allows the existing users to access the application. The login page is the first page of FoetusCare application. It is shown in Fig. 2.

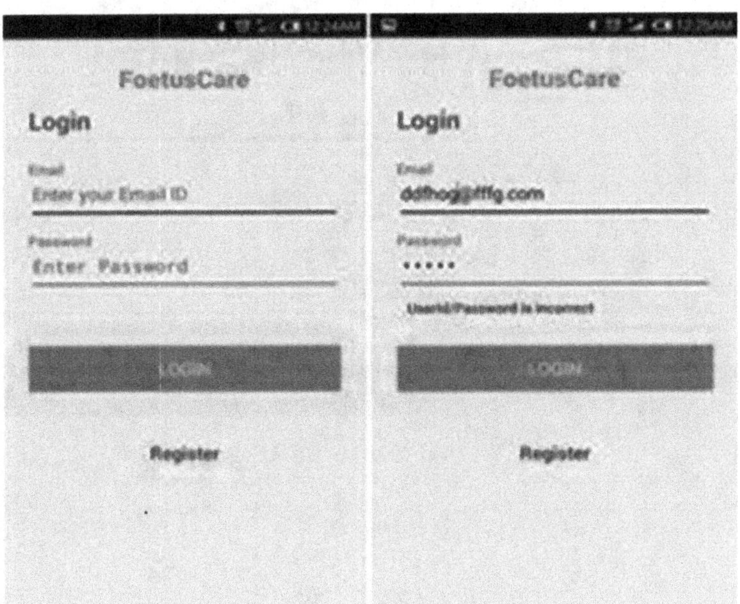

Fig. 2 Login page

It gets the user id and password. After typing the details, the user has to click the "Login" link. Validation of the user id and password against the database are verified by invoking the Restful web services hosted in the cloud. Password is encrypted with MD5 sum. Users with invalid credentials will not be allowed to view the internal screens of the app. If the entered login or password is wrong, it will show the error messages like "User ID/Password is incorrect." The valid users can be directed to the next (pregnancy data) page.

2. ***Registration***: In this module, the user id and password of the new user is collected and stored. New users will be able to register on their own with the "Registration" link present in the login screen. They have to give their user id and password. It is shown in Fig. 3. Upon the successful registration, users will be redirected to the first screen (pregnancy data page) of the application.
3. ***Pregnancy Data Management***: This module collects and stores all the basic pregnancy details (such as mother's name, last menstruation date, pre-pregnancy weight, height, age and etc) of the mother. When the user is logged on to the system for the first time, she should enter the above-mentioned pregnancy details. The user has to enter the data into the provided area with the mentioned units as shown in Fig. 4. Entered details will be stored in the database through Restful web service. One user can store multiple pregnancy data, but only the latest pregnancy data will be active. User can update the basic details of the

Fig. 3 Registration

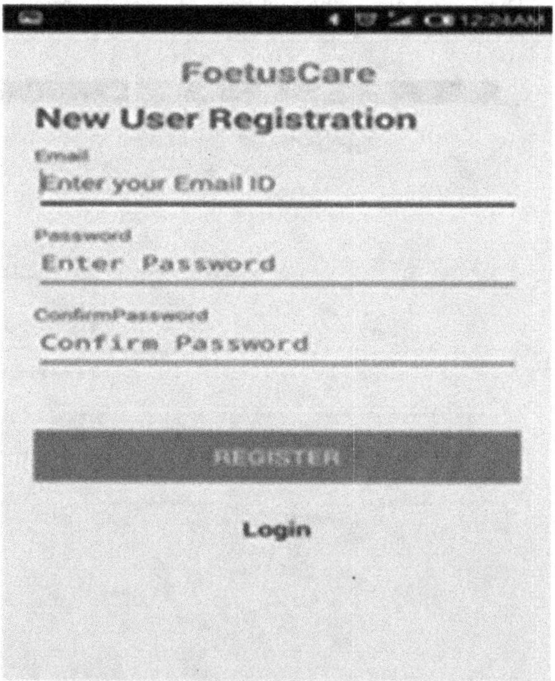

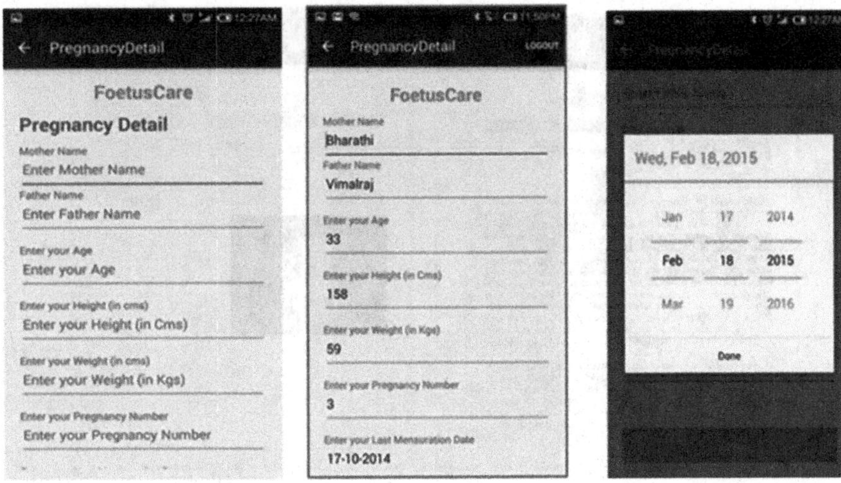

Fig. 4 Pregnancy data management

active pregnancy data at any time. Previous pregnancy data can be viewed any time.

4. ***Fetus Growth and Symptoms***: This page displays the due date, current week of pregnancy, fetus growth, and the symptoms of the mother each week. Fetus growth details are shown in detailed pictures. Fetus growth and symptoms that the mother may experience in the current week will be shown by default. Normal pregnancy duration is 40 weeks. Among the 40 weeks details, the user can view any week's information by navigating to that particular week. This page is shown in Fig. 5. These weekly data are also stored in the cloud. This data can be accessed and shown to the user through Restful web services.
5. ***Weight Gain and Calorie Needs***: The recommended weight gain for each week is calculated and shown. The amount of calories needed by the mother in each trimester is calculated and shown on this page [8]. Figure 6 shows the weight and calorie information. This page gets the weight of the mother in the current week. She is allowed to modify the previous week's weights also. Based on the height and pre-pregnancy weight data, the body mass index (BMI) value is calculated and suggests her to keep her weight as recommended. The formula used to calculate the BMI is given below (height in meters and weight is in kilograms).

$$\text{Body Mass Index} = \text{Weight/Height} * \text{Height} \quad (1)$$

Table 1 summarizes the range of BMI categorization. By referring the table "Weight Gain range During Pregnancy" (which is re-examined by Institute of Medicine of the National Academies on May 28, 2009) the recommended weight gain of each week is calculated, displayed and compared with the actual weight of mother each week.

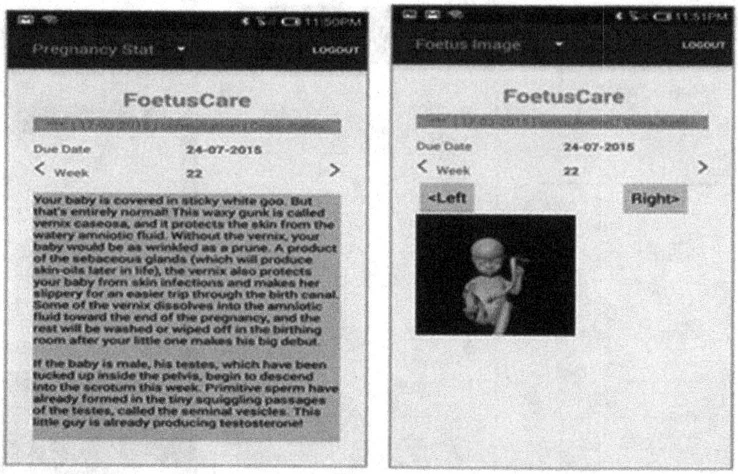

Fig. 5 Fetus growth and symptoms

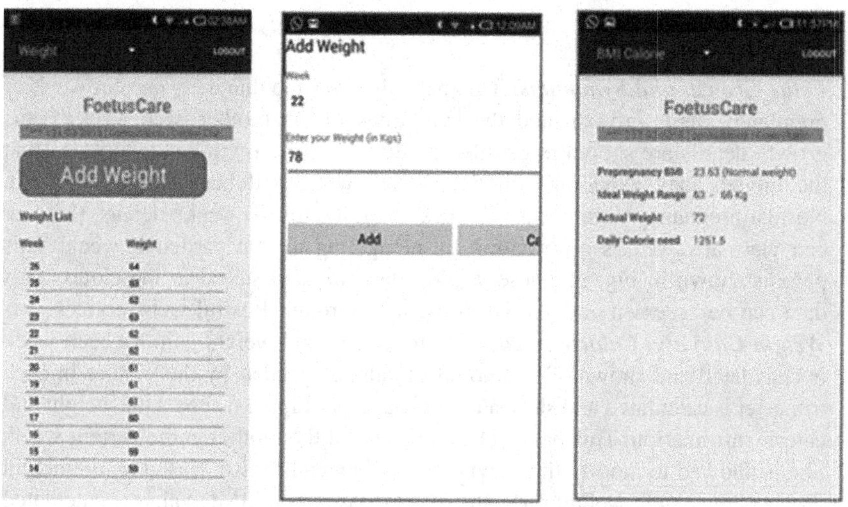

Fig. 6 Weight and calories needs

Table 1 BMI categorization

Pre-pregnancy BMI value (kg/m^2)	Category
<18.5	Underweight
18.5–24.9	Normal weight
25.0–29.9	Overweight
>30.0	Obese

A recommended body weight range table for a whole pregnancy period is summarized in Table 2.

The amount of calories needed by the mother in each trimester is calculated and shown on this page. To calculate calorie need "Mifflin–St. Jeor" formula is used. The ADA (American Dietetic Association) published a comparison of various equations for calorie requirement calculation and found that the "Mifflin–St. Jeor"

Table 2 Recommended weight range and weight gain

Week	Recommended weight range (kg)	Recommended weight gain (kg)
Week 1	50.0–50.0	0–0
Week 2	50.0–50.2	0.04–0.2
Week 3	50.1–50.3	0.08–0.3
Week 4	50.1–50.5	0.1–0.5
Week 5	50.2–50.7	0.2–0.7
Week 6	50.2–50.8	0.2–0.8
Week 7	50.2–51.0	0.2–1.0
Week 8	50.3–51.2	0.3–1.2
Week 9	50.3–51.3	0.3–1.3
Week 10	50.4–51.5	0.4–1.5
Week 11	50.4–51.7	0.4–1.7
Week 12	50.5–51.8	0.5–1.8
Week 13	50.5–52.0	0.5–2.0
Week 14	50.9–52.5	0.9–2.5
Week 15	51.3–53.0	1.3–3.0
Week 16	51.7–53.5	1.7–3.5
Week 17	52.1–54.1	2.1–4.1
Week 18	52.5–54.6	2.5–4.6
Week 19	52.9–55.1	2.9–5.1
Week 20	53.3–55.6	3.3–5.6
Week 21	53.7–56.1	3.7–6.1
Week 22	54.1–56.6	4.1–6.6
Week 23	54.5–57.1	4.5–7.1
Week 24	54.9–57.7	4.9–7.7
Week 25	55.3–58.2	5.3–8.2
Week 26	55.7–58.7	5.7–8.7
Week 27	56.1–59.2	6.1–9.2
Week 28	56.5–59.7	6.5–9.7
Week 29	56.9–60.2	6.9–10.2
Week 30	57.3–60.7	7.3–10.7
Week 31	57.7–61.2	7.7–11.2

(continued)

Table 2 (continued)

Week	Recommended weight range (kg)	Recommended weight gain (kg)
Week 32	58.1–61.8	8.1–11.8
Week 33	58.5–62.3	8.5–12.3
Week 34	58.9–62.8	8.9–12.8
Week 35	59.3–63.3	9.3–13.3
Week 36	59.7–63.8	9.7–13.8
Week 37	60.1–64.3	10.1–14.3
Week 38	60.5–64.8	10.5–14.8
Week 39	60.9–65.4	10.9–15.4
Week 40	61.3–65.9	11.3–15.9

formula is most accurate. The "Mifflin–St. Jeor" formula for calculating calorie needs of the mother in normal days is

$$10 * \text{Weight} + 6.25 * \text{Height} - 5 * \text{Age} - 161 \quad (2)$$

In the above formula weight is in kilograms, height is in centimeters, and age is in years.

In our FoetusCare application, the pre-pregnancy calorie requirement is calculated using the above formula and the extra calories required for growing a healthy baby is added to that. Based on common recommendation, it is best to increase calorie intake at each trimester is given below:

- First Trimester—85 Extra Calories
- Second Trimester—285 Extra Calories
- Third Trimester—475 Extra Calories

6. ***Alerts Module***: This module provides alerts to the user for doctor visit, scan and other tests. The alert page is shown in Fig. 7. This alert page shows the list of predefined alerts. User can override the alert dates and also they can add additional alerts to suit her need. The predefined alert dates are computed based on the week. The user will be alerted 2 days in advance. The common consultation dates, test, and scan dates recommended are summarized in Table 3.
7. ***Communication with Doctor***: This module adds the communication facility between the mother and deoctor. Through this user can send her queries to her doctor. For each and every new discomfort, she can get a solution from doctor without visiting in person. This communication is achieved via email. By default, the system will fetch and attach the basic pregnancy details and weight data while sending each email. So that the doctor can easily recall the mother's reports and can reply her without any delay. The weight data of the mother for each week is sent to the doctor automatically. If there is any abnormal change in mother's weight, doctor can revert to her immediately. The reply to the corresponding

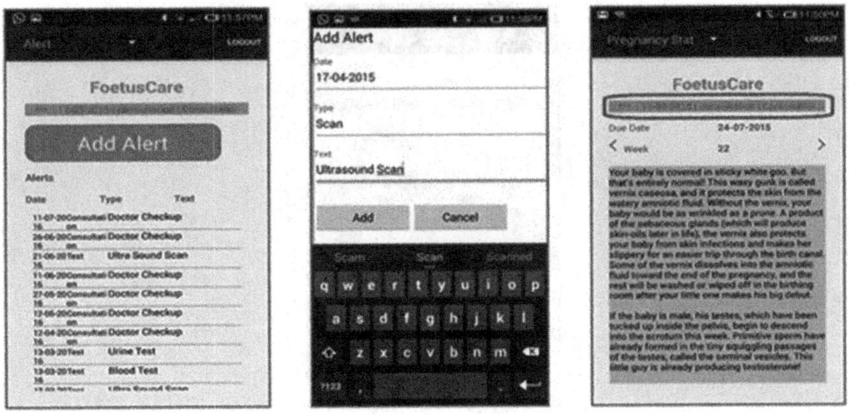

Fig. 7 Alert

Table 3 Alert information

S. No.	Consultation/Test/Scan	Day
1	Doctor consultation	30/60/90/120/150/180/210/225/240/255/270
2	Ultrasound scan	55
3	Blood test	55
4	Urine test	55
5	Ultrasound scan	150
6	Blood test	150
7	Urine test	150
8	Ultrasound scan	250

query from doctor can be viewed on the same screen. User can see all sets of queries and the corresponding responses by clicking front or back buttons. The queries and responses are displayed with date information. This is illustrated step-by-step in Figs. 8, 9, and 10.

5 Results and Discussions

BMI calculation and weight gain calculation are compared with the results of the Web Pregnancy Weight Gain Calculator [9]. Results are given in Table 4. Height and weight are given in the metric scale, i.e., height is measured in centimeters and weight is measured in kilograms.

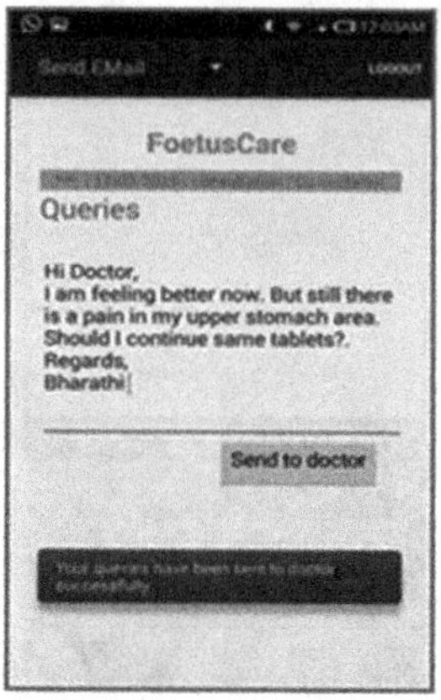

Fig. 8 Physician connect

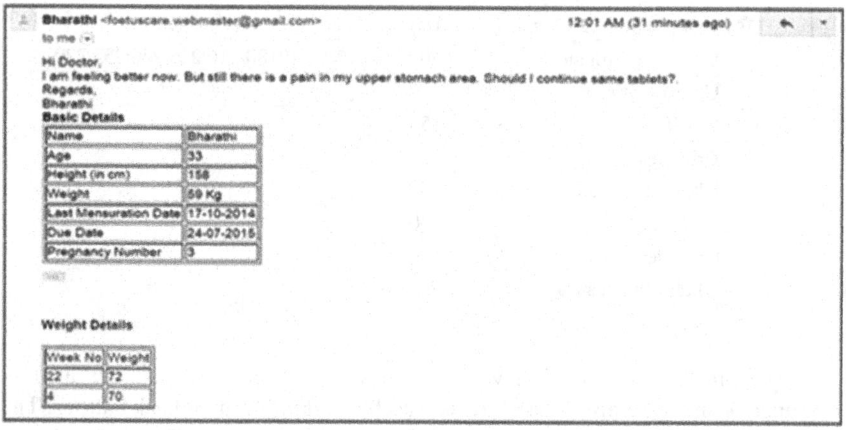

Fig. 9 Physician mail

Due date calculation results are compared with the results of the Baby center website due date calculator. Results are given in Table 5.

The results of the BMI, weight gain calculation, and due date calculation between our application and frequently used websites are similar.

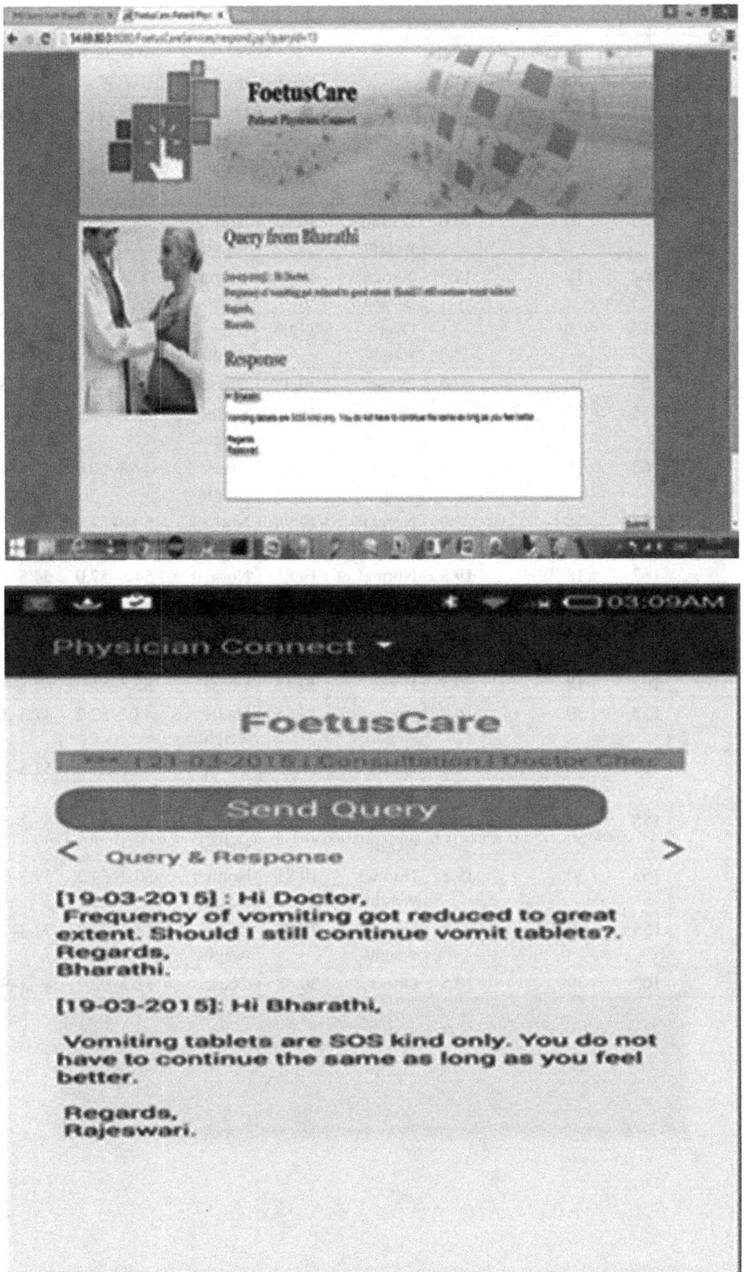

Fig. 10 Response to the patient

Table 4 Comparison of BMI calculation and weight gain calculation

Pre-pregnancy weight	Height	Pregnancy week	BMI				Weight gain	
			PWGC		FoetusCare		PWGC	FoetusCare
40	155	10	16.7	Under weight	16.65	Under weight	40.4–41.5	40.4–41.5
45	162	10	17.2	Under weight	17.15	Under weight	45.4–46.5	45.4–46.5
48	155	10	20.0	Normal weight	19.98	Normal weight	48.4–49.5	48.4–49.5
52	162	10	19.8	Normal weight	19.81	Normal weight	52.4–53.5	52.4–53.5
65	155	10	27.1	Over weight	27.06	Over weight	65.4–66.5	65.4–66.5
80	162	10	30.5	Obese	30.48	Obese	80.4–81.5	80.4–81.5
40	155	18	16.7	Under weight	16.65	Under weight	42.8–45.0	42.5–44.6
45	162	18	17.2	Under weight	17.15	Under weight	47.8–50.0	47.5–49.6
48	155	18	20.0	Normal weight	19.98	Normal weight	50.8–53.0	50.5–52.6
52	162	18	19.8	Normal weight	19.81	Normal weight	54.8–57.0	54.5–56.6
65	155	18	27.1	Over weight	27.06	Over weight	67.8–70.0	67.5–69.6
80	162	18	30.5	Obese	30.48	Obese	82.8–85.0	82.5–84.6
40	155	30	16.7	Under weight	16.65	Under weight	48.2–52.2	47.3–50.7
45	162	30	17.2	Under weight	17.15	Under weight	53.2–57.2	52.3–55.7
48	155	30	20.0	Normal weight	19.98	Normal weight	56.2–60.2	55.3–58.7
52	162	30	19.8	Normal weight	19.81	Normal weight	60.2–64.2	59.3–62.7
65	155	30	27.1	Over weight	27.06	Over weight	73.2–77.2	72.3–75.7
80	162	30	30.5	Obese	30.48	Obese	88.2–92.2	87.3–90.7

Table 5 Due data calculation

Last menstruation date	Due date	
	Baby center	FoetusCare
11-Jan-14	18-Oct-14	18-Oct-14
22-Feb-14	29-Nov-14	29-Nov-14
08-Mar-14	13-Dec-14	13-Dec-14
05-Apr-14	10-Jan-15	10-Jan-15
20-May-14	24-Feb-15	24-Feb-15
13-Jun-14	20-Mar-15	20-Mar-15
02-Jul-14	08-Apr-15	08-Apr-15
17-Aug-14	24-May-15	24-May-15
10-Sep-14	17-Jun-15	17-Jun-15
24-Oct-14	31-Jul-15	31-Jul-15
09-Nov-14	16-Aug-15	16-Aug-15
29-Dec-14	05-Oct-15	05-Oct-15

6 Conclusion

During pregnancy, women may experience a wide range of common symptoms. Still, pregnancy can be confusing and wondering. This application explains all such symptoms of a mother during each stages of her pregnancy. And it shows the fetus growth stage by stage, helps the mother to keep her weight by calculating needed weight gain for each week, and needed calories in each trimester. In addition to that, mother can send queries to her gynecologist. This application will act as a portable manual for all pregnant mothers. This can be accessed and used by any of the android devices.

References

1. My Pregnancy Today Android application by BabyCenter, 8 Dec 2014
2. Pregnancy Disc Android app developed by HipoApps, 10 Dec 2014
3. Sinthanayothin C, Wongwaen N, Bholsithi W, Xuto P (2014) ZBaby: android application for pregnancy due date, fetus development simulation and weight gain during pregnancy. ICSEC, 30 July–1 Aug 2014
4. Marjorie Greenfield "Dr. Spock's Pregnancy Guide", 2003
5. Reto M (2009) Professional android application development
6. Android Studio overview. http://developer.android.com/tools/studio
7. Amazon Elastic Cloud Computing (EC2). http://aws.amazon.com/ec2
8. Daily Calorie Needs. http://www.freedieting.com/calorie_needs.html
9. Pregnancy Weight Gain Calculator. http://www.calculator.net/pregnancy-calculator.html
10. Spring framework documentation. http://docs.spring.io/spring/docs/current/spring-framework-reference/htmlsingle
11. Andriod JSon Parser. http://www.tutorialspoint.com/android/android_json_parser.htm

Mobile Forensic Investigation (MFI) Life Cycle Process for Digital Data Discovery (DDD)

S. Rajendran and N.P. Gopalan

Abstract The process of gathering digital data is called digital data discovery (DDD) is done by using digital forensic software. The digital forensic experts can investigate and gain data from the acquired electronic devices. The electronic devices may be a storage media, accessories of the mobile phone, smart phones, and add-on parts like SD RAM, camera, audio, video recorders, etc. The existing digital forensics investigation (DFI) procedures are followed fundamentally with computers and standard file systems, but in recent days the use of smart phones and new mobile operating systems and new file systems presents more challenges for DFI. This can be addressed by introducing new policies, methodologies and life cycle process for effective Mobile Forensic Investigation (MFI). This paper will contribute potential evidence from DFI to MFI. It also deals with the technical compliance of all smart phone hardware and software relatively. The MFI life cycle process of forensic digital data discovery (DDD) is befitting to professional forensic community to understand the ecosystem of MFI.

Keywords Digital data discovery (DDD) · Mobile forensic investigation (MFI) · On-site device · On-site data · Data image

1 Introduction

The digital data discovery (DDD) is an emerged discipline and it has very high demand in digital forensic research in the present era. Accessing electronics gadgets becomes order of our daily life and these are indirectly helping to find 95 % of criminal occurrences. DDD plays a key role in digital forensic investigation.

S. Rajendran (✉) · N.P. Gopalan
National Institute of Technology, Tiruchirappalli, India
e-mail: sraj@nitt.edu

N.P. Gopalan
e-mail: npgopalan@nitt.edu

L.P. Suresh and B.K. Panigrahi (eds.), *Proceedings of the International Conference on Soft Computing Systems*, Advances in Intelligent Systems and Computing 398, DOI 10.1007/978-81-322-2674-1_37

The procedure for the forensic lifecycle process [1, 2] is a re-evaluating policy in nature. Forensic investigation starts from the incident zero level by interviews, interrogatories, and prevention of the original data device. The requirements are kept in secret and secured. This can be done by depositions of on spot-seized materials like data card, mobile phones, and any other electronic data storage media hard drive, spy camera, CDs, DVDs, flash drive, and similar storage devices. The digital crime data analysis, data reporting, documentation, assembling is to be done by expert authentication. This data is to be present in jurisdictions by suitable format prescribed by cyber law [3, 4]. Several cyber law specialists are used to prove the proficiencies by using the software to do the forensic examination.

This paper reveals exclusively on mobile forensics and it is organized with necessary introduction and available literature then followed by digital forensics investigation methodology, potential factors contributing to mobile forensics and proposed a lifecycle process for mobile forensics investigation (MFI).

2 Review of Literature

Mobile forensics can be achieved by some techniques and scientific tests are applied in the DDD investigation process. Another word the forensic science is an application of digital science to answer the questions of interesting to a legal system, this may be used a digital forensic opinion for the particular investigation case.

Mobile equipments are required for specialized [5, 6] interface, storage media, software, and hardware. Mobile operating system has its own file system and file structure. Mobile phone operating system performs it function from volatile memory but the computers are in RAM and ROM only.

The very important challenges in mobile devices and forensics are variations from design to design, model to model, time to time, manufacturer to manufacturer, and the adaptation of the technology. At this point, all the manufactures keep evaluating new technology for making own marketing by means of new invention and applications. The investigator must be aware of up-to-date information and latest technology. To have the collective knowledge of almost all types and its facilities build within it. This is more essential qualification for expert in forensic investigation.

The smart phone manufacturers will introduce a new version of operating system with new model mobile phones. This is the benchmark report for the present, leading, branded smart phone compared its responses. When these scores are compared, the iPhone 6 ends up having a clear lead over the Samsung Galaxy Note 4. The versions of the software are different in the benchmarks above. Even in Sun Spider 0.9.1, the score for the iPhone 6 is 353.4. Apple's software optimization on the iPhone can be expected to have worked in this case.

3 Methodology for DFI

The forensic methodology is always repeatable and defensible. This repeatable process must require to create a log by manually [2, 4]. It will avoid making a similar mistake more than once. It confirms proper chain of investigating procedure and lead to influences of successful. This helps in summarizing evidence, assured efficiency of forensic case administration.

The lifecycle process of forensics is in which DDD is done by various methods. The best practice will involve the following important seven progressive stages. Each step comprises several responsibilities and restrictions. At the time of investigation some cases may possibly permit to avoiding certain phases. Basically DFI is very complex in nature, involves many risk and it needs to severe observation. To all tasks in the procedure is suggested that the forensic investigations is complicated by the time to time.

The Fig. 1 shows to technically discover the data from any kind of digital media for procedure to collect/detects/recover the data from the identified/seized/captured electronic storage devices by legally to electronic data discovery.

The following table shows that comparison of different manufacture with similar model, facility smart phone specifications. It will be useful for forensic digital investigating device parameters. The parameters are the mobile phones, computers, laptops, storages media, etc (Table 1).

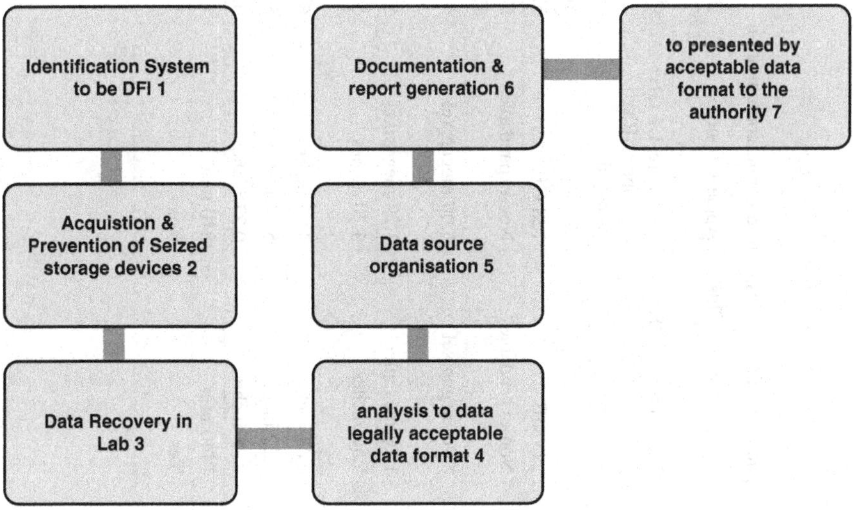

Fig. 1 Technical step for DDD

Table 1 Smart phones similar facility with different configuration

Description data	Apple iPhone 6 Plus	Apple iPhone 6	HTCOne (M8)	Samsung Galaxy S5	Samsung Galaxy Note 4
CPU	Dual-core 1.4 GHz Cyclone-ARM v8-based	Dual-core 1.4 GHz Cyclone (ARM v8-based	Quad-core 2.3 GHz (US/EMEA)/	Quad-core 2.5 GHz Krait 400	Quad-core 2.7 GHz Krait 450, Quad-core 1.3 GHz and Quad-core 1.9 GHz Cortex-A57
Data type	4G	4G	4G	4G	4G
Memory	128 GB	128 GB	32 GB	32–128 GB	32 GB
Rear camera	Non-expandable	Non-expandable	Expandable to 128 GB	16.00 megapixels	16.00 megapixels
Front camera	8.00 megapixel	8.00 megapixel	4.00 megapixel	22.00 megapixel	33.70 megapixel
Display size	1.20 megapixel	1.20 megapixel	5.00 megapixel	5.10 inches	55.70 inches
Mobile hot spot	5.50 inches	4.70 inches	5.00 inches	5.11 inches	6.21 inches
Height	6.07 inches	4.55 inches	5.64 inches	4.59 inches	6.04 inches
Thickness	6.22 inches	5.44 inches	5.76 inches	0.32 inches	0.33 inches
Talk time	0.28 inches	0.27 inches	0.37 inches	1260 min	910 min
NFC	1440 min	840 min	1200 min	4G	4G
Video comparisons	Yes	Yes	Yes	Yes	Yes

4 Mobile Forensics

The utilization of smart mobile devices is occupying a major role in our lives; almost all are carrying at least one type of electronic device. Using electronic mobile devices becomes inevitable for every one. During the past decade, it indicates the criminal incident and victims are finalized based on the digital evidence that plays an important role. Mobile forensics is the practice of recovering digital evidence from a mobile equipment by under the forensically to complete.

The basic-type mobile phones can store limited amount of data on the phone and additional storage media. This could be easily examined by the forensic specialist. The major difficulties begin from the smart phones [5]. It has a substantial amount of data storage. This will store verity of information with different file formats. All the information is recovered from smart devices by a forensic investigator by the predetermined procedures used to collect the stored and deleted information. It will lead to different kind of technical difficulties and issues will increase the level of investigation complexity.

The digital forensics, particularly data from the mobile phone is used to store, transmit, and receive, in the form of electronic signal to communicate with each other. This communication is mutually shared and exchanging data. This data may be personal or corporate information and in the form of voice, video, and text, or a combination of more than one type of data format. Presently, mobile phone utilization is rapidly increased and more new apps—applications are available on the internet at free of cost. By using this apps—money transactions, online booking, trading, reservations are possible. The mobile devices are used for more activities like sales, communications, marketing, and other online applications. Simultaneously this will create more illegal activities like a crime. It gives warning to us to focus more on this DFI area. Those areas may be as follows

- Economic crimes involving false, credit card scams, money laundering, etc.
- Cyber pornography involving production and spreading of pornographic material.
- Illegal marketing articles such as weapons, wild life etc.
- Intellectual property crimes such as theft of computer source code, software piracy, copyright violation, trademark abuses, etc.
- Harassments such as cyber nuisance, cyber offense, indecent and ill-treating mails, SMS, MMS, etc., and this list continues.

The Law implementation and digital forensics detectives will make more struggles to the investigator. In order to efficiently produce the result in digital data, evidence from the acquired mobile devices is required more concentration.

5 Mobile Forensics Investigation Life Cycle Process

Actual forensics begins with sorting based on the descriptions, data recovery, keyword searches, hidden data review, communicated by the devices and service provider. Particularly detecting data from the mobile phone is different from the computer data investigation. Here various kinds of data format will be available. The availability of data will be classified into three ways. They are: (1) From device (mobile phone), (2) From base station (BTS) and main switch controller (MSC), and (3) Service provider. The methodology of digital forensic investigation sequential procedures for DDD is shown in the Fig. 2

5.1 From Device

In the mobile phone device, also called as mobile station, it is possible to collect data in the variety of formats like text, contact database, SMS, MMS, audio, video, photo images with a combination of the mentioned format of information are possible.

1. Log for incoming, outgoing, missed call details of time, call is established duration of connectivity and some phone can have facility of recording audio as well as video.

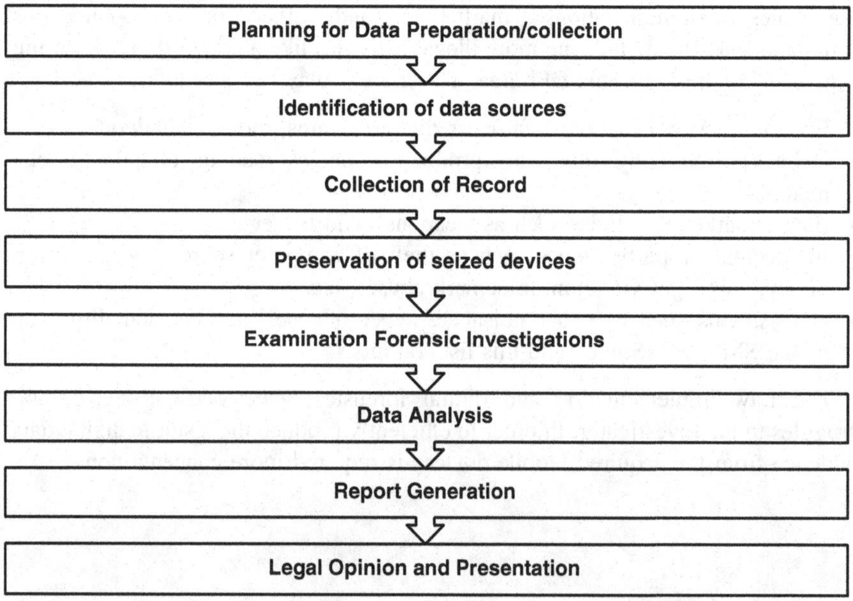

Fig. 2 Sequential procedures for DDD

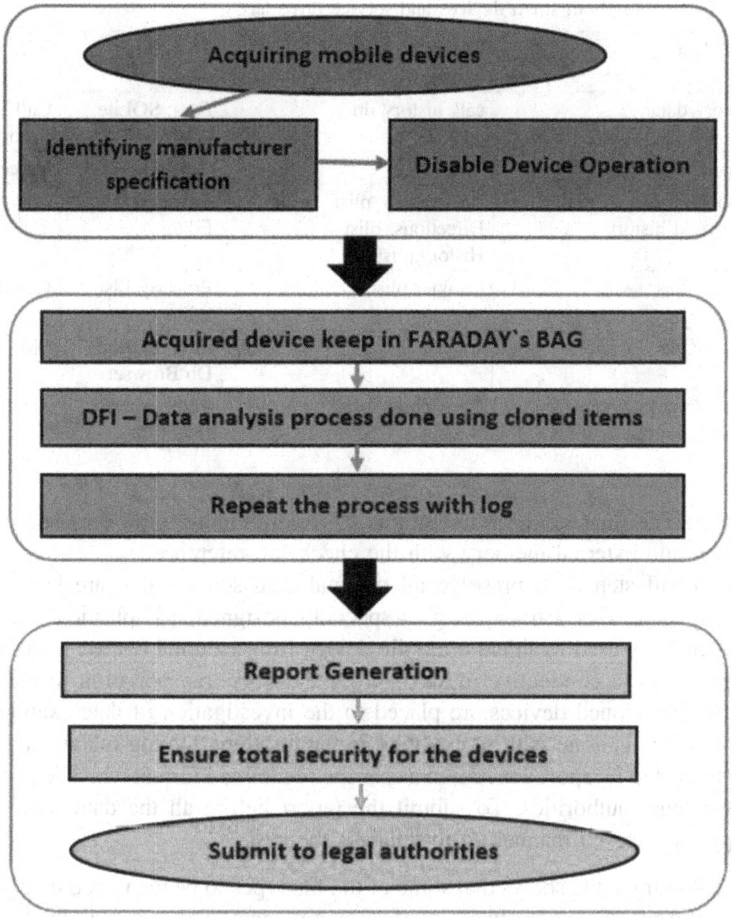

Fig. 3 MFI process

2. Report generation: Oral-based, written, scientific searching technic, statistic reports procedures, case logs reporting, and kind of paper work.
3. Legal opinion and presentation: This is almost the final step and will have proof of preparation and presentation of the detected data in the cyber law as prescribed format.

The mobile forensic investigation (MFI) can broadly be categorized into three steps as shown in Fig. 3

1. Data gathering from in and around the mobile device. The data include identification, features and manufacturer specification of device, model, make, IMEI, SIM, network info, and owner information. The next immediate action is to disable the network, internet, Bluetooth, tethering, and all kinds of external

Table 2 This data in both smart devices and service provider

Software Title	File(s)	Tool to be used	File directory
Call history data	call_history db	Froq, SQLite Database	Call history
Voicemails	.amr	QuickTime	Voicemail
Map bookmarks, map route directions and history	Bookmarks .plist, Directions. plist History.plist	Property List Editor	Maps
Internet cookies	Cookies.plist	Property List Editor	Cookies
SMS and MMS	sms.db	Froq, SQLite Db Browser	SMS
Safari, bookmarks, internet history web pages suspends	Bookmarks. plist	Property List Editor	Safari

activity. The final stage of first step is to make cloning of all data source like internal and external memory with the checksum reference.

2. The second step is to preserve all original data sources that are kept in the Faraday bag. This Faraday bag is specially designed, RF plastic-coated and shielded bags used to shield a mobile device from external contact. This bag is coupled with a conductive mesh to provide secure transportation to the laboratory. The cloned devices are placed to the investigation of data examination and analysis is done with re-evolving in nature along the log maintained.
3. The final step is report generation as per the predefined format which is provided by the legal authorities. To submit the report before all the data sources are keeping preserved manner up to finalizes the case.

The following table shows that some of the file types to be recovered by different type of software. There are two classifications of data. They are: (1) Data on the device alone and (2) Data partially on device and partially data from the service provider log. This both data can be categorized and placed on the table as shown in Tables 2 and 3, respectively.

5.2 *Spoliation of Data*

There is possibility to get or making damage of seized items unfortunately or intentionally. The original data may be a document, spreadsheet, scheduler, calendars, and database files associated with the particular case or applications to the event. The storage media such as SD ram, DVDs, CDs, pen drives, ZIP drives. It may/may not be as compatible as present technology.

Table 3 This data will be available in smart device alone

Software title	File(s)	Tool to be used	File directory
Celltower data, screenshot images	Consolidated.db	Froq, SQLite Db Browser	Caches
Contact information	Address Book. sqlit db	Froq, SQLite Db Browser	Address book
Passcode history	Password History. plist	Property List Editor	Configuration profiles
Event data	Calendar.sqlitedb	SQLite Database	Calendar Froq, browser
Application usage	ADDataStore. sqlit edb	Froq, SQLite DbBrowse	Logs
Keyboard logger	Dynamic-text.dat	Text Edit, Lantern	Keyboardt
Wallpaper background	-Image	Preview	LockBack ground.jpg
Notes	Notes.db	Froq, SQLite Db Browser	Notes
Web icons	Png info.plist	Preview, Property List Editor	Webclip.
System/app data	Numerous property lists	Property List Editor	Preferences

5.3 *Chain of Custody for DDD*

The mishandling evidence, ignorance of custody procedure will add more complication to the case. Similarly documentation of raw evidence sources of data, the use of [6, 7] "write-blocking" devices to ensure no data changes take place inadvertently. An initial forensic screening of disk drives for relevant data and making bit-by-bit soft copies of hard drives in the form images. The protection that digital-form file hashes to compare and checksum in all phases of investigation. Some documents like photography and serial number inventory of all evidence and maintain each and every movement of investigation step.

5.4 *Experience and Efficiency*

The highly convenient forensic software tools with high performance computer will create an excellent report at short period. This tool can produce the exact and accurate result. The net result depends on investigators working with field knowledge and involvement will produce the accurate and quick result.

5.5 *Data Sources of Electronic Evidence*

Commonly, computer forensics means searching for email, text, audio, video, messages chatting, and word processing documents. Data from the mobile phone devices will have the text messages, pictures, video, call history, contacts list, schedules, reminders, text chats, video chats, emails, personal digital assistants, reminders, schedulers, web sites visited, phone service provider company records, GPS internet site history logs, internet "cookie" artifacts, instant messaging logs, webmail messages, voicemail and PBX systems, and handheld PDAs. The professional legal services depend on the type of incidents and its created impacts will decide the level seriousness.

5.6 *The Forensic Tools*

The forensic software [6, 8] should have high complexity and gentleness in nature. An appropriate electronic evidence need to have proper procedures, well-developed protocols, luxurious hardware and software tools, along with knowledge to achieve best results. The most cases, by this will produce fine addressing by associating with a third party expert forensic firm. Consuming your private data may abolish relevant data and costs. Accurately managed forensic software can lead the forensic expert to the relevant electronic discovery targets in less time, at low cost and manpower. The ultimate process technology will uplift the possibility of a perfect result.

6 Conclusion

The DDD approach with MFI combined approach will support the forensics investigation result by a scientifically proven one. The MFI process will not replace the DFI rather it complements to modern investigation. It will be able to discover positively to get more information/data and it ensures all sources of digital evidence are under the discovery. This paper suggested better techniques and methods in applying forensic investigating operations with a trendy approach. It also revealed information from all additional device data synchronized with smart phones. MFI addresses the contribution of improving competitiveness and safe society through modern crime investigation technique.

References

1. Carroll JM (2000) Making use: scenario-based design of human computer interactions. The MIT press, USA
2. Rowlingson, R (2004) Ten step process for forensic readiness. Int J Digit Evid Winter, 2(3):1–28
3. Introduction to NISTIR 7628 guidelines for smart grid cyber security. The Smart Grid Interoperability Panel Cyber Security Working Group, Sept 2010
4. Ciardhuain SO (2004) An extended model of cyber crime investigations. Int J Digit Evid 3(1)
5. www.phonearena.com/phones/Samsung-Galaxy-Note-4_id8577/benchmarks
6. RSA2012 cybercrime trends report http://www.rsa.com/products/consumer/whitepapers/11634_CYBRC12_WP_0112.pdf
7. Garfinkel SL (2010) Digital forensics research: the next 10 years. Naval Postgraduate School, Monterey, USA, Digit Invest 7:S64–S73
8. Palmer GL (2001) A road map for digital forensic research. Technical report DTR-T0010-01, DFRWS. Report for the first digital forensic research workshop (DFRWS)
9. Chang Y-H, Yoon K-B, Park D-W (2013) Technology for forensic analysis of synchronized smartphone backup data 978-1-4799-0604-8/13/$31.00 ©2013 IEEE
10. Maurer U (2004) New approaches to digital evidence. In: Proceedings of the IEEE 92(6):933–947
11. Palmer G (2002) Forensic analysis in the digital world. Int J Digit Evid 1(1)
12. Baryamureeba V, Tushabe F (2004) The enhanced digital investigation process model. In: Proceeding of digital forensic research workshop. Baltimore, MD

MIMO-Based Efficient Data Transmission Using USRP

M. Lenin, J. Shankar, A. Venkateswaran and N.R. Raajan

Abstract In full-duplex wireless communication, the major constraint is to utilize the bandwidth in an efficient way and to transmit and receive data simultaneously at the same time using a single channel. We introduce a concept of MIMO system for avoiding self-interference cancelation by using an USRP device to transmit data packets through channel. In this paper, the USRP will act as a transmission medium for the process and is interconnected through the MIMO channel. The generated data are processed through the USRP and reach the destination without any interference or without any loss of packets during transmission. The packets are transmitted using the LABVIEW software.

Keywords MIMO · Full-duplex · USRP · Packets · Interference

1 Introduction

In wireless communication, one of the major technologies proposed is Multiple In and Multiple Out (MIMO) process [1]. MIMO plays a major role in wireless and achieves communication effectively within a limited band. The MIMO system is combined with full-duplex to utilize the available bandwidth as much as possible for effective communication [2, 3]. In full-duplex communication, data can be

M. Lenin (✉) · J. Shankar · A. Venkateswaran · N.R. Raajan
Department of Electronics and Communication Engineering, School of Electrical and Electronics Engineering, SASTRA University, Thanjavur, Tamil Nadu, India
e-mail: leninmkarthi@gmail.com

J. Shankar
e-mail: shankarece123@gmail.com

A. Venkateswaran
e-mail: arangulavan.venkates@gmail.com

N.R. Raajan
e-mail: nrraajan@ece.sastra.edu

L.P. Suresh and B.K. Panigrahi (eds.), *Proceedings of the International Conference on Soft Computing Systems*, Advances in Intelligent Systems and Computing 398, DOI 10.1007/978-81-322-2674-1_38

transmitted and received in both directions simultaneously using a different frequency. It will act as a bidirectional communication process. The full-duplex single channel will be the most useful technique in the future wireless communications as it doubles the spectrum within the limited system [2]. While using full-duplex communication, we obtained drawbacks such as self-interference which occurred during processing and consumption of power also increased. These impacts can be eliminated by using a concept of MIMO system obtained in the full-duplex process implementation [1–4].

MIMO wireless communication process covers a wide coverage area, high energy efficiency, and high system capacity [3]. The design and analysis of MIMO channel is a critical factor in theoretical models represented as a wireless fading channel. In spatial correlation, a lot of attention was gained on channel configuration in realistic [5]. The self-interference problem over data transmission is eliminated by applying a MIMO system in full-duplex single channel [6–8].

USRP is a hardware used to interface analog signal with digital software. It has a separate transmitter and receiver allotted in a single hardware. It is interconnected through the gigabyte ethernet cable with the computer [1, 9–12].

2 Full-Duplex Process

Full-duplex communication process can be obtained simultaneously on two sides either by time or frequency. We represent this form of duplexing as Time Division Duplexing (TDD) and the other form as Frequency Division Duplexing (FDD) [1–4, 13, 14]. TDD can operate based on division of time to each node. So TDD has half-duplex communication. Then the other duplexing known as FDD can operate based on frequency domain; this duplexing process assigns different carrier frequency with every single data [14]. By assigning different frequency for data over transmission it requires more data to be processed at the same time [15]. These processes will increase the availability of user allotted for system at the same time. While using this concept more number of packets are transmitted within a limited frequency, and will cause interference in the required system; at the same time the power consumption is also increased during packet transmission [16].

3 USRP Platform

USRP is used for analog signal processed with a digital software. USRP N210 kit has components such as DSP 3400 FPGA, 100 MS/s dual ADC, 400 MS/s dual DAC, and gigabyte ethernet connectivity [9, 17]. N210 has a basic transmitter and receiver, components contain motherboard and many daughter boards obtained with transmitter and receiver to generate radio frequency. If daughter board tries to RF front end of system to be associated with transmission and receptions of signal.

Then the motherboard is the central part of processing which can perform analog-to-digital converters (ADC) and digital-to-analog converters (DAC) on transceiver similarly [12]. The receiving side consists of digital down converter (DDC) with decimator (Decim) and similarly transmit side has digital up converter (DUC), and interpolator (Interp) [11]. Finally, baseband signal is converted into streamed data using the data streaming function block.

4 Process Implementation

Initially, the data are processed by LABVIEW software and the USRP is interlinked with the required IP address allotted for the specific node [17]. The frequencies are assigned and modulation technique for transmission is selected [7, 18]. The messages are typed in the required area of the text box. There are a couple of transmitters and receivers proposed for this process, named as T × 1, T × 2, R × 1, and R × 2 [1, 2, 4, 8].

Initially, the transmitter T × 1 parameters are assigned with sampling rate and allotted frequency. It increases the strength of the message signal using a technique of modulation [2]. Here different forms of PSK modulation techniques are assigned for the packet data transmission process. The constellation graph of input signal is plotted during transmission [7].

After selection of modulation process, set the parameters for a specific packet. Packet parameters such as guard bits, synchronization bits, message bits, and number of packets in the message are assigned. On receiver side, the same IP address is assigned for the T × 1 to get the data [7, 14]. Then select the frequency and sampling rate on the receiver side to select the demodulation technique to retrieve the transmitted signal [19].

5 Experimental Analysis

5.1 Transmitter Section

In the transmitter part, the USRP 1 contains an IP address obtained as 192.168.11.4 and USRP 2 is obtained with the IP address 192.168.10.4. The USRP has a pair of transmitter and receiver present in each kit [17]. Kit 1 has a transmitter and receiver named as T × 1 and R × 1; the same is present in kit 2 named as T × 2 and R × 2 [8–13]. Here a full-duplex operation obtained with IP address for transmission is performed. Initially the LABVIEW is interlinked with kit 1 with a specific IP address, then the modulation technique and packet parameters assigned to transmit in the transmission are selected [20, 21]. Finally, select the required message to be transmitted through a channel with the USRP 1 [14, 22]. The same set of processes must be followed in another part of kit and made ready to transmit in USRP 2. Every signal is transmitted through a single allotted channel of MIMO cable [2, 3]. The data are transmitted from transmitter 1 and received at receiver 2. The couple of transmitters 1 and 2 are shown in Figs. 1 and 2.

The constellation graph for transmitters 1 and 2 are obtained with the message as plotted in Figs. 3 and 4.

The assigned packet parameters allotted in the block of USRP 1 and USRP 2 transmitter should be the same for the entire transmission process as shown in Fig. 5 [7, 18]. Here the input messages are converted into packets based on the packet parameters [5].

The modulation technique of transmitters 1 and 2 is shown in Figs. 6 and 7.

The messages are typed in the required location in LABVIEW block [9, 12, 13]. Here the typed messages of both the transmitters are shown in Figs. 8 and 9.

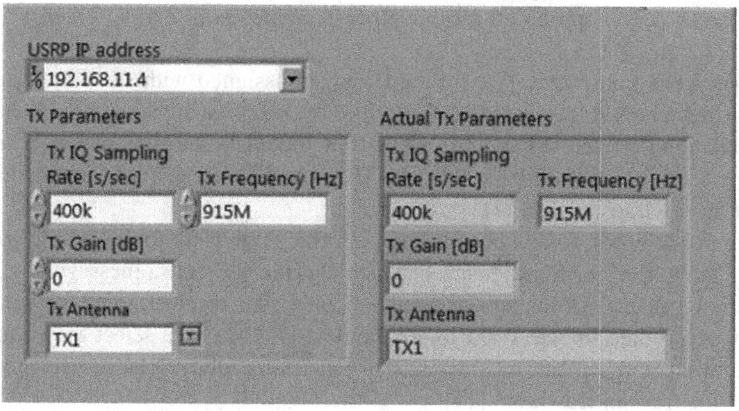

Fig. 1 USRP 1 transmitter specification

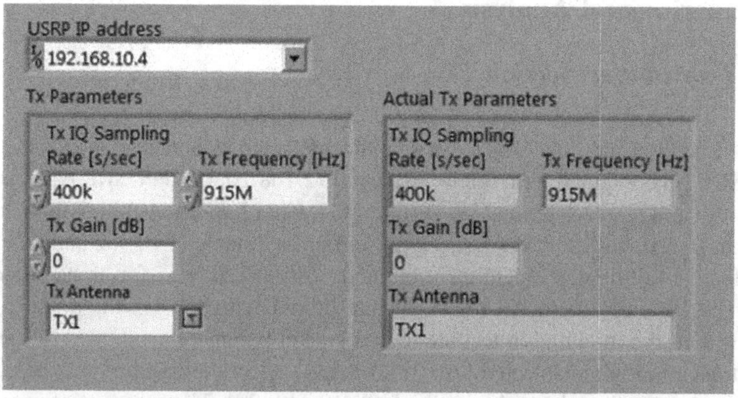

Fig. 2 USRP 2 transmitter specification

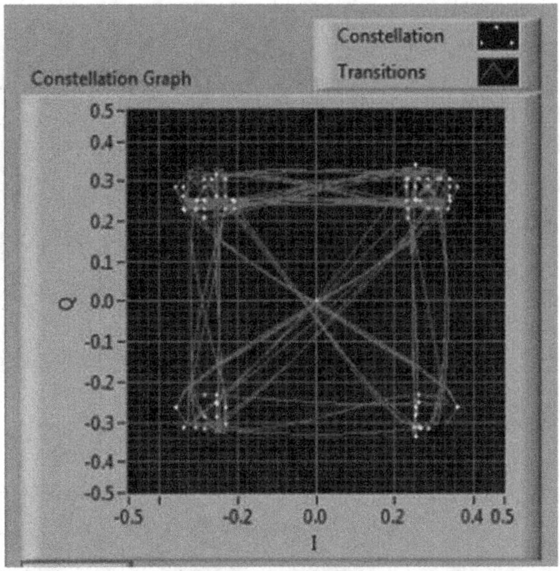

Fig. 3 Transmitter 1 constellation graph

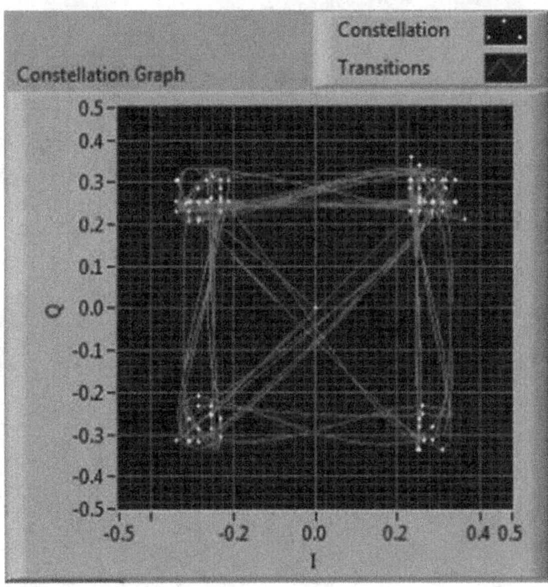

Fig. 4 Transmitter 2 constellation graph

5.2 *Receiver Section*

On receiver side, the operations performed in LABVIEW are the reverse process of transmitter section. Here the same IP address is allotted in the transmitter as mentioned in this section [6, 11, 13]. Based on IP address, the transmitted message

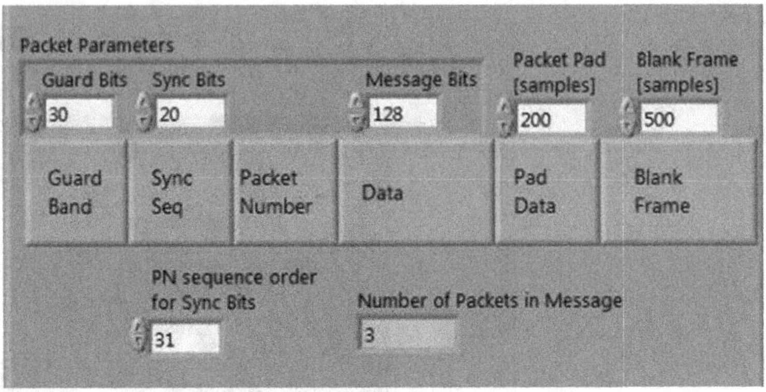

Fig. 5 Packet parameter

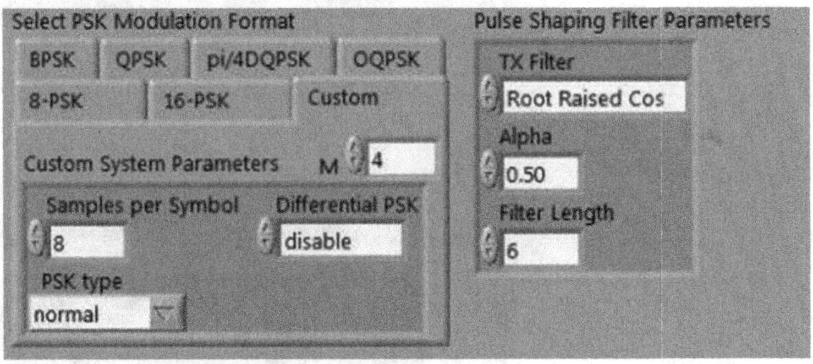

Fig. 6 Transmitter 1 modulation

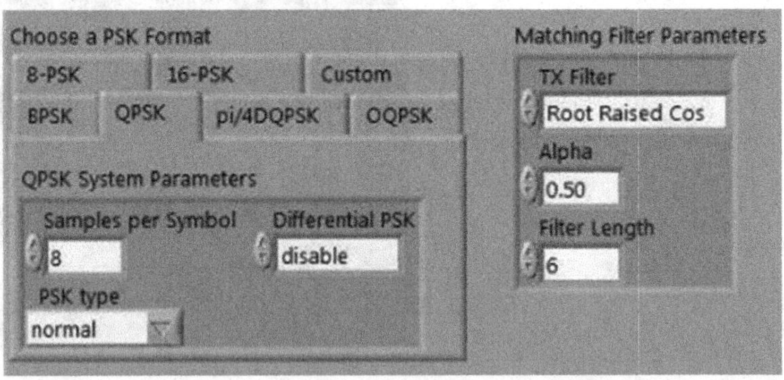

Fig. 7 Transmitter 2 modulation

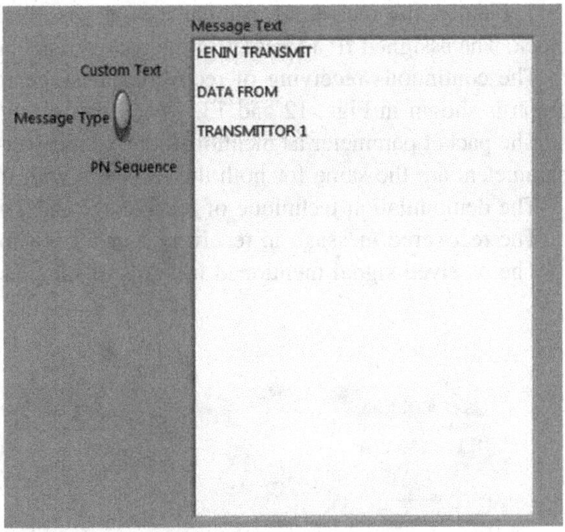

Fig. 8 Transmitter 1 message

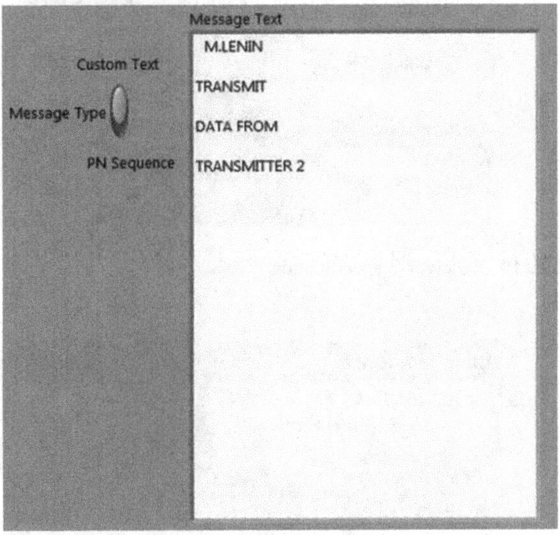

Fig. 9 Transmitter 2 message

from the transmitter side is retrieved. The couple of receivers are allotted in the transmission such as R × 1 and R × 2 which uses the same set of IP addresses. To receive a message such as from transmitter 1 to receiver 2 and transmitter 2 to receiver 1, a format can be assigned in the specific function. Initially, the receiver USRP block assigns the IP address as 192.168.10.4 and 192.168.11.4 as mentioned in R × 1 and R × 2 [12, 16, 18, 23]. Then select the demodulation and packet parameter process which is suitable to recover the message from the transmitter

[22]. Finally, the recovered original message is obtained as shown in the output block. The assigned IP addresses in the receiver side are shown in Figs. 10 and 11.

The continuous receiving of recovered message indicated by the constellation graph is shown in Figs. 12 and 13.

The packet parameter as mentioned on the receiver side is shown in Fig. 14, the parameters are the same for both the receivers with the same band [18, 24].

The demodulation technique of receivers 2 and 1 is shown in Figs. 15 and 16.

The recovered message in receivers 2 and 1 is shown in Figs. 17 and 18.

The received signal mentioned in terms of samples plotted is shown in Fig. 19.

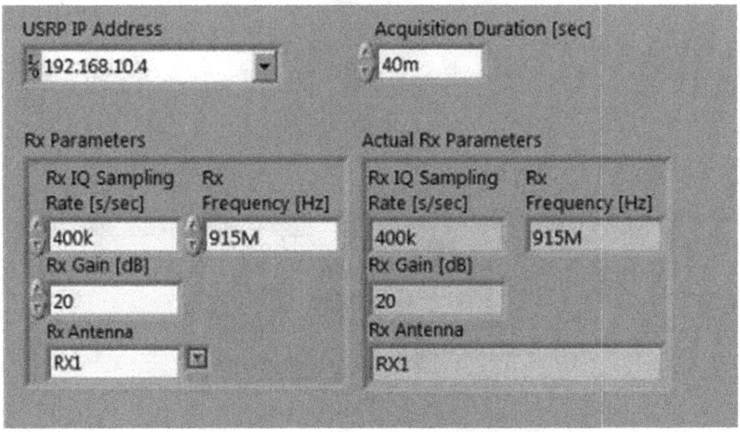

Fig. 10 Receiver 2 specification

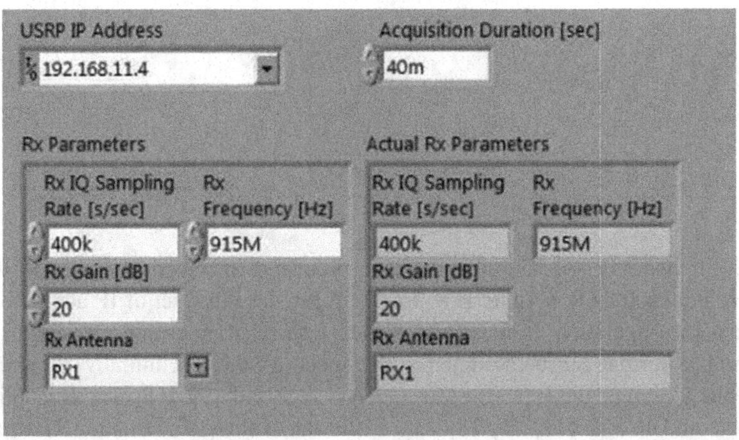

Fig. 11 Receiver 1 specification

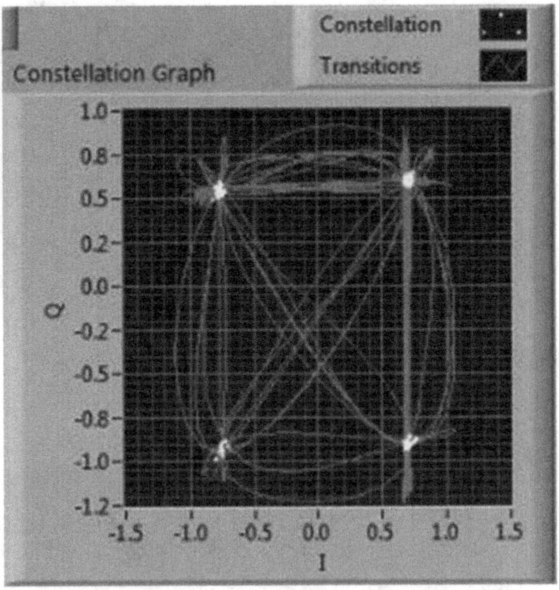

Fig. 12 Receiver 2 constellation graph

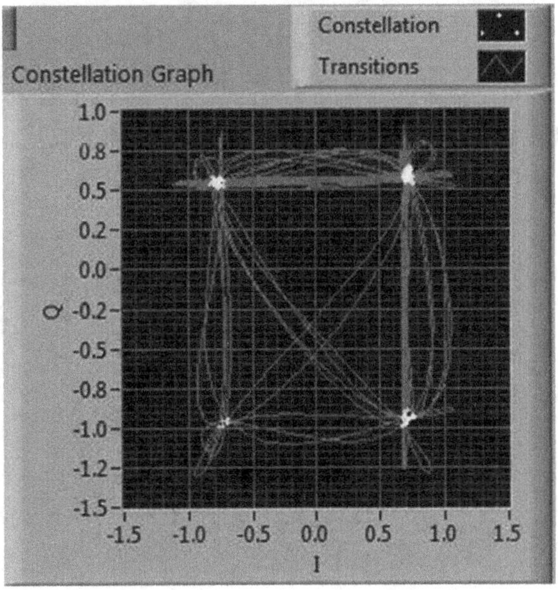

Fig. 13 Receiver 1 constellation graph

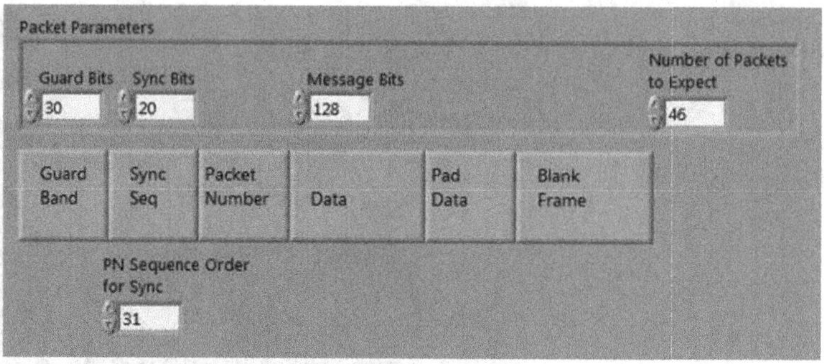

Fig. 14 Receiver packet parameter

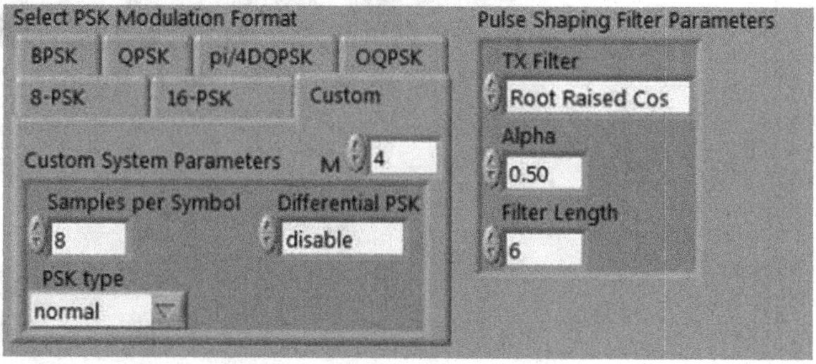

Fig. 15 Receiver 2 demodulation

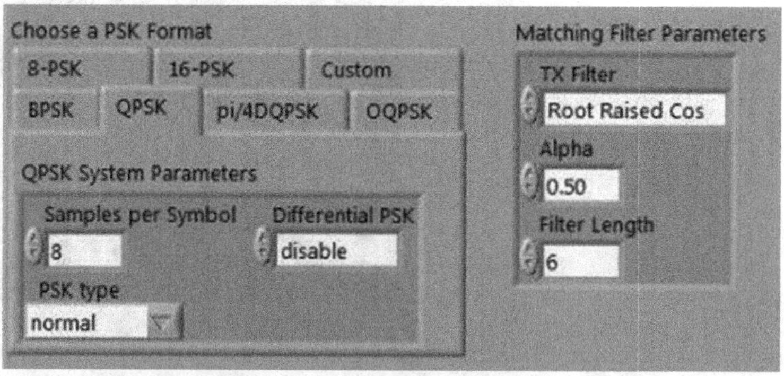

Fig. 16 Receiver 1 demodulation

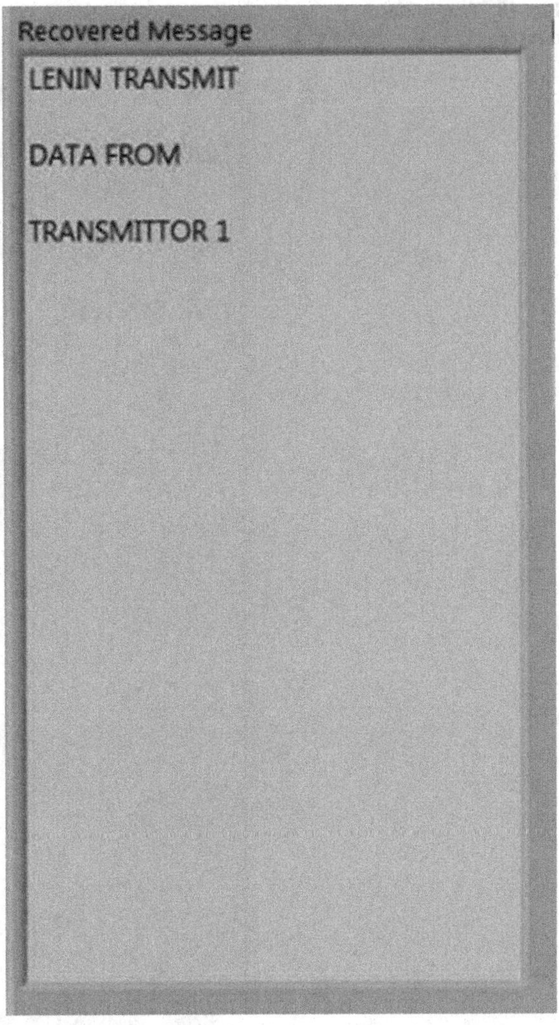

Fig. 17 Transmitter 1 message

In the USRP hardware, the signals are received successfully and the required graphs are analyzed in this experiment. Then the full-duplex formation of a plotted result is obtained on receiver side. To analyze the constellation graph by full-duplex, compared result are without any interference.

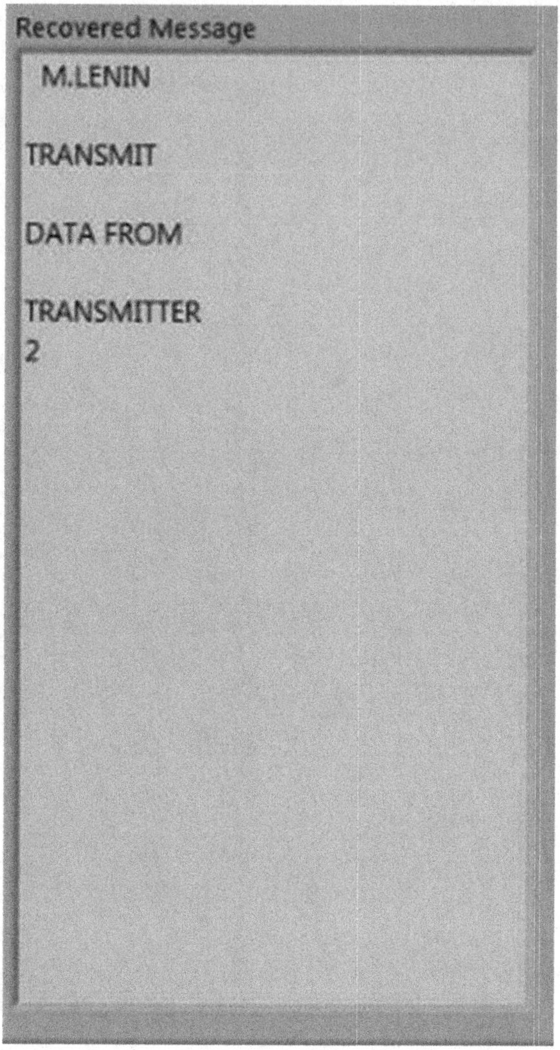

Fig. 18 Transmitter 2 message

5.3 *BER Comparison*

Comparison of various forms of modulation techniques performed in the MIMO system is shown in Fig. 20. The comparison results show better efficiency of transmission data obtained during different modulation [6, 8].

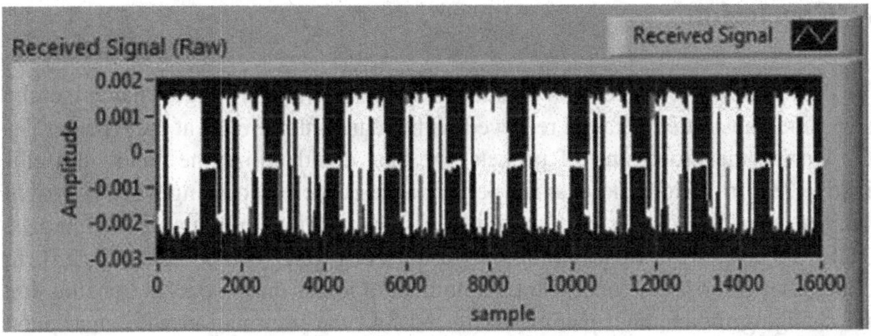

Receiver 2 signal

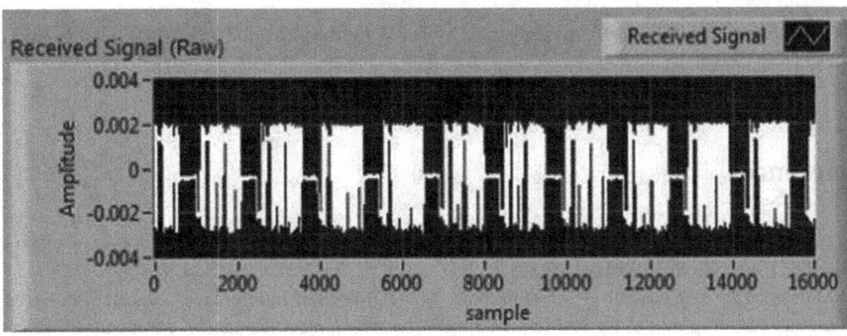

Receiver 1 signal

Fig. 19 Received signal

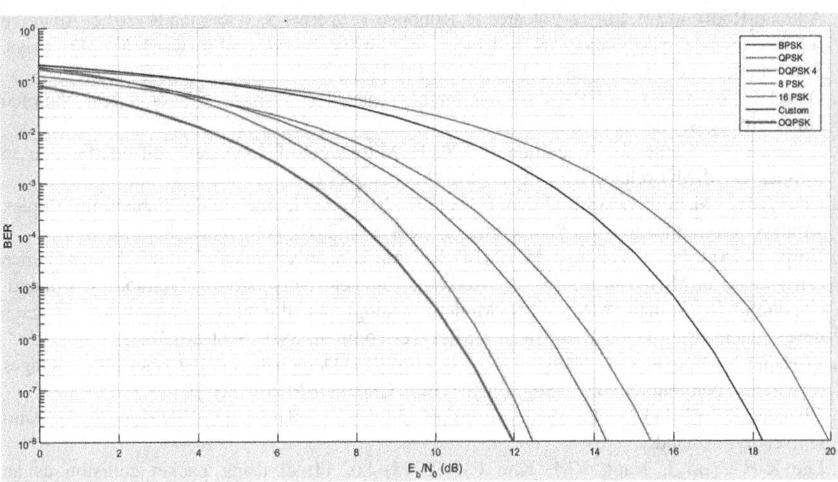

Fig. 20 BER comparison

6 Conclusion

The data packets are successfully passed through an available single channel; at the same time transmitted data are retrieved without any interference at the receiver [9–12]. During the processing of packets, the data could utilize the bandwidth efficiently [18]. In USRP, the channel performance is increased using MIMO system obtained in packet data transmission process with various modulation techniques, finally we collect the better efficiency result using BER comparison [13, 25]. In future processing, the device with high number of nodes during packet transmission and changing the channel parameters to transmit the information with encryption standards can be implemented [2].

Acknowledgments The work was supported by R&M fund (R&M/0027/SEEE-011/2012–2013) of SASTRA University.

References

1. Hua Y, Liang P, Ma Y, Cirik AC, Gao Q (2012) A method for broadband full-duplex MIMO radio. IEEE Sig Process Lett 19(12)
2. Meerasri P, Uthansakul P, Uthansakul M (2014) Self-interference cancellation-based mutual-coupling model for full-duplex single-channel mimo systems. Int J Antennas Propag
3. Lenin M, Shankar J, Raajan NR (2014) Hardware based full duplex wireless communication. In: Conference on emerging electrical systems and control (ICEESC 2014), vol 9(24). ISSN 0973-4562
4. Alexandris K, Balatsoukas-Stimming A, Burg A (2014) Measurement-based characterization of residual self-interference on a full-duplex MIMO Testbed, Switzerland
5. Antonio-Rodriguez E, Lopez-Valcarce R, Riihonen T, Werner S, Wichman R (2012) Adaptive self-interference cancellation in wideband full-duplex decode-and-forward MIMO relays. Smart radios and wireless research (SMARAD), Finland
6. Ong LT (2012) An USRP based interference canceller. Temasek Laboratories, National University of Singapore, IEEE
7. Riihonen T, Werner S, Wichman R (2011) Mitigation of loopback self-interference in full-duplex MIMO relays. IEEE Trans Sig Process 59(12)
8. Choi J, Jain M, Srinivasan K, Levis P, Katti S (2010) Achieving single channel, full duplex wireless communication. In: Proceedings of MobiCom2010, New York, Sept 2010
9. Korpi D, Anttila L, Valkama M (2014) Reference receiver aided digital self-interference cancellation in MIMO full-duplex transceivers. Tampere University of Technology, Finland
10. Gandhiraj R, Soman KP (2013) Modern analog and digital communication systems development using GNU Radio with USRP. Telecommun Syst (Sept 2013)
11. El-Hajjar M, Nguyen QA, Maunder RG, Ng SX (2014) Demonstrating the practical challenges of wireless communications using USRP. IEEE Commun Mag (May 2014)
12. Dhamala UP (2011) On the practicability of full-duplex relaying in OFDM systems. Aalto University, Finland
13. Lee K-H, Yoo J, Kang YM, Kim CK (2015) 802.11mc: using packet collision as an opportunity in heterogeneous MIMO-based Wi-Fi networks. IEEE Trans Veh Technol 64(1)
14. Duarte M, Sabharwal A (2010) Full-duplex wireless communications using off-the shelf radios: feasibility and first results. In: Proceedings of ASILOMAR

15. Widrow B, Duvall KM, Gooch RP, Newman WC (1982) Signal cancellation phenomena in adaptive antennas: causes and cures. IEEE Trans Antennas Propag AP-30:469–478
16. Duarte M, Sabharwal A (2010) Full-duplex wireless communications using off-the-shelf radios: feasibility and first results, pp 1558–1562
17. Widrow B et al (1975) Adaptive noise cancelling: principles and applications. Proc IEEE 63 (12):1692–1716
18. Chen X, Einarsson BP, Kildal P-S (2014) Improved MIMO throughput with inverse power allocation—study using USRP measurement in reverberation chamber, antennas and wireless propagation letters, vol 13. IEEE
19. Sahai A, Patel G, Sabharwal (2011) A pushing the limits of full-duplex: design and real-time implementation. Rice University, Technical Report TREE, July 2011
20. Akbar N, Khan AA, Butt FQ, Saleem MS, Riffat S (2013) Performance analysis of energy detection spectrum sensing using MIMO-OFDM VBLAST Test Bed. In: WMNC'13, National University of Sciences and Technology, IEEE
21. Zhou W, Villemaud G, Risset T (2014) Full duplex prototype of OFDM on GNU Radio and USRPs. IEEE radio and wireless symposium (RWS), pp 217–219
22. Syrjala V, Valkama M, Anttila L, Riihonen T, Korpi D (2014) Analysis of oscillator phase-noise effects on self-interference cancellation in full-duplex OFDM radio transceivers. IEEE Trans Wireless Commun
23. Cox C, Ackerman E (2013) Demonstration of a single-aperture, full duplex communication system. In: Proceedings of radio and wireless symposium, Jan 2013
24. Ettus Research. http://www.ettus.com
25. Howells P (1965) Intermediate frequency side-lobe canceller, U.S. Patent 3202990, Aug 1965

Automatic Road Sign Detection and Recognition Based on SIFT Feature Matching Algorithm

P. Sathish and D. Bharathi

Abstract The paper presents a safety and comfort driving assistance system to help driver in analyzing the road sign boards. The system assists the drivers by detecting and recognizing the sign boards along roadside from a moving vehicle. Signboards are detected using color and shape detection techniques, and they are recognized by matching the extracted scale invariant features with the features in the database. The proposed method based on SIFT feature detects 90 % of road sign boards accurately. Experimental analysis carried out using C/C++ and Open CV Libraries shows better performance in detecting and recognizing sign boards when real-time video frames are given as input.

Keywords Road traffic accidents (RTA) · Hue · Saturation · Value · Polygonal approximation · Hough transform · Scale invariant feature transformation

1 Introduction

Road Traffic Accidents (RTA) are one of the leading causes of death worldwide. 1.2 million people fall as victim every year. The most common cause of RTA is improper driving. Proper presents a novel driving assistance system which takes over this task, thus reducing the accident counts. In our approach an onboard camera on the vehicle monitors the road sign boards and assists the driver accordingly. Detecting Road sign boards have challenging problem like different illumination conditions, view point variation, aging of traffic signs, dislocation, and even vandalism. Other problems like occlusion, shadow occurrence existence of similar objects leads to false sign detection. Hence, the system developed should be

P. Sathish (✉) · D. Bharathi
Department of CSE, Amrita Vishwa Vidhyapeetham, Coimbatore, India
e-mail: sathishatm27@gmail.com

D. Bharathi
e-mail: d_bharathi@cb.amrita.edu

L.P. Suresh and B.K. Panigrahi (eds.), *Proceedings of the International Conference on Soft Computing Systems*, Advances in Intelligent Systems and Computing 398, DOI 10.1007/978-81-322-2674-1_39

robust to invariance under rotation and translation. Based on the shape, the sign boards are classified into three: circular, rectangular, and triangular.

The proposed method assumes that the vehicles are moving along the direction of road and the road the road sign boards are orthogonal to the road direction. Hence, the shape (rectangular or circular panels of different color) is detected without any distortion. In case of danger zone, the system alerts the driver, thus saving lives from the clutches of death.

2 Literature Survey

YCbCr color model: Color segmentation using YCbCr color space. Shape filtering is carried out template matching candidates detected with color. So the sign board from images as color and shape, thus distinguishes a sign board from its background. Multilayer perception neural networks are used to determine the classification module, determining the type of road sign board present.

RGB color model: In Broggi et al. (2007),author proposes a road sign board detection and classification system based on a three-step algorithm comprising of color segmentation, shape recognition using a neural network. To save computational time author prefers RGB color model, it helps to identify the sign boards based on color properties. Two methods are used for shape detection, one is based on pattern matching with simple models and the other is based on edge detection and geometrical cues. Finally, for each detected sign board, neural network is created and trained.

BRISK feature detection and TLD framework: In Piccioli et al. (1996), author propose a new algorithm that uses RGB to combine ratios-based color segmentation, automatic white balance preprocessing, and Douglas-Peucker shape detection for establishing ROI's. Scale and rotation invariant BRISK features are used for recognition and matching with the features of the template images in the database. Tracking-learning-detection (TLD) framework is adopted for tracking the recognized sign board in real-time scenario assisting the driver for better functioning.

3 Proposed Method

The proposed method consists of three categories to detect the sign board: color segmentation, shape detection, and classification based on shape as shown in Table 1. The recognition of sign board is done by template matching using SIFT features. The overview of road sign board detection and recognition is shown in Fig. 1.

Table 1 Proposed system

Color segmentation	HSV color model
Shape detection	Draw contours (Approximate shape)
	Hough transform (Detect lines and circles)
Recognition	Template match using SIFT features

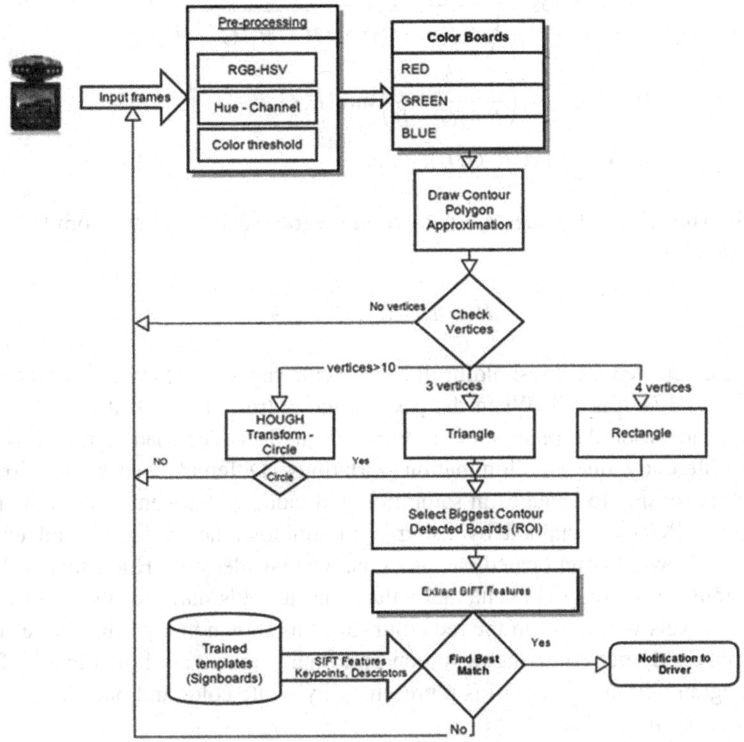

Fig. 1 Flowchart for sign board detection and recognition

3.1 Color Segmentation

The algorithm proposed detects and segments the sign board based on color region extraction in the varying illumination conditions. The input RGB image from the camera is enhanced by using true color and histogram equalization technique. The sign board is segmented from the input frame using HSV color space. HSV works much better in handling varying lighting conditions and it is easy to access the color values from Hue channel. HSV is used instead of RGB, because of different lighting conditions RGB values vary and it is difficult to access color values. While in RGB color space, an image is treated as an additive result of three base colors (blue, green, and red). HSV is a rearrangement of RGB in a cylindrical shape. Hue is represented as a circular angle and colors has extended in the angle between 0° and 360°. Being a circular value means that 360° is the same as 0°. A Hue of 0° is red, a Hue of 90° would be green, 180° is blue, 270° is pink, and again Hue is red at 360°.

This is the conversion of RGB to HSV,

$$H = \cos^{-1}\left(\frac{0.5(R-G)+(R-B)}{\sqrt{(R-G)^2+(R-B)(G-B)}}\right)$$
$$S = 1-\left(\frac{3}{R+G+B}\right)\min(R,G,B)$$
$$V = \max(R,G,B)$$

Then the Hue channel values are converted to gray scale of range from 0 to 255. And it done by,

$$H = H/20 < H < 179$$

Colors are selected by thresholding the hue scale range. If Hue has a pixel value from 160 to 179 it is red. When the pixel value is from 40 to 60 it is green. Blue color appears when the pixel value is from 100 to 125. The road sign boards may appear differently due to illumination variations. Different shades of color are obtained according to changes in saturation and value components where the pixel of interest (POI) is obtained by using optimum tolerance values found experimentally. To avoid complicated calculations, we consider only Hue channel. Here, we are considering only Hue channel values and it holds purity of the color it will select the pixels which lies in the red color range as shown in Fig. 2b. The resultant image obtained after color segmentation is a binary image as shown in Fig. 2c in which region within sign board is represented by white color and outside region of sign board is made black.

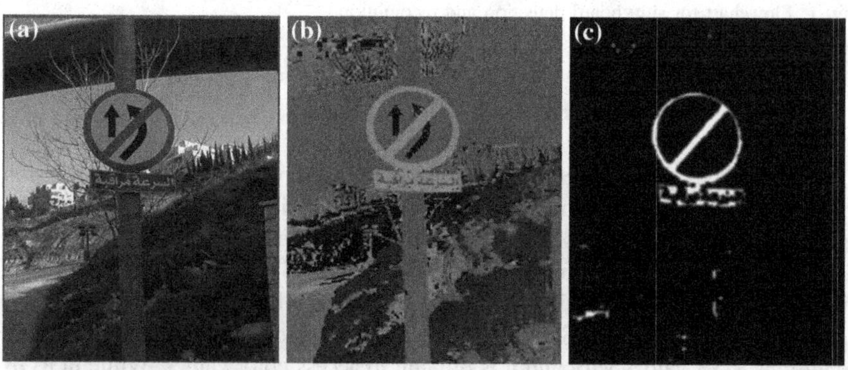

Fig. 2 **a** Input image. **b** Hue channel. **c** Red color extracted

Algorithm 1 Signboard Detection

Input: Frames from Dashboard camera

Output: Detect signboard ROI

1: Convert RGB→HSV Color model

2: Select Hue channel

3: Do Color Segmentation

set REDBOARD = if(intensity >160 && 179 < intensity)

set BLUEBOARD = if(intensity >100 && 125< intensity)

set GREENBOARD = if(intensity > 60 && 85< intensity)

4: Extract Contours

Do Polygon approximation

Detect Triangle = if(vertices=3)

Detect Rectangle = if(vertices=4 && angle 90°)

Detect Circle = if(vertices>10) apply Hough Transform detect Circle

5: Find Contour size

Select max.Contoursize()&Set ROI

3.2 Shape Detection

The contour is extracted from the binary image using canny edge detector. From the extracted contour, the polygonal curve approximation is done, which is the key for detecting the shape of the contour. For example, the shape is determined to be a rectangle if the polygonal curve meets the following conditions: It is convex, has 4 vertices and all angles are ~90 (degree as shown in Fig. 3b. Given the approximation curve, we can quickly determine a contour is a triangle if the curve has 3 vertices. Similarly for pentagon, hexagon, and circle which has 5 vertices, 6 vertices, and even more vertices, respectively. To obtain the degree of the corners, we detect square, rectangle, pentagon, and hexagon. We will use these values to determine the contour's shape by using the shape properties. The proposed method

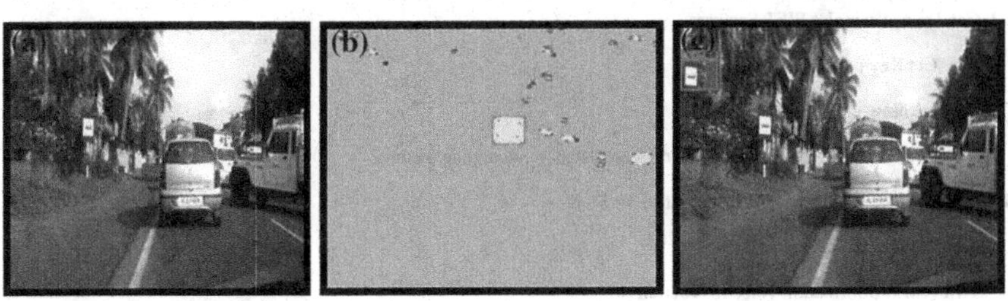

Fig. 3 **a** Input frame. **b** Signboard segmentation. **c** Signboard detection

is a lossless approximation that encodes horizontal, vertical, and diagonal contour segments with ending points. The Freeman chain code method is used to find the contour of an image as a chain code which is then converted into points. These points are then compressed to produce the fine contours of the object in the image. The triangular sign board is detected by using a polygonal approximation of edge chains inside the search region, which is performed to eliminate the chains that strongly depart from a straight segment. The Circular sign board is hard to find using polygonal approximation algorithm since there are no edges found in the circle. Hence, we use Hough transform to detect the circular shaped road sign board. Using Hough transform the circular shape is recovered by finding the cluster of points in the 3D parameter space. It is computed by identifying the circumference passing through the current point and the other edge points that lie at a certain distance on the same chain.

3.3 *Recognition of Road Signs*

Recognition of Road Sign board is a hard multi-class problem. Many traffic signs are not standardized in terms of color, shape, and symbol contained. A good example is the signs giving direction, which may vary in terms of size, shape, and the background color of the board. Such sign boards also contain various symbols, like arrows, as well as variable-font characters. Road Sign board recognition involves two critical stages: feature extraction and classification. In most of the object recognition problem, the feature exaction step is considered as most important one where a discriminative representation of an object provides a mapping from the high-dimensional original image space to the low-dimensional feature space in which different class patterns are separated from one another. The SIFT algorithm is used to detect key points and describe them because the SIFT features are invariant to scale and rotation and are robust to changes in viewpoint and illumination. SIFT is divided into two stages, key point detection and key point description.

Algorithm 2 Signboard Recognition

Input: Signboard ROI from each frame

Output: Instruction to Driver by Text/Voice

1: Extract SIFT Features from ROI

2: Get Trained samples SIFT Features

Get Key points & Descriptor

3: Do Matching

Find good matches By Using Euclidean distance calculation method

If (distance (2nd neighborhood distance ratio) <0.8)

Mark as Good Match

4: Retrieve text Information/voice Instruction

Each stage consists of two substages respectively.

1. *Scale-space detection*: Overall scales and image locations are searched during the initial stage of computation. The next task is to identify visual interest points, which is invariant to scaling and orientation changes. Laplacian of Gaussian is used to find such visual interest points. The image then is convolved with the Gaussian filter at specified scales and orientations that results in octaves as shown in Fig. 4.

$$\frac{\partial G}{\partial \sigma} = \sigma \Delta^2 G[\text{Heat equation}]$$

$$\sigma \Delta^2 G = \frac{\partial G}{\partial \sigma} = \frac{G(x, y, k\sigma) - G(x, y, \sigma)}{k\sigma - \sigma}$$

$$G(x, y, k\sigma) - G(x, y, \sigma) \approx (k - 1)\sigma^2 \Delta^2 G \quad \text{Typical values} : \sigma = 1.6; \; k = \sqrt{2}$$

$$G(x, y, k\sigma) = \frac{1}{2\pi (k\sigma)^2} e^{\frac{-(x^2 + y^2)}{2k^2\sigma^2}} [\text{Gaussian equation}]$$

Iteratively, the different versions of the same image are filtered by multiplying σ and k. Then for each octave the image is scaled down by half. In each octave we find a difference in values which helps in finding visual interest points. The key points are the maxima/minima, obtained as a result of Difference of Gaussian operation at multiple scale.
Figure 4b Compare a pixel (x) with 26 pixels in current and adjacent scales. It will select a pixel (x) if it is larger/smaller than all 26 pixels.

2. *Key point localization*: The selection of each key points is based on the magnitude of stability. There can be changes of outliers, which occurs due to noise, low contrast, and improper location of candidates along edges. To localize prominent key points, an outlier rejection process is carried out initially. Taylor

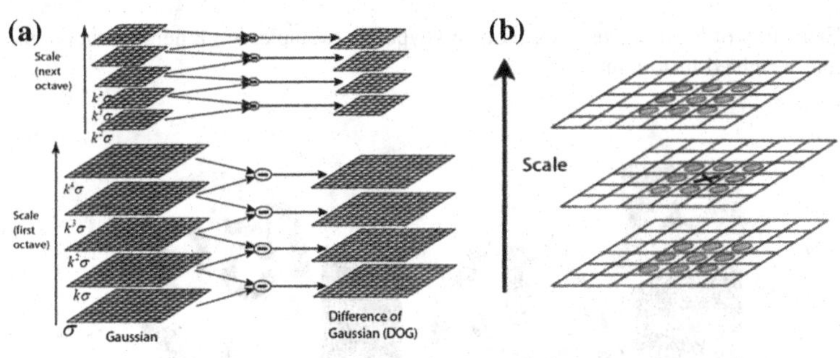

Fig. 4 Key Point selection: **a** building a scale-space difference of Gaussian and **b** scale-space peak detection

series expansion of Difference of Gaussian on image is used to accomplish the above task of finding improved maxima/minima. Thus more stable and better key points are obtained.

3. *Orientation assignment*: For every key point orientations are being assigned based on local image gradient directions. The image data is transformed according to the assigned orientations, and then further operations are performed by providing invariance to the assigned orientation, scale, and location of the each feature.
4. *Key point descriptor*: It will describe the key point as a high-dimensional vector. The local image gradients are measured at the selected scale in the region around each key point. These are transformed into a representation that allows for significant levels of local shape distortion and change in illumination. SIFT feature is a local feature. Compared with traditional overall features, it improves the efficiency of the method. After selecting key points, it will select 16 × 16 neighborhoods around key points, and it divides 16 × 16 neighborhoods by 4 × 4 block as shown in Fig. 5. In each 4 × 4 block it will compute 8 bin histogram using the gradient orientation and scale it by the magnitude and it will get 16 histograms, each has eight dimensions put them together, we will get vector of 128 dimensions, which is called SIFT descriptor.

The features are extracted from the trained dataset and XML file is created. From which the road sign board is matched and recognized from the input video frame as shown in Fig. 6. A descriptor from the detected signboard ($D1$) is matched to a

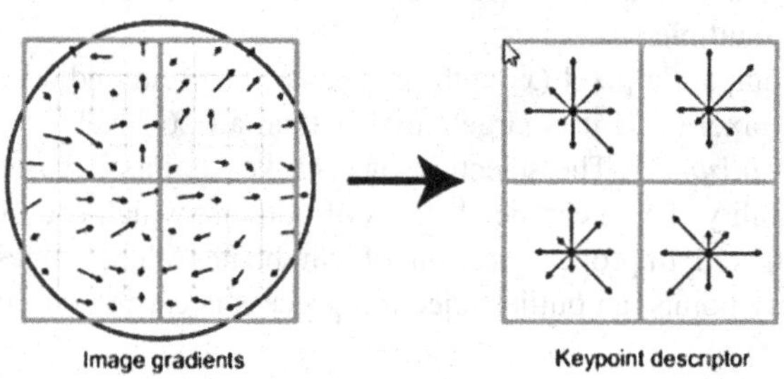

Fig. 5 Extraction of local image descriptors at keypoints example: 8 × 8 neighborhood and 2 × 2 block 8 bin weighted histogram

Fig. 6 SIFT keypoints extraction and descriptor matching

descriptor from the trained templates ($D2$) only if the distance $d(D1, D2)$ multiplied by threshold is not greater than the distance of detected signboard ($D1$) to all other descriptors. The default value of the threshold is 1.5.

4 System Implementation

4.1 Training the Dataset

Indian traffic sign boards are collected and classified into three groups based on shapes like circle board, triangle board, and rectangle board. The dataset is trained and the key point features are extracted from the trained dataset and descriptor of each key points with 128 dimensional vectors are stored in XML file.

4.2 Hue Channel Extraction and Thresholding

The Hue channel from the input frame is extracted which gives the purity of color within the color space range from 0° to 359° when measured in degrees and then threshold. To segment each color, the color ranges have been set to gray scale values and each color filtered by threshold values (Table 2).

4.3 Polygonal Approximation and Shift Feature Matching

For a given image, the contours are extracted from the binary image by applying polygonal approximation to the binary image followed by Hough transform. After polygon approximation, color is filled in the contour region and the largest contour is segmented out separately. From the segmented region, number of vertices is found in order to narrow down the shape of the sign board. If contour has 3 vertices it is a triangle sign board. If it has 4 vertices, then it is a rectangle sign board, and if it has more than 10 vertices, select that contour, and send it to Hough transform function to find out whether detected part is circle board. Then Region of Interest (ROI) will be set and key points are extracted and matched with the trained sets as shown in Fig. 7.

Table 2 Color Segmentation-Hue range

Color	Hue range ($H = H/2$) 0–179
Red	$160 < H < 179$
Green	$60 < H < 85$
Blue	$100 < H < 125$
Yellow	$40 < H < 50$

Fig. 7 Analysis of different dataset forms: **a** input frame, **b** color threshold, **c** shift matching, **d** recognized board contour extraction and polygonal approximation, **e** scene1-sign board, **f** scene2-signboard

Experimental Results:

Test videos are taken randomly from the manually collected dataset for processing and the results are shown in the table with the true positive, false positive, and false negative and precision. This system is built using Visual studio 2010 with OpenCV2.4.9 Libraries. In order to analyse the efficiency of the system the results are shown in the table. The performance measured in our approach is in terms of average precision rate, which is 85.6 %. This shows that the proposed system is robust and works with better accuracy. Eight manually recorded videos are considered for experimentation and their prediction analysis is in Table 3.

Table 3 Images and their corresponding predictions

#Recorded video	True positive	False positive	False negative	Sign board detection rate/precision
1	3	0	1	100
2	4	1	3	80
3	5	1	2	83.3
4	4	0	3	100
5	5	1	2	83.3
6	6	2	4	75
7	7	2	2	77.7
8	6	1	3	85.7

Precision = (true positive/true positive + false positive) × 100
Average precision = total detection rate of N videos/N
Here N stands for total number of videos used for experimental analysis ($N = 8$)

5 Conclusion

In this paper, we propose to detect and recognize the signboards from a moving vehicle. On comparing with the existing approaches, the proposed approach is invariant to rotation, translation, and size of the sign board. The SIFT features used in our approach makes the system robust to occlusions, thus performing better in real-time scenario. Experimental analysis carried out by a set of manually recorded real-time videos shows that the system is robust and gives better accuracy.

Bibliography

1. Zhihui Z, Hanxizi Z, Bo W, Zhifeng G (2012) Robust traffic sign recognition and tracking for advanced driver assistance systems. In: Proceedings of the 15th international IEEE conference on intelligent transportation systems, Anchorage, Alaska, USA, 16–19 Sept 2012
2. Broggi A, Cerri P, Medici P, Porta PP, Ghisio G (2007) Real time road signs recognition. In: proceedings of IEEE intelligent vehicles symposium, Istanbul, pp 981–986
3. Adam A, Ioannidis C (2014) Automatic road-sign detection and classification based on support vector machines and hog descriptors. ISPRS annals of the photogrammetry, remote sensing and spatial information sciences, vol II-5
4. Fleyeh H, Dougherty M (2005) Road and traffic sign detection and recognition. Advanced OR and AI methods in transportation
5. Dean HN, Jabir KVT (2013) Real time detection and recognition of indian traffic signs using Matlab. Int J Sci Eng Res 4(5)
6. Piccioli G, De Michelib E, Parodi P, Campani M (1996) Robust method for road sign detection and recognition. Image Vis Comput 14:209–223
7. Fang C, Chen S, Fuh C (2003) Road-sign detection and tracking. IEEE Trans Veh Technol 52:1329–1341
8. de la Escalera A, Armingol J, Mata M (2003) Traffic sign recognition and analysis for intelligent vehicles. Image Vision Comput 21:247–258
9. Vitabile S, Gentile A, Sorbello F (2002) A neural network based automatic road sign recognizer. In: International joint conference on neural networks, Honolulu, HI, USA
10. Miura J, Kanda T, Shirai Y (200) An active vision system for real-time traffic sign recognition. In: ieee intelligent transportation systems, Dearborn, MI, USA
11. Douville P (2000) Real-time classification of traffic signs, real-time imaging. 6(3):185–193

A Performance Analysis of Black Hole Detection Mechanisms in Ad Hoc Networks

V. Manjusha and N. Radhika

Abstract A continuous infrastructure-less, self-configured collection of mobile devices connected as a network without wires is termed as mobile ad hoc network (MANET). Dynamic topology, shared physical medium, and distributed operations are some of the characteristics of MANETs due to which they are highly vulnerable to security threats. Black hole attack is a kind of attack that compromises security by redirecting the entire traffic to a node that is not actually present in the network. This paper aims to detect black hole using three detection schemes, i.e., Time-based threshold detection schemes, DRI table with cross checking scheme, and distributed cooperative mechanism. Comparative analyses on all three schemes are done to find out which one detects black hole more accurately. The measurements are taken on the light of packet delivery ratio, throughput, energy consumption and end-to-end delay. Simulation is performed using Network Simulator-2 (NS2).

Keywords MANETs · AODV · Black hole attack · Malignant node

1 Introduction

A group of mobile nodes that disseminate with each other are known as mobile ad hoc networks (MANET). They do not have any centralized authority. Every [1] node in the network act as a router. They are dynamic, distributed, and infrastructure-less networks that works regardless of any access points or base stations. The access to information and services are provided in MANETs irrespective of their position. They tend to be self-healing, self-configured, peer-to-peer networks. They have

V. Manjusha (✉) · N. Radhika
Department of Computer Science & Engineering, Amrita Vishwa Vidyapeetham, Coimbatore, India
e-mail: manju.2302@gmail.com

N. Radhika
e-mail: n_radhika@cb.amrita.edu

L.P. Suresh and B.K. Panigrahi (eds.), *Proceedings of the International Conference on Soft Computing Systems*, Advances in Intelligent Systems and Computing 398, DOI 10.1007/978-81-322-2674-1_40

improved flexibility and are scalable due to their dynamic topology. A vital aspect of MANETs is its mobility. The setting up of network can be done at any place and time. The main challenges that MANETs face are topology that changes dynamically, bandwidth constraints, cooperative algorithms, and limited security.

One of the major concerns for the proper functioning of any network is security. Currently wired networks are more secure than MANETs against attacks. MANETs does not have a centralized monitoring server. The absence of this server makes the detection of attacks more difficult. MANETs does not have a predefined boundary too. Nodes work in an environment where they are allowed to join and leave the network at anytime. The network has only limited power supply, where the nodes may behave in a selfish manner which may cause several problems. It is very difficult to detect the malignant node as nodes may freely enter/leave the network. Thus, these attacks prove to be more hazardous than other external attacks.

One among the main security attacks in MANETs is black hole attack in which a malignant node broadcast itself as the one which has the shortest distance to the destination node or to the packet it wants to interrupt. Even without checking the routing table, this node can advertise the availability of routes which enables the attacker node to reply to the route request and hence intercepting the packet. This paper is organized into five sections. Section 2 presents Literature Survey. Section 3 suggests the proposed system. Section 4 presents the analysis. Section 5 presents the conclusion.

2 Literature Survey

2.1 Routing Protocols in MANETs

Communication between nodes facilitates the propagation of information. A routing protocol [2] is used to cite how routers inseminate this information, and how the paths between two nodes are chosen. A routing algorithm selects a particular route based on the routing protocol. In prior, every router will be aware of the networks associated to it precisely. This protocol exchanges information initially with its immediate neighbors, and thereby increases the networks throughput. A demanding area in MANETs, to which researchers pay keen consideration is routing protocols. Based on their functionality, the MANETs routing protocols [3] are widely classified into three different classes.

2.2 Reactive Protocols

These protocols do not begin route identification by themselves, until a source node requests to find a route. Also called as demand-driven protocols, they set up routes on demand. They establish a path from source to destination whenever communication is to be set up. Ad hoc on-demand distance vector (AODV) is a reactive protocol.

2.3 *Proactive Protocols*

Proactive routing [4] protocols keep record of the most recent network topology. Each node in the network has knowledge about the other node in advance. The routing information is kept in tabular form. If there are any changes, the tables are altered corresponding to the change. The topological information is shared among nodes and they can also access the information regarding routes whenever required. DSDV, OLSR, and OSPF are some examples of this kind of protocols.

2.4 *Hybrid Protocols*

Hybrid protocols utilize the good aspects of both proactive and reactive protocols and merge them for obtaining improved results. The networks are divided into two zones, and while one protocol can be used within, the other can be used between zones. Zone Routing Protocol (ZRP) is a kind of Hybrid Routing Protocol.

2.5 *Ad Hoc On-Demand Distance Vector (AODV)*

Ad hoc on demand distance vector routing (AODV) protocol [5] is the protocol used in the proposed work due to its rapid route creation and establishment properties. It [6] facilitates discovery of route by making use of control messages, route request (RREQ) and route reply (RREP) whenever a node wishes to transmit packets to the destination. Upon receipt of the route error (RERR) message, the source re-establishes route. Information about the nearby nodes is received from broadcasted hello packets and need not have a centralized system to administer the routing process. AODV increases the cost of latency so as to avoid the overhead of control traffic messages. This protocol is a loop free protocol that adopts sequence numbers to avoid the so-called "infinity counting problem."

2.6 *Attacks in MANETs*

In black hole attack, [7] a malignant node broadcasts that it has the shortest distance to destination node or to the data packet it wants to interrupt. This causes Denial of

Fig. 1 Black hole attack

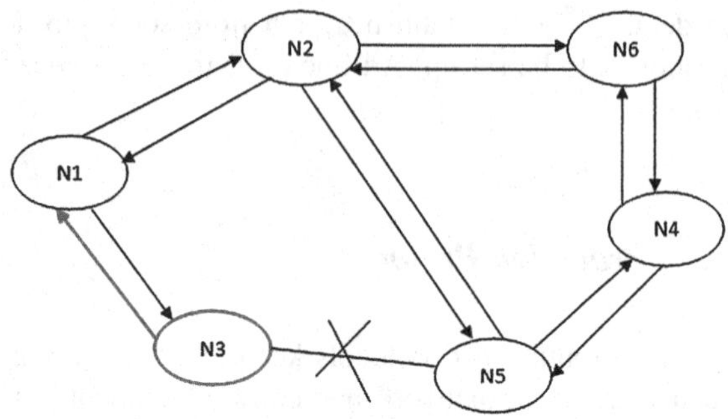

Service (DoS) attack by losing the obtained packets. Here, the malignant node endorses the availability of its route without checking its routing table. As soon as the source node gets the route reply from this node, it forwards its data packet to the malignant node. In flooding-based protocol, reply from the malignant node is received even before the reply is received from true node. Hence a malignant route is formed. After the route has been established, the malignant node decides whether to forward the packet to an unknown address or to drop all packets. Methods on how malignant node is inserted into the data routes vary. Figure 1 indicates the formation of black hole. Node N1 wants to transmit packets to node N4 and initiates the path selection process. If the node N3 is assumed to be malignant, it will declare that it has discovered the active route to the destination node as it gets the RREQ packets. The RREP packets are then sent to the node N1 prior to any other node. Thus, the node N1 takes this as the active route and the active route discovery is complete. Now node N1 ignores further replies and starts sending packets to node N3. Likewise all the packets are lost.

3 Related Works

In [3] AODV there are two types of black hole attacks. They are as follows:

(i) Internal Black Hole Attack
In this attack, the malignant node is inserted into the path of source and destination. Internal malignant nodes places itself an active route element and attacks the network at the beginning of data transmission. Since the malignant nodes are present within the network, the internal nodes are more prone to attack.

(ii) External Black Hole Attack
In this attack, malignant node first finds out the active route to the destination along with the destination address, then transmits a route reply packet (RREP) along with the destination address field to an unidentified destination address. Here, the hop count is set to a minimum value and the sequence number to a

maximum value. Next, an RREP is transmitted to the nearest neighbor in the active route and this is transmitted straight to the source node. The nearest neighbor node that receives the RREP transmits it through the recognized inverse route of source node. The updated information obtained in the route reply will be incorporated in the source nodes routing table.

3.1 Glimpse of Existing Methods for Detection of Black Hole Attack with Pros and Cons

MANETS [8] has attracted the attention due to the widespread use of mobiles. They can be deployed easily, as they are infrastructure less. Hence, they can be used in personal area networks, home area networks, and so on. MANETS are widely used in military operations and disaster rescue. Since MANETS are infrastructure-less, they are more prone to security attacks. There are several security attacks and detection schemes in MANETs. Some of the detection schemes in MANETs are depicted in Table 1.

4 Proposed System

A black hole attack is considered to be a severe security issue in MANETs. Here, [9] a malignant node broadcast itself having the shortest path to the destination. It captivates all data packets from the source by fraudulently pretending that it has the efficient path. To establish a route and to transfer data, each node in the MANET acts as a router. Intermediate nodes are used to transfer packets, whenever a node chooses to transmit data to another node. Thus a vital task in MANET is to search and establish a path from source to destination. Hence routing is a significant element in MANETs. In this paper, a comparative study on three black hole attack detection schemes in MANETS have been done. The three detection techniques are as follows.

4.1 Time-Based Threshold Detection Scheme

In this method, the fundamental idea is to ensure the time obtained for the initial route request along with the value of timer threshold. Each node after obtaining the first request has to setup a timer in the so-called timer expired table and the succeeding requests will be established when the timer expires. It will store the arrival time of the route request along with the sequence number. After the timeout, it first checks its collect route reply table (CRRT), to see if there are any next hops. Whenever the next hop is repeated, the path is assumed to be safe, i.e., the path does not carry any malignant node.

Table 1 Methods for detection of black hole attack with pros and cons

Sl. no	Schemes	Routing protocol	Simulator	Pros	Cons
1	BDSR (Baited Black Hole DSR)	DSR	Qualnet	Provides better packet delivery ratio	Does not overcome network overhead
2	Neighborhood based routing recovery	AODV	NS2	The probability if detecting a single attacker is high	Failed when attacker cooperate to cast the false reply packets
3	Redundant route and unique sequence number scheme	AODV	NS2	Increased route verification	Attacker can listen to channel and updates table for last sequence number
4	Time-based threshold detection scheme	AODV	NS2	Provides high packet delivery ratio and reduced overhead	High end to end delay when the malignant node is away from the source node
5	Random two-hop ACK and Bayesian detection scheme	DSR	GloMoSim	Effective detection of malignant node and reduced overhead	Fails when there are multiple malignant node
6	DRI table with cross checking scheme	AODV	NS2	Detects collaborative black holes and minimum packet loss	Reduced end to end delay
7	Distributed cooperative mechanism	AODV	NS2	High packet delivery ratio and detection rate more than 98 %	Higher control overhead than AODV
8	Resource-efficient accountability scheme (REAct)	DSR	Qualnet	Reduces communication overhead	Binary search method easily exposes audit nodes information
9	Detection, prevention and reactive AODV (DPRAODV)	AODV	NS2	High packet delivery ratio than AODV	High routing overhead and end to end delay

4.2 DRI (Data Route Information) Table and Cross Checking Scheme

In [10] this method, the route entry to the destination for the source node is not present. So it will transmit a RREQ message to determine the same. Any node that receives this request either replies or broadcasts it again. When the destination replies, each and every intermediate node get updated or the entry gets inserted in the routing table. Based on the reply originated a trust is formed between the source and destination and data is send along the path and the DRI table will get updated along with the entire intermediate nodes linking source and destination. The benefits of the method are that it can effortlessly make out the collaborative black hole nodes. However, the main drawback is the safeguarding of supplementary DRI table in the memory of each node. Also it wants the nodes to accumulate the previous routing information which is very costly. The more the size of the DRI table, the more the computational overhead of the cross checking process.

4.3 Distributed Cooperative Mechanism (DCM)

In DCM method, a four step procedure is used for detecting black hole attack. In the initial phase of this mechanism, each node consists of an additional table which is known as estimation table. This table comprises of the evaluation of credibility of each node according to the overhearing of packets. When a node is said to be suspicious, it enters the second phase called as local detection where it is verified with its partner cooperative node. If the value of inspection is negative, the node enters third phase where all one-hop neighbors take part in broadcasting about the credibility of that particular suspicious node. The final phase which is the global reaction phase, the information is exchanged with all nodes in the network and thus the black hole node is detected.

5 Analyses

5.1 Packet Delivery Ratio

Packet delivery ratio is defined as the ratio of total number of packets transmitted by the CBR (Constant Bit Rate) source to the total number of packets obtained by CBR sink at the corresponding destination nodes. In Fig. 2, *x*-axis indicates the packet delivery ratio and *y*-axis indicates the different detection schemes chosen under study. Here, we can analyze that distributed cooperative mechanism scheme has higher packet delivery ratio, when compared with the other two schemes.

5.2 Throughput

Throughput is defined as the total number of packets transmitted over network in one second time. In Fig. 3, *x*-axis indicates the throughput and *y*-axis indicates the

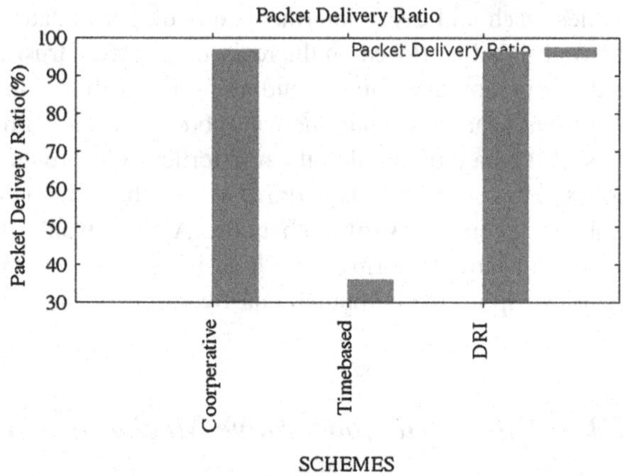

Fig. 2 Packet delivery ratio

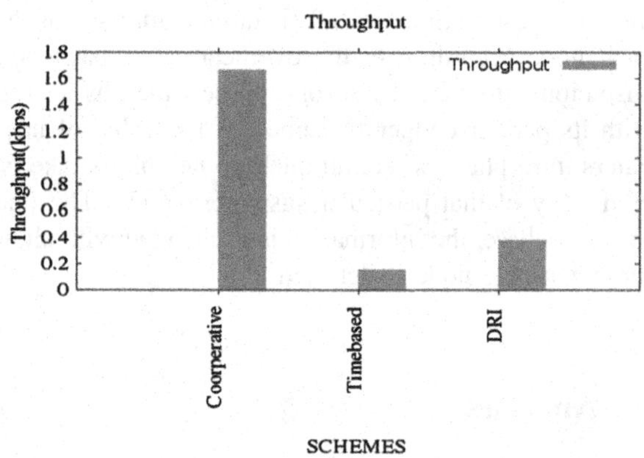

Fig. 3 Throughput

different detection schemes chosen under study. Here we can analyze that distributed cooperative mechanism has higher throughput, when compared with the other two schemes.

5.3 End-to-End Delay

The total time taken to transmit a packet between source and destination across a network is known as end-to-end Delay. In Fig. 4, x-axis indicates the end to end delay and y-axis indicates the different detection schemes chosen under study. Here we can analyze that Time-based threshold detection scheme has higher end-to-end delay, when compared with the other two schemes.

Fig. 4 End to end delay

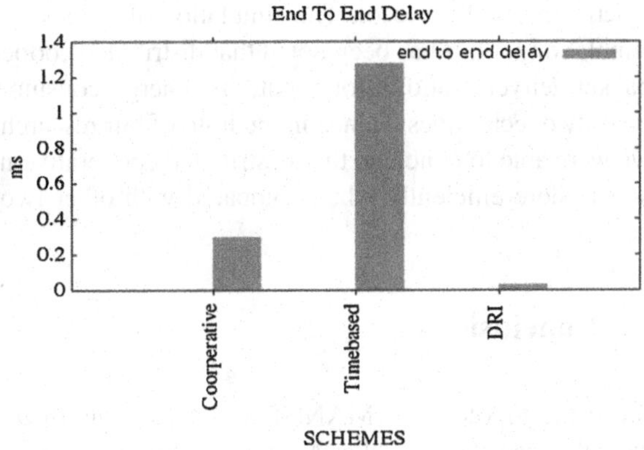

5.4 Energy Consumed

In Fig. 5, x-axis indicates the energy consumption and y-axis indicates the different detection schemes chosen under study. Here, we can analyze that distributed cooperative mechanism has higher energy consumption, when compared with the other two schemes.

In our study, the black hole attack has been analyzed using three different detection techniques evaluated by the performance parameters like throughput and packet delivery ratio. We have demonstrated three different techniques to detect black hole attack: (1) Time-based threshold detection scheme, (2) DRI table and cross checking scheme, and (3) distributed cooperative mechanism and based on their performance in trace of throughput, end-to-end delay, packet delivery ratio, and energy consumed. It can be stated that black hole attacks causes many security issues in a network. The detection techniques have made use of AODV which is a

Fig. 5 Energy consumed

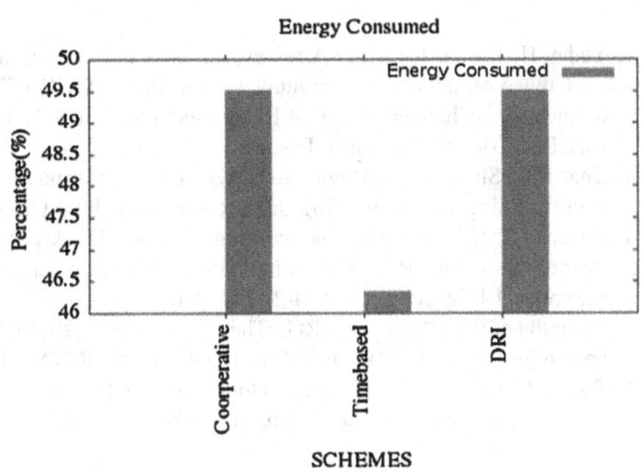

reactive routing protocol. The simulation of the above work was done using NS-2 simulation tool. It has been noted that distributed cooperative mechanism has higher packet delivery ratio, throughput, and energy consumed, when compared with the other two techniques. Hence in the light of our research and analysis of simulations, we were able to conclude that Distributed cooperative method can detect black Hole attack more efficiently when compared with other two techniques.

6 Conclusions

Since the topology of MANET is dynamic, its distributed operation and limited available bandwidth makes MANET more prune to many security attacks. A black hole attack is an attack in which a malignant node pretends to be the destination by transmitting false RREP to the source which starts route finding, and there after rejects data traffic. In this work, comparative surveys on different existing methods employed for the black hole attack detection schemes in MANETs with their demerits are presented. In the light of the above experiments, we arrived at a conclusion that black hole attacks seriously affect a network and cause security issues. Therefore, there is an imminent necessity for detection and prevention mechanisms. The detection and prevention of Black Holes are considered as challenging tasks in MANETs.

The future research includes the proposal of an efficient Black Hole attack detection and elimination algorithm with minimum possible delay and overhead can be employed in ad hoc networks. The precise location where a black hole is present can be identified, and many more parameters can be used for the comparison between the three detection schemes.

References

1. Yadav H, Kumar R (2011) A review on black hole attack in MANETs. In: 13th international conference on advanced communication technology (ICACT)
2. Chanchal A, Chander D (2013) Black hole attack in AODV routing protocol: a review. Int J Adv Res Comput Sci Softw Eng
3. Bhat MS, Shwetha D, Devaraju JT (2011) A performance study of proactive, reactive and hybrid routing protocols using qualnet simulator. Int J Comput Appl 28
4. Perkins CE (2008) Ad hoc networking. Addison-Wesley Professional
5. Sharma S, Gupta R (2009) Simulation study of blackhole attack in the mobile ad hoc networks. J Eng Sci Technol 4(2):243–250
6. Vasanthavalli S, Bhargava RG, Thenappan S Dr (2014) Peruse of black hole attack and prevention using AODV on MANET. Int J Innov Res Sci Eng Technol
7. Tsou P-C, Chang J-M (2011) Developing a BDSR scheme to avoid black hole attack based on proactive and reactive architecture in MANETs. ICACT

8. Agarwal R, Arora K, Singh RR (2014) Comparing various black hole attack prevention mechanisms in MANETs
9. Kaur G, Singh M (2013) Packet monitor scheme for prevention of black-hole attack in mobile Ad-hoc network. Int J Comput Sci Commun Eng
10. Anita EM, Vasudevan V (2010) Black hole attack prevention in multicast routing protocols for mobile ad hoc networks using certificate chaining. Int J Comput Appl 1(12):21–22

Sensor Web Enablement for an Indoor Pervasive Network Using RESTful Architecture

M.V. Babitha and A.K. Sumesh

Abstract Integration of various sensors into observation systems is complicated due to the increase in the number of sensor manufacturers and the various protocols associated with them. In order to address the complexity, the used infrastructure must be interoperable and independent of underlying platform and must consider sensor observations in a uniform manner. This paper focuses on adapting a RESTful architecture for OGC's SWE services. This is done in order to overcome the limitations of SWE. The major limitations are associated with the architecture design and data format and this paper addresses these issues by implementing the different SWE services using RESTful architecture. The reason for this proposal is the fact that sensor nodes can be observed as RESFful resources and they are made attainable through SWE services and can be queried over SWE services. Another reasonable fact behind this proposal is that JSON data format which is lightweight can be used instead of XML in order to exchange messages; which can be a better extension for SWE. The performance analysis shows the advantage of using JSON notation instead of XML in terms of file size reduction.

Keywords Sensor web · SWE · REST · SOS · SAS · SPS · WNS · SDS · JSON · O&M · SML

1 Introduction

In the today's fastest world, to make things easier we depend on different types of applications in an indoor space, In turn these applications depend upon the information's obtained from a pervasive network, which results in an imminent need for

M.V. Babitha (✉) · A.K. Sumesh
Department of CSE, Amrita Vishwa Vidyapeetham, Ettimadai, Coimbatore
Tamil Nadu, India
e-mail: babithamv999@gmail.com

A.K. Sumesh
e-mail: ak_sumesh@cb.amrita.edu

L.P. Suresh and B.K. Panigrahi (eds.), *Proceedings of the International Conference on Soft Computing Systems*, Advances in Intelligent Systems and Computing 398, DOI 10.1007/978-81-322-2674-1_41

easily attaining those information's. For the easy access of data, an approach put forward is to make the sensed data accessible via the World Wide Web. This led to the idea of Sensor web. Sensor Web is an infrastructure for automatically finding and accessing sensor data from different types of sensor devices via the web. But integrating these sensors into the web is a difficult task. But using the different types of sensor data from the wireless sensor network is difficult because of the incompatible services and data formats which can cause interoperability issues between different sensor nodes in the same WSN (Wireless Sensor Network). So the OGC (Open Geospatial Consortium) started for developing SWE (Sensor Web Enablement) standards. SWE standards allow the accessing of sensor information with the usage of a set standard services and encodings.

However, SWE standards have shortcomings in terms of data format used for the obtained observations and the architectural mode used for the implementation of the SWE services. REST (Representational State Transfer) is a lightweight approach for implementing the web services that can be adopted to overcome the above mentioned shortcomings of SWE [1]. This paper proposes a system that implements the different services like SOS (Sensor Observation Service), SAS (Sensor Alert Service), WNS (Web Notification Service), and SPS (Sensor Planning Service) using REST concept and compare their performance with the implementation using SOAP (Simple Object Access Protocol) in terms of the parameters like data buffer size and data transmission duration.

This paper is divided into four portions. The related works and the comparisons of the existing approaches are given in the second part whereas part 3 gives a brief introduction to the sensor web. Section 4 tells about the details of sensor web enablement. Implementation overview and performance analysis are given in Sect. 5. Section 6 concludes the paper and it layout the future works that can be done in this area.

2 Related Works

There are so many standard as well as nonstandardized approaches for implementing Sensor Web. Some nonstandardized approaches like Hourglass [2], an approach that creates a software infrastructure, which is able to address some of the challenges with the application development that make use of sensor network. But they were able to solve only a few research problems in this area. Second comes the global sensor network (GSN) [3], which acts as a flexible middleware that resolves the challenges that may occur when the sensor network is deployed in a real-world application. GSN [3] "hides arbitrary data sources behind its virtual sensor abstraction and provides simple and uniform access to the host of heterogeneous technologies available through powerful declarative specification and query tools which support on-the-fly configuration and adaptation of the running system."

Another nonstandardized approach includes the sensor network services Platform (SNSP) [4].

The SOCRADES [5], is also a nonstandardized approach that tells about a service-oriented architecture that acts as a connecting medium between the applications and shop floor devices. A much more better and promising integration platform was developed by modifying and extending the existing XMII functionalities. In this rich platform, the web service enabled devices were integrated providing access via XMII.

All the above-mentioned approaches are using their own proprietary mechanisms for implementing Sensor Web. The vision of sensor web is incomplete when proprietary mechanisms were used for implementation. Due to the challenges in sensor web OGC took an initiative for sensor web enablement. Since OGC's SWE is a standard, many researchers have worked on it to improve that. One such approach is GinisSense-system [6], where they focus to improve the services through the implementation of SAS and WNS. The future researches in this approach include improving the eminence of alert and notification services and to implement observation and planning service using the SWE standards so as to develop an operational system that can be used in diverse applications.

For the similar background, Chu [7] have developed a platform independent middleware called Open Sensor Web Architecture (OSWA). OSWA is an open-source system that makes use of grid technologies in order to overcome challenges that may occur while collecting sensor data for observation and making decisions from those observations. OSWA is an implementation of SWE. OSWA developed a software infrastructure for simplifying the application development for heterogeneous wireless sensor networks. It also conducted an analysis of the standards defined by OGC for sensor web by integrating them into their OSWA. OSWA is developed based on a group of uniform and standard operations and information encodings as specified in the SWE standards by the OGC. 52 North Sensor Web framework [8] is a middleware implemented as per the definitions given in the SWE standards. They formed a collection of open-source software consisting of the SWE services. They encoded the output in a uniform format for making the clients easily accessing the data.

2.1 Comparison of Existing Approaches

See Table 1.

Table 1 Comparison Table

Approaches	Strength	Weakness
Hourglass	Standard approach for Sensor Web	Uses interfaces and encodings as per the specifications developed by them for requesting data
Global sensor network	Standard approach for Sensor Web	Need to standardize their services
Sensor Network Services Platform	Standard approach for Sensor Web	Require the standardization of their services for planning the sensors and for data subscription by the users
GinisSense-system	They developed services that enhance the observation capability	Their main concentration was on notification and alert services

3 Sensor Web

It is an infrastructure for incorporating the sensor networks into the World Wide Web and thus making the sensed data easily available over the internet. In other words a sensor web can be considered as a distributed system of interconnected sensor nodes connected with the web. Simple whether station sensors to complex virtual sensors were used in the sensor web infrastructure. It is used not only for getting the sensor data but it can be also used for getting alerts and also for tasking the sensors.

3.1 Applications of Sensor Web

1. Disaster management
2. Building automation system
3. Earth observation system
4. Smart home
5. Health sector for old people
6. Traffic management

3.2 Challenges of Sensor Web

The main challenge of sensor web [9] is the hindrance related with the need of using data in different formats obtained from different sensors in a sensor network. So it is difficult for an application to integrate these resources for making it usable for the application. Also proprietary mechanisms were used for integrating these different sensor resources, i.e., there is no standard way to create a sensor web and no standard way of representation for sensor data which may cause the interoperability issues between the sensors.

4 Sensor Web Enablement

The development of this standard is initiated by OGC to solve the challenges associated with the sensor web. More specifically, it defines a set of standard interfaces and encodings to handle the heterogeneous sensor resources and making them easily available for the required applications [1] (Fig. 1).

4.1 SWE Services

See Fig. 2.

4.1.1 SOS (Sensor Observation Service)

This is a standard service interface developed according to the specification given by OGC for querying and accessing the obtained data [10]. The sensors can be a stationary one or mobile, As well as it comes under the category of remote or in situ sensors. The SOS provides high interoperability between the different sensors

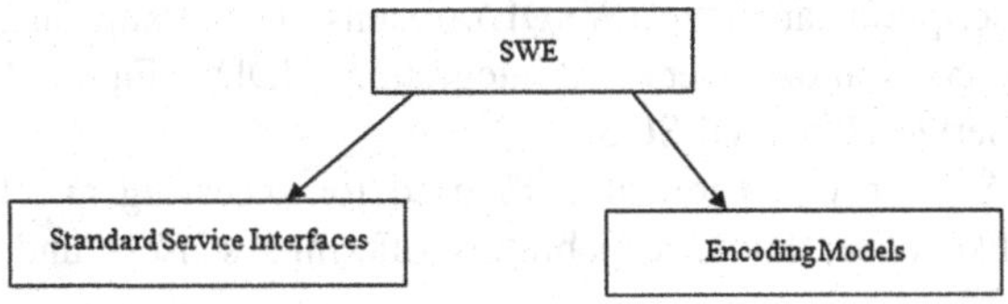

Fig. 1 SWE framework

Fig. 2 SWE services

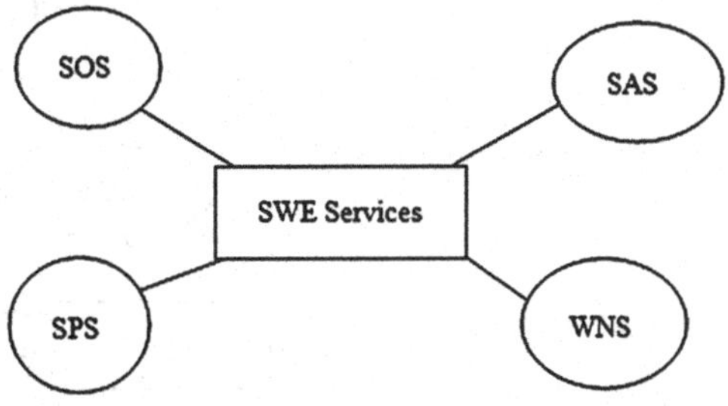

because of the usage of SWE standards. The Observations obtained from this service is O&M encoded format. Because of this standard output format any client application can make use of this observation as per their need. Along with the sensor observation information about the sensors is also obtained in the SML language.

SOS Operations

- GetCapabilities—allows for requesting descriptions about the services provided.
- GetObservation—allows for requesting sensor data in observation and measurement model (O&M).
- DescribeSensor—this procedure is used for requesting information about the sensor itself, i.e., the sensor Meta data in Sensor Model Language.
- GetFeatureOfInterest—this procedure is used for requesting the features if interest of the required observation in GML encoded format.
- GetObservationById—allows for requesting single instance of an observation by its id.
- InsertSensor—this procedure is used for adding a new sensor into service.
- UpdateSensorDescription—for modifying information about a sensor which is already added into the service.
- DeleteSensor—used for deleting every information's and observations of a selected sensor from the observation service.
- InsertObservation—for inserting a new observation of an already inserted sensor.
- InsertResult—for inserting results for a before registered result template.
- GetResult—for getting the result alone without the metadata of the information's.

.

Service-oriented SOS [7] makes use of simple object access protocol (SOAP), web service description language (WSDL), extensible markup language (XML), and universal description discovery and integration (UDDI). Figure 3 illustrates the architecture of Service Oriented SOS.

RESTful SOS is an extension of SOS used for accessing and handling SOS resources just like observations capabilities offerings sensors and features in a

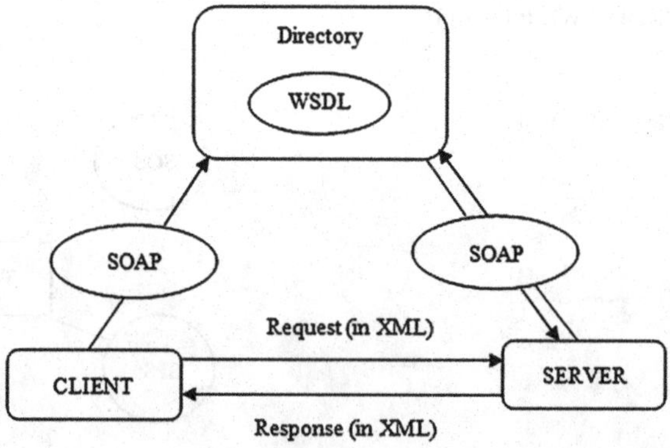

Fig. 3 Architecture of Service Oriented SOS

RESTful way. So that the plain HTTP methods like (GET, DELETE, POST, and PUT) will be able to combine with those resources. REST brings about client development simple and weightless.

4.1.2 SPS (Sensor Planning Service)

It is a standard that contains specification given by the OGC for the tasking of sensors or, for requesting the collection feasibility.

4.1.3 SAS (Sensor Alert Service)

It is a standard specification given by the OGC for subscribing the sensor data by the users from the sensor web. The subscription of alerts provided by the SAS is based on the condition specified by the users. Filtering condition can be sensor location or calculated value. A notification, using asynchronous, push-based communication mechanism, is sent to the users when the conditions are matched.

When the conditions are satisfied, the subscribed users will receive a notification using asynchronous, push-based communication techniques. Alerts are sent using XMPP (Extensible Messaging and Presence Protocol).

4.1.4 WNS (Web Notification Service)

WNS service is used for getting notification from the sensor network using the SWE standards. It is of two types namely one-way and two-way communication. It uses HTTP, Email, XMPP, SMS, Phone and Fax for sending notifications.

5 Information Modeling Language

The three encoding language given in the SWE standard are O&M, SML, TML (Fig. 4).

Fig. 4 Standard Encoding Languages

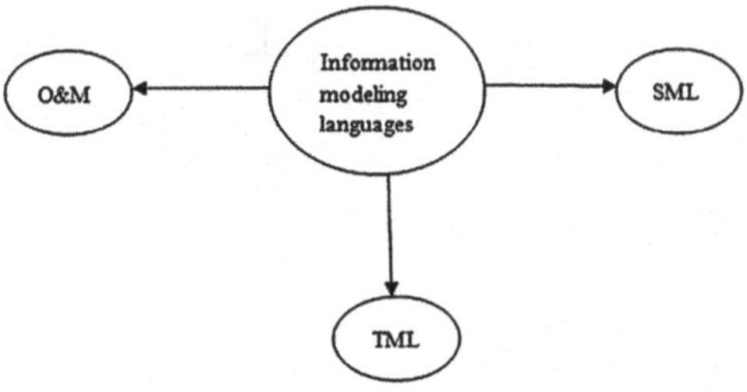

We need an observation model for creating a unified model for observational data obtained from any field, because for analyzing the collected data we need the data to be in a uniform model. The O&M standard is developed by OGC which contains a conceptual model for observation and measurement data and their xml implementation also. SML standard is used for the describing the sensors and TML is used for describing the transducers.

6 Implementation Overview

6.1 System Architecture

The SWE client will send a feasibility request to SPS service to check whether the required service is available or not. The SPS service will check in the service registry and give the reply. Then the SPS request for the service. Then the called service will give the reply to SPS. Finally the client will get a feasibility response from SPS and the required information from the service which is invoked (Fig. 5).

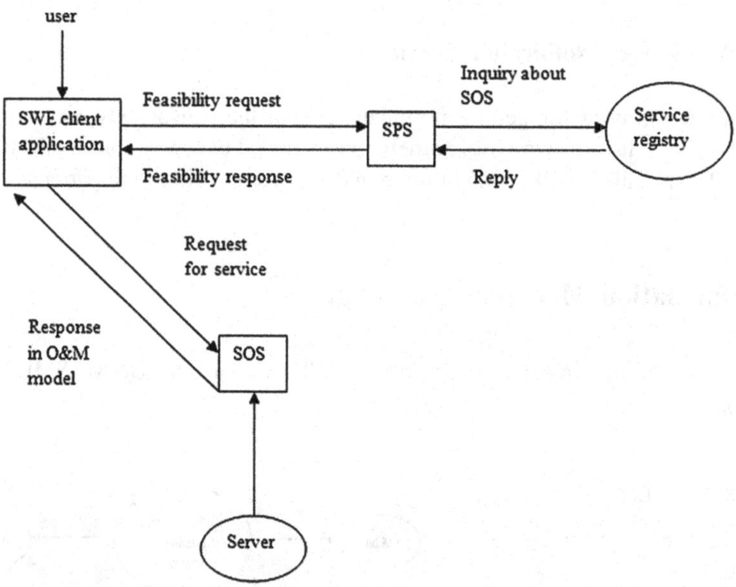

Fig. 5 System Architecture

6.2 Procedure

The data from different sensors is collected and stored in a database. When the client requests for the sensor data for its application, the data is obtained in the O&M encoding format. So that client can easily make use of the data for its application. Along with the observation, information about the selected sensor is also obtained in the SML encoding format. The web service is implemented as per the SWE standards using SOAP protocol and REST architecture.

6.3 Sample Output

See Figs. 6 and 7.


```
json ×
[{"name": "S1", "value": 5.0, "station": {"geometry": {"type": "Station", "coordinates": [0.0, 0.0]}, "id": 1, "description": "Station 1
contains threee sensors Sensor1, Sensor2, Sensor3"}, "time": "2015-03-27 14:47:29.752933+00:00", "type": "sensor", "id": 1}]
```


Fig. 6 Output in JSON notation


```
xml.txt ×
<om:CompositeObservation xsi:schemaLocation="http://www.opengis.net/om/1.0 http://www.psos.noaa.gov/gml/psos/0.6.1/schemas/
ioosObservationSpecializations.xsd" gml:id="AirTemperaturePointCollectionTimeSeriesObservation">
    <om:observedProperty xlink:href="http://mmisw.org/ont/cf/parameter/air_temperature"/>
    <om:featureOfInterest xlink:href="urn:cgi:feature:CGI:EarthOcean"/>
    <om:result>
        <psos:Composite gml:id="Temperature Observations" recDef="/temparature-record-definition/">
            <gml:valueComponents>
                <psos:Count name="NumberOfObservationPoints">1</psos:Count>
                <psos:Array gml:id="temperature single observation">
                    <gml:valueComponents>
                        <psos:Composite gml:id="Station1TimeSeriesRecord">
                            <gml:valueComponents>
                                <psos:Count name="Station1NumberOfObservationTimes">1</psos:Count>
                                    <gml:valueComponents>
                                        <psos:Composite gml:id="Station1T1Point">
                                            <gml:valueComponents>
                                                <psos:CompositeContext gml:id="Station1T1ObservationConditions" processDef="#Station1Info">
                                                    <gml:valueComponents>
                                                    </gml:valueComponents>
                                                </psos:CompositeContext>
                                                <psos:CompositeValue gml:id="Sensor 1 observations "processDef="#Sensor1Info">
                                                    <gml:valueComponents>
                                                        <psos:Quantity name="AirTemperature" uom="C">5.0</psos:Quantity>
                                                    </gml:valueComponents>
                                                </psos:CompositeValue>
                                            </gml:valueComponents>
                                        </psos:Composite>
                                    </gml:valueComponent
                            </gml:valueComponents>
                        </psos:Composite>
                    </gml:valueComponents>
                </psos:Array>
            </gml:valueComponents>
        </psos:Composite>
    </om:result>
</om:CompositeObservation>
```


Fig. 7 XML encoding of O&M

Table 2 Data buffer size gain in bytes

XML	JSON	GAIN
306	156	150
698	298	400
1022	422	600
1098	306	792
1678	1046	632
2098	1052	1046
2377	1310	1067

6.4 Performance Analysis

The output obtained by using SOAP protocol and REST architecture are compared using the parameter data buffer size. And found out that there is a gain when comparing the SOAP with XML output and REST with JSON output (Table 2).

7 Conclusion

This paper specifies the usage of REST architecture for implementing SWE services for temperature monitoring application using an indoor pervasive system. The adaptation of REST architecture improves the limitations of OGC's SWE services. The performance analysis shows the advantage of using JSON notation instead of XML in terms of file size reduction. The future work include implementing the SDS (Sensor Discovery Service) using the new Meta data language STARFL (Starfish fungus language) for providing more interoperability between the sensors.

References

1. Mohsen R, Baccar S, Abid M (2012) RESTful sensor web enablement services for wireless sensor networks. In: IEEE eighth world congress on Services (SERVICES). IEEE
2. Shneidman J, Pietzuch P, Ledlie J, Roussopoulos M, Seltzer M, Welsh M (2004) Hourglass: an infrastructure for connecting sensor networks and applications. In Harvard Technical Report TR-21-04
3. Aberer K, Hauswirth M, Salehi A (2006) A middleware for fast and flexible sensor network deployment. In: Proceedings of the 32nd international conference on very large data bases, VLDB '06. VLDB endowment, pp 1199–1202
4. Cecílio J, Furtado P (2011) Distributed configuration and processing for industrial sensor networks. In: Proceedings of the 6th international workshop on middleware tools, services and run-time support for networked embedded systems, MidSens '11, ACM, New York, pp 4:1–4:6
5. de Souza LMS, Spiess P, Guinard D, Köhler M, Karnouskos S, Savio D (2008) Socrades: a web service based shop floor integration infrastructure. In: Internet of things 2008 conference, Zurich, Switzerland, Lecture Notes in Computer Science, Springer, pp 50–67 26–28 Mar 2008

6. Veljkovic N, Bogdanovic-Dinic S, Stoimenov L (2010) Ginissense-applying ogc sensor web enablement. In: 13th AGILE international conference on geographic information science, Guimaraes, Portugal
7. Chu X (2005) Open sensor web architecture: core services. Dissertation, University of Melbourne, Australia
8. Bröring A et al (2009) Development of sensor web applications with open source software. In: First open source GIS UK conference (OSGIS 2009), vol 22
9. Chu X, Buyya R (2007) Service oriented sensor web, sensor networks and configuration. Springer, Berlin, pp 51–74
10. Mokhtary M (2012) Sensor observation service for environmental monitoring data

Bio-sensor Authentication for Medical Applications Using WBAN

Gomathi Venkatesan and Chithra Selvaraj

Abstract Advances in sensor technologies, such as Bio-sensors, along with recent developments in the embedded computing area are enabling the design, development, and implementation of Body Area Networks (BAN). A typical BAN consists of a number of wearable and implanted sensors through wireless communication (WBAN) can be used to monitor the parameters of the human body continuously. The unstable physiological regulatory system has to report about the status of the patient to the physician. There are various issues exist associated with WBAN. One of the significant process of WBAN is the communication of the sensor nodes. A key requirement for WBAN applications is to provide a secure communication channel between the body sensors and the communication system which are prone to malicious attacks. Lack of adequate security features may not only lead to a breach of patient privacy resulting in wrong diagnosis and treatment. This paper aims for implementing a secure authentication mechanism using Elliptic Curve Digital Signature Algorithm technique (ECDSA) for securing the physiological signal from the human body during inter-sensor communications.

Keywords WBAN · Secure communication · Secure authentication · ECDSA

1 Introduction

The recent development in Physiological sensors, integrated circuits consuming less power, and wireless communication have given rise to the development of new generation of wireless sensor networks named Wireless Body Area Network (WBAN) otherwise called as Body Sensor Networks (WBSN). WBAN is a network

G. Venkatesan (✉) · C. Selvaraj
Department of Information Technology, SSN College of Engineering, Chennai, India
e-mail: gomathi631@gmail.com

C. Selvaraj
e-mail: chithraS@ssn.edu.in

L.P. Suresh and B.K. Panigrahi (eds.), *Proceedings of the International Conference on Soft Computing Systems*, Advances in Intelligent Systems and Computing 398, DOI 10.1007/978-81-322-2674-1_42

of wireless bio-sensor nodes that are attached or implanted on the surface of human body to monitor and report the physiological status of the person periodically to the caretakers. Recently after the standardization of WBAN IEEE 802.15.6, the research focus on WBAN has evolved considering the improvements in its major domains like communication, sensor design, physical layer issues like channel modeling, security, MAC layer issues like QoS support, synchronization and power supply.

WBAN can be used for communication in medical applications to continuously monitor the health status of an individual from any location. One of the important issues that have to be considered is the process of securing the crucial health information. Lack of security measures may breach the patient data by allowing the illegitimate users to eavesdrop, modify, and impersonate the original data of the person under health risk. In this paper, a secure WBAN system has been proposed to monitor the health conditions of a patient.

The security mechanism has been proposed to be applied at two levels, one at the Intersensor communication level. The patients health status is continuously monitored using the physiological signals measured from the patients' body using various bio-sensors. The parameters like body temperature and heart beat has to be communicated to the communication system. The bio-sensor nodes have to be authenticated prior to communication by utilizing the measured physiological signals. This is used for generating the asymmetric cryptographic keys that are used for Elliptic Curve Digital Signature Algorithm (ECDSA) scheme [1]. The ECDSA uses compact key sizes that are best suitable for the low power and memory constraint devices. The sensors have less memory space; hence ECDSA algorithm is suitable for applying security which is a public key cryptographic algorithm.

At the next level, the data has to be communicated from the WBAN to the care taker. Hence, the authentication mechanism is applied at the network level where the authentication of users is verified in the WBAN network and the caretakers' network before the original data is communicated. This reduces the possibility of impersonation attacks.

2 Related Work

The WBAN is used for monitoring the individuals' physiological status usually deployed in health care units. The health status of the individual has to be secured at any communication level [2]. The WBAN security issues like authentication, integrity, confidentiality and access control for the bio-sensor parameters like EEG, ECG, and PPG signals are being actively developed in academic research and industry [2]. Security protocols for Sensor Networks (SPINS) a protocol developed for the authenticity, integrity of Sensor Nodes which uses symmetric keys for encrypting and decrypting the data [3–5]. It has been implemented only for general sensor networks and is not suitable for WBAN. The Authentication in WBAN can be implemented using mechanisms like Biometric authentication, Channel Based Authentication, Proximity Based Authentication [6].

The channel Based Authentication has been implemented by comparing the variation between the Received Signal Strength (RSS) of the On-Body sensors and the Off-Body sensors [6, 7]. In the channel based Authentication the RSS gets weakened when the object is far apart from the subject, so additional hardware setup is needed to compromise this issue. The directional antenna can be deployed to the WBAN nodes from deviation of the RSS signal strength [6].

The proximity based authentication is based on the distance computation between the sensor nodes deployed on the individual [8]. The proximity based authentication is measured between the physically paired devices using the RSS variation and channel characteristics of the WBAN sensor nodes [7]. The proximity Authentication has the same issue as channel based authentication i.e. as the proximity of the sensor increases the on-body sensor node is assumed wrongly to be off-body sensor node.

Recently research on WBAN focuses on using physiological signal based key agreement schemes [9–11] have been proposed in order to adapt the resource constraint behavior of WBAN by implementing plug and play and transparency properties. These physiological signal based key agreement schemes uses the biometric signals sampled by bio-sensors deployed on the patient body. This symmetric key generated from the patient body is used for securing the sensor communication within WBAN. The symmetric key agreement is less secure for the WBAN system and thus hybrid security system which uses the potential of symmetric and asymmetric schemes have been implemented [3].

3 System Model

WBAN is a network of Bio-sensor devices connected wirelessly to monitor and sense the physiological conditions of the individual under supervision. These sensor devices are attached to the clothes or implanted on a subject or individual to monitor the health status of the individual at regular intervals and forward it wirelessly over a multi-hop network to a sink node.

The block diagram of WBAN security model is shown in Fig. 1. This block diagram gives the overview of the components used in achieving security in the system. The components used for providing a secure intersensor communication channel are bio-sensor nodes for measuring the health status of the individual, control unit for digitalizing the analog signals and sink node for aggregating the data for further processing to the external network.

This paper focuses on securing the intersensor communication within WBAN and the external network communication using suitable authentication mechanisms. The system model provides basic security at both intersensor level and the external network level. The implementation of Security in WBAN model in which the

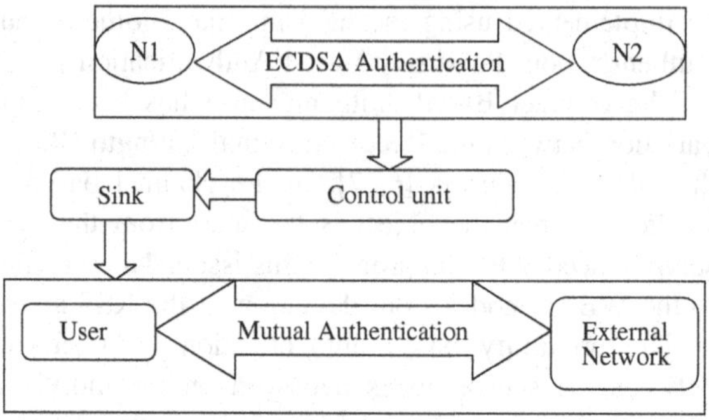

Fig. 1 WBAN security model

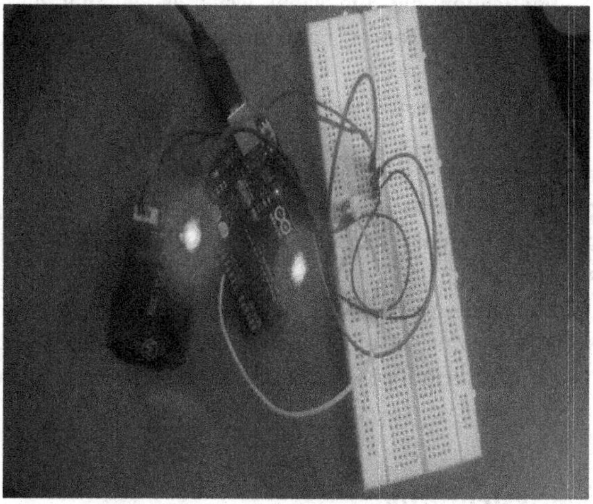

Fig. 2 System model

intersensor node communication is authenticated using ECDSA algorithm and the external network communication is authenticated mutually (Fig. 2).

The bio-sensors used are heart beat and temperature sensors which are deployed on the control unit (Arduino-UNO) for converting analog signals into digital format. This data from the control unit is sent to the sink node or a personal device capable of advanced processing.

4 Implementation

The two implementation phases involved in the Secure Authentication of WBAN are intersensor Authentication and Network Authentication.

The intersensor Authentication involves Data Extraction, Key generation, Key Comparison, Communication at the BAN network, i.e. data exchange between the sensors in a multi-hop environment. The communication and authentication involved within WBAN is dynamic and real time process. In order to secure a stream of data, the WBAN Authentication should be effective for the memory and power constraint sensor devices for which ECDSA Algorithm has been implemented to secure the sensor data. The modules like Data Extraction, Locking and Message Exchange have been implemented in every sensor node. These modules are responsible for secure transfer of messages from the sensor nodes to the sink node.

The secure intersensor communication is achieved between the sender and receiver nodes using public key cryptographic technique. A random number is generated based on the physiological signal based features. The sensor data is extracted as the physiological signal and is used for key generation. The random number generated is used as input for generating Public and Private keys using Elliptic Curve Cryptography.

The sender node would have an elliptic curve E denoted over finite field Fq, where q is a prime power. The elliptic curve equation from which different elliptic curves would be obtained for different values of a and b. All the points which satisfy the equation plus a point at infinity would lie on the elliptic curve. The private key, known only to the sender, is a random number s, such that, s € Z where Z is a set of integers. The public key is obtained by multiplying the private key with the generator G, G € E, in the elliptic curve. The value of x in the Elliptic curve equation was varied from 0 to 4 to generate the (x, y) pairs, namely (1,0), (2,2), (2,3), (3,1), (3,4), (4,3), (4,2). Any of these points can be chosen as Generator point G. The value of G was randomly chosen for each execution cycle. The public key P is computed such that $sG = P$.

The private key is used for locking phase and the public key is used at the receiver node for unlocking. This is illustrated in Fig. 3.

In the Data Extraction phase the data is extracted from the nodes of sender and receiver considered as Node A and Node B. The data extracted from the sender and

Fig. 3 Sensor node authentication

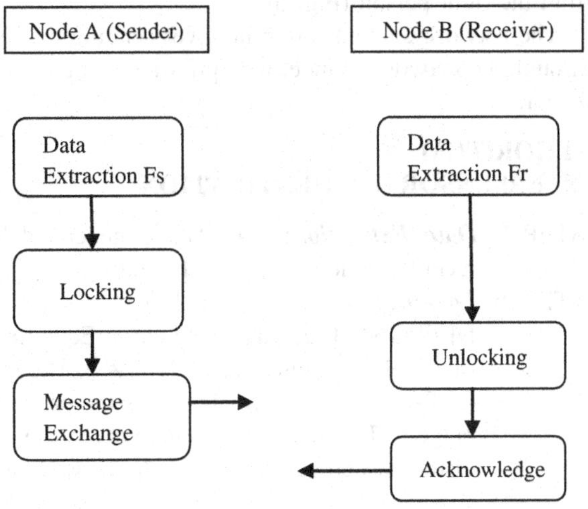

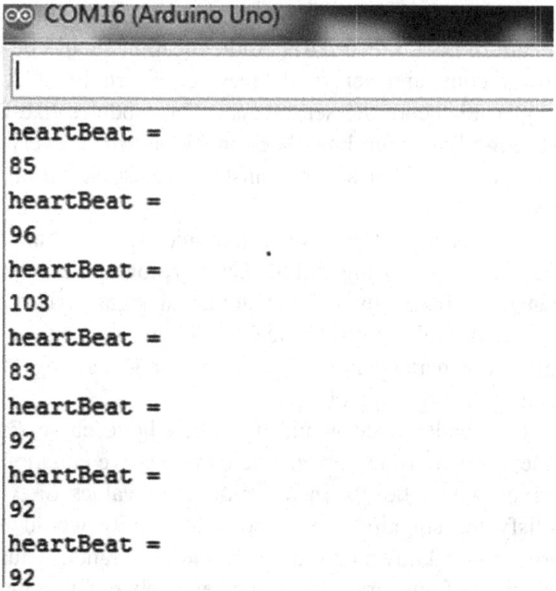

Fig. 4 Data extraction

the receiver node is utilized for generating secret keys. The secret keys thus generated is used for locking and unlocking of the messages in which the signature of the sender is verified for its authenticity at the receiver end and acknowledgement is send to the sender by the receiver (Figs. 4 and 5).

The Network Authentication involves Users login and authentication, Upload/Download file, i.e., communication between the BAN network and the Physician/Doctor. The authentication details of the physician would be verified using his login Id and password before allowing him to access the data sent by the user. The user will also be mutually authenticated to ensure that the data sent is from the right person (Fig. 6).

The network authentication involves sending the recorded data to the physician through the network channel with prior authentication of the network users and the doctor.

ALGORITHM
INTERSENSOR AUTHENTICATION

STEP 1: ***Data Extraction***: The Data is generated by the sender node and the receiver node denoted by F_s and F_r.

STEP 2: ***Locking***:
DEFINING CURVES: The curve Equation is $Y^2 = X^3 + aX + b$ must satisfy the condition $4a^3 + 27b^2 \neq 0$. The curve parameters used are P = (q, FR, a, b, G, n, h) where q is the prime field (FR) integer within the range n. The point on the curve is represented as G and h is the cofactor of the hashed parameter, h = hash (E (q)/n).

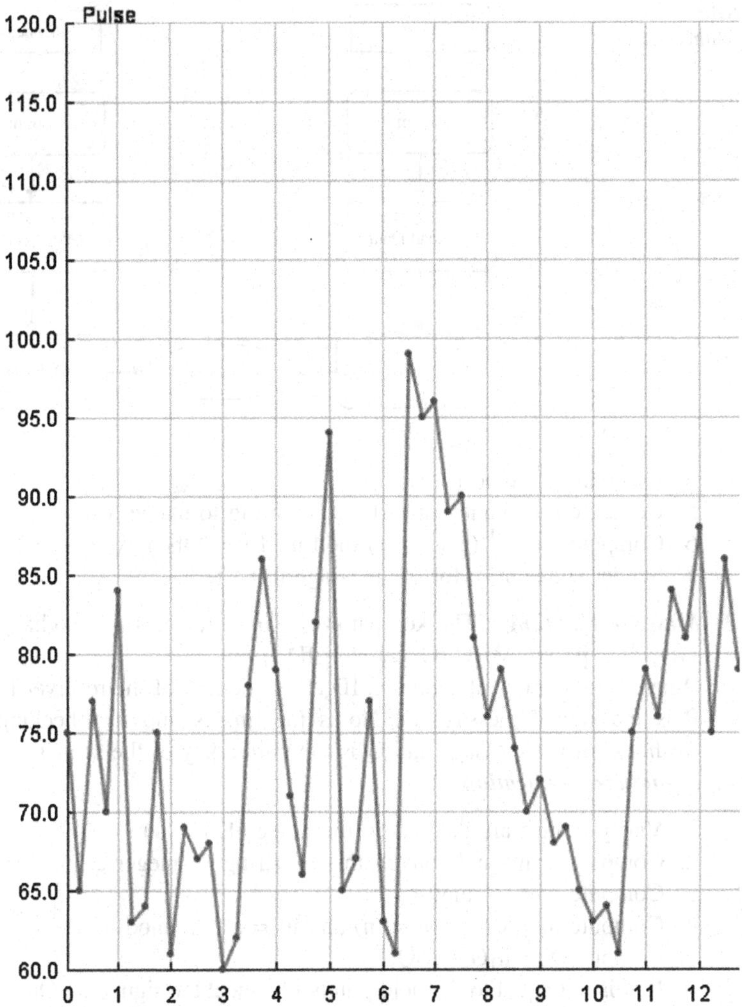

Fig. 5 Graph generation

KEY GENERATION: The public and private key pair is generated utilizing the stream data.

1. Select a random or pseudorandom integer d in the interval [1, n − 1].
2. Compute Q = d * G.
3. The public key of sender is Q, and private key is d

SIGNATURE: To sign a message m with key pair (d, Q),

1. Select a random integer k, $1 \leq k \leq n - 1$.
2. Compute k * G = (x_1, y_1).
3. Compute r = x_1 mod n. If r = 0 then go to 1.

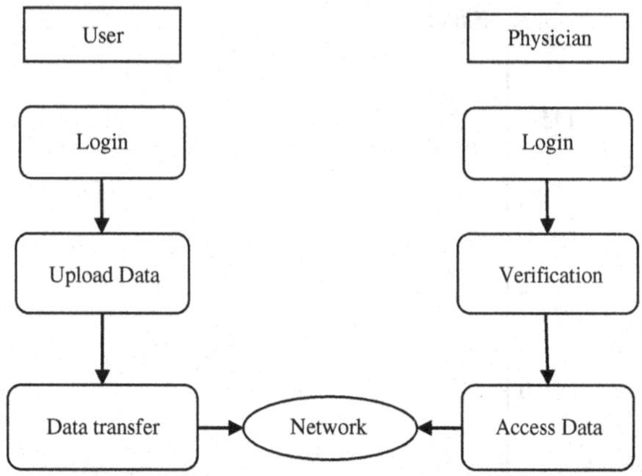

Fig. 6 Network authentication

4. Compute k^{-1} mod n.
5. Compute h (m) and convert the bit string to integer e.
6. Compute $t = k^{-1}(e + d * r)$ mod n. If t = 0 then go to step 1.
7. Sender signature p for the message m is (r, t).

STEP 3: ***Message Exchange***: The key parameter used for message exchange is, $\{ID_s, ID_r, p, No, MAC(Q, p \mid No \mid ID_s)\}$
Where, the Identity of sender is ID_s. The identity of the receiver is ID_r. P is the signed message m. No is the time stamp for checking the freshness of the message and Q is the public key of the sender.

STEP 4: ***Signature Verification***:

1. Verify r and t are integers in the range (1, n − 1).
2. Compute h (m) and convert this bit string to integer e.
3. Compute $w = t^{-1}$ mod n.
4. Compute $u_1 = e * w \pmod n$ and $u_2 = r * w \pmod n$.
5. Compute $X = u_1G + u_2Q$.
6. If X is not equal to the field points FR reject the signature. Otherwise convert the x- coordinate x_1 of X to an integer $\bar{x}_1^{-1}$.
7. Compute $v = x_1$ mod n.
8. Accept signature if v = r.

STEP 5: *Acknowledgment*: The acknowledgement is sent by the receiver to the sender in order to verify the authenticity of the communicating parties.

$(ID_r, ID_s, (v), No)$ where, v is the status of the verified signature sent to the sender by the receiver.

EXTERNAL NETWORK AUTHENTICATION

STEP 1: Login of Authenticated users with user name and password.
STEP 2: Upload the file recorded at abnormal conditions to the care taker's network.
STEP 3: Authenticated Doctor accesses the uploaded file in the network.
STEP 4: Checks for abnormality in the data and proceeds for further diagnosis.

5 Results

5.1 *Intersensor Authentication*

The Authentication at the intersensor communication involves providing security by implementing ECDSA algorithm which is proved to have a strong and compact key size that is suitable for sensor nodes utilizing its memory and power effectively when compared to other asymmetric algorithms. The strength of the public key algorithms ECC and RSA are compared based on the key sizes used in the implementation and the results have shown that the performance ratio of ECC algorithm is better compared to RSA. The observed results have been tabulated in Table 1 and shows that ECC is the best suitable public key algorithm for sensor nodes.

The hardware implementation of sensor setup is tested with the Arduino software tool and the results of ECC testing phase is shown in the Fig. 7. The testing has been performed for comparing the key value generated at the sender and receiver nodes.

The results of implementing ECDSA algorithm for verifying the authenticity of the sender and the receiver nodes is shown in Table 2. The authenticity is achieved by verifying the similarity between the senders secret key and the receivers secret key. The time taken for generating the secret and public keys for a sample of 2 sensor nodes is given in Table 2.

5.2 *External Network Authentication*

The authentication being carried out between the sensor nodes and a node representing the external network is said to be external network authentication. This

Table 1 Comparison of RSA and ECC

Security level	RSA key length	ECC key length	Performance ratio
80	1024	160–223	4.6:6.4
112	2048	224–255	8.0:9.1
128	3072	256–283	10.8:12.0
192	7680	384–511	15.0:20.0

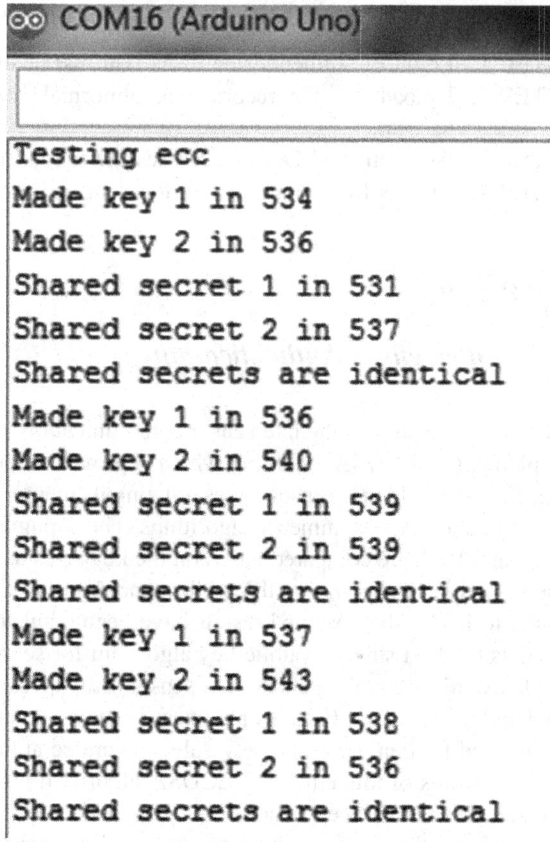

COM16 (Arduino Uno)
Testing ecc
Made key 1 in 534
Made key 2 in 536
Shared secret 1 in 531
Shared secret 2 in 537
Shared secrets are identical
Made key 1 in 536
Made key 2 in 540
Shared secret 1 in 539
Shared secret 2 in 539
Shared secrets are identical
Made key 1 in 537
Made key 2 in 543
Shared secret 1 in 538
Shared secret 2 in 536
Shared secrets are identical

Fig. 7 Key generation

Table 2 Implementation results of ECDSA algorithm

ECDSA verification	Public key (ms)	Secret key (ms)
Node 1	534	531
Node 2	536	537
Time taken (ms)	2	6
Results	Shared	Identical

authentication mechanism is applied between the sensor nodes and the physician. A key requirement for such communication of patient data through network is Security. The security is provided through mutual authentication of the registered users in the network that can prevent illegitimate users from accessing patient's data. The mutual authentication is implemented by the process of login of registered patient to the network and the details about the patient are uploaded in the form of file in case of assumption of abnormality. The authenticated doctor after receiving the message can download the file and acknowledges to the patient for receipt of the data. The doctor can further analyze the received patient data for further diagnosis.

6 Conclusion

In this paper, the implementation of sensor nodes for medical application has been carried out. The heart rate and temperature sensors have been utilized to detect the health status of the patient. The status of the patient has been sent in a secure way to the physician to identify the instability in the status of the patient's health condition. The system has been modeled for implementation of secure authentication for WBAN. As a part of the work to provide security for the data generated by the WBAN sensor nodes, ECDSA has been implemented. This algorithm verifies the keys of both the sender and the receiver nodes during intersensor communication and the external network authentication where the legitimacy of the users accessing the data is verified. The results have shown that the sensor data of the patient has been forwarded to the physician in an authenticated way.

References

1. Kodali RK (2013) Implementation of ECDSA in WSN. ICCC
2. Chen M, Gonzalez S, Vasilakos A, Cao H, Leung V (2011) Body area networks: a survey. Mobile Networks Appl 16(2):171–193
3. Jang CS, Lee DG, Han J-W, Park JH (2011) Hybrid security protocol for wireless body area networks. Wirel Commun Mob Comput 11:277–288. doi:10.1002/wcm.884
4. Sharmilee KM, Mukesh R, Damodaram A, Bharathi VS (2008) Secure WBAN using rule-based IDS with biometrics and mac authentication. In: 2008 10th IEEE international conference on health networking applications and services, IEEE, pp 102–107. doi:10.1109/HEALTH.2008.4600119
5. Ramli SN, Ahmad R (2011) Surveying the wireless body area network in the realm of wireless communication. In: 2011 7th international conference on information assurance and security (IAS), IEEE, pp 58–61
6. Shi L, Li M, Yu S, Yuan J (2012) Bana: body area network authentication exploiting channel characteristics. In: Proceedings of the fifth ACM conference on Security and privacy in wireless and mobile networks, WISEC '12, ACM, New York, pp 27–38
7. Shi L, Yuan J, Yu S, Li M (2013) ASK-BAN: authenticated secret key extraction utilizing channel characteristics for body area networks. In: WiSec'13, Budapest, Hungary, ACM, 17–19 Apr 2013
8. Mathur S, Miller R, Varshavsky A, Trappe W, Mandayam N (2011) Proximate: proximity-based secure pairing using ambient wireless signals. In: Proceedings of the 9th international conference on mobile systems, applications, and services. ACM, New York, pp 211–224
9. Venkatasubramanian K, Banerjee A, Gupta S (2010) PSKA: usable and secure key agreement scheme for body area networks. IEEE Trans Inf Technol Biomed 14(1):60–68
10. Hu C, Cheng X, Zhang F, Wu D, Liao X, Chen D (2013) OPFKA: secure and efficient ordered-physiological-feature-based-key agreement for wireless body area networks. In: Proceedings IEEE INFOCOM
11. Zhou J, Cao Z, Dong X (2013) BDK: secure and efficient biometric based deterministic key agreement in wireless body area networks. BODYNETS, pp 488–494

PSO-Based Multipath Routing in Wireless Sensor Network

T. Vairam and C. Kalaiarasan

Abstract Multipath routing is one of the major challenges to be considered in wireless sensor network (WSN). This can be used to increase the network lifetime, ease the need for update the route table frequently, balance the traffic load, and improve the overall network performance. The proposed approach cost-based multipath routing (CBMR) is meeting the above criteria based on the cost of sensor nodes. On the other hand, selecting multipath is an NP-complete problem. In this paper, the number of path produced by CBMR are optimized using particle swarm optimization (PSO) algorithm is proposed. Simulation results show that CBMR-PSO routing approach has better performance as compared with CBMR and EECA.

Keywords WSN · CBMR · PSO · Throughput and delay

1 Introduction

Wireless sensor networks (WSN) are randomly disseminated independent sensors to observe the physical or environmental conditions, such as motion, heat, pressure, speed, sound, temperature, light, etc. and to considerately transfer their data all the way through the network to a Base Station (BS)/destination. WSN [1, 2] is a compilation of sensor nodes with controlled power supply and restricted computation capability. WSN is an infrastructure less networks which serves a vital task in monitoring. In WSN, there exist a restricted communication range between the sensor nodes and there will be a high volume of sensor nodes. Moving packets

T. Vairam (✉)
Department of IT, PSG College of Technology, Coimbatore, Tamil Nadu, India
e-mail: tvairam@gmail.com

C. Kalaiarasan
SNS College of Technology, Coimbatore, Tamil Nadu, India
e-mail: ckalai2001@yahoo.com

L.P. Suresh and B.K. Panigrahi (eds.), *Proceedings of the International Conference on Soft Computing Systems*, Advances in Intelligent Systems and Computing 398, DOI 10.1007/978-81-322-2674-1_43

across the network (i.e., from source to sink) is usually performed through multi-hop data transmission. Therefore, routing in WSN has been considered an imperative field of research over the past decade.

In a single-path routing infrastructure, only one path exists between sources and sink nodes in the networks. This will make simpler the routing tables and the packet flow paths, but any one of the node in the path gets failed the entire path will be affected. Multipath routing schemes distribute traffic among multiple paths. There exists more than one path between source and sink. The multipath routing scheme has been widely utilized for different network management purposes, such as improving data network life time, transmitting data in reliable manner, reliability, providing efficient routing, and Quality of Service in wired and wireless networks.

Many issues in WSNs are considered as multidimensional optimization problems and being approached through bio-inspired techniques. Swarm intelligence [3] is a fairly novel field that was originally defined as "The endeavor we put for designing an algorithm for distributed problem-solving strategies are inspired by the collective behavior of social insects." Nowadays sensor appliances require networking substitutes that minimize the cost and complexity when improving the overall reliability. The routing problem can also be formulated as multidimensional optimization problem where there is a need of maximizing the throughput and minimizing the delay. The objective of the proposed system is to transmit data between sources and sink efficiently in wireless medium with increased throughput and reduced delay, routing overhead, and energy consumption. Work presented here is organized as follows: Sect. 2 presents the related work with respect to multipath routing and PSO. Section 3 presents a description of the proposed system. Section 4 discusses the implementation environment and the result analysis. Finally, the conclusions and the future work are provided on Sect. 5.

2 Literature Survey

This section confers some of the existing multipath routing protocols. The comparison of basic multipath routing protocol has been done in [4]. IAMR [5] protocol is proposed to improve the trustworthiness of data transmission, Quality of Service, fault-tolerance. In this, the traffic intersection spread out among the multiple paths. Path cost has been calculated for all nodes. Based on the cost parameter, the path will be chosen. Energy efficient collision aware (EECA) [6] is a node-disjoint multipath routing algorithm for WSNs. This builds two collision free paths using controlled and power adjusted flooding and then transmits the data with minimum power. It fixes the constraints for route discovery flooding to the neighbor's of current node repeatedly added to the route which is being discovered: Low Energy Adaptive Clustering Hierarchy with Deterministic Cluster-Head Selection [7]. The main objective of this protocol is also to increase the network lifetime. They have an assumption as WSN is considered to be a dynamic clustering method. Here the network is made up of set of nodes and cluster heads. The job of cluster head is to collect data from its neighbor node and send them to the base station. Cluster head is dynamically changing based on QoS parameters.

EEEMRP [8], in this paper the optimal path along with maximum energy was achieved by incorporating Cuckoo Search algorithm in to AOMDV. PSOR [9], the objective of this paper is to maximize the network life time as much as possible. They proposed a novel algorithm called particle swarm optimization (PSO)-based routing (PSOR) which consider the nodes energy level and the lengths of the routed paths.

3 Proposed Model

The flow of the system is represented in Fig. 1. Initially the nodes are deployed randomly. The source and sink are identified when an event is being generated. The proposed approach includes two methods, first method is finding multiple paths between source and sink using cost-based multipath routing (CBMR) approach and the second method is finding optimized path from the paths obtained from the CBMR approach using PSO. The process of CBMR has alienated into two different phases namely initialization phase, route discovery, and establishment phase. The neighborhood information is acquired by each node in the initialization phase. This information will be utilized in the next phase to find the next hop node toward the sink. The route discovery and establishment phase is triggered whenever an event is detected. The second method is choosing the optimized multipath using PSO. Due to the ecological match between WSN and swarm intelligence, proficient routing techniques can be achieved. The algorithm for CBMR is given in Table 1.

The process of PSO algorithm in finding optimal values follows the work of the animal society [10]. PSO consists of a swarm of particles, where particle signifies each node in a path.

$$V_{(l+1)}(\text{new}) = W * V_l(\text{old}) + \text{con}_1\text{rnd}_1(\text{Pb}_l - X_l) + \text{con}_2\text{rnd}_2(\text{Gb}_l - X_l) \quad (1)$$

$$X_{l+1} = X_l + V_{l+1}(\text{new}) \quad (2)$$

Where con_1 and con_2 are constants, and rnd_1 and rnd_2 are random numbers uniformly distributed between 0 and 1, Pb is personal best, and W is weight. In each iteration l, velocity V, and position X are updated using Eqs. 1 and 2. The update process is iteratively repeated until an acceptable global best (Gb) is achieved.

Table 2 shows the number of path chosen by CBMR for 25 nodes. And this will be given to PSO as input to optimize results of CBMR. Our objective is to increase the performance of WSN. Performance includes decreasing the delay and increasing the throughput with low cost. So fitness function includes delay, throughput of the path as a parameter.

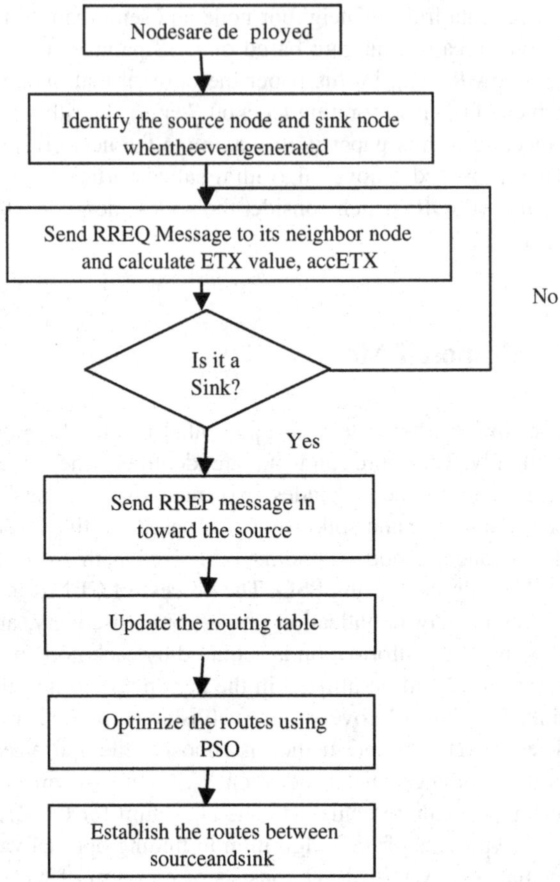

Fig. 1 System flow

Table 1 Algorithm for CBMR

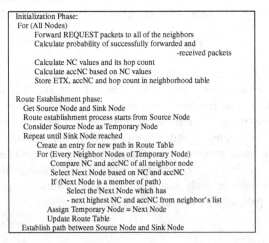

```
Initialization Phase:
 For (All Nodes)
        Forward REQUEST packets to all of the neighbors
        Calculate probability of successfully forwarded and
                                                    -received packets
        Calculate NC values and its hop count
        Calculate accNC based on NC values
        Store ETX, accNC and hop count in neighborhood table

Route Establishment phase:
    Get Source Node and Sink Node
    Route establishment process starts from Source Node
    Consider Source Node as Temporary Node
    Repeat until Sink Node reached
         Create an entry for new path in Route Table
         For (Every Neighbor Nodes of Temporary Node)
              Compare NC and accNC of all neighbor node
              Select Next Node based on NC and accNC
              If (Next Node is a member of path)
                    Select the Next Node which has
                     - next highest NC and accNC from neighbor's list
             Assign Temporary Node = Next Node
             Update Route Table
  Establish path between Source Node and Sink Node
```

Table 2 Path chosen by CBMR

S.No	Path	Nodes involved in a path
1	Path 1	1, 4, 11, 13
2	Path 2	14, 9, 19, 16
3	Path 3	22, 23, 12, 24, 15
4	Path 4	7, 18, 16, 25
5	Path 5	10, 20, 5, 21, 2

Table 3 Particles

S.No	Particles
1	4, 11, 13
2	14, 9, 19, 16
3	22, 23, 12, 24, 15
4	7, 18, 16, 25
5	10, 20, 5, 21, 2

$$\text{Fitness(s)} = W1 \times \text{Delay} + W2 \times \text{Throughput} \qquad (3)$$

The value of $W1$ and $W2$ are 0.5. The delay and the throughput are calculated based on the number of nodes and the energy of each node in path.

Table 3 shows the particle formed for this problem. Each path is considered as a particle and each node involved in a path is considered as gene. Fitness function of each chromosome is calculated using Eq. 3.

Finally, the route maintenance phase handles path failures during data transmission. The load balancing algorithm is considered for distributing the traffic over the multiple established paths. When a route is established the node starts transmitting data. Figure 2 shows the path established between source and sink. Solid line (→) represents the paths chosen by the CBMR. Whereas the dotted line (—) represents the optimized path chosen by CBMR-PSO.

Fig. 2 Path chosen by CBMR and CBMR-PSO

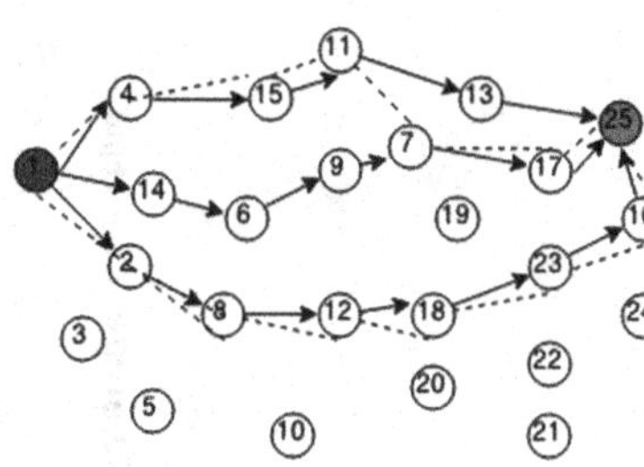

4 Implementation

In this section, we describe the performance metrics, simulation environment, and simulation results. We used MATLAB to implement and conduct a set of simulation experiments for our algorithms and did a comparative study with the EECA and normal CBMR protocol. Our simulation environment consist of various set of nodes starting from 30 to 250 nodes which are randomly deployed in a field of 1000 × 1000 m all nodes are identical with a transmission range 250 m [11, 12]. Table 4 shows the simulation parameters.

In all graphs shown in Figs. 3, 4, 5, 6, and 7 has *x*-axis as number of nodes and the *y*-axis varies according to its performance metrics. We evaluate the performance of CBMR-PSO based on the performance metrics such as throughput, end-to-end delay, routing overhead, energy, and number of paths.

The average throughput of CBMR-PSO, CBMR, and EECA has given in Fig. 3. It shows that the performance of CBMR-PSO is good than the other two approaches as only the optimized path is used for transferring data packets. The average delay of CBMR-PSO, CBMR, and EECA are given in Fig. 4, here also, the CBMR-PSO

Table 4 Input parameters

Parameters	Value
Number of sensors	30–250
Traffic type	Constant bit rate
Bandwidth	2 Mb/s
Transmission range	250 m
Initial energy in batteries	10 J
Energy threshold	0.001 mJ
Fitness value	Highest value

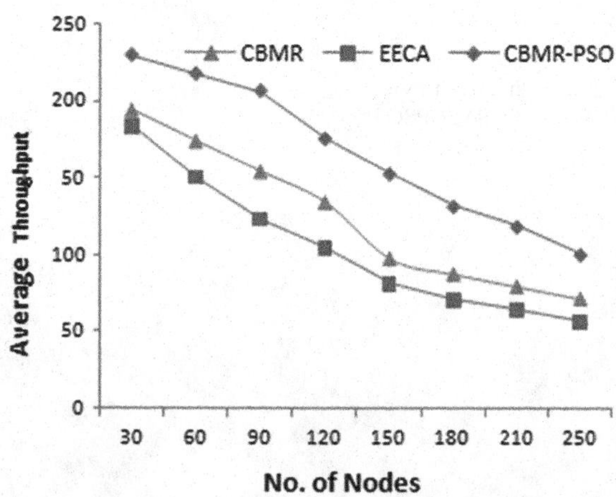

Fig. 3 Average throughput

Fig. 4 Average delay

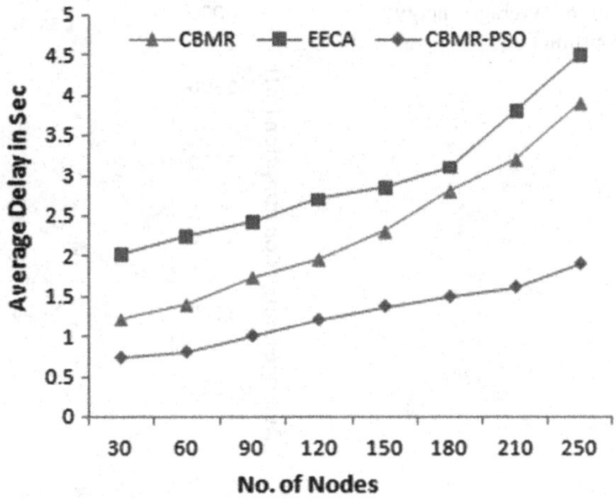

Fig. 5 Average routing overhead

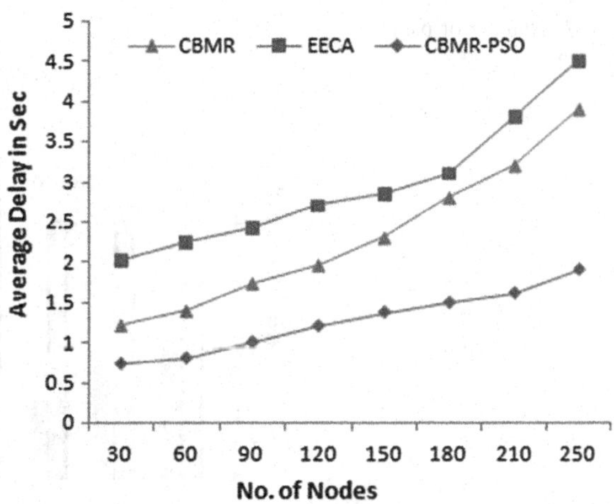

performs well compared with other two approaches. Figures 5 and 6 show analysis of message overhead and energy, respectively.

Figure 7 shows the comparison of number of paths selected in CBMR and CBMR-PSO. From the Fig. 7, we infer that CBMR-PSO produced the optimized path than CBMR. Number of path selected may vary depends upon the location of source node and sink node. Since we are using less number of paths in CBMR-PSO, we can avoid number of packets dropped and the life time of network will also increase.

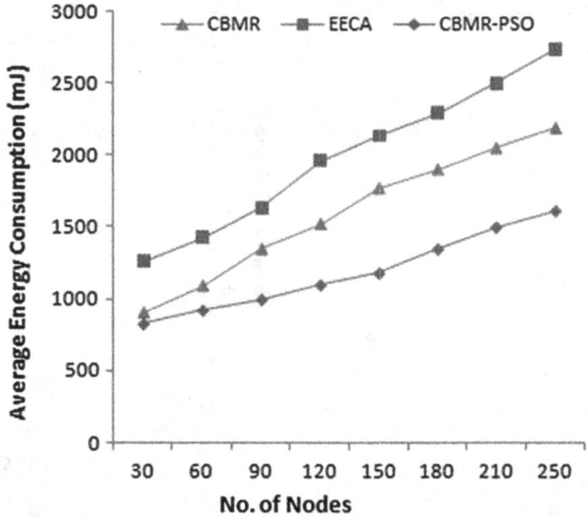

Fig. 6 Average energy consumption

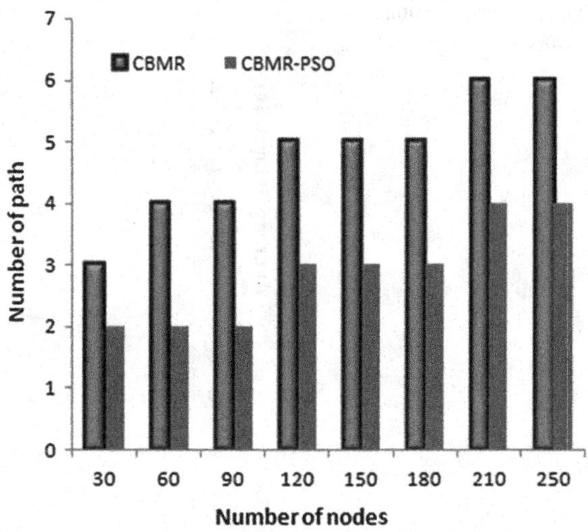

Fig. 7 Number of path chosen

5 Conclusion

In this paper, we have proposed a novel approach to enhance the performance of multipath routing in WSN. The proposed technique performs routing based on cost metrics and bio-inspired algorithm. CBMR protocol was implemented to improve performance of WSNs. The proposed system includes CBMR with PSO algorithm in order to optimize the results. Swarm Intelligence algorithm like PSO was used to maximize the throughput and minimize the delay. By using PSO the average percentage of throughput is increased and delay is decreased than CBMR Protocol and EECA. In future, instead of sending normal text data the multimedia data could also be considered.

References

1. Li S, Neelisetti RK, Liu C, Lim A (2010) Efficient multipath protocol for wireless sensor networks. Int J Wirel Mob Netw 2(1)
2. Akkava K, Younis M (2005) A survey on routing protocols for wireless sensor networks. Ad Hoc Netw J 3:325–349
3. Sharawi M et al (2013) Routing wireless sensor networks based on soft computing paradigms: survey. Int J Soft Comput Artif Intell Appl 2(4):21–36
4. Yamuna D, Vairam T, Kalaiarasan C (2012) Efficient comparison of multipath routing protocols in WSN. In: Proceedings of international conference on computing, electronics and electrical technologies
5. Vairam T, Kalaiarasan C (2013) Interference aware multi-path routing wireless sensor network. Int J Emerg Sci Eng 1(10)
6. Wang Z, Bulut E, Szymanski BK (2009) Energy efficient collision aware multipath routing for wireless sensor networks. In: Proceedings of international conference on communication, ICC09, Dresden Germany, pp 1–5, 14–18 June 2009
7. Arab E, Aghazarian V, Hedayati A, Motlagh NG (2012) A LEACH-based clustering algorithm for optimizing energy consumption in wireless sensor networks. In: Proceedings of ICCSIT
8. Raj DAA, Sumathi P (2014) Enhanced energy efficient multipath routing protocol for wireless sensor communication networks using cuckoo search algorithm. Wirel Sens Netw 6:49–55
9. Sarangi S, Thankchan BA (2012) A novel routing algorithm for wireless sensor network using particle swarm optimization. IOSR J Compu Eng (IOSRJCE) 4(1):26–30. ISSN:2278-0661
10. Rini DP, Shamsuddin SM, Yuhaniz SS (2011) Particle swarm optimization: technique, system and challenges. Int J Compu Appl 14(1):19–26
11. Bredin JL, Demaine ED, Hajiaghayi MT, Rus D (2010) Deploying sensor networks with guaranteed fault tolerance. IEEE/ACM Trans Netw 18(1):216–228
12. Bettstetter C, Reseta G, Santi P (2003) The node distribution of the random way point mobility model for wireless ad hoc networks. IEEE Trans Mob Comput 2(3):257–269
13. Gao Y, Shi L, Yao P (2000) Study on multi-objective genetic algorithm. IEEE, June 28–July 2
14. Marina MK, Das SR (2006) Ad hoc on-demand multipath distance vector routing. Wirel Commun Mob Comput 6:969–988
15. Perkins CE, Royer EM (1999) Ad Hoc on-demand distance vector routing. In: Proceedings of second IEEE workshop mobile computing systems and application, pp 90–100
16. Kulkarni RV, Venayamoorthy GK (2011) Particle swarm optimization in wireless sensor networks: a brief survey. IEEE Trans Syst Man Cybern Part C Appl Rev 41(2):262–267
17. Radi M, Dezfouli B, Bakar KA, Razak SA, Nematbakhsh MA (2011) Interference-aware multipath routing protocol for Qos improvement in event-driven wireless sensor networks. Tsinghua Sci Technol 16(5):475–490
18. Lou W, Liu W, Zhang Y (2006) Performance optimization using mulitpath routing in mobile ad hoc and wireless sensor networks. Proc Comb Optim Commun Netw 18:117–146
19. Averkin AN, Belenki AG, Zubkov G (2007) Soft computing in wireless sensors networks. Conf Eur Soc Fuzzy Logic Technol (EUSFLAT) (1):387–390
20. Kulkarni RV, Forster A, Venayagamoorthy GK (2011) Computational intelligence in wireless sensor networks: a survey. Commun Surv Tutorials IEEE 13(1):68–96

Information Hiding in H.264, H.265, and MJPEG

S. Priya and P.P. Amritha

Abstract Steganography refers to the process of inserting information into a medium to secure the communication. Video steganography, which is the focus of this paper, can be viewed as an extension of image steganography. In fact, a video stream consists of a series of consecutive and equally time-spaced still images, sometimes accompanied with audio. There are many image steganographic techniques that are applicable to videos as well. In this paper, data hiding is done in H.264, H.265, and MJPEG using LSB- and PVD- based steganographic algorithms. A comparative study is also done for the above mentioned video codecs. Here, we address the issues of the least detectable steganographic algorithm when performing steganalysis in terms of MMD and the best channel and location for embedding the secret information. We also discuss the applications that work on these compression formats that can take advantage of this work.

Keywords Data embedding · H.264 · H.265 · MJPEG · Compression · PSNR · MMD

1 Introduction

Video streams have high degree of spatial and temporal redundancy in representation and have all encompassing applications in daily life, thus they are considered as good candidates for hiding data. We can use video steganography for many useful applications. Military and intelligence agencies' communications are an example. Tew and Wong in [1] discuss more about the applications of H.264 compression framework. Image steganography [2] is the basis of video steganography [3]. If one

S. Priya (✉) · P.P. Amritha
TIFAC CORE in Cyber Security, Amrita VishwaVidyapeetham, Coimbatore, India
e-mail: firstpiya@gmail.com

P.P. Amritha
e-mail: ammuviju@gmail.com

L.P. Suresh and B.K. Panigrahi (eds.), *Proceedings of the International Conference on Soft Computing Systems*, Advances in Intelligent Systems and Computing 398, DOI 10.1007/978-81-322-2674-1_44

is well versed with image steganography and its various techniques, they can easily work with video steganography. A video stream can be considered as sequence of images and frames. The secret data can be hidden in these frames. But there are many different things that make video steganography different from the area of image steganography. The chances for the detection of the secret message are less in video steganography when compared to that of image. Various information hiding techniques are included in steganography, whose aim is to embed a secret message into a frame. There are different mediums where we can do steganography. Here, in video steganography, we are comparing the hiding techniques in three compression formats H.264, H.265/High-Efficiency Video Coding (HEVC) and Motion Joint Photographic Experts Group (MJPEG). The main problem that we are addressing here is the performance analysis of the compression formats H.264, H.265, and MJPEG, when video steganography is done on them. Also we can see that, in video the options for embedding location are richer. Therefore, finding the locations best suited for embedding is also considered. As there are a wide variety of applications for each of the compression formats, another issue is to rate the performance analysis on the basis of the applications of each compression frameworks. There are a number of embedding operations like Least Significant Bit (LSB) embedding and Pixel Value Differencing (PVD). Finding the embedding operations and channels that are least detectable while doing steganalysis is also discussed.

Section 2 describes about the related works. Section 3 features the proposed system and system design which is efficient than the existing system described in related works. Section 4 presents the experimental results and discussion. Section 5 ends with conclusion.

2 Related Works

Many papers have discussed the various embedding locations and operations that are being done on H.264. H.265 has not been used in information hiding as H.264. Sullivan et al. in [4] gives an overall idea of H.265 coding standard. Neufeld and Ker in [5] discusses and analyzes the most efficient embedding algorithm in various embedding locations in H.264 using benchmarking techniques like Maximum Mean Discrepancy (MMD) in their paper. They determine the best embedding option from a list of embedding operations and locations chosen, by using MMD as the benchmark. A number of videos with different sizes are selected for this purpose. Since the numbers of videos that are used are high, a stable MMD can be formed. MMD [6] is a benchmark that can be used to validate the least detectable embedding operation from a set of operations, in the context of steganography. The basic idea of MMD is to compare the value from two sample sets. In this case, the original video frame and distorted video frame become the two sets. The comparison between the pixel values of the above mentioned frames are taken into consideration while calculating MMD. There are mainly a few findings from the work done in this paper. The main embedding operations that are used in this paper are LSB Matching [7], LSB Replacement [8], and F5 [9]. The least detectable embedding operation is found to be F5. Varying payloads are used in this case and it was found out that when the payload is equally spread between Luma and Chroma the operation is least detectable. The location where the embedding operation is least detectable is the combination of P and B frames. Grois et al. in [10] analyzes performance of the two latest video coding standards H.264/MPEGAVC, H.265/MPEG-HEVC (High-Efficiency Video Coding), and VP9 (a scheme introduced by Google). After doing a lot of experimentations it was found that, the coding efficiency of VP9 is very much less when compared to H.264/MPEG-AVC and H.265/MPEG-HEVC. Xu et al. in [11] proposes a hiding method that could be employed in compressed video streams like MPEG. Using this method, in each Group of Pictures (GOP), the control information to facilitate data extraction was embedded in I frame, P frames and B frames. Motion vectors of macroblocks are used as the location to embed the data repeatedly and this data is transmitted. A threshold was calculated using an algorithm and using the threshold, the data were embedded into the frames.

3 Proposed System

H.264 and H.265 or HEVC are the latest standards for compression in the industry now. MJPEG is the lowest form of compression framework. Compression efficiency is doubled in the case of H.265 when compared to H.264. An encoder converts video into a compressed format and a decoder converts compressed video back into an uncompressed format. There are a series of steps to process a video and perform steganography. To be clear about the process, first the parts that constitute a video needs to be studied. A video contains several number and types of frames. They are I (Intrapredicted) [12], P (Predicted), and B (Bidirectional) frames. I frames are intrapredicted frames, which means they depend on only the current frame for prediction purpose. P and B frames are interpredicted frames meaning, they depend on other frames too. Chen et al. in [13] analyzes and gives a comparison of interprediction of H.264 and H.265 coding. P frames depend on past frames for prediction whereas B frames depend on both previous and next I or P frames for prediction purposes. A frame can again be divided into macroblocks and macroblocks into slices. I frame uses a single motion vector where as P and B uses two or more motion vectors for prediction. The phase angle and magnitude of the motion vectors are calculated. Using a threshold and according to the corresponding values of motion vectors [11], the secret message is embedded into it. Table 1 gives the comparison between the three compression schemes that are being used.

Table 1 Comparison of MJPEG, H.264, and H.265

Category	MJPEG	H.264	H.265
Names	Motion JPEG	MPEG-4 Part 10 AVC	MPEG-H, HEVC, Part 2
Major changes	Digital video sequence is made up of a series of individual JPEG images	40–50 % bit rate reduction compared to MPEG-2. Content of HD content delivery broadcast online	40–50 % bit rate reduction compared to H.264. Potential to realize UHD, 2 K, 4 K for broadcast and online
Progression	MJPEG was first used by the QuickTime Player in the mid 1990s	Successor to MPEG-2 Part	Successor to MPEG-4 AVC, H.264
Compression model	Compressed sequentially, frame by frame is transmitted through the network. Each transmitted frame is processed independently	Hybrid spatial Temporal prediction model –Flexible partitioning of macroblock (into a maximum size of 16 × 16) and sub macroblock for motion estimation. –Intra prediction –9 directional modes –Entropy coding is CAVLC and CABAC	Enhanced hybrid Spatial temporal Prediction model –Partitioning into Coding, Prediction and Transform (64 × 64) units –35 directional modes –Entropy coding is CABAC
Specification	Up to 30 fps	Up to 4 K Up to 59.94 fps 21 profiles, 17 levels	Up to 8 K HDTV Up to 300 fps 3 approved profiles, draft for 5 more and 13 levels
Disadvantages	–Makes no use of any video compression techniques to reduce the data since it is a series of still, complete images –Results in a relatively high bit rate or low compression ratio for the delivered quality compared with video compression standards such as MPEG-4 and H.264	–Requires more processing power than MJPEG which means that H.264 IP cameras tend to be slightly more expensive than MJPEG IP cameras –Unrealistic for UHD content delivery due to high bit rate requirements	Computationally expensive due to larger coding units and motion estimation techniques

3.1 Architectural Model

Proposed system implementation is developed as in Fig. 1. For primary test, a video from the DSLR of Nikon, namely Nikon D3100, with 14 s duration, 23 frames per second (fps), 1080 × 1920 frame width and frame height and frame rate 23 fps is taken. The data rate and total bit rate of the respective video are 20,509 and 20,917 kbps respectively. Initially without any compression, the video was of size 35.2 MB. Then the video was encoded using H.264 and HEVC compression formats. For this purpose, the softwares named H.264 encoder and DivX converter were used. H.264 encoder comes with many options to compress the input video in various formats. There are slots to input video and we can change the encoding setting as well as aspect ratio. The various encoding settings available in it are Full High Definition (1920 × 1080), Commonly Used High Definition (1280 × 720), Standard Definition (640 × 480), Internet-size Content (320 × 240), 3G Content (176 × 144), High Quality (Same size with Source), Normal Quality (Same size with Source), and Fast Speed (Same size with Source).

The variants available in aspect ratio are 5:4, 16:10, and 16:9. The raw uncompressed video was then compressed to H.264 format with the encoding, setting high quality and with no change in aspect ratio. The output video formed was of size 1.69 MB. The data rate changed to 814 kbps and total bit rate to 942 kbps. MJPEG conversion was done using another software named Oxelon Media Converter. DivX Converter also comes with a variety of options to encode and compress videos in various formats. It has a drag and drop interface for convenience of use and the various encoding options available in it are HEVC UHD K, HEVC 1080 p, HEVC 720 p, PLUS 4K, PLUS HD, HD 1080 p, HD 720 p, home theatre, iPad, iPhone, and mobile phone. Here, since the original video was of

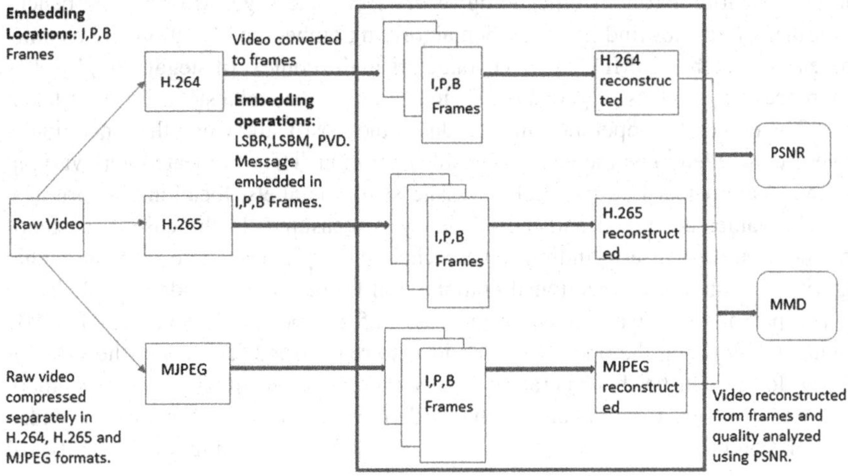

Fig. 1 Architecture of the proposed system

dimension 1920 × 1080 p, the option HEVC 1080 p was chosen. The output video was of size 6.40 MB. Thereafter H.264, HEVC and MJPEG videos are converted into frames using MATLAB code. From a series of frames formed, the embedding is done in B and P + B frames. The embedding operations that are used here are LSB embedding, PVD, and TPVD. Afterward, the video is reconstructed using MATLAB code. Then again, the parameters like data rate and total bit rate are checked. There are a number of benchmarks used for measuring video quality like PSNR. We consider PSNR here for the purpose of quality checking. MMD has been used here for comparison of steganographic methods and for finding least detectable algorithm. For finding out whether two datasets were obtained from the same distribution, MMD was used. Filler and Fridrich in [14] discusses about MMD in detail.

3.2 Design and Implementation

Initially a raw video with 23 fps is taken and is separately compressed using H.264, H.265, and MJPEG formats. The video is then converted into frames. We generally do not choose I frames as they are rare. To know about the detectability options we use, B frames and a combination of both P and B frames separately as embedding locations. A secret message (with varying payloads) is embedded into frames using three different embedding algorithms. Here we use LSB embedding [15], Pixel Value Differencing (PVD) [16], and Tri-way Pixel Value Differencing (TPVD) [17]. After embedding of the message, the video is reconstructed from the frames and the quality difference between the non-embedded and embedded video sequences, is measured using PSNR (Peak Signal-to-Noise Ratio). We use PSNR and MMD as the estimates here. PSNR is used to measure the quality difference in the video before and after embedding of the secret message. MMD is the benchmark that we use for finding out which algorithm; embedding location and channel are least detectable. MMD has two parameters in the context of steganography. The difference in the values of pixels between a cover image and a stego image is taken here. The embedding operation that produces the least value when this operation is done is concluded to be the least detectable operation. This is repeated with varying different parameters like payload (message size), channels, locations, algorithms, etc. The values are noted and the efficiency is measured. In the case of steganographic security, we are finding out which is the least detectable steganographic algorithm in the above mentioned compression frameworks of video (Table 2).

The payload size was varied from 25 to 125 characters. The values of MMD obtained after doing the operations on video were recorded. The lower the value of MMD, lower will be that operation's detectability. Video quality is a benchmark that can be used to analyze the video when passed through a video transmission/processing system. PSNR is used to measure the quality of

Table 2 PSNR values

PSNR value after embedding in decibels (dB)			
Compression frameworks	B frames	P + B frames	I, P, B frames
H.264	85.1	81.3	80.1
H.265	90.9	89.9	87.4
MJPEG	70.1	69.1	67.8

reconstructed images. Given a noise-free $M \times N$ monochrome image X and its noisy approximation Y, MSE is defined as:

- **Mean-Squared Error (MSE)**:

$$\mathrm{MSE} = \frac{\sum_{j=1}^{N}\left(\sum_{i=1}^{M}(X_{i,j} - Y_{i,j})^2\right)}{MN}$$

- **Peak Signal-to-Noise Ratio (PSNR)**:

$$\mathrm{PSNR} = 10\log\frac{\mathrm{MAX}^2}{\mathrm{MSE}}$$

Here, MAX is the maximum possible pixel value of the image. When the pixels are represented using 8-bits per sample, this is 255.

4 Implementation Results

Embedding was done in H.264, H.265, and MJPEG frames. The frames were reconstructed to form video. The performance comparison was done using PSNR between the original and reconstructed video, before and after data extraction. The embedding operation was done in all the three I, P, B frames as well as in B frames and P + B fames separately. After embedding, the value was showing a marginal improvement in all the three compression frameworks. The PSNR values after embedding are given below. H.265 framework was found to have better PSNR.

The payload was equally spread between the Luma and Chroma channels. The lower the value of MMD for an operation, it will be least detectable. In the case of steganographic security, from the above mentioned operations, TPVD was found to have lower value of MMD. The table showing the MMD values of the operations in corresponding frameworks with varying payloads is shown in Table 3.

Table 3 MMD values

Compression framework	Frames	Payload (character)	MMD values of embedding operations		
			LSB	PVD	TPVD
H.264	P + B	25	2.77	2.52	2.32
		125	3.08	2.96	2.06
	B	25	4.02	4.15	3.92
		125	5.84	4.24	4.13
H.265	P + B	25	2.32	2.12	1.97
		125	2.61	2.34	2.16
	B	25	3.22	3.06	2.93
		125	4.34	3.82	3.42
MJPEG	P + B	25	6.45	5.76	4.82
		125	5.86	5.36	4.61
	B	25	5.32	4.84	4.05
		125	4.76	3.93	2.04

5 Conclusion

A variety of applications like digital television, DVD-Video, mobile TV, video-conferencing and internet video streaming were found to have using H.264 standard compression framework. H.265 can also work in similar applications. But the main advantage was found to be that, it has 50 % more coding efficiency than H.264 and can support for resolutions up to 8192 × 4320. H.265 was found to have good performance than that of H.264 and MJPEG. TPVD is least detectable since it has the lowest value of MMD among the three embedding operations.

References

1. Tew Y, Wong KS (2014) An overview of information hiding in H.264/AVC compressed video. IEEE Trans Circuits Syst Video Technol 24:305–319
2. Cheddad A, Condell J, Curran K, McKevitt P (2010) Digital image steganography: survey and analysis of current methods. Sig Process 90:727–752
3. Sadek MM, Khalifa AS, Mostafa MGM (2014) Video steganography: a comprehensive review. Multimedia Tools Appl 1–32
4. Sullivan GJ, Ohm J, Han WJ, Wiegand T (2012) Overview of The High Efficiency Video Coding (HEVC) Standard. Circuits Syst Video Technol IEEE Trans 22:1649–1668
5. Neufeld A, Ker AD (2013) A study of embedding operations and locations for steganography in H.264 Video. In: IS&T/SPIE Electronic imaging, International Society for Optics and Photonics
6. Gretton A, Borgwardt KM, Rasch M, Schölkopf B, Smola AJ (2006) A kernel method for the two-sample-problem. In: Advances in neural information processing systems, pp 513–520
7. Mielikainen J (2006) LSB matching revisited. IEEE Signal Process Lett 13:285–287

8. Kekre HB, Mishra D, Khanna R, Khanna S, Hussaini A (2012) Comparison between the basic LSB replacement technique and increased capacity of information hiding in LSB's method for images. Int J Comput Appl 45:33–38
9. Westfeld A (2001) F5 A steganographic algorithm, in information hiding. Springer, Berlin, pp 289–302
10. Grois D, Marpe D, Mulayoff A, Itzhaky B, Hadar O (2013) Performance comparison of H.265/MPEG-HEVC, VP9, and H.264/MPEG-AVC Encoders. In: PCS, vol 13, pp 8–11
11. Xu C, Ping X, Zhang T (2006) Steganography in compressed video stream. In: ICICIC'06 first international conference on innovative computing, information and control. IEEE, vol 1, pp 269–272
12. Liu Y, Li Z, Ma X, Liu J (2012) A robust data hiding algorithm for H.264/AVC video streams without intra-frame distortion drift. In: Proceedings of the 2012 second international conference on electric information and control engineering, IEEE Computer Society, vol 01, pp 182–186
13. Chen ZY, Tseng CT, Chang PC (2013) Fast inter prediction for H.264 to HEVC transcoding. In: 3rd international conference on multimedia technology, ICMT
14. Filler T, Fridrich J (2011) Design of adaptive steganographic schemes for digital images. In: IS&T/SPIE electronic imaging, International Society for Optics and Photonics
15. Provos N, Honeyman P (2003) Hide and Seek: An Introduction to Steganography. Security & Privacy, IEEE 1:32–44
16. Wu DC, Tsai WH (2003) A Steganographic method for images by pixel value differencing. Pattern Recogn Lett 24:1613–1626
17. Sherly AP, Amritha PP (2010) A compressed video steganography using TPVD. Int J Database Manage Syst 2

Intelligent Data Prediction System Using Data Mining and Neural Networks

M. Sudhakar, J. Albert Mayan and N. Srinivasan

Abstract Stock market prediction is a significant area of financial forecasting; this is of great interest to stock investors, stock dealers, and applied researchers. Important issues in the development of a fully automated stock market prediction system are: attribute extraction from the metadata, attribute selection for highest forecast accuracy, and minimize the dimensionality of selected attribute set and the precision, and robustness of the prediction system. In this paper, we are using two methodologies; data mining and neural networks for achieving an intelligent data prediction system. Data mining is a specific step in this method, which involves the application of new learning algorithms for extracting the patterns (models) from the existing data. In the proposed system, accurate results were predicted by adjusting the advanced neural network parameters. Neural network extracts valuable information from an enormous dataset and data mining helps to presume future trends and behaviors. Therefore, a combination of both the methods could make the calculation much reliable.

Keywords Data mining · Artificial neural network · Data prediction · Stock index

1 Introduction

In general, stock market system is a nonlinear one. A prediction is a report about the things that will happen in near future. Many investors in stock market aim to earn more profit, by predicting. Prediction is one of the effective ways to view the future

M. Sudhakar (✉) · J. Albert Mayan · N. Srinivasan
Department of Computer Science and Engineering, Sathyabama University, Chennai, Tamil Nadu, India
e-mail: sudhakar_jetsetgo@yahoo.co.in

J. Albert Mayan
e-mail: albertmayan@gmail.com

N. Srinivasan
e-mail: srinijyothish@gmail.com

L.P. Suresh and B.K. Panigrahi (eds.), *Proceedings of the International Conference on Soft Computing Systems*, Advances in Intelligent Systems and Computing 398, DOI 10.1007/978-81-322-2674-1_45

happenings. Prediction in stock market system increases the trust of investors in understanding the day to day changes, up's and down's in the stock market. Various computing techniques are required and to be clubbed together to predict the stock market system.

Data mining and neural networks can be efficiently used to learn about the nonlinearity in stock market prediction system. Data mining is a part of knowledge discovery process, which involves data preprocessing, selecting mining algorithm and processing the results of mining.

The ultimate aim of the investors is to attain a dependable prediction method for the stock market; this objective can be achieved with the support of econometric concepts, statistical methods, and technical study. These factors motivate many researchers to think about the new models and approaches, towards forecasting. Data mining is the computational process of discovering patterns in large data sets. The main goal of the data mining process is to extract the information from an existing dataset and convert it into an understandable form. A large collection of data is required for generating the information. Along with data collection, it requires an automatic aggregation of data, information abstraction and recognition of discovery pattern.

Extraction which was formerly unidentified, implicit and potentially useful information from data in databases, is an effective way of data mining. It is commonly known as knowledge discovery in databases [1]. Neural networks is an important method in forecasting the stock market data, due to its ability in dealing with fuzzy, undefined and inadequate data which may fluctuate promptly in period of time.

Neural networks are used in many financial applications, such as stock market prediction system, gold price prediction and so on. Many researches on neural networks, in prediction system have addressed about the various applications of neural networks in different ways. However, the required standard parameters were still not addressed clearly. Moreover, most of the researches and application of neural networks has been proved about its efficiency over existing traditional methods.

2 Literature Survey

Most of the prediction system uses artificial neural networks, which uses the historical time-series data. A backpropagation neural networks was introduced to predict the stock market data, and then applied in real-time data to check about the further improvements in accuracy. A survey of various articles that uses neural networks and neuro-fuzzy models for predicting the stock markets were taken and it was detected that soft computing technique overtakes conventional models [2].

It was proven that neural networks outperform statistical technique in forecasting stock market prices [3]. It was shown that, it was through a process to prediction the everyday stock rate using neural networks and the final outcome is compared with the statistically forecasted outcome. The neural networks along with set of well-trained data set, correct inputs, and suitable architecture will be able to predict the stock market data accurately. Henceforth, neural networks may be used as better system in predicting the stock market rate.

3 Data Mining and Neural Networks

Neural Networks are one of the efficient data mining methods. Neural networks have the ability to analyze the model, after the processes of learning the intellectual system and the neural functions of the stock market and able to predict the new experiments from other observations, after completing the process of learning from the former data set.

In the initial stage, it is required to setup a model of specific network architecture. The network architecture consists of the number of layers and neurons in it. The overall dimension and hierarchical structure has to be designed. The newly designed neural networks is then undergoes into the training process. In this stage, neurons were given an iterative process to total number of inputs to regulate the load of the network, this results in predicting the data from the training data. After the learning process (from the existing dataset) the new network model is organized and it can be used to predict the data (Fig. 1).

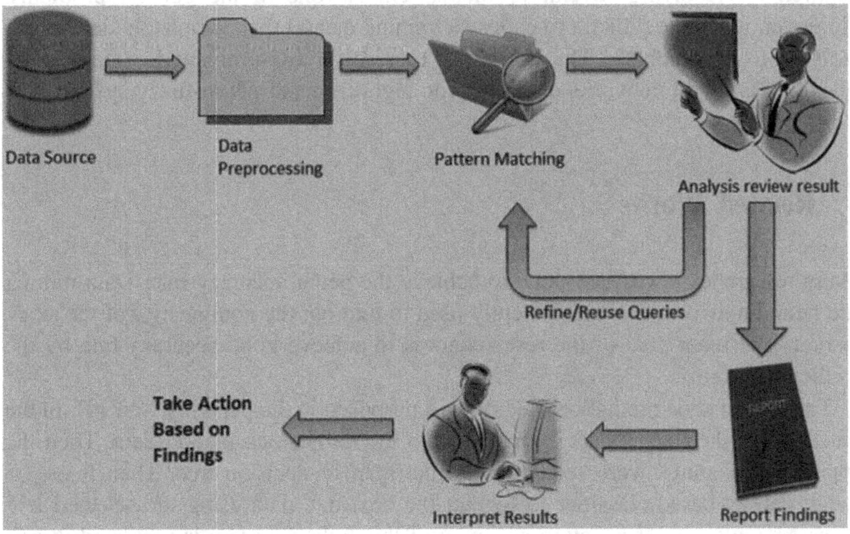

Fig. 1 Data mining process

3.1 Advantages of Neural Networks

Accuracy: They have the ability to calculate most approximate data in complex nonlinear cases.
Flexible: Generally very flexible on incomplete, missing, and noisy data.
Independence from existing predictions: Neural networks are free from historical time series data and expectations about the distribution of the data.
Ease of maintenance: Neural networks can be updated with new data, and used effectively in dynamic environments.
Parallel processing: Even though an element fails in the neural networks, it has ability to continue the process, without any problem (since it is parallel, in nature).

4 Problem Description

One of the major problem in forecasting the share market data is that, it is an unordered system. Many variables are involved and that might affect the share market directly, sometimes indirectly. Moreover, there are no substantial relations between the variables and their rate. Thereby, we are unable to pull any exact calculated bondage between the variables. There are no rules of forecasting the share rate, with the help of these variables.

There are no common approaches to determine the optimal number of neurons required for resolving several problems. This is one of the general problems. Moreover, it is quite difficult to choose a training dataset that completely defines the problem to be resolved. The only way to resolve this problem is achieved by designing a neural networks using genetic algorithm and neuro-fuzzy system.

5 Related Works

Many researches have been done to achieve the better accuracy rate. Data mining and neural networks can be efficiently used to find out the nonlinearity of the stock market. The main goal of the researchers is to achieve good accuracy rate by the predicted system.

The hybrid style prediction method was proposed in the paper by Nair [4]. In the initial stage, the features are extracted from the daily stock market data. Then the appropriate features were selected with the help of decision tree. Then it uses a rough dataset-based classifier to predict the next day data using the selected features. Moreover, the prediction accuracy of this hybrid system is validated in real time stock market data, Bombay Stock Exchange Sensitive Index data (BSE-SENSEX). The input of this system is daily open and close values, high and low values of SENSEX and technical index. Feature extraction and feature selection are the two major things that were used.

Statistical studies, technical investigation, essential analysis, time series analysis, chaos theory, and linear regression are some of the methods that have been assumed to forecast the stock market direction [5]. Moreover, these techniques will not able be to produce consistent prediction of the stock market, and many experts remain unsure about the effectiveness of these approaches. However, these methods represented a first-level standard where neural networks are expected to outperform and command significance in the stock market forecasting system. There are no set of standard rules or sequential steps involved in simplifying the patterns of data. In general, generalization is used to predict probable result of specific task, which involves learning part and prediction part. Backpropagation networks may be used for prediction of stock market.

6 Proposed Works

The main goal of the proposed system is to achieve better accuracy, and check the results with the real-time stock market data. Data mining and neural networks were used in this system. The proposed system was designed in .Net framework—to make the application user-friendly and GUI (Graphical User Interface). Historical data is collected from the stock market, and data preprocessing has to be applied into the collected data to convert the raw data into understandable form of data. Regression algorithm was used to select the required parameters. Then backpropagation algorithm was tossed to train the selected parameters (concept of neural networks). In order to predict the future parameters, we are implementing fuzzy rules in this system. Then neurons were trained based on the new requirement of the newly proposed model.

Optimization algorithm also may be used in this process to minimize the dimensionality of the data. The above procedure was repeated couple of times to find out the best value for optimization parameters.

7 System Architecture

In the first step of this system, stock market data were collected based on the technical index. Then the collected data were preprocessed in order to convert the raw input data into understandable data. For the process of selecting the parameters regression algorithm is used, then backpropagation algorithm is used to predict the stock market data. Finally, performance of the system was evaluated (Fig. 2).

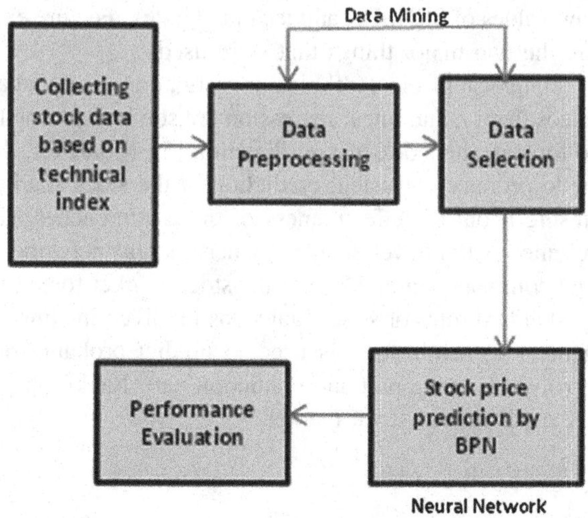

Fig. 2 System architecture

8 System Implementation

Four major phases are involved in the implementation methodology of the newly proposed model. All these four phases play vital role in the proposed system.

They are:

- Data pre-processing
- Regression algorithm
- Backpropagation algorithm
- Fuzzy rules

8.1 Data Preprocessing

It is the process of converting the raw input data into clear understandable data, before the training process. This minimizes the dimensionality of the input data to optimize the generalization performance.

8.2 Regression Algorithm

Regression algorithm is used to select the required parameters. Lot of parameters are required and referred in the process of data prediction. Initially, the required parameters needs to be selected (which is mandatory). Regression algorithm is effectively used for the process of parameter selection, based on the parameters and their values, final data were predicted by the system.

8.3 Backpropagation Algorithm

Training of the data is one of the very important phases in the process of data prediction. For this process, backpropagation algorithm is efficiently used. In order to train the selected parameters and to process the overall stock market data with the trained dataset, backpropagation algorithm was applied in the proposed system. All the selected parameters were well-trained in this phase of newly proposed system.

8.4 Fuzzy Rules

Fuzzy rules are set of rules that are created in order to predict the stock market data in the final stage of processing. A set of rules were created and applied into the proposed system, the result of this leads to prediction of future stock market price. In general, for the process of predicting the future stock market price, a set of rules are created. These rules were also referred as fuzzy rules.

9 Feature Extraction Process

Feature extraction is a process that involves decreasing the total amount of resources required to define a large dataset. More number of variables is involved in the complex data, which is one of the main drawbacks in feature extraction. Large amount of memory and computational power are required for handling a large number of variables, also it requires training sample for training process.

In general, analyzing with large number of variables require a huge amount of memory and computational power to handle the training sample and generalizes poor new outcomes. Feature extraction is a process of creating combinations of the required variables to get around the problems, while describing the data with enough accuracy.

During machine learning, pattern recognition and image processing, feature extraction process starts from a basic set of measured data that forms derived values proposed to be more informative, non-redundant, enabling the successive learning and generalization stages. Feature extraction is closely associated to dimensionality reduction.

In case if an input data to a process is huge to be processed and doubted to be redundant, then it can be converted into a reduced data set of features. This process is called as feature extraction. Extracted features are then expected to have the required information from the input data, so that the preferred task can be performed, with the help of reduced representation (Fig. 3).

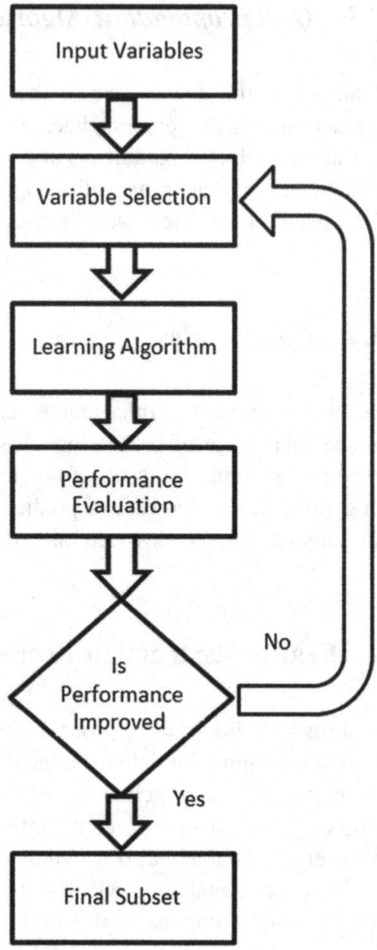

Fig. 3 Variable selection and feature extraction

Most of the researchers and analysts go with future extraction process, because of the following reasons,

- Reduced dimensionality of data
- High relevance to the objective
- Data visualization
- Low redundancy

10 Feature Selection Process

Feature selection is the process of selecting a subclass of relevant features (variables) to use in constructing the model. The data may contain many redundant features during feature selection process. A feature selection system is to be

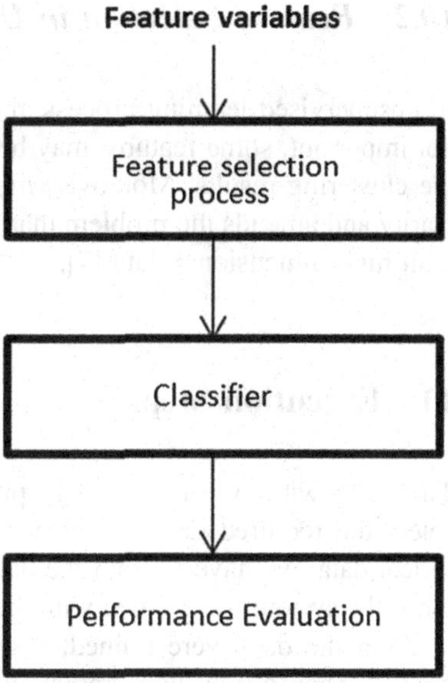

Fig. 4 Feature selection

successful from feature extraction process. Feature extraction generates new features from the functions of original features; however, the feature selection gains a subset of the new features. Also, feature selection is a key in data analysis process, as it shows which features are important for data prediction (Fig. 4).

Feature selection process supports the researchers to create a perfect predictive model. By selecting the features it gives us a better level of accuracy. Feature selection methods are also used to recognize and eliminate the redundant and unwanted attributes. Those attributes may minimize the level of accuracy in the predictive model.

10.1 Feature Selection in Supervised Learning

It is a new approach for the investors targeting the combined evolutionary algorithms and artificial neural networks [6]. In supervised learning, feature selection algorithms increase the functions of accuracy. Since, the given class labels were given, it is to keep the features alone that are closely related to the classes.

In specific, this study expects to address the multi-objective mindset of the purchaser targeting application—increasing the hit rate and reducing the complexity of the new model through feature selection process.

10.2 Feature Selection in Unsupervised Learning

In unsupervised learning process, there was no label for class. All the features are not important, some features may be redundant, irrelevant, and they may misguide the clustering results. Moreover, minimizing the number of features increases the clarity and amends the problem that few unsupervised learning algorithms interrupt with high dimensional data [7].

11 Execution Steps

The main window of the newly proposed prediction system shows the users to select the required dataset (select the historical dataset). After selecting the historical data, we have to train the dataset based on our proposed system algorithm. Once the system was trained with the data, then it was stored in our system (Fig. 5).

After the data were trained, a set of recent data has to be selected again and predict. Then our system predicts the new data and shows us the predicted result and error rate. Stock market open value and close value are taken into account for predicting the new data.

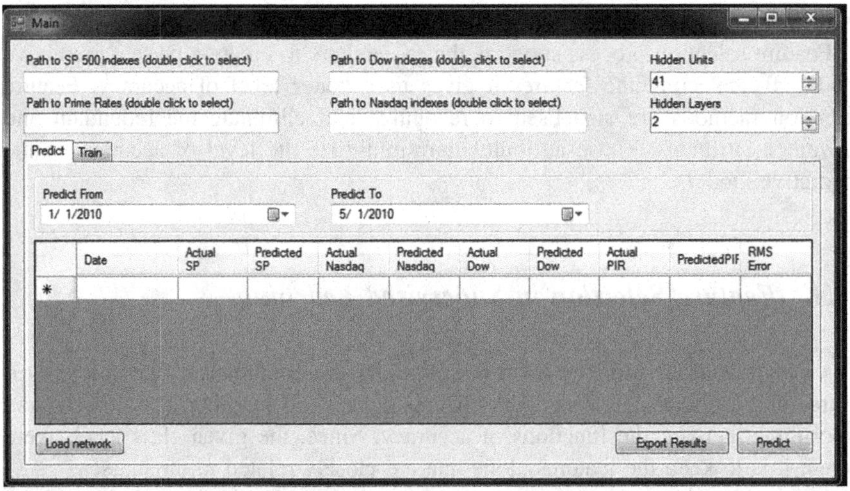

Fig. 5 Historical data selection and training

12 Future Works

In general, stocks are usually predicted on the basis of daily stock data, although few researches use weekly and monthly data. Moreover, in future this should focus more on the investigations of other prediction system also. Almost all the researchers highlight the incorporation of NN with additional methods of artificial intelligence as best solutions for improving the restrictions. In order to improve the accuracy rate of the stock market prediction system, other new techniques may be applied along with neural networks.

Moreover, neural networks are comparatively new method and not frequently examined, they open up many possibilities for joining their methods with new technologies, such as intelligent agents, Active X, and others. Therefore, those technologies could help in intelligent collection of data that involves searching process, selection process, and designing of the huge input patterns. Neural networks investigators improve their restrictions, and that is the valued contribution in the practical significance.

13 Results and Discussion

After several experiments and researches were conducted with different system architectures, the newly designed predictive model results us the best accurate stock rate prediction using the new learning approach that associates the variables of technical and essential analysis. The best outcomes of the hybrid and technical analyses are compared with the existing system. It would be help us to improve the proposed system and its technology, which may results in accuracy. The outcome of the neural network model was examined by comparing the projected values with the real values over a sample time period. The derived output is to be considered useful for exchanging decision support; global success rate of level of accuracy should be extensively high.

14 Conclusion

Growth and development in the field of telecommunication technology, information technology, and internet contributes to improvement of computer science methods. Most of the stock market data are highly time-variant and they are generally in a nonlinear design, forecasting the future rate of a stock is challenging task. Prediction technique provides educated information concerning the present status of the stock price. Various data mining methods for stock market prediction system is reviewed under the literature survey. It is observed that artificial neural network technique is very useful in predicting the stock market price.

Data selection methods for data mining and neural network are one of the important parts in this process; also it needs the knowledge of the field. Several efforts have been made for designing and developing a good system for stock market prediction. Unfortunately, there was no such system found. This paper could not exactly present the situation in neural network applications in stock market. Many researches were done and implemented by the experts and some of them were proven. From the previous researches and studies we conclude (1) Neural networks were effectively used in the field of stock market prediction systems. (2) Backpropagation algorithm is frequently used in NN with various artificial intelligence methods. (3) Neural network is having the ability to predict the stock rates accurately under critical conditions, such as uncertain data. In near future, it can be investigated with real time stock market data, and try to improve accuracy.

References

1. Dunham MH, Sridhar S (2006) Data mining: introductory and advanced topics, 1st edn. Pearson Education, New Delhi. ISBN 81-7758-785-4
2. Atsalakis GS, Valavanis KP (2009) Surveying stock market forecasting techniques—part II: soft computing methods. Expert Syst Appl 36:5932–5941
3. Vaisla KS, Bhatt AK (2010) An analysis of the performance of artificial neural network technique for stock market forecasting. (IJCSE) Int J Comput Sci Eng 02(06):2104–2109
4. Nair BB, Mohandas VP, Sakthivel NR (2010) A decision tree—rough set hybrid system for stock market trend prediction. Int J Comput Appl 6(9)
5. Ravichandran KS, Thirunavukarasu P, Nallaswamy R, Babu R (2005) Estimation of return on investment in share market through ANN. J Theor Appl Inf Technol 44–54
6. Kim Y, Street WN, Menczer F (2003) Feature selection in data mining. Data Min: Oppor Chall 80–105
7. Dy JG, Brodley CE (2004) Feature selection for unsupervised learning. J Mach Learn Res 5:845–889

Resource Allocation for Wireless Network Environment

V. Noor Mohammed, Siddharth Jere and Vora Sulay Kaushikbhai

Abstract This paper presents a radio resource allocation scheme which can be used for OFDM-based systems to mitigate Inter-Cell Interference (ICI). A graph-based framework is created for the wireless cellular system and an algorithm is applied to allocate Physical Resource Block (PRB) to the system. This effectively reduces the ICI which improves the overall network performance.

Keywords OFDMA · Radio resource allocation · Interference mitigation

1 Introduction

Next generation wireless networks aim for high data-rates, efficient resource usage and economical networks. With the increasing number of users, the radio spectrum will become a scarce resource in the near future. To overcome this scarcity, Orthogonal Frequency Division Multiple Access (OFDMA) has been proposed. Using this technique, high spectral efficiency can be obtained and frequency selective fading can be effectively reduced. The current technologies like Long-Term Evolution (LTE) [1] and IEEE 802.16 m [2] are adopting OFDMA due to its advantages. In order to effectively use OFDMA, a proper subcarrier allocation scheme as well as power allocation scheme has to be decided. All this can be collectively included in Radio Resource Management (RRM) scheme. The performance of OFDMA is highly dependent on the choice of a good RRM scheme. If it is not chosen properly, a high system performance cannot be obtained [3–6]. At present, we utilize the frequency reuse technique. But in the future denser techniques with lower frequency reuse factor will have to be used. To effectively use a technique with a lower frequency reuse factor also, an efficient RRM technique is needed.

V. Noor Mohammed (✉) · S. Jere · V.S. Kaushikbhai
SENSE, VIT University, Vellore, Tamil Nadu, India
e-mail: vnoormohammed@vit.ac.in

L.P. Suresh and B.K. Panigrahi (eds.), *Proceedings of the International Conference on Soft Computing Systems*, Advances in Intelligent Systems and Computing 398, DOI 10.1007/978-81-322-2674-1_46

The frequency reuse technique is agreed upon to be used which gives rise to another problem, namely inter-cell interference (ICI). As the same spectrum is utilized in all the cells, users can encounter a severe degradation of performance. This problem is more dominant for the users operating at the edge of the cells. This gives rise to the need of a good RRM scheme which can reduce ICI for a multicell scenario as a single-cell scenario cannot be compared with the practical scenario [7–9].

For ICI-aware RRM Scheme, we consider Signal-to-Interference and -Noise Ratio (SINR) in place of Signal-to-Noise Ratio (SNR). Due to the complexity of this problem, the solution is divided into two phases [10]. In the first phase, a subcarrier allocation scheme is created which reduces the ICI. This technique is known as ICI Coordination (ICIC). The second phase deals with efficient power allocation to increase the network throughput. We concentrate on the first phase which is a resource allocation technique used to mainly reduce ICI. The subcarrier allocation approach is again divided into two parts: Developing a graphic framework using graph theory which reduces ICI and then taking the framework as an input and allocating the subcarriers according to the algorithm.

Thus, the major contributions of this paper are

1. The two-part subcarrier allocation technique used to address the problem that the users at the cell-edge suffer from high ICI while the users at the cell-center suffer from comparatively less ICI. The graphic framework is created so that the same subcarrier is not allocated to two users operating at the edge of the cell as, such users can interfere with each other. Then using the algorithm, the subcarrier is allocated.
2. The performance of this technique is then analyzed for a multicell network. The simulation shows that the performance of the users at the edge of the cell is significantly improved while the performance of the users at the center of the cell is not affected.

2 System Model and Problem Formulation

2.1 System Model

We consider a multicell network using OFDMA technique for downlink. One example of such network is given in figure.

As shown in the Fig. 1, a base-station (BS) using an omni-directional antenna is placed at the center of the cell. It serves users who are randomly located in the cell. In OFDMA systems, the frequency resource is divided into subcarriers while the time resource is divided into time slots. For wireless communication, a traffic bearer is defined as the smallest radio resource unit that can be allocated to a user for the transmission of data in each transmission time. For LTE systems, this traffic bearer is termed as a Physical Resource Block (PRB). In general, the PRB contains twelve

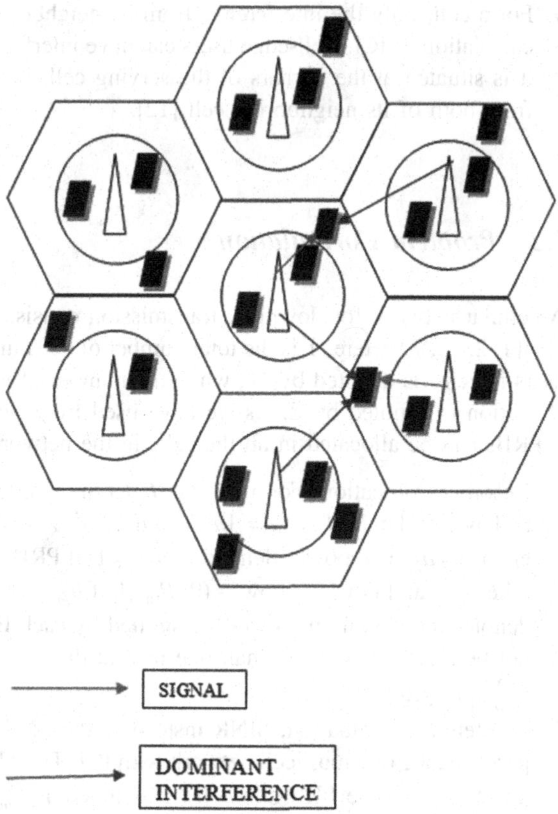

Fig. 1 Multicell network using OFDMA technique

consecutive subcarriers in the frequency domain and one slot duration in time domain. One slot is 0.5 ms [11]. Thus, we can define a PRB as a group of subcarriers that can be allocated to any user in a given time. From now on, PRB is used to denote a single unit for radio allocation in OFDMA network. In addition, the assumptions made in this paper are:

1. For each cell, users are divided into cell-center and cell-edge users depending on their geographical location and straight-line distance from the BS. The boundary separating the cell-center and cell-edge region is decided as a design parameter. The location of the mobile station is updated at the BS periodically using the uplink channel.
2. In every Transmission Time Interval (TTI), the BS has to decide the assignment of PRB to its users. The duration of one TTI is equal to one time slot of the PRB.
3. The transmission power is assigned independently with each PRB. The sum of the total allocated power cannot be greater than the maximum power allocated to the BS.

4. For a cell, only the interference from its neighboring cell is considered for the calculation of ICI. Cell-edge users can have interference from atmost two cells if it is situated at the corners of the serving cell, i.e., it is having equal distance from both of its neighboring cell [12].

2.2 Problem Formulation

A cellular network for downlink transmission consists of a set of BSs denoted by $T = \{1, 2, \ldots, T\}$. Here, T is the total number of cells in the network. The number of users in cell t is denoted by M^t, while the number of PRBs which are available for allocation is denoted by Z. As we have used frequency reuse—1 concept, all the Z PRBs can be allocated in all the cells in the network.

1. Resource Allocation: For a cell $t \in T$, let the matrices for the allocation of PRB and power be $B^t M^t \times Z = [b^t_{mz}]$ and $P^t M^t \times Z = [\text{pow}^t_{mz}]$ respectively, with elements b^t_{mz} and pow^t_{mz} defined as $b^t_{mz} = 1$, if PRB z is allocated to user m and 0, otherwise and $\text{pow}^t_{mz} = \text{pow} \in (0, P_{\max}]$, if $b^t_{mz} = 1$ and 0, otherwise. Here, $P_{\max}$ denotes the maximum power transmitted by each BS. Since the same PRB will not be assigned to more than one user at the same time in each cell, we have $\sum_{m=1}^{M^t} b^t_{mz} = 1$.
2. Interference Evaluation: SINR instead of the SNR can be used to evaluate the performance of a multicell network with ICI. The SINR at a particular instant for a user m with the PRB z in cell t is denoted as β^t_{mz} and it can be expressed as

$$\beta^t_{mz} = \frac{b^t_{mz}\text{pow}^t_{mz} g_z^{(t \to m)} L(d^{t \to m})}{\sum_{t^* \in T, t^* \neq t} p^{t*}_{m*z} g_z^{(t* \to m)} L(d^{t* \to m}) + N_0}$$

where t^* denotes the neighbor cell in which PRB z is allocated to user m^*, $g_z^{(t \to m)}$, $g_z^{(t* \to m)}$ represent the gain of the channel from BSs of cell t and neighboring cell t^* for user m on PRB z, respectively, $L(d^{t \to m})$, $L(d^{t* \to m})$ represent the path loss which is dependent on the distance (independent of z) from BSs of the cell currently serving and the cell which is causing interference to user m, respectively, and N_0 is the variance of the thermal noise. Subsequently, the bit rate that can be achieved by a user m of cell t can be expressed using Shannon's formula as

$$R^t_m = \sum_{z=1}^{Z} BW * (1 + \beta^t_{mz}) \text{ [bits/s]}$$

where BW denotes the bandwidth of a PRB.

3 Proposed Radio Resource Allocation Scheme

In this section, we develop an algorithm for coarse ICIC and fine PRB assignment in order to achieve an efficient radio resource allocation in the network.

3.1 Phase I: ICIC

In the multicell scenario, the first phase of our proposed radio resource allocation is to develop an ICIC scheme using a simple graph-based framework. Our aim is to construct a graph that reflects major interference that occurs in the real-time network environment. According to the graph theory, the corresponding interference graph is denoted by $G = (V, E)$, where V is a set of nodes which represents each of the users in the network, and E is a set of edges connecting those users that can cause heavy mutual interference when they are allotted the same PRB. For further simplicity, the interference intensity for edge connections is determined only by the geographical location and proximity of users in the network, which means severe interference can be suffered by the cell-edge users due to the shorter distances to the adjacent BSs. For building the link, we define that the link between user $a1$ and $a2$ exists when $E(a1, a2) = 1$, otherwise, $E(a1, a2) = 0$.

The rules for the construction of the interference graph are as follows.

1. Users in same cells share a link.
2. For a cell-edge user, link is established with other cell-edge users of its dominant interfering cells (Fig. 2).

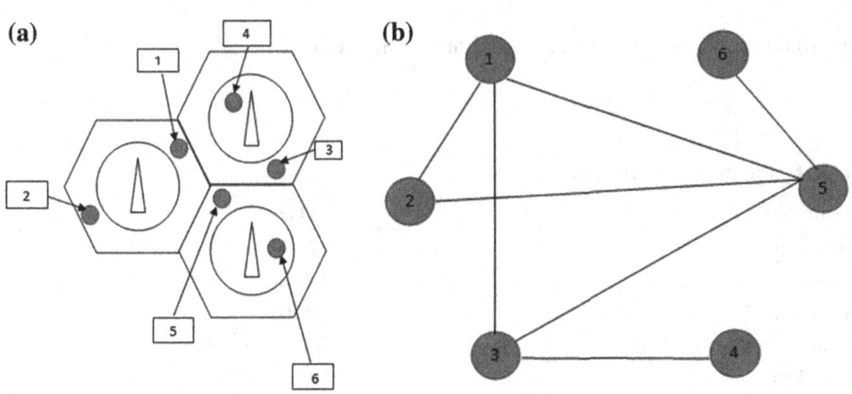

Fig. 2 **a** 3-cell Scenario. **b** Interference graph

3.2 Phase II: Fine PRB Assignment

After the first phase, strategic planning for ICIC has been carried out but the actual allocation of PRBs has not been done yet. In the second phase, hence, we decide how to practically assign the PRB in the network from the interference graph.

Since majority ICI has been appropriately considered in the first phase only SNR is considered instead of SINR.

SNR for user m on PRB z is given as

$$\mathrm{SNR} = (P_{\max}/Z)g_z^{(t\rightarrow m)}L(d^{t\rightarrow m})/N_0$$

4 Algorithm

1. Take the interference graph as input.
2. Select a node A and allocate a PRB to it.
3. Select another node B and check if there is any link between A and B.
4. If no link exists, check whether a link exists between a node that has been allocated same PRB as A. If not, then allocate a PRB to B.
5. If a link exists in both cases, do not allocate the PRB to B and consider a new node C and repeat the above two steps.
6. After this PRB has been allocated to all the nodes take the next PRB and repeat the process for the remaining users.

5 Simulation Parameters

The main simulation parameters used are as in Table 1.

Table 1 The main simulation parameters

Parameters	Values
Total cells	7
Radius of cell	500 m
Bandwidth	5 MHz
Ratio of area of cell-edge	1/3 * total cell area
Total PRBs	10
Maximum power that can be allocated in a cell ($P_{\max}$)	43 dBm
Model for path loss	131.1 + 42.8 $\log_{10}(d)$ dB, d in km
Thermal noise power	−106 dBm

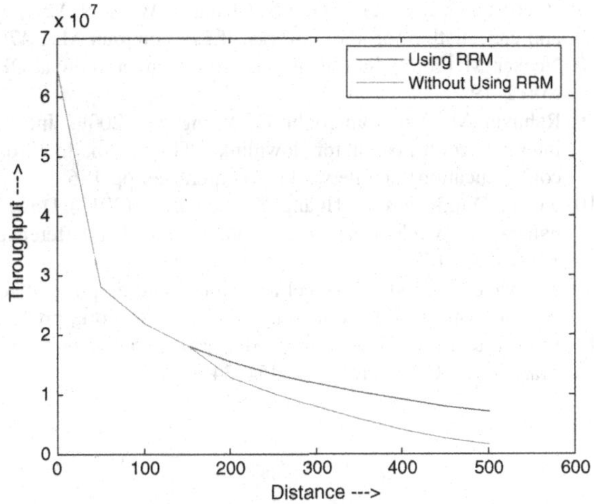

Fig. 3 RRM scheme is applied as compared to the normal scenario in which no RRM is used

6 Simulation Results

From the graph given below, we can see that the channel rate (throughput) is more for the scenario in which our proposed RRM scheme is applied as compared to the normal scenario in which no RRM is used (Fig. 3).

7 Conclusion

Hence, we have proposed a downlink resource allocation scheme for OFDMA-based wireless communication systems which can effectively suppress the ICI.

References

1. Astely D, Dahlman E, Furuskar A, Jading Y, Lindstrom M, Parkvall S (2009) LTE: the evolution of mobile broadband—[LTE part II: 3GPP release 8]. IEEE Commun Mag 47 (4):44–51
2. Etemad K (2008) Overview of mobile WiMAX technology and evolution. IEEE Commun Mag 46(10):31–40
3. Shen Z, Andrews JG, Evans BL (2005) Adaptive resource allocation in multiuser OFDM systems with proportional rate constraints. IEEE Trans Wirel Commun 4(6):2726–2737
4. Lin Y, Chiu T, Su Y (2009) Optimal and near-optimal resource allocation algorithms for OFDMA networks. IEEE Trans Wirel Commun 8(8):4066–4077
5. Jang J, Lee K (2003) Transmit power adaptation for multiuser OFDM systems. IEEE J Sel Areas Commun 21(2):171–178
6. Song G, Li Y (2005) Cross-layer optimization for OFDM wireless networks—part I: theoretical framework. IEEE Trans Wirel Commun 4(2):614–624

7. Boudreau G, Panicker J, Guo N, Chang R, Wang N, Vrzic S (2009) Interference coordination and cancellation for 4G networks. IEEE Commun Mag 47(4):74–81
8. Necker M (2008) Interference coordination in cellular OFDMA networks. IEEE Netw 22 (6):12–19
9. Rahman M, Yanikomeroglu H, Wong W (2009) Interference avoidance with dynamic inter-cell coordination for downlink LTE system. In: Proceedings on 2009 IEEE wireless communications and networking conference, pp 1–6
10. Yu Y, Dutkiewicz E, Huang X, Mueck M (2013) Downlink resource allocation for next generation wireless networks with inter-cell interference. IEEE Trans Wirel Commun 12(4):1783–1793
11. Narayanan L (2002) Channel assignment and graph multicoloring. In: Handbook of wireless networks and mobile computing. Wiley, New York, pp 71–94
12. Li G, Liu H (2006) Downlink radio resource allocation for multi-cell OFDMA system. IEEE Trans Wirel Commun 5(12):3451–3459

Design of PSO-Based PI Controller for Tension Control in Web Transport Systems

N. Hari Priya, P. Kavitha, N. Muthukumar, Seshadhri Srinivasan and K. Ramkumar

Abstract Web transport systems (WTS) are widely employed in industries to transport materials as a long sheet (web) from one section to another. Maintaining web tension along the transit is imperative as it accounts for material integrity and quality. Hence, an efficient web tension controller is needed to provide performance optimization and reduced production down time. Conventional controllers such as PI controllers have several shortcomings, such as performance distortion in noisy environment, inability to handle interactions among the process variables and lack of optimality. This investigation presents an offline optimization-based PI controller tuned using particle swarm optimization (PSO) for WTS to control the web tension. Simulation studies indicate that the proposed controller strategy exhibits good performance optimization by maintaining constant web tension without violating physical constraints.

Keywords Particle swarm optimization · Web transport system · Offline optimization · Proportional integral controllers · Web tension controllers · Evolutionary algorithms

---

N. Hari Priya (✉) · P. Kavitha
School of Electrical Engineering, Vellore Institute of Technology, Vellore, India
e-mail: haripriya.n2013@vit.ac.in

P. Kavitha
e-mail: kavithaslvm@gmail.com

N. Muthukumar · K. Ramkumar
Department of Electronics and Instrumentation Engineering, School of EEE, SASTRA University, Thanjavur, India
e-mail: muthukumar.n@sastra.ac.in

K. Ramkumar
e-mail: ramkumar@eie.sastra.edu

S. Srinivasan
International Research Centre, Kalaslingam University, Srivilliputhur, India
e-mail: cpscourse@klu.ac.in

L.P. Suresh and B.K. Panigrahi (eds.), *Proceedings of the International Conference on Soft Computing Systems*, Advances in Intelligent Systems and Computing 398, DOI 10.1007/978-81-322-2674-1_47

1 Introduction

Material processing industries like paper, iron, polystyrene, etc. transports materials as sheet (called as web) over long distances during manufacturing and finishing stages. The web is subjected to tension variations and disturbances that affect the integrity and quality of the web. Further, the web tension is affected by multiple rollers that makes web tension regulation a nonlinear and multivariable process [1]. Web tension controllers (WTC) is used to regulate and maintain proper web tension. Performance improvement in the WTC can lead to significant improvement in productivity, reduction in product wastage and energy consumption. However, the nonlinear and multivariable nature of WTS poses a big challenge in designing high efficiency WTC.

WTC design to regulate web tension has received considerable attention from the researchers. Traditionally, PI controllers are used for web tension regulation as they are cheap, reliable and are simple to design (see, [2–9]). Although these controllers are simple and reliable, PI controllers have several drawbacks, such as performance degradation, susceptibility to noise and lack of optimality [10]. To overcome the drawbacks of PI controllers, model-based adaptive controllers [11, 12] and robust controllers (e.g., [5, 13–17]) are introduced to achieve performance optimization in WTS. However, these controllers are computationally intensive and require sophisticated process for real-time implementation. The limitations of WTC, from these results are (i) The advanced controller strategies do not guarantee performance optimization at low cost and (ii) Inability of the conventional controllers to handle the nonlinearities and interactions in the process. The objective of this investigation is to provide performance optimization in WTS using simple, efficient and optimal control design with reduced cost expenditure.

This investigation proposes to use offline optimization using evolutionary algorithms (EA) to tune PI controllers for WTC performance enhancement. The EA are preferred as they reach towards global optimization in a small time span. Among the available EAs, particle swarm optimization (PSO) show faster convergence rate towards global minimum when compared to other EA, such as genetic algorithm (GA). Hence, this investigation proposes to use PSO as an offline optimizer to tune the PI controllers to regulate the web tension in WTS.

The paper is structured as follows. Section 2 explains about the mathematical model of WTS. The PSO algorithm to tune PI controller is given in Sect. 3. Results and discussion are provided in Sect. 4, followed by conclusion in Sect. 5.

2 Mathematical Model of WTS

This section explains about WTS and derives the mathematical model to implement PSO-PI controller. The variables and constants involved in the mathematical model are given in Table 1. The WTS consists of winder, un-winder and dancer

Table 1 Parameters used in WTS

Symbols	Description
L_N	Length of the web in m
$F_{t,2}$	Tension in the web in N
E	Young's modulus in N/m
A	Cross sectional area of the web in m^2
V_1	Velocity of the un-winder roller in m/s
V_2	Velocity of the winder roller in m/s
V_d	Velocity of the roller in m/s
R	Radius of the roller in m
N_g	Gear ratio
J_m	Moment of inertial of the motor in Nm^2
B_m	Viscous friction constant of the motor in Nm s/rad
τ_m	Torque of the motor in Nm
ω_m	Angular velocity of the motor in rad/s
M_d	Mass of the dancer in kg
B_d	Viscous friction of the dancer in N s/m
K_d	Spring constant in N/m
d	Dancer position in m
C	Damping ratio of the material Pa/s

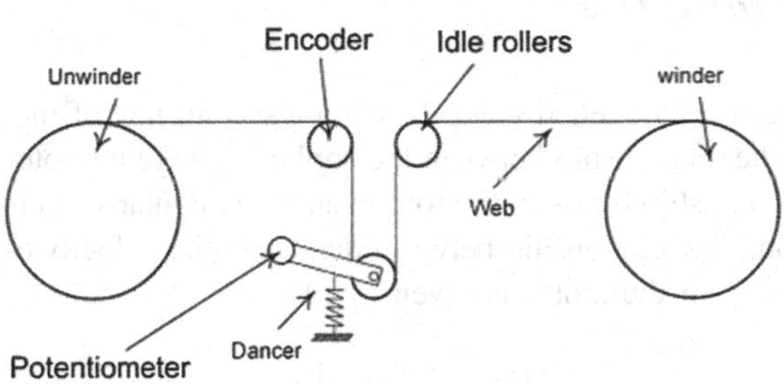

Fig. 1 Sketch of a web handling system copied from [18]

arrangement, over which the web travels. The schematic of WTS is shown in Fig. 1. The web tension is maintained by suitably adjusting the velocity of the winder, un-winder and position of the dancer. The winder and un- winder are coupled to an electric motor through a gear system. This arrangement induces nonlinearity and interactions among the rollers that results in a multivariable system. The tension of the web is measured using load cell. The multivariable mathematical model of WTS can be modelled as three subsystems, they are: (i) The web material, (ii) The dancer and (iii) The winder drive train.

2.1 Web Material

This section provides a mathematical relationship between web tension and the roller velocities. The mathematical model can be obtained from law of conservation of mass and Hooks law [6, 19] and the dynamics of the web material is given by [3].

$$\frac{F_{t,2}}{-V_1 + V_2 - 2V_d} = \frac{\frac{EA}{L_N} + \frac{CAs}{L_N}}{s + \frac{V_1}{L_N}}. \tag{1}$$

2.2 Dancer

The mechanical model of the dancer can be modelled with Newton's second law of motion and it is given as

$$2(F_{t,2} + F_{t,1}) - M_d g = (M_d s^2 + B_d s + K_d)d. \tag{2}$$

2.3 Winder Drive Train

The drive train can be modelled using Newton's second law of motion. The train model provides the relationship between the applied voltage to motor and the roller velocity. The relationship between the torque and the angular velocity of the motor is given by (3) and the relationship between the tangential velocity of the roller and the angular velocity of the motor is given by (4).

$$\frac{\omega_{\mathrm{m}}}{\tau_{\mathrm{m}} - \tau_{\mathrm{cou}}} = \frac{1}{J_{\mathrm{m}} s + B_{\mathrm{m}}}. \tag{3}$$

$$V_2 = \frac{R}{N_{\mathrm{g}}} \omega_{\mathrm{m}}. \tag{4}$$

The complete web winder model is as shown in Fig. 2 by combining Eqs. (1)–(4).

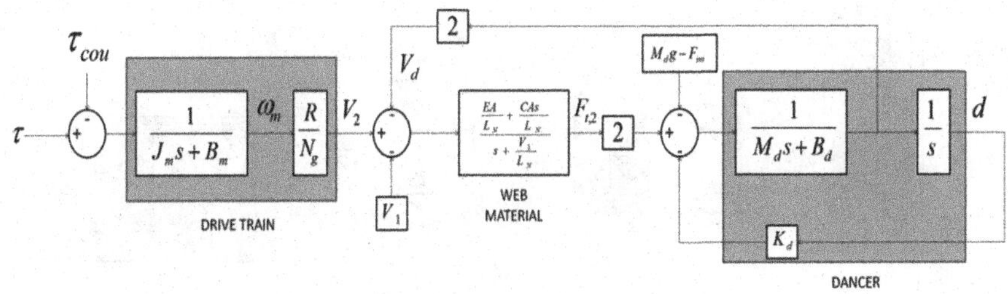

Fig. 2 The complete system of the web winding process

3 PI Controller Tuning Using Offline Optimized Particle Swarm Optimization (PSO-PI)

PSO is an agent based optimization technique that involves particles. Several Particles are grouped to form swarms. Each particle has a position and velocity component. Each swarm moves in the search space to find a best solution [20]. The values of the individual particle decision variables P_i, are randomly chosen within the range $[P_{\max}, P_{\min}]$. Each particle is associated with a velocity V_i that are randomly distributed within the range. $[-V_{\max}, V_{\min}]$. Initially, the particle is allocated with a search direction and velocity components. The particles move in a search space with allocated position and velocity to reach a new fitness value p_i. This fitness value is compared with the best particle fitness value pbest and the group fitness value gbest to update the velocity component of the particle given by (5)

$$v_i^{k+1} = wv_i^k + a_1 \text{rand}_1 * \left(\text{pbest}_i - s_i^k\right) + a_2 \text{rand}_2 * \left(\text{gbest}_i - s_i^k\right). \tag{5}$$

where v_i^k is the velocity of particle i at kth iteration, w is the weighting function, a_1, a_2 is weighting factor, p_i^k is the current position of ith particle at iteration k, pbest_i is the best position of particle i and gbest is the best of particles of the group. Based on the new velocity updates, the particle position is modified by (6).

$$p_i^{k+1} = p_i^k + v_i^{k+1}. \tag{6}$$

This process continues until a best solution is found. In PSO-PI controller illustrated in Fig. 3, the proportional and integral gains of the PI controller are determined by minimizing the fitness function framed by PSO algorithm.

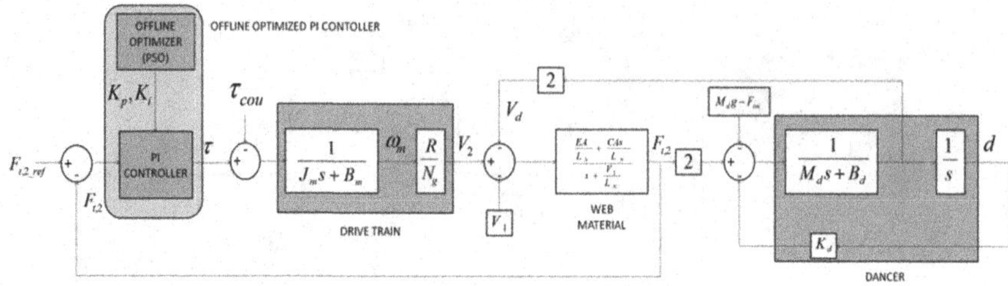

Fig. 3 PSO tuned PI controller for WTS

4 Results and Discussion

SIMULINK is used for the simulation of PSO-PI controller for the mathematical model obtained in Sect. 2, the response of WTS for reference tension of 170 N is shown in Fig. 4. The dancer position in regulating the web tension is shown in Fig. 5. The plant parameters and PSO parameters are provided in Tables 2 and 3, respectively. The gain values of PI controllers tuned using PSO algorithm is provided in Table 4. The performance of the WTS using PSO-PI controller is given in Table 5 and the performance is found to be satisfactory.

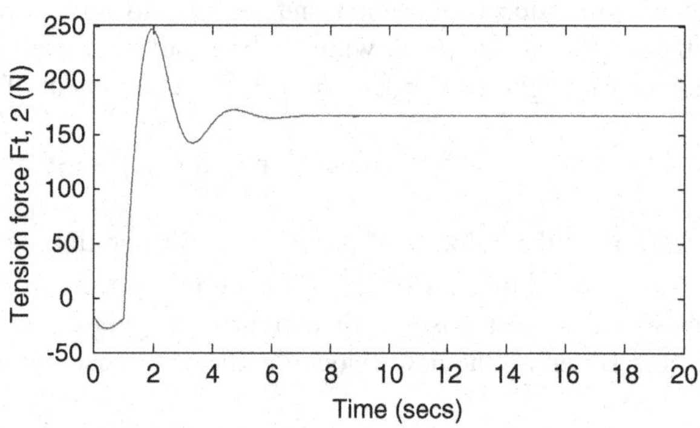

Fig. 4 Web tension response for a reference value of 170 N using PSO-PI controller

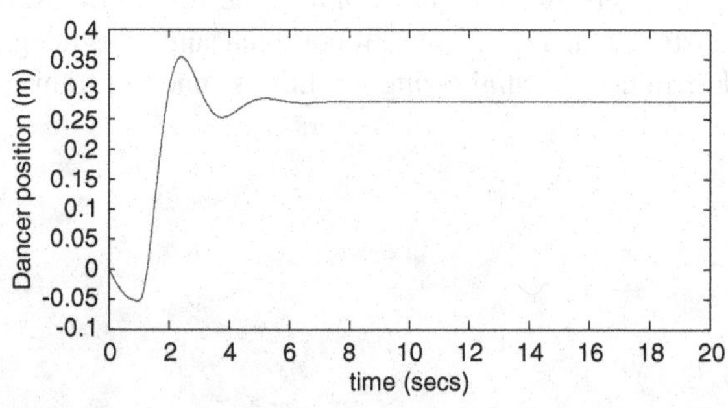

Fig. 5 Dancer position obtained in tracking a reference tension of 170 N using PSO-PI controller

Table 2 WTS model parameters

Symbol	Value	Unit
J_m	3.1e−3	Kg m^2
B_m	0.55e−3	Nm s/rad
T_{cou}	85e−3	Nm
E	4e9	Pa
A	4.35e−6	m^2
L_N	0.61	M
R	57.3e−3	M
N	10.5	–
K_d	1131	N/m
M_d	0.69	kg
B_d	500	Nm s/m
F_{ini}	12	N

Table 3 PSO algorithm parameters

Parameter	Value
Population size	100
No of steps	100
PSO momentum	0.9
Dimension	2
PSO parameter $c1$ and $c2$	0.12, 1.2

Table 4 PI controller parameters obtained from PSO algorithm

Control parameters	Value
P	2.025
I	1.563

Table 5 Performance of PSO-PI controller

Controller	Rise time (s)	Settling time (s)	Peak overshoot (%)
PSO-PI	0.61	5.61	26.4

5 Conclusion

For a WTS, the offline optimized PI controller using PSO algorithm has been designed. Simulation indicates that the PSO-PI controller provides better performance optimization. The PI controller provides exhibits excellent transient characteristics. Designing model-based controllers to address the parameter variations will be the future scope of the investigation.

References

1. Muthukumar N, Seshadhri S, Ramkumar K, Kavitha P, Balas VE (2015) Supervisory GPC and evolutionary PI controller for web transport systems. Acta Polytech Hung 12(5):135–153
2. Hou Y, Gao Z, Jiang F, Boulter BT (2001) Active disturbance rejection control for web tension regulation. In: Proceedings of the 40th IEEE conference on decision and control, vol 5
3. Jeetae K (2006) Development of hardware simulator and controller for web transport process. J Manuf Sci Eng 128(1):378–381
4. Lin KC (2002) Frequency-domain design of tension observers and feedback controllers with compensation. In: 28th annual conference of the industrial electronics society, IECON 02, vol 2. IEEE, pp 1600–1605
5. Ku Chin Lin (2003) Observer-based tension feedback control with friction and inertia compensation. IEEE Trans Control Syst Technol 11(1):109–118
6. Sakamoto T, Fujino Y (1995) Modelling and analysis of a web tension control system. In: Proceedings of the IEEE international symposium on industrial electronics, ISIE'95, vol 1
7. Valenzuela M, Bentley JM, Lorenz RD (2002) Sensorless tension control in paper machines. In: Conference record of the 2002 annual pulp and paper industry technical conference, vol 44 (53), pp 17–21
8. Liu W, Davison EJ (2003) Servomechanism controller design of web handling systems. IEEE Trans Control Syst Technol 11(4):555–564
9. Young GE, Reid KN (1993) Lateral and longitudinal dynamic behavior and control of moving webs. J Dyn Syst Meas Control 115(2B):309–317
10. Garelli F, Mantz RJ, De Battista H (2006) Limiting interactions in decentralized control of MIMO systems. J Process Control 16:473–483
11. Wang B, Zuo J, Wang M, Hao H (2008) Model reference adaptive tension control of web packaging material. In: International conference on intelligent computation technology and automation (ICICTA), vol 1, pp 395–398
12. Pagilla P, Dwivedula R, Siraskar N (2007) A decentralized model reference adaptive controller for large-scale systems. IEEE/ASME Trans Mechatron 12:154–163
13. Claveau F, Chevrel P, Knittel D (2005) A two degrees of freedom H_2 controller design methodology for multi-motors web handling system. In: Proceedings of the 2005 American control conference, vol 2(8–10), pp 1383–1388
14. Gassmann V, Knittel D, Pagilla P, Bueno M (2012) A. Fixed-order H_∞ tension control in the unwinding section of a web handling system using a pendulum dancer. IEEE Trans Control Syst Technol 20:173–180
15. Knittel D, Laroche E, Gigan D, Koc H (2003) Tension control for winding systems with two-degrees-of-freedom H ∞ controllers. IEEE Trans Ind Appl 39:113–120
16. Sakamoto T, Izunihara Y (1997) Decentralized control strategies for web tension control system. In: Proceedings of the IEEE international symposium on industrial electronics, ISIE '97, vol 3, pp 1086–1089
17. Koc H, Knittel D, de Mathelin M, Abba G (2002) Modeling and robust control of winding systems for elastic webs. IEEE Trans Control Syst Technol 10:197–208
18. Adaptive control with self-tuning for centre—driven web winders: Graduation report in Electro-Mechanical System Design, Aalborg University
19. Whitworth DPD, Harrison MC (1983) Tension variations in pliable material in production machinery. Appl Math Model 7(3):189–196
20. Kennedy J, Eberhart RC (1995) Particle swarm optimization. In: Proceedings of IEEE international conference neural networks IV, pp 1942–1948

Distributed Service Level Agreement-Driven Dynamic Cloud Resource Management

M.K. Gayatri and G.S. Anandha Mala

Abstract The size and complexity of Cloud systems are growing more rapidly than expected, and hence, the management of these resources is a major research area. Resource provision with respect to SLA (Service Level Agreement) is directly tied up with customer satisfaction. Failure management is a real-time metric which needs to be addressed by providing continuous availability of resources to the users. This paper contributes to these issues by handling the client without intervention by a human. Continued availability of the client is successfully accomplished by deploying distributed sets of Orchestrators and SLA manager. Ability to deploy nodes remotely and restart nodes is the major area of analysis in this thesis. Monitoring of SLA from a single point for each cloud resource management system is the bottleneck in times of SLA manager failure, which is the brain of the system, and hence we need to overcome the same. Distributing the SLA manager is one of the approaches and which is also being proposed as a solution to the SLA manager failure. SLA Provides a lot of potential in managing cloud resources. In this paper, we build a model to show how WS-Agreement works using JSON. Then, we implement Load Balancing via Grid Gain to make a virtual copy of the same infrastructure as we obtained in SLA. This achieves the jobs to run on virtual machines by eliminating disaster recovery using particle swarm optimization algorithm.

Keywords Cloud · SLA · Orchestrator · JSON · Gridgain

M.K. Gayatri (✉) · G.S. Anandha Mala
Computer Science and Engineering, Easwari Engineering College, Chennai, India
e-mail: mkgayatri@gmail.com

G.S. Anandha Mala
e-mail: gs.anandhamala@gmail.com

L.P. Suresh and B.K. Panigrahi (eds.), *Proceedings of the International Conference on Soft Computing Systems*, Advances in Intelligent Systems and Computing 398, DOI 10.1007/978-81-322-2674-1_48

1 Introduction

Cloud computing refers to the using of computing resources over the net for several demands. It is an intelligent mode of computing, which employs the resources on the internet for saving our data than on our own systems. Spatially located in a different place our needs are serviced by the latest and updated applications. Privacy is compromised though, but the accessibility of our data is increased. We will be able to access our data from anywhere which is not possible if it is only present in our home computer. Privacy needs are handled as required using the authentication services available again on the cloud. In an organization, its use is extensive. An example is the online invoicing service that is a service available on the cloud which is more efficient and robust compared to the in house service.

Infrastructure planning is becoming equally soft as a child play today with the advent of cloud computing platform. It uses the plug and play model where in the resources are pooled as and when needed. Major advantages of this model are that they are really comfortable to use and comparatively low price. The concept of reusability is the foundation for cloud computing models that we practice presently. Grid computing, distributed computing, utility computing, autonomic computing, are the deviations of cloud computing.

Major features of cloud computing are broader network services, on-demand computing, polling of resource, rapid elasticity, and services to measure for cost effectiveness. On-demand computing means that the user or an organization can request and manage their own resources required to achieve their computing task. A broader network service is that it caters to the users' needs over the private networks or the internet. Pooling of resources means that customers pick up their resources from a pool of computing resources, which are usually located in remote data centers. Services can vary and the scale may be larger or smaller; and these are measured by time of usage, type of resource usage, type of service usage, and users are billed accordingly.

Popularity of the cloud services comes from their basic characteristic and most expected nature of human which is the reduced cost and complexity. A new user does not require investing the time in setting up the infrastructure for the computing service that is to be accomplished. Major overheads that an infrastructure set up has been buying of hardware, checking for software licenses, invest in the information technology hardware. These present us the benefits, such as expeditious ROI (Return on Investment), low cost to set up, overnight deployability, instantaneous customizability, flexibility. New innovations in the industry can be included on the go like making advances in a particular service that a single company cannot afford to invest in the research. Scalability is a consequential benefit by which resources can be increased or decreased as the need arises and thus reliability and efficiency come along with the benefits.

Cloud computing models are classified based on service and deployment. Service models are Infrastructure as a Service (IaaS), Platform as a Service (PaaS), and Software as a Service (SaaS). The models are based on the hierarchy from hardware to application. The Infrastructure model is the hardware provider and its necessary network connections. In Platform as a Service model, required hardware, OS, and required connectivity are provided, and the users deploy their own apps. An application that is already ready to use and its required hardware and software are provided to be used by the customers for their need in a SaaS model.

Deployment models are classified based on the level of access the clouds have to the users. Major types are the private cloud, community cloud, public cloud, or hybrid cloud. A third-party cloud provider offering services on the cyberspace is called as public cloud. E-mail services, Facebook, Twitter are the real-time examples of public clouds. Organizations can also be part of using the services of public clouds. In some cases, a cloud provider infrastructure is developed and maintained for the use of a single or a group of organizations which is managed by the cloud service provider which is called as a private cloud. Sharing of services with a group of organizations is termed as community cloud in which each service is provided between the systems to achieve a particular computing task.

2 Related Work

Islam et al. [1] in their paper titled Empirical prediction models for adaptive resource provisioning in the cloud (2012) printed in Science Direct have proposed a technique that offers more adaptive resource management for applications hosted in the cloud environment. It determines the efficient mechanism to achieve resource allocation as needed in the swarm. They have developed prediction-based resource measurement, provisioning strategies using neural network and linear regression to satisfy upcoming resource demands. The report also compares various algorithms to obtain a better technique to allocate better resources, keeping in mind SLA Agreements.

Kertesz et al. [2] discuss in their report titled An interoperable and self-adaptive approach for SLA-based service virtualization in heterogeneous Cloud environments published in Science Direct on how to give a self-manageable architecture for SLA-based service virtualization that provides a means to ease interoperable service executions in a varied, differentiable, scattered types of services. They were able to conclude that heterogeneous Cloud environments need business-oriented autonomic service executions. The concept is built on three areas: agreement negotiation, brokering, and service deployment. Principles of autonomic computing are used to cope with failures in the Clouds. The attack used is Unified systems with heterogeneous, distributed service-based environments, such as Grids, SBAs, and Clouds. SSV architecture is validated in a simulation environment.

Ardagna et al. [3] were able to integrate workload prediction models into capacity allocation techniques that are able to coordinate multiple distributed resource controllers working in geographically distributed cloud sites in their paper titled Dual time-scale distributed capacity allocation and load redirect algorithms for cloud systems (2012). The aim of this paper was to find solutions for performance guarantee that are able to dynamically adapt the resources of the cloud infrastructure in order to satisfy SLAs and to minimize costs. They were able to prove that costs were minimized by using an IaaS. Capacity allocation and load redirect distributed algorithms are represented as the foundation of the solution when the job traffic is high.

Maurer et al. [4] proposes an approach to manage Cloud infrastructures by means of MAPE and MAPE-K (Protocols) published in Science directly under the title Adaptive resource configuration for Cloud infrastructure management (2013). It discusses about autonomic computing and examines the autonomic control loop and adjusts it to govern cloud computing infrastructures. Compares case-based approach and rule-based approach that is the feasibility of the CBR approach and major improvements by the principle-based approach.

Gao et al. [5] proposed a scheme that incorporates various techniques like queuing theory, integer programming, timing analysis, and control theory techniques in their paper titled Service level agreement-based energy-efficient resource management in cloud data centers (2013). To provide with strict QoS standards, they developed effective mathematical models with a strong theoretical base. Both DVFS and server consolidation techniques are exploited in the proposed scheme to achieve energy efficiency and desired application-tier performance in cloud data centers.

Zapater et al. [6] analyzed on how to (i) provide the required computing and sensing resources (ii) allow the population-wide diffusion (iii) exploit the storage, communication, and computing services provided by the Cloud (iv) tackle the energy-optimization issue as an exclusive requirement, which is often revisited issued by them in their paper titled A novel energy-driven computing paradigm for e-health scenarios (2013) published in Science Direct. The Major aim of the report was to thin out the storage of cloud data and to increase speed of live monitoring data.

Kaur and Chana [7] published a paper in Computer Methods and Programs in Biomedicine titled Cloud-based intelligent system for delivering health care as a service (2014) discussed about real-time monitoring of user health data for diagnosis of chronic illness such as diabetes. The paper focused on utilizing data mining techniques for delivering intelligent service behavior and stronger focus on QoS parameters like traffic handled by the application, usage of the devices on cloud, level of vulnerability to attacks, and return on investment.

García et al. [8] propose a paper titled SLA-driven dynamic cloud resource management (2014) published by Elsevier, Future Generation Computer Systems that represent cloud resources using SLAs and WS-Agreement specification. They introduce Cloudcompaas, an SLA-aware PaaS Cloud platform that manages the complete resource life cycle. They were able to achieve improvement in performance in terms of cost and number of failed user requests, of the proposed architecture using elasticity (upscaling and downscaling) rules.

3 Proposed System

System architecture is the conceptual model that defines the structure or behavior of the system.

The system architecture of the proposed system is given in Fig. 1. A description about the architecture is explained in the subsequent paragraphs.

3.1 Schedule Cloud Feed and Store the Resource Information

In this module, we implement the catalog of the connected devices. It tracks the current information about every system like IP Address, total memory, free memory, CPU count, RAM Speed, etc. This information is stored in server node in a database. The database contains the information about the availability for allocation job in next phase.

3.2 Distributed SLA Manager

SLA Manager allows service providers to easily map contractual Service Level Agreements to service level performance metrics and track those SLAs against configurable conformance periods.

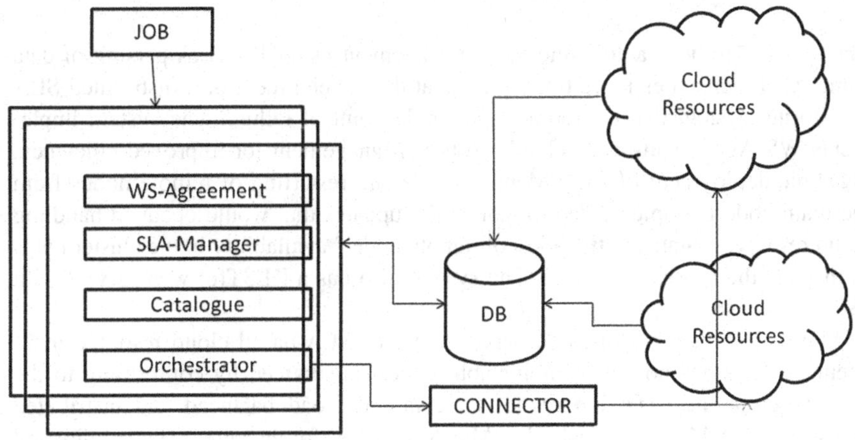

Fig. 1 System architecture

3.3 Orchestrator and Connector

The Connector is a tool that facilitates hybrid cloud computing for organizations. The Cloud Connector essentially helps to orchestrate and administer the migration of Virtual Machines across different data centers and clouds. A cloud orchestrator is the unit that manages the interconnections and interactions among cloud-based physical units that will be assigned to do the jobs. Products on the Cloud that are designed with orchestrators use a sequence of processes to coordinated the different processes and resources that are required to complete the jobs.

3.4 Setting up of Real Cloud

Setting up of real cloud involves configuring of resources and each resource should communicate about its available resources to the catalog. Each resource IP Address is used to identify it separately.

3.5 Load Balancing at the Infrastructure Connector

This module is designed to check the load distribution algorithm that is deployed and check for the results of how successfully the Service level.

4 Implementation

The system provides autonomic SLA management using JSON as parsing of data using XML consumes more time. It aims at the development of a distributed SLA monitoring system that overcomes the single point of failure. The system implements WS Agreement-based cloud resource Management for improved efficiency. Dynamic deployment of SLA Manager code and restarting of remote nodes from the main node is implemented to handle the updates that would occur in handling cloud resources. Status notification of remote nodes' availability in the cluster helps to manage the availability issue. The system also has a RESTful web service with JSON response.

Figure 2 shows the functional architecture of SLA-based cloud resource management. It shows how a incoming job request is serviced by conforming to the SLA requirements. The job requests are initially load balanced and distributed across the SLA Manager. Each SLA Manager checks for the agreement validity and its adherence to the existing agreement. If the agreement is in conformance the job is sent to the orchestrator. Orchestrator assigns the task to different service provides

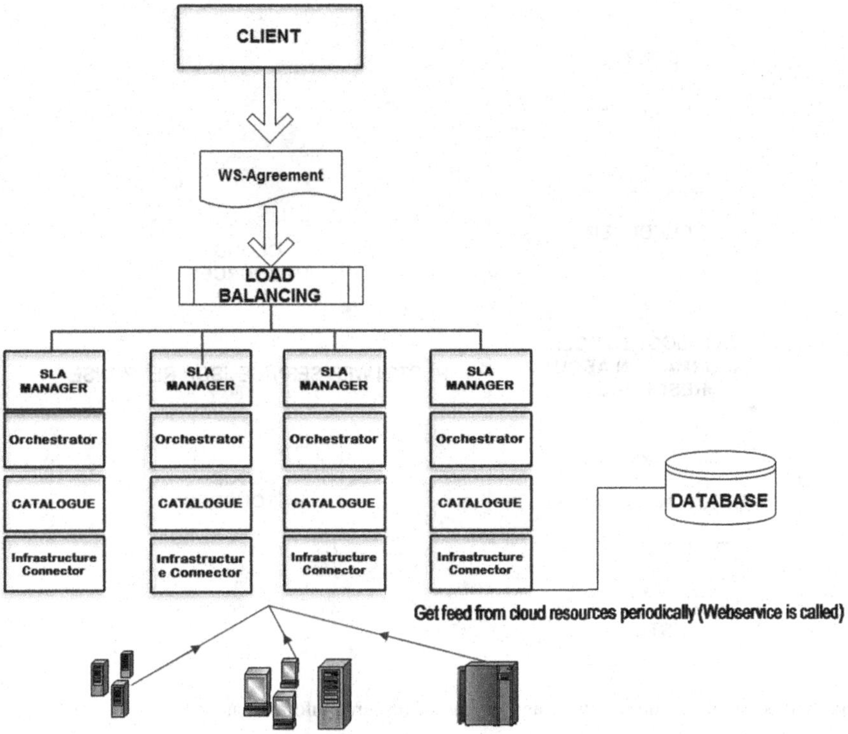

Fig. 2 Overall process of managing SLA

after checking for the availability with the catalog. Job is serviced using the infrastructure connector.

4.1 Catalog Information System

The Catalog implements an Information System, by means of a distributed and replicated database accessible through a RESTful API. The other components use this Information System for recovering and storing of many types of information like agreements, correspondence templates, and runtime and monitoring data. It delivers the current data about every system like IP Address, total storage, free storage, CPU count, RAM Speed, etc. This info is stored in server node in a database. The database holds the information about the availability for an allocation job in the following form. Figure 3 depicts the workflow of how the feed is received by the catalog. A timer class is implemented to invoke the methods in the Catalog class that get the system information about the system tied to the web. The data gathered is stored in a database server.

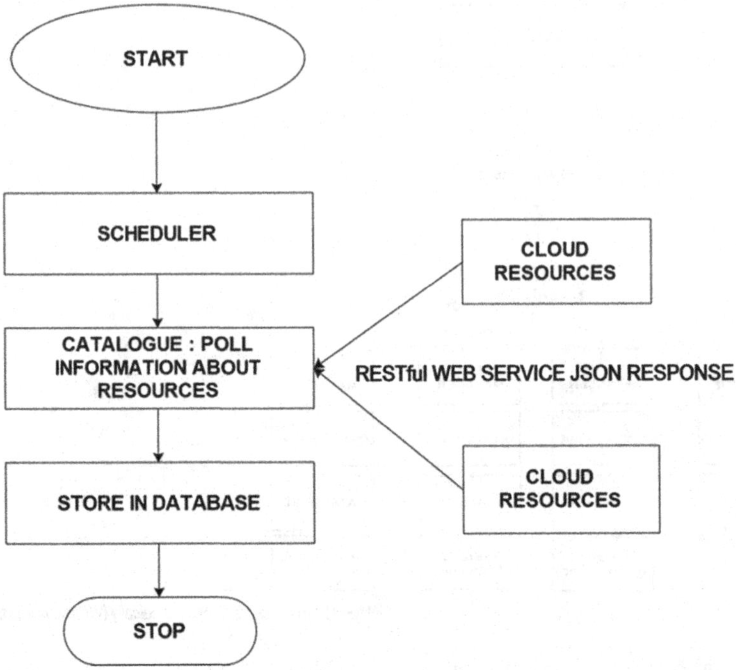

Fig. 3 Process of scheduling timer and polling of resource information

4.2 Distributed SLA Manager

The SLA Manager is the entry level to the program. The SLA-driven nature of the program implies that every interaction between components is done by means of correspondences. Consequently, any external interaction must occur through the SLA Manager component. The SLA Manager is responsible for building up of agreement documents, checking for correctness and registering the agreement that has given the correct standards. There are four operations that are the basic responsibilities of this component which are querying, searching, creating, and deleting. SLA Manager allows service providers to easily map contractual service level agreements to service level performance metrics and track those SLAs against configurable conformance periods. The creating operation sends the SLA Manager an agreement offer. The component checks that the offer complies with the accord template. If this operation fails, the offer is spurned. If the offer is well determined, the SLA Manager sends it to the Orchestrator to schedule its deployment. If this operation goes bad, for example, because no free resources are usable, the offer is spurned. After an agreement has been assumed and its resources have been allocated, the SLA Manager registers the agreement in the Monitor component.

The query operation enables users to retrieve the state of agreements that they have committed to the platform (including the rejected ones) and to delete an active

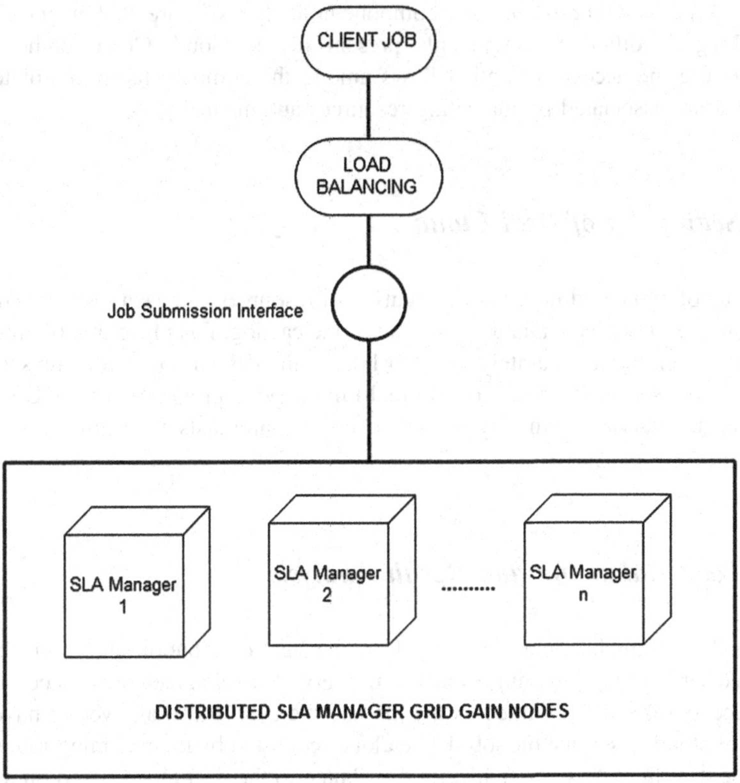

Fig. 4 Distributed SLA manager

correspondence. The deleting operation provides users with a mean to deallocate the resources associated with an agreement and stop its monitoring. The SLA Manager checks if the appointed agreement is presently alive and if the user has rights to edit it before interacting with the Orchestrator and the Monitor to delete the agreement.

Figure 4 shows the working of distributable SLA Manager in which the grid gain nodes are created and restarted as required. Nodes availability is updated and information about the node unique ID and the URL are listed in the monitoring screen.

4.3 Orchestrator and Connector

The Connector is a component that makes the hybrid cloud computing possible for organizations. The Cloud Connector essentially helps to form and control the movement of computing machines virtually across different information centers and clouds. A cloud orchestrator is a component in the software that manages the networking of different components present in the cloud. Cloud orchestrator products use the access controlled flows among the components to complete the tasks that are associated by allocating resources automatically.

4.4 Setting Up of Real Cloud

Setting up of real cloud involves configuring of resources and each resource should communicate about its available resources to the catalog. Each resource IP Address is used to identify it separately. A web host with domain space for our service, Cloud Server set up from one of the real-time service providers, and a URL for accessing the services remotely are the major requirements for setting up a real cloud.

4.5 Load Balancing and Result Analysis

This module is implemented to check how the load distribution algorithm that is deployed services the job with respect to the service level agreements successfully. A FB access token is submitted as the job and the user is also allowed to have the choice of cloud to service the job. If the cloud requested by the incoming job is not available then the job is automatically distributed to the peer cloud based on a load balancing algorithm. Particle swarm optimization algorithm is implemented for load balancing the incoming job requests.

5 Experimental Results

Implementation of a monitoring system using grid gain helps to analyze the number of resources that are connected to the cloud. It also helps to expedite the process of allocating the resources to the incoming job requests which is proven using the Grid gain implementation. The user will be able to view and update the SLA Manager code just on a click without the need to update each and every node as we do in traditional systems.

6 Conclusion and Future Work

Acquisition of information from the set of resources that are connected to the network and how that is updated with respect to the IP Address when the information about the same resource changes in the course of time is done using the catalog code. Distributing the SLA Manager organizes the work of the incoming job effectively. Using of Grid gain platform to implement the distributed functionality is a new approach that is handled in this paper.

Future work on this paper is planned on developing a Orchestrator that manages the job servicing between the SLA Manager and the Infrastructure Connector. Setting up of Infrastructure connector and job servicing using real-time cloud is also in the scope of enhancement. Load Balancing and Result Analysis is to be taken up after the previously mentioned components are completed.

References

1. Islam S, Keung J, Lee K, Liu A (2012) Empirical prediction models for adaptive resource provisioning in the cloud. Future Gener Comput Syst (Elsevier) 28(1):155–162
2. Kertesz A, Kecskemeti G, Brandic I (2014) An interoperable and self-adaptive approach for SLA-based service virtualization in heterogeneous cloud environments. Future Gener Comput Syst (Elsevier) 32:54–68
3. Ardagna D, Casolari S, Colajanni M, Panicucci D (2012) Dual time-scale distributed capacity allocation and load redirect algorithms for cloud systems. J Parallel Distrib Comput (Elsevier) 72(6):796–808
4. Maurer M, Brandic I, Sakellariou R (2013) Adaptive resource configuration for cloud infrastructure management. Future Gener Comput Syst (Elsevier) 29(2):472–487
5. Gao Y, Guan H, Qi Z, Song T, Huan F, Liu L (2014) Service level agreement based energy-efficient resource management in cloud data centers. Comput Electr Eng (Elsevier) 40 (5):1621–1633
6. Zapater M, Arroba P, Ayala JL, Moya JM, Olcoz K (2014) An interoperable and self-adaptive approach for SLA-based service virtualization in heterogeneous cloud environments. Future Gener Comput Syst 32:54–68
7. Kaur PD, Chana I (2014) Cloud based intelligent system for delivering health care as a service. Comput Methods Program Biomed (Elsevier) 113:1346–359
8. García AG, Espert IB, García VH (2014) SLA-driven dynamic cloud resource management. Future Gener Comput Syst (Elsevier) 31:1–11

Efficient Route Discovery in VANET Using Binary Bat Approach

D. Saravanan, S. Janakiraman, S. Sheeba Roseline, M.P. Sharika, U. Madhivadhani and J. Amudhavel

Abstract Vehicular ad hoc network is an engrossed environment of exploration as folks always keep moving from one location to another. Hence, the traffic compactness increases due to the increasing number of locomotives. This shows that there is a ceaseless claim for the conveyance report for the current location, the information about the local traffic, path, etc. The course of conveying the message plays a crucial role in the domain of VANET. This paper gives the employment of the Binary Bat Algorithm (BBA) for the thrifty breakthrough of the routes in order to transmit or broadcast the required data.

Keywords Congestion control · Bio-inspired · Position vector · Velocity vector · Frequency vector

---

D. Saravanan (✉) · S. Sheeba Roseline · M.P. Sharika · U. Madhivadhani · J. Amudhavel
Department of CSE, SMVEC, Pondicherry, India
e-mail: svsaravana1@gmail.com

S. Sheeba Roseline
e-mail: roselinesheeba08@gmail.com

M.P. Sharika
e-mail: shari.mptly@gmail.com

U. Madhivadhani
e-mail: madhianju@gmail.com

J. Amudhavel
e-mail: info.amudhavel@gmail.com

S. Janakiraman
Department of Banking Technology, School of Management, Pondicherry University, Pondicherry, India
e-mail: jana3376@yahoo.co.in

L.P. Suresh and B.K. Panigrahi (eds.), *Proceedings of the International Conference on Soft Computing Systems*, Advances in Intelligent Systems and Computing 398, DOI 10.1007/978-81-322-2674-1_49

1 Introduction

Vehicular ad hoc networks is an increasing area of research due to the need for the motion from one location to another in a reduced amount of time. VANET, which is a sub category of MANET, considers each node as vehicles and is in a highly mobile state. Ad hoc networks refer to that the communication [1] takes place without any intermediate devices such as routers, etc. Hence, VANET communication is enabled without any intermediate devices. The VANET communication [1] is of two types such as the Vehicle to Vehicle (V2V) and Vehicle to Infrastructure (V2I). The communication [1] occurs by means of IEEE 802.11p, which is designed mainly for the vehicular networks. The two main components involved in the communication are the Road Side Units (RSU) and the On-Board Units (OBU). The RSUs consist of the transmitters and the receivers while the OBUs consist of the sensors that are used to either look for other sensors in the OBUs of the other vehicles or the transceivers of the RSUs. The communication takes place mostly between the OBU and the RSU [1].

There are several issues [2] while the vehicles are communicating between each other to pass the information about the obstacles in the path of travel. Some of the issues are flooding in the initial phase of the route discovery, delay in the transmission of the packets, increasing congestion in the networks, and bad performance due to the distance between the vehicles and also duplicate packets may be forwarded in and around the cluster that is formed. Link breakage may also occur such that the packets that are available with the information may not reach the destination from the source. In order to overcome all these disadvantages and to efficiently discover a route to forward the packets, several protocols have been devised which has both advantages and disadvantages.

Bio-inspired algorithms [3] have also been employed in VANETs in various areas of research. These algorithms are usually derived based on the behavior of the nature. The bio-inspired computing [3–5] is associated with the standards of the artificial intelligence such that the method of learning for the machines is made easier. The bio-inspired computing [5–7] is mostly used for the optimization problems which act as a function between the nature and the technology. These algorithms are used to solve the most complicated problems in the areas of computer science. In this paper, we have proposed to implement BBA for the efficient discovery of the route in order to transmit the packet in such a way that there occurs minimal traffic congestion.

We have organized the paper as follows: Sect. 2 contains the related works that had already been done in VANETs. Section 3 contains the proposed work using the BBA. The last Sect. 4 concludes the paper.

2 Related Works

2.1 Route Discovery

The acquiring of the path is the first phase in VANET. The traffic compactness increases due to the increasing number of vehicles and the information about the traffic density has to be passed out to all the vehicles in the network in order to rule out the traffic congestion. There are various protocols and bio-inspired algorithms [3] used in order to exalt a path to traverse the information packets in and around the vehicles.

2.2 Protocols in Route Discovery

2.2.1 Topology-Based Routing Protocol

This genre of routing protocol [8] is relied on the coupling information between the source and the destination nodules in order to transmit the packet. This is categorized into three classes as

- Proactive routing
- Reactive routing
- Hybrid routing

.

2.3 Proactive Routing

These protocols hinge on the shortest path algorithm and the required data are cached in a tabular form and thereby also known as table driven. Each nodule amends their routing table at regular intervals [8].

2.4 Fisheye State Routing

Each node contains the information about the neighboring nodes in the nexus. The advantage of the FSR is that the sizes of the updated messages are small and hence, reduces the complexity of the messages. The disadvantage is that it cannot be used over a large network as it cannot maintain the information of a complex structure.

2.5 Reactive Routing

This is also summoned as the on-demand routing protocol as it periodically brings the table up to date [8]. Hence, this is very commonly used in a highly mobile environment.

2.6 Ad Hoc on Demand Distance Vector

This protocol establishes a path between the source and the destination nodes only when the source node wants to forward the packet to the destination node. This protocol updates the table only when there is a link breakage or looking for a new path. This takes place by means of Route Request [RREQ] and Route Reply [RREP] messages. The disadvantage of this protocol is that the bandwidth of the messages are high as there is a periodic update in the routing table.

2.7 Geographic-Based Routing Protocol

This category of routing protocol [6, 8, 9] is used to indicate the vehicles about its current position and the positions of also its neighboring vehicles.

2.8 Greedy Perimeter Stateless Routing

When the source node wants to transmit the packets to the destination, it forwards the packets to the nodes that are closest to the destination [6, 8, 9]. It uses two types of forwarding strategies such as the greedy forwarding and the perimeter greedy forwarding technique. The disadvantage of GPRS is that it is not suitable in areas of dense traffic and if there are no nodes that are closest to the destination, then the face routing occurs which causes a very long delay in the transfer of messages.

2.9 Geographic Source Routing

This is a combination of both the position-based routing [6, 9] and the topological-based routing as it updates the position of the nodes periodically and stores it in the format of the reactive routing. It uses the Dijkstra algorithm to formulate the shortest path between the start and the end nodules. It forwards the control beacons before the actual beacon is sent. The disadvantage of this protocol is that it is not suitable for very low traffic density and the routing overhead is also high.

2.10 *Bio-Inspired Algorithm Used in Route Discovery*

2.10.1 Ant Colony Optimization

The ant colony optimization algorithm [3] is used to effectively discover the route to forward the packets. This bio-inspired algorithm is based on the behavior of the ants. The hunting behavior of the ants is that it first randomly searches for the prey in every direction. It secrets a hormone called the pheromones, which leaves a mark as the ant moves around. Once the ant catches the prey it takes the food back to its nest following the patch that is formed by the pheromones. It traces the path again and again until the food gets out of bound. Then the ant starts searching for the other route to find the prey. The same process is carried out in VANET. A random search is done such that all the possible routes are found and made a mark. When a source wants to forward a packet to the destination, it traces the path that was already found to forward the packet. The pheromone that is secreted by the ants disappears after a while. In the same way, as the nodes are highly mobile, the path that is traced gets depleted which forms a disadvantage of the ant colony optimization. The route has to be tracked again and again in order to transmit the packet which creates a delay of transfer.

3 Proposed Work Methodology

3.1 *Behavior of the Bats*

The specialized behavior of the bats is that they are blind, indeed they fly without hitting any obstacles. This is by means of the ultrasonic sounds produced by the bats. When bats fly across they produce the ultrasonic sound and based on the echoes produced, they react to the obstacles on the path. This phenomenon is known as the echolocation. The echolocation behavior includes the pulse rate, the position, the velocity of the bats and the frequency, and the loudness produced by the bats. Each of the location of the bats is indicated as the position of the bats and the velocity depicts the speed at which the bat moves in search of the prey. The frequency refers to the rate of sound produced by the bats at regular intervals of time. The bats from their hive moves in search of the prey by producing the ultrasonic sound and when it finds a prey it reduces its loudness and increases the pulse rate to indicate its closeness toward the prey.

Based on the echolocation technology and the fore aging behavior of the bats, the bat algorithm has been devised. The BBA [10] has been applied to those with the binary search spaces. The BBA is proved to have better performance than the other optimization algorithms such as the Genetic algorithm, Particle Swam algorithm, and Harmony Search algorithm, etc. Hence, the BBA has been implemented for efficient route discovery in VANETs.

3.2 *Binary Bat Algorithm in VANET*

In order to find an efficient route to traverse the information in VANET, the BBA is implemented. Here, the bats are considered to be the vehicles. The echolocation behavior is compared to the communication that occurs between the RSUs and the OBUs or between OBUs and OBUs. The parameters that are considered are the position vector, velocity vector, and the frequency vector.

3.2.1 Position Vector-Based Updation

When the vehicles enter into the clusters, it is taken to be the initial position. In order to reach the final destination, the vehicle senses for the information either from the other vehicles or from the infrastructure. When another vehicle that is either close to the destination or in a route with minimal traffic [5, 10] is found, then the position of the vehicles are updated. It is updated in such a way that the new position is added up to the initial position. In the next iteration, the position is again updated in accordance with the other solutions that are available with respect to the position. When the position of vehicles comes closer to destination, the rate of sensing reduces and the information is passed out to the destination.

3.2.2 Velocity Vector-Based Updation

The velocity refers to the speed at which the vehicles travel. The initial velocity of any vehicle is 0. Depending on the speed of the vehicle [11, 12] that is moving ahead of the current vehicle, it can be determined that the route in which the vehicle is traveling is of minimal congestion and it can be assumed that there is no obstacle in the path of travel.

3.2.3 Frequency Vector-Based Updation

Frequency in bats refers to the rate at which the ultrasonic sound is emitted for an interval of time. The frequency of the vehicles refers to the velocity of the vehicle either in a cluster or for a period of time. When the position or the velocity of the vehicle changes, the frequency rate of the vehicle also gets updated periodically. The frequency rate reduces when the current vehicle reaches the destination.

3.2.4 Route Discovery Combining Position, Velocity, and Frequency Vector

The initial position and velocity of vehicles are assigned to be 0. As the vehicle starts accelerating from the initial spot, the position and the velocity of that particular vehicle gets updated. Each updated value is preserved in the Gbest solution [13, 14]. The velocity value is added to the previous value when it is updated every time. The position renewal helps in discovering the route that has been traced. As the destination is nearer, the signal emission of the vehicle gets reduced. The signal emission contains information about the velocity and the position [12, 15] traced at different spots by the vehicle. Thus, the path has been traced and this information may be passed to any other vehicle in the cluster. This is also stored in the database of the RSU such that the information may be used for later retrieval. If any obstacles occur in the path of travel, then an alternate route is taken and the information is passed to all other vehicles in the cluster in the same fashion.

4 Conclusion

The BBA has many relevant features and it is applied to almost all the fields. The usage of BBA in VANET for the efficient route discovery is found to be better than the other optimization algorithms. The implementation is done using the network simulators as a combination of veins, sumo, and omnetpp and is thus found to have a better performance.

References

1. Sharef BT, Alsaqour RA, Ismail M (2013) Vehicular communication adhoc routing protocols: a survey. J Netw Comput Appl 40:363–396
2. Bilal SM, Bernardos CJ, Guerrerro C (2012) Position based routing in vehicular plexus—a survey. J Netw Comput Appl 36:685–697
3. Vengattaraman T, Abiramy S, Dhavachelvan P, Baskaran R (2011) An application perspective evaluation of multi-agent system in versatile environments. Int J Expert Syst Appl 38 (3):1405–1416
4. Paul B, Ibrahim M, Naser A, Bikas M (2011) VANET routing protocols: Pros and Cons. Int J Comput Appl 20(3)
5. Raju R, Amudhavel J, Kannan N, Monisha M (2014) Interpretation and evaluation of various hybrid energy aware technologies in cloud computing environment—a detailed survey. In: International conference on green computing communication and electrical engineering (ICGCCEE), 6–8 Mar 2014, pp 1–3. doi: 10.1109/ICGCCEE.2014.6922432
6. Amudhavel J, Vengattaraman T, Basha MSS, Dhavachelvan P (2010) Effective maintenance of replica in distributed network environment using DST. In: International conference on advances in recent technologies in communication and computing (ARTCom), 16–17 Oct 2010, pp 252–254. doi: 10.1109/ARTCom.2010.97

7. Raju R, Amudhavel J, Kannan N, Monisha M (2014) A bio inspired energy-aware multi objective chiropteran algorithm (EAMOCA) for hybrid cloud computing environment. In: International conference on green computing communication and electrical engineering (ICGCCEE), 6–8 Mar 2014, pp 1–5. doi:10.1109/ICGCCEE.2014.6922463
8. Raju R, Amudhavel J, Pavithra M, Anuja S, Abinaya B (2014) A heuristic fault tolerant MapReduce framework for minimizing makespan in Hybrid Cloud Environment. In: International conference on green computing communication and electrical engineering (ICGCCEE), 6–8 Mar 2014, pp 1–4. doi:10.1109/ICGCCEE.2014.6922462
9. Shinjde SS, Patil SP (2010) Various issues in vehicular adhoc Plexus. Int J Comput Sci Commun l(2):399–403
10. Fonseca A, Vazao T (2012) Applicability of position based routing for VANET in highways and urban environment. J Netw Comput Appl (23 Mar 2012)
11. Venkatesan S, Dhavachelvan P, Chellapan C (2005) Performance analysis of mobile agent failure recovery in e-service applications. Int J Comput Stand Interfaces 32(1–2):38–43. ISSN:0920-5489
12. Zeadally S, Hunt R, Chen YS (2010) Vehicular ahdoc networks (VANETs): status, result and challenges. Springer, Berlin
13. Lochert C, Hartenstein H, Tian J, Fubler H, Hermann D, Mauve M (2000) A routing strategy for vehicular adhoc networks in city environments acknowledged in 2000
14. Dhavachelvan P, Uma GV, Venkatachalapathy VSK (2006) A new approach in development of distributed framework for automated software testing using agents. Int J Knowl Based Syst 19(4):235–247
15. Binitha S, Sathya SS (2012) A survey of bio inspired optimization algorithm. Int J Soft Comput Eng 2(2). ISSN:2231

Mining Ambiguities Using Pixel-Based Content Extraction

B.S. Charulatha, Paul Rodrigues, T. Chitralekha and Arun Rajaraman

Abstract Internet and mobile computing have become a major societal force in that down-to-earth issues are being addressed and sorted out whether they relate to online shopping or securing driving information in unknown places. Here the major concern of communication is that the Web content should reach the user in a short period of time. So information extraction needs to be at a basic level and easier to implement without depending on any major software. The present study focuses on extraction of information from the available text and media-type data after it is converted into digital form. The approach uses the basic pixel map representation of data and converting them through numerical means, so that issues of language, text script and format do not pose problems. With the numerically converted data, key clusters similar to keywords used in any search method are developed and content is extracted through different approaches making it computation-intensive for easiness. One approach is that statistical features of the images are extracted from the pixel map of the image. The extracted features are presented to the fuzzy clustering algorithm. The similarity metric being Euclidean distance and the accuracy is compared and presented. The concept of ambiguity is introduced in the paper, by comparing objects like 'computer,' which have explicit content representation possible to an abstract subject like 'soft-computing,' where vagueness and ambiguity are possible in representation. With this as the objective, the approach used for content extraction is compared and how within certain bounds it could be possible to extract the content.

B.S. Charulatha (✉)
JNTUK, Kakinada, Andhra Pradesh, India
e-mail: charu2303@yahoo.co.in

P. Rodrigues
DMI College of Engineering, Thiruvallur, Chennai, Tamil Nadu, India

T. Chitralekha
Central University Puduchery, Puduchery, India

A. Rajaraman
IIT Madras, Chennai, Tamil Nadu, India

L.P. Suresh and B.K. Panigrahi (eds.), *Proceedings of the International Conference on Soft Computing Systems*, Advances in Intelligent Systems and Computing 398, DOI 10.1007/978-81-322-2674-1_50

Keywords Web content · Content mining · Heterogeneous · Unstructured · Statistical features · FCM · Euclidean distance

1 Introduction

The Web and internet are becoming a major source of large data as they have access to millions of sites and billions of pages providing information in all fronts from ancient history to futurology dealing with 22nd century and different disciplines. The content of the Web page is in the form of flat files, data bases and other repositories. The indexed WWW has more than 13 billion pages, more than 630,000,000 sites, (Netcraft, Feb 2013) 2 million scholarly articles published per year. And 4 % growth rate each year with more than 50 million scholarly articles so far [1]. Content extraction is the process of identifying the main content and/or removing the additional contents [2]. The Web page/document may be structured or unstructured and homogeneous or nonhomogeneous. The form of the Web page and type of data can be simple, relational or multimedia-based and if one considers the script, they are generated from different sources. So to get tangible and coherent information from such Web pages requires data processing and extraction and the research towards this started with data mining. Next advancement in mining is consideration of different forms of data like the streaming audio and video files, which are embedded. This brought in wider and more complex media mining. This is gaining popularity, as the use of internet for sharing of knowledge or entertainment or community development, is on the rise. The use of the Internet has spread across the nook and corner of the world. Since earlier developments in mining are built around English, research and communication/entertainment industry give importance to the English and even data or text mining is built around this. Hence extracting information and then knowledge from these documents calls for preparing, converting, and modifying the data and develop strategies for mining. A Web page can consist of text with images, streaming videos, scrolling news which makes the user seeing lot of information on a single Web page. Web pages are often cluttered with distracting features around the body of an article that distract the user from the actual content they're interested in. These "features" may include pop-up ads, flashy banner advertisements, unnecessary images, or links scattered around the screen [3]. The focus of present work is to extract the content from the Web pages and understand what actually the Web page considered tells about. Since generality of Web pages can be in different ways like language, script, form, and format, the method should preferably be not translation as in text mining and should be computer-understandable and not software-dependent. So a generic approach using pixel-based processing to assess overall content is the focus of the study. The objective of the present study is to extract the features dealing with Web documents and to conclude the relevancy of the content with reference to the subject of interest [4]. So the approach in this paper is based on data preparation, feature extraction, and development of 'key clusters' which will form the basis to assess the content through algorithms. Here the content may relate an '*object*' which may be real or a '*subject/discipline*', which is only possible to comprehend. Hence the clusters which may contain crisp or fuzzy data content are reflective of crispness and vagueness of content and the clustering algorithm and fuzzy C-Means are the methodologies used to extract content. Later case studies for sample Web pages used from the Web if hyperlink is provided or with author-generated pages, are shown to demonstrate the effectiveness of the approach.

2 Approaches for Content Extraction

Content representation and content extraction are two essential needs for Web surfing. A simple, robust, accurate and language-independent solution for extracting the main content of a HTML formatted Web page and for removing additional content such as navigation menus, functional and design elements, and commercial advertisements was presented by Moreno et al. [5]. This method creates a text density graph of a given Web page and then selects the region of the Web page with the highest density. The results are comparable or better than state-of-the-art methods that are computationally more complex, when evaluated on a standard dataset. But methods like these depend on a structured document or Web page (HTML formatted) and when present day Web pages are unstructured and non-homogeneous, content extraction from Web pages is largely needed when searching or mining Web content. Hence a method more generic and independent of language form or media type is needed. The proposed method is different from conventional mining approaches like translation and easily computer-understandable and software-independent and the approach uses the basic representation of data as pixel maps. Internally data are represented only by pixels. Hence the pixel representation is considered for processing. Since pixel maps are large datasets different reduction methods to extract features and attributes are considered using salient features of the pixel matrix, which may be binary or grayscale or even colored. Using the attributes clusters are formed and different methods to infer like Euclidean distance fuzzy clustering and statistical assessment are done for extracting the content. The pixel map of the image is of higher dimension. Hence dimension reduction is done to the pixel map of the image using the nonzero elements in the matrix representation. The matrix is reduced to a scalar or a vector or even a matrix size is 2×2 and 3×3. The features extracted are the statistical properties of the images. Now the pixel map of the text or image is represented by the various attributes like mean, standard deviation, eigenvalue, determinant, diagonal, rank, and norm. Normalization of the attributes is done to so that data dependence of the attributes is taken care of as discussed in Ref. [4, 6]. The influence of the attribute on distance and clustering algorithm is used for better classification. Attribute is the basis of all further inferences and since the choice of attribute and its closeness to actual matrix of the data are not clear, one can study the influence of each one, the subsequent formulations which finally result in grouping. So the attribute which makes the groups more or less similar irrespective of the methods used in the process can well be the basis for future classifications. This could be one attribute or a cluster with features forming the basis for content extraction and is called as '*key cluster*,' similar to key words, popular in Web search. Key clusters can be used to search any kind of data whether structured or unstructured and homogeneous or nonhomogeneous. Finding key clusters and checking with more data is the core of this method. For demonstration, the case studies are chosen for 'object' and 'subject' and here again the definition of 'crisp' or 'unambiguous' for the two are chosen for with different data. The method is tested with actual content and how far they are effective in extracting the content.

3 Case Studies

The method explained earlier is tested for four different data sets, with two belonging to 'object' and the other two to a 'subject.' Even in the two data sets, concept of ambiguity is brought by choosing words which have crisp representation and another set which is vague.

The discussion is given separately for 'object and 'subject' datasets.

3.1 Results for 'Object' Data

For this two objects one a clear one like 'cat' and another, an abstract one 'computer' are chosen and the words describing them are given in Fig. 1. While object 'cat' is represented with image and three language words, 'computer' is represented by one image and three words related to it.

Similarly, more words can be chosen from Web page and using the method the centroidal distances are calculated and membership values for nearly 113 words the graph looks alike that as shown in Fig. 2.

One can see clearly that this variation can be either represented by max. or min. or equivalent values and here max. value is taken for inference. With any new data, if the centroidal distance falls within certain threshold then the chances are the content could be 'cat' or 'computer'.

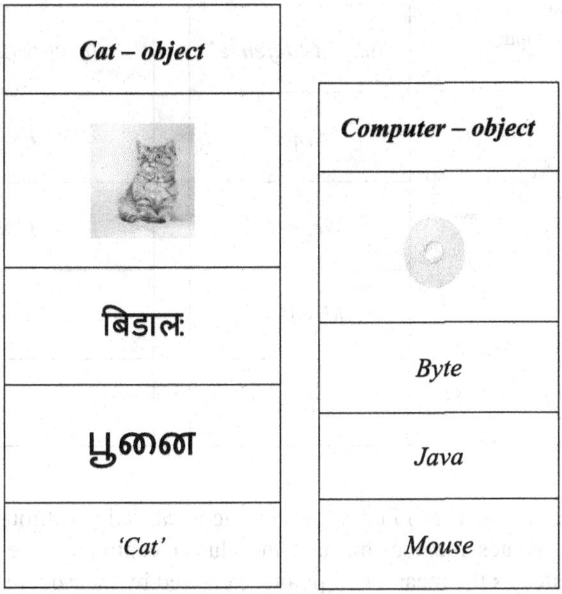

Fig. 1 Example of two object words with input attributes

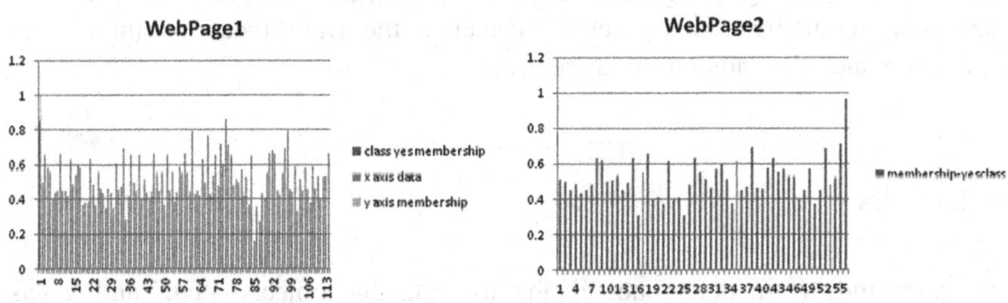

Fig. 2 Membership values for data for content 'cat' and 'computer'

3.2 Results for 'Subject' Data

Two subject data sets for 'defense' and 'soft-computing' are chosen as shown in Fig. 3.

4 Fuzzy C-Means

Fuzzy C-means (FCM) is a method of clustering which accepts data to belong to more than one cluster. Each data is a member of every cluster but with a certain degree known as membership value. This method (developed by Dunn in 1973 and

Fig. 3 Example of two subject words with input attributes

*Subject-'**defense**'*	*Subject-'**softcomputing**'*
Army	*Fuzzy*
Weapon	*Oops*
Missiles	*Genetic*
battlefield	*Parallel*

improved by Bezdek in 1981) Fuzzy partitioning is carried out through an iterative procedure that updates membership and the cluster centroids. The centroid of a cluster is computed as the mean of all points, weighted by their degree of belonging to the cluster. The degree of being in a certain cluster is related to the inverse of the distance to the cluster. In this paper the distance metric used is Euclidean distance. By iteratively updating the cluster centers and the membership grades for each data point, FCM iteratively moves the cluster centers to the 'right' location within a data set. Performance depends on initial centroids.

5 Results

Two case studies are done one relating to realizable objects—'***cat***' and '***computer***'—with attributes describing them and another two subjects—'***defense***' and '***softcomputing***'—with words describing closely the content are chosen to demonstrate the applicability. For the training phase a set of twenty words are taken as shown in Table 1 for both the categories of subject and object in the class of ambiguity and non ambiguity words. Ten words are taken altogether for creating key clusters related to subject. Ten images are taken for creating key clusters related to object. For testing, three different Web pages are taken. The Web pages are preprocessed by removing the tags, special characters and stop words. The preprocessed pages are tokenized. The tokens are converted to pixel map for finding the fuzzy membership. With this conclusion can be drawn whether the Web page is relevant or irrelevant. The first Web page is taken from the link which has an article titled 'Bridging Software Communities through Social Networking' the abstract of which is chosen. The second Web page is that of http://in.mathworks.com/help/

Table 1 Data set chosen for Training

Object		Subject	
Non-Ambiguous	Ambiguous	Non-Ambiguous	Ambiguous
Hardware	Compute	Addition	Fuzzy
Software	Language	Arithmetic	Oop
CPU	Mouse	Binary	Neural
JAVA	Byte	Decimal	Genetic
Objectcode	Interpreter	Solution	Parallel

Table 2 Comparison of maximum values for different Web pages

Web page	Max value with class 1	Web page	Max value with class 1	Web page	Max value with class 1
1	0.515478	1	0.514333	1	0.545585
2	0.517929	2	0.517286	2	0.557233
3	0.532939	3	0.508842	3	0.580049

matlab/ref/urlwrite.html which is relevant to computer industry. The third Web page is that of http://ibnlive.in.com/news/pistol-tamil-actor-vijay-to-shake-a-leg-in-the-hindi-remake-of-thuppakki/376479-71-180.html. This Web page is that of cinema [7]. The code is developed in matlab. The result from the matlab source code is found to be correct with first two pages and false positive with the third page since this page is about a movie remake.

A comparison of maximum values for different Web pages for the content is shown in Table 2, demonstrating the closeness of the values in assessing the content.

6 Conclusion

A method based on generic pixel maps for any kind of data is developed for assessing the content and the performance is tested for both 'object' and 'subject' words using a well known method and the results are encouraging when applied to different Web pages. The concept can be expanded to other languages and other concept of object and subject. The images are restricted to black and white and the text is restricted to computer-generated ones. This needs to be extended to various other images and hand-written formats. The algorithm used is FCM and the metric used is Euclidean. It can be tried with other fuzzy algorithms and metrics. Number of key clusters can be extended beyond two.

References

1. Ross Mounce Content Mining University of Bath, Open Knowledge Foundation Panton Fellow PPT presention
2. Gottron T (2008) Content code blurring: a new approach to content extraction, DEXA '08: 19th international workshop on database and expert systems applications. IEEE Computer Society, pp 29–33
3. Gupta S, Kaiser G, Neistadt D, Grimm G (2003) DOM based content extraction of HTML documents, WWW '03: Proceedings of the 12th international conference on world wide web. ACM Press, New York, pp 207– 214
4. Charulatha BS, Rodrigues P, Chitralekha T, Rajaraman (2014) A clustering for knowledgeable web mining, ICAEES2014, a Springer international conference
5. Moreno J, Deschacht K, Moens M (2009) Language independent content extraction from web pages. In: Proceeding of the 9th Dutch-Belgian information retrieval workshop, pp 50–55
6. Charulatha BS, Rodrigues P, Chitralekha T, Rajaraman A (2014) Heterogeneous Clustering, ICICES 2014 published by IEEE. ISBN 978-1-4799-3835-3
7. Charulatha BS, Rodrigues P, Chitralekha T, Rajaraman A (2015) Content Extraction in traditional web pages related to defense, bilingual international conference on information technology: yesterday, today and tomorrow, 19–21 Feb 2015 organised by DRDO India

Intelligent Collision Avoidance Approach in VANET Using Artificial Bee Colony Algorithm

S. Sampath Kumar, D. Rajaguru, T. Vengattaraman, P. Dhavachelvan, A. Juanita Jesline and J. Amudhavel

Abstract Clustering seems to be the most desired process in the network arena, especially in vehicular ad hoc network (VANET). Several algorithms were proposed for the optimization of the routing in VANET that enables efficient data transfer through better manipulated clustering. In spite of using those algorithms, the major impact lies in the clustering method, so it necessarily depends upon the effective manipulation of analysis of dynamic clustering (Kashan et al. in DisABC: a new artificial bee colony algorithm for binary optimization, 12(1):342–352, 2012). Bee colony optimization is another vibrant bio-inspired methodology that is being used in solving all the complex problems in the network sector. Since bee colony optimization is highly heuristic in nature we adhere to it to obtain good degree of clustering.

Keywords Clustering · ABC algorithm · Cluster patterns · Road side unit (RSU) · Application unit (AU) · Collision avoidance

S. Sampath Kumar (✉)
Department of ECE, MIT, Pondicherry, India
e-mail: sampathsara@gmail.com

S. Sampath Kumar · D. Rajaguru
Department of IT, PKIET, Karaikal, India
e-mail: raja.guru42@gmail.com

T. Vengattaraman · P. Dhavachelvan
Department of CSE, Pondicherry Unviersity, Pondicherry, India
e-mail: vengattaraman.t@gmail.com

P. Dhavachelvan
e-mail: dhavachelvan@gmail.com

A. Juanita Jesline · J. Amudhavel
Department of Computer Science and Engineering, SMVEC, Pondicherry, India
e-mail: jcnr93@gmail.com

J. Amudhavel
e-mail: info.amudhavel@gmail.com

L.P. Suresh and B.K. Panigrahi (eds.), *Proceedings of the International Conference on Soft Computing Systems*, Advances in Intelligent Systems and Computing 398, DOI 10.1007/978-81-322-2674-1_51

1 Introduction

In vehicular environment, any information from the road side unit (RSU) unit or between the vehicles is passed through the wireless network individually. In order to transmit the common information to a group of nodes the clustering [1] has been done. Clustering is the process in which certain number of nodes will be grouped together with certain criteria such as grouping by similarity, position, speed, pattern, etc., and there are two types of clustering algorithm. They are hierarchical and partitional clustering algorithms [1]. The hierarchical clustering algorithm is based on the similarity between the data points to be clustered. Data are grouped by the similarity pattern either through, bottom-up or bottom-down approach. But the partitional algorithm allows the values to be updated which are not possible in the other approach. Clustering can be applied in two ways: crisp and fuzzy [1]. The crisp technique assumes that each pattern [2] should be assigned to a unique cluster where the cluster should be disjoint and it should not overlap with other clusters. In fuzzy technique, the patterns may be applied to any number of clusters provided with a specific membership function. The artificial bee colony algorithm [3, 4] has been used in the paper to find the optimized solution. In artificial [5, 6] algorithm, the algorithm consists of three kinds of bees: employee bee, onlooker bee, and scout bee. In this paper, the bees are considered as vehicles and the nectar is considered as the indicating message. Initially, a vehicle will be present in a cluster and it receives the necessary details from the RSU. Depending upon the information from the RSU, the vehicle moves to either the next cluster or the destination. If any obstacles are faced by the vehicle, then the information is immediately passed to the corresponding RSU, which transmits the information to other nodes. In ordinary vehicular environment, the message has to be sent to individual vehicles whereas in clustered environment, the information has been sent to the cluster, which means once the information has been transmitted to the cluster, the nodes which are present in the cluster will receive the information.

1.1 Motivation

Many approaches have been proposed for clustering of nodes dynamically in the vehicular environment, however, with some difficulties. Clustering may be done based on some similarity conditions like pattern, location, velocity, frequency, etc., achieving the clustering dynamically with these conditions was thorny in the previously proposed works. This motivates us to put forward the new approach for clustering the nodes dynamically using the artificial bee colony algorithm [3, 4]. This algorithm has been used in various fields for finding an optimized path but it is new to the vehicular environment. The idea which is going to be discussed further will help to cluster the nodes dynamically in an efficient manner using the bee algorithm, which helps to find an optimized path with efficient data transmission.

1.2 Contribution

In this work, we are going to propose an approach for clustering the nodes dynamically using the binary artificial bee colony algorithm [7]. Many approaches have been introduced for clustering the nodes dynamically but this approach is new to the vehicular network [8]. We begin our work by discussing few points of the previous approach, the clustering techniques, and about the artificial bee colony [5, 6] algorithm. This approach is used in finding an optimized path and clusters the nodes dynamically. We are going to discuss diverse tactics like the introduction to clustering and artificial bee colony techniques, the technique which we are going to propose, the challenges that are to be met, etc., to the best of our knowledge this is considered to be the first approach in the vehicular network for dynamic clustering using artificial bee colony algorithm.

1.3 Organization

We have organized our work as follows: Sect. 1 provides a brief introduction about clustering of nodes and their corresponding algorithms, the artificial bee colony algorithm, and their functioning. Section 2 imparts the detailed description of our proposed work and the works that are going to be carried out. Section 3 describes the challenges that are to be met and finally, Sect. 4 concludes our proposed work.

2 Proposed Work

As clustering seems to be an efficient way for providing communication between vehicles, we deem in establishing a trusted dynamic link [4] among vehicles through clusters. The vehicles or say nodes are characterized by the dynamic shift behavior in the environment. The proposed model includes three main stages for clustering of nodes and information exchange, (1) node selection process, (2) pattern formation, (3) cluster performance. The enactments of clusters are measured using the compactness and separation factors.

- Compactness: the working efficiency of a cluster will depend on how close the patterns are arranged with each other within the cluster.
- Separation: the dynamic interval amid the cluster heads of different clusters should be as large as possible.

1. **Node Selection Process**
   For selecting a particular set of nodes for a cluster formation, all possible combinations of node behaviors are analyzed in a prospective approach. The cluster formation requires a certain equality measure [1] for determining the

relevant vehicular nodes in accordance with the existing neighbors. The similarities between the nodes provide the data about how far they are from each other and the similarities present among them. Based on the similarity and dissimilarity of the nodes the selection process is made for a cluster formation.

- Modified random selection of vehicles
  The selection of clusters is done based on a random value. The initial state of the cluster is allotted as zero. As the vehicles move further, based on the previous results, random clusters are determined and their values are updated. So, each stage involved in the progress of the vehicle toward the destination contains a set of clusters that are formed based on the random values.
- Modified greedy selection of vehicles
  In this selection process, the clusters are determined based on the greedy values. Based on the randomly generated clusters, a particular cluster is extracted from it using the greedy approach to acquire an optimal path to reach the destination. The transfer of data is performed based on the greedy cluster that is found randomly. This greedy clustering is performed dynamically based on a random selection procedure. The major advantage is, it provides an optimal output.

The modified selection practice of vehicles is performed based on the application of two vital operators.

- Crossover operator
  Crossover operator is applied in areas of high density of vehicles such as an urban region to enhance the probability of obtaining an optimal solution. In this stow, two-point crossover is followed, where any two positions of vehicles are randomly selected from two neighboring clusters, and the nodes that lie within the range of these nodes are identified and interchanged with the other cluster. This may be performed during the dynamic movement of vehicles.
- Swap operator
  Swap operation is performed within a single cluster, i.e., it is performed on a single RSU unit that coordinates a particular cluster. Swap is performed between vehicles that come under the same cluster. In a cluster, say, if swapping is to be performed, then a specific behavior of the vehicle is considered. For example, the swap operation can be made based on the velocity, like low-velocity vehicles can be exchanged with high-velocity vehicles.

2. **Pattern formation**
   The efficiency of a cluster is based on the patterns that are designed within it. The structure of each pattern, entrenched inside the huddle follows a certain vector, based on which the vehicles are gathered. In a single hurdle, several patterns may be formed based on utility-based vectors such as speed, distance, neighbor, etc. The RSU groups the vehicles based on the utility factor into a single pattern with the help of application units (AU) [2].

- Speed-based cluster pattern: The velocity of each vehicle is obtained from the AU [2]. Based on this factor, a pattern is framed to accommodate a particular set of node in it.
- Distance-based cluster pattern: The vehicles or nodes present at a specific distance are clutched as a separate pattern. Here the nodes tend to be closer to each other with a constant interval.
- Neighbor-based cluster pattern: The formation of this pattern is merely based on the destination of each vehicle. If the destination of vehicles is in the neighborhood, then they are said to be clubbed as a separate pattern.

3. **Cluster Performance**
The behavioral and structural patterns of clusters depend on the cluster head's performance and the intervals between the vehicles. There are two components that are used to analyze the outcome of a cluster.

- Intra-cluster distance:
The mean value of all the intervals between the cluster head in a huddle and their patterns is calculated to determine the compactness of the constructed cluster. For a cluster to be compact and efficient, the intra-cluster distance needs to be minimized.
- Inter-cluster distance:
To obtain well separated and distinct clusters, the inter-cluster distance should have a maximum value. The inter-cluster distance is defined as the gap between a pair of cluster heads in a network [8]. This is used in measuring the separateness of a huddle.

3 Open Challenges

Some common issues faced are patterns for cluster cannot be predetermined as dynamic cluster [4] formation plays a major role here. Also due to these factors, attributes-related cluster will also get affected coincidently. The mainly concentrated factors in ABC algorithm [3] in general are the position of the food source and the amount of nectar in that food source. So, these factors, when related to VANET, are the direction of movement of the vehicles toward the destination and possible number of ways from a node to the destination node. Finding the initial node in any network architecture to start the traversing to reach the destination is another challenge to be faced. Various services that we provide to the user are to be maintained and monitored, whenever a new cluster forms or any changes occur in the existing clusters. Load that a cluster maintains plays a major role, i.e., number of nodes that the cluster can handle and hold should be maintained, in case of over load and less concentration of nodes' cluster design should be reconstructed. Advanced soft computing [9–12] of vehicular information is also a part of this technology. Possibility of more than one cluster in a VANET network [2, 8] is high so that each cluster will perform its own processes. Thus, parallel processing happens whenever a cluster-based network.

4 Conclusion

Clustering is a mandatory process that has to be done on any VANET-based network architecture. Forming such clusters can be achieved through various methods. Our proposed idea is to cluster the nodes dynamically [2] using the artificial bee colony algorithm [3, 4], which includes some processes like modified random and greedy selection [7] of vehicles in VANET, which helps in choosing the best node that will lead to the destination node in an optimized path. The inter- and intra- cluster distances are used to find the efficiency of the clusters. This approach to the best of our knowledge is considered to be an effective approach and forms the clusters efficiently and the data transmission is done in a successful manner. Future work considerations will be more on numerous clustering parameters like load balance, density, and incoming, ongoing, and outcoming speed.

References

1. Dhavachelvan P, Uma GV (2005) Complexity measures for software systems: towards multi-agent based software testing. In: Proceedings-2005 international conference on intelligent sensing and information processing, ICISIP'05 2005, art. no. 1529476, pp 359–364
2. Ozturk C, Hancer E, Karaboga D (2015) A novel binary artificial bee colony algorithm based on genetic operators, information sciences, vol 297, 10 Mar 2015, pp 154–170. ISSN 0020-0255. http://dx.doi.org/10.1016/j.ins.2014.10.060
3. Amudhavel J, Vengattaraman T, Basha MSS, Dhavachelvan P (2010) Effective maintenance of replica in distributed network environment using DST. International conference on advances in recent technologies in communication and computing (ARTCom) 2010, pp 252, 254, 16–17 Oct 2010. doi: 10.1109/ARTCom.2010.97
4. Kashan MH, Nahavandi N, Kashan AH (2012) DisABC: a new artificial bee colony algorithm for binary optimization. Appl Soft Comput 12(1):342–352, Jan 2012. ISSN 1568-4946. http://dx.doi.org/10.1016/j.asoc.2011.08.038
5. Karaboga D, Okdem S, Ozturk C (2010) Cluster based wireless sensor network routings using artificial bee colony algorithm, autonomous and intelligent systems (AIS), 2010 international conference on, pp 1, 5, 21–23 June 2010. doi: 10.1109/AIS.2010.5547042
6. Krishnamoorthi M, Natarajan AM (2013) A comparative analysis of enhanced artificial bee colony algorithms for data clustering, computer communication and informatics (ICCCI), 2013 international conference on, pp 1, 6, 4–6 Jan 2013. doi: 10.1109/ICCCI.2013.6466275
7. Ozturk C, Hancer E, Karaboga D (2015) Dynamic clustering with improved binary artificial bee colony algorithm. Applied Soft Computing, vol 28, Mar 2015, pp 69–80. ISSN 1568-4946. http://dx.doi.org/10.1016/j.asoc.2014.11.040
8. Raju R, Amudhavel J, Pavithra M, Anuja S, Abinaya B (2014) A heuristic fault tolerant MapReduce framework for minimizing makespan in hybrid cloud environment. International

conference on green computing communication and electrical engineering (ICGCCEE) 2014, pp1, 4, 6–8 Mar 2014. doi: 10.1109/ICGCCEE.2014.6922462
9. Dhavachelvan P, Uma GV (2005) Multi-agent based integrated framework for intra-class testing of object-oriented software. Int J Appl Soft Comput 52(2):205–222
10. Raju R, Amudhavel J, Kannan N, Monisha M (2014) A bio inspired energy-aware multi objective chiropteran algorithm (EAMOCA) for hybrid cloud computing environment. International conference on green computing communication and electrical engineering (ICGCCEE) 2014, pp 1, 5, 6–8 Mar 2014. doi: 10.1109/ICGCCEE.2014.6922463
11. Dhavachelvan P, Uma GV (2004) Reliability enhancement in software testing: an agent-based approach for complex systems, 7th ICIT 2004, Springer Verlag—lecture notes in computer science (LNCS), vol 3356, pp 282–291. ISSN 0302-9743
12. Raju R, Amudhavel J, Kannan N, Monisha M () Interpretation and evaluation of various hybrid energy aware technologies in cloud computing environment—a detailed survey. International conference on green computing communication and electrical engineering (ICGCCEE) 2014, pp 1, 3, 6–8 Mar 2014. doi: 10.1109/ICGCCEE.2014.6922432
13. Chen Y, Liu J (2012) A new method for underdetermined convolutive blind source separation in frequency domain, computer science and network technology (ICCSNT), 2012 2nd international conference on, pp 1484, 1487, 29–31 Dec 2012. doi: 10.1109/ICCSNT.2012.6526201

Enhancing the Interactivity in Distributed Interactive Applications

D. Saravanan, S. Janakiraman, P. Niranjna Devy, T. Sophia, M. Mahalakshmi and J. Amudhavel

Abstract Interactivity is a prior step of distributed interactive applications (DIAs). In distributed server architecture, the synchronization delay meets with requirements of fairness and consistency of DIAs, as well as the performance of interactivity not only depends on latencies on the client and server but also on the latencies of inter-server network (Zhang and Tang in IEEE/ACM Trans Net 20(6):1707–1720, 2012; Delaney et al. in Presence: Teleoperators Virtual Environ 15(2):218–234, 2006; Marzolla et al. in ACM Comput Entertainment, 2012; Webb et al. in Enhanced mirrored servers for network games, pp 117–122, 2007). The above-mentioned aspects are affected directly based on by which the client is assigned to the server in a network (Dhavachelvan in Int J Knowl–Based Syst 9 (4):235–247, 2006; Brun et al. in Comm ACM 49(11):46–51, 2006) . We look over the complications of client assignment to server for improving the enhancement of DIAs. We mainly concentrate on endless DIAs that are changed due to the time aspect. We evaluate the small amount of feasible interacting time for DIAs to sustain consistency and fairness among the users. The heuristic algorithms are

D. Saravanan (✉) · P.N. Devy · T. Sophia · M. Mahalakshmi · J. Amudhavel
Department of CSE, SMVEC, Pondicherry, India
e-mail: svsaravana1@smvec.ac.in

P.N. Devy
e-mail: niranjdhoni7@gmail.com

T. Sophia
e-mail: sophiathamiz@gmail.com

M. Mahalakshmi
e-mail: mahamsundaram@gmail.com

J. Amudhavel
e-mail: info.amudhavel@gmail.com

S. Janakiraman
Department of Banking Technology, School of Management, Pondicherry University, Pondicherry, India
e-mail: jana3376@yahoo.co.in

L.P. Suresh and B.K. Panigrahi (eds.), *Proceedings of the International Conference on Soft Computing Systems*, Advances in Intelligent Systems and Computing 398, DOI 10.1007/978-81-322-2674-1_52

proposed and we are going to prove that *Greedy Assignment* and *Distributed-Modify Assignment* algorithms are going to produce near-optimal interactivity by reducing the interaction time among the clients related to the immediate algorithm that assigns each user to its adjacent server.

Keywords Consistency · Fairness · Interactivity · Distributed interactive application · Client assignment

1 Introduction

Distributed interactive applications (DIAs) such as online multiplayer games which allow members at different sites through networks [1, 2] to interact with one another. Thus, due the interactivity of DIAs in the network, participants can have enjoyable experiences. Interactivity is based on the time at which the participant or member issues an action to the time from one site and when the outcome of the action is granted to the same participant or other participant in different sites [3]. This indicates the duration as the interaction time between the participants.

A good interactivity in DIAs is provided by a network latency [4–6] which acts as major obstacles in networks [1, 7, 8]. Latency cannot be rejected from the interaction among participants in network and it has a lower limit when compared to the speed of light. We mainly targeted on minimizing the latency [5, 9] for a good interactivity in DIAs. The geographical spread area is increased in network by the deployment of distributed server. There are no limitations on the server resources in specific site or locations. Distributed server architecture, there are "n" number of nodes called as client assignment assigned to server in a specified geographical location. Each participant in the network called as user-client which is allocated to its adjacent server and it associates to its allocated server for initiating actions and receiving updates of a specified location or site. While pointing to an action, the user-client initiates its request to its allocated server. For requesting updates, the server forwards its request action to all servers. After, receiving the action each server calculates the new state of action and sends the updates to its assigned client. Each client interacts with one another through their assigned server. The interaction time between each client pair and assigned server should include their network latency [3, 8, 10]. To the requirements of DIAs, the consistency and fairness determines the interaction time.

To the users of DIAs, consistency is one of the key factors to provide interactive experience. All participants should ideally see the same information at the same time independently in network. This is called as consistency. Fairness on the other hand, it implies that there must be a chance of participation of all client in network [11]. Due to diverse network latency there is a delay in interaction among the clients. So, there is a problem in maintaining consistency and fairness [1, 11, 12].

This delay depends on client to their assigned server. For a better interactive performance in DIAs based on how the clients are assigning to the servers is an important one.

We found out the complications while allocating the user-clients to the servers for improving the enhancement in DIAs. Examples of endless problem in assigning clients to servers are multiplayer online games and distributed virtual environments. It is achieved by consistency and fairness requirements. First, we have to find minimum achievable time interaction in DIAs when the user-clients are allocated to the servers by consistency and fairness. Based on the analyses, we define that it could be NP-Complete by combination optimization problem. For that we are going to propose some heuristic algorithms such as Greedy Assignment and Distributed-Modify Assignment algorithms, which are going to produce near-optimal interactivity by reducing the communicating time between user-clients related to the unmediated algorithm that allocates each user-client to its adjacent server.

The contribution is formulated as follows: Part 2 Proposed system of client assignment problem. Part 3 Discussion of heuristic algorithms such as Greedy Assignment and Distributed-Modify Assignment algorithms. Finally, Sect. 4 concludes the paper of continuous client assignment problem in network by DIAs.

2 Proposed Work

In our proposed work, there are several modules, which is used to reduce network latency and to maximize interactivity in DIAs. The several modules are:

2.1 Model of System

System model consists of set of nodes. The nodes are assigned here as clients. Each and every user-client is allocated to server in a specified location. The link between a node to another node in network is called as routing path. As shown in Fig. 1, different clients are named here as c1, c2, c3. The clients are assigned to different servers such as s1, s2. The connection between client-to-server or server-to-server and client-to-client is called as routing path in network. It also determines the interaction path between the above-mentioned factors.

Each participant in the network is called as a user-client and it is allocated to its server and it bridges to its allocated server for initiating actions and receiving updates of a specified location or site. While pointing to an action, the user-client initiates its request to its allocated server. For requesting updates, the server forwards its request action to all servers. After receiving the action each server calculates the new state of action and sends the updates to its assigned client. Each client interacts with one another through their assigned server. The interaction time between each client pair

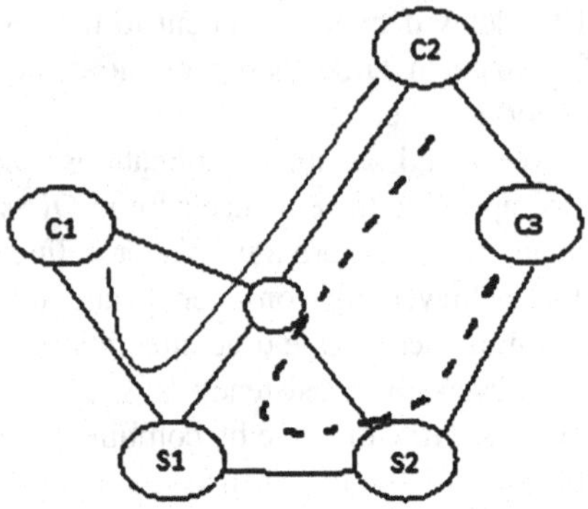

Fig. 1 Interaction paths in networks. *Arrow* Interaction path between c1 and c2. *Dashed lines* Interaction path between c2 and c3

and assigned server should include their network latency. To the requirements of DIAs, the consistency and fairness determines the interaction time.

Suppose clients c1 and c2 are allocated to server s1, then the communication path between c1 to c2 via s1 is indicated by straight line. And the client c3 allocated to server s2, then the communication path between c2 and c3 via server s2 is indicated by dotted lines. The dimension of the communication path between any two user-clients such as c1, c2 and c2, c3 represents the network latency. This helps to minimize feasible interaction time between clients to its assigned servers in network.

2.2 Flexibility and Integrity Models

The process of simulation which represents the action of a system over time is called as simulation time. The simulation time and wall clock time are not the same. They differ from one another based on readings. The reproduction time of user-client stays behind the reproduction time of allocated server due to its network latency by delivering the updates from server to its user-client.

To the users of DIAs, the consistencies is one of the key factors to provide interactive experience. In any point, all participants should ideally see the same information at the same time independently in network. This is called as consistency. Fairness on the other hand, it implies that there must be a chance of participation of all client in network. Due to diverse network latency there is a delay in interaction among the clients. So, there is a problem in maintaining consistency and fairness. These delays depend on client to their assigned server.

2.3 Further Consideration

We mainly focused on minimizing latencies and time delays involved in communication between allocated user-clients. The latencies are mainly raised by delays on the server side. For reducing those delays in network the server should not be lagged to response back to its assigned client or server. Thus, a real-world system often makes a certain percent of network latencies cause arise in unfairness.

3 Heuristic Algorithms

In this section, we are going to deal about three heuristic algorithms. These three algorithms are dependent on latencies in between the allocated user-clients and servers.

- Nearest-Server Assignment algorithm
- Greedy Assignment algorithm
- Distributed-Modify Assignment algorithm

3.1 Nearest-Server Assignment Algorithm

The prior algorithm is represented as Nearest-Server Assignment algorithm which allocates each client to its possible adjacent server. The implementation of this algorithm is done for the purpose of finding network latency among clients and servers. Here, every user-client measures the latencies by itself and between all the servers. This assignment algorithm minimizes user-client to server latencies, which could be seriously increasing the latency between the allocated servers of different user-clients, and this makes the communication too poor than its excellence. Thus, to overcome these problems, we mainly focused on Greedy Assignment and Distributed-Modify Assignment algorithms.

3.2 Greedy Assignment Algorithm

The next algorithm is Greedy Assignment algorithm which approaches to assign clients constantly that is initiating with empty assignment. The algorithm first considers all possible unassigned clients in network. Then, it assigns unassigned client to nearest possible servers. For example, if the user-client "c" is considered to be allocated to the servers, then all unallocated user-clients are assigned to another server other than the previous assigned server. Thus, it could not increase or maximize the interaction path length.

To calculate efficiency, the path lengths from all assigned clients to each server are categorized for preprocessing stages. These ordered lists are then updated incrementally by deleting newly allocated user-clients at the end point. Finally, we can obtain unallocated user-client in the ordered list. On the other side, by matching the maximum communication path dimension we have to assign a new user-client "c" to server "s". This is mainly done by Greedy Assignment algorithm.

3.3 Distributed-Modify Assignment Algorithm

The third algorithm is Distributed-Modify Assignment algorithm. At any single server without the global knowledge of the network is functioned in a wide manner. It is initiated with an initial allocation. For reducing maximum communication path is continuously modified until it cannot be reduced further. These all are done in a distributed server. Thus, this process is called as Assignment Modification algorithm or Distributed-Modify Assignment algorithm. In a distributed network, one server is selected or elected as a coordinator to calculate the distance of interaction path length. Further, the server is selected to perform assignment modification. Consider, the interaction path length as L to measure the initial allocation, each and every server measures its latencies to all servers in network to maximize more interactivity experience in DIAs. It also measures the distance of all the assigned clients and it is managed in a categorized list or sorted list. The coordinator calculates L based on the received message where each server "broadcasts messages to all servers and its longest span to all its allocated user-clients in network." Finally, it sends the inter-server measurement of path to the coordinator. The assignment modification in distributed server is done by the coordinator. To function assignment modification, the coordinator selects a server s which is involved in a longest interaction path is performed by current L value. Then, server s pursuit to reduce L by modifying the assignment of client "c" which introduces the protracted communication path. Initially, server s transmits to all servers the modification of client's "c" and protracted span to its allocated user-clients by ruling out "c". On accepting the message, each of the servers considered to be s measures its span to "c" and figure out the maximum dimension of the communication path associated with "c" is allocated to it. Finally, assignment modification is done by the coordinator in distributed server to maximize interactivity in DIAs.

3.4 Dealing with Server Capacity Limitations

We have not assumed any capacity constraints at the server-side of our proposed work. Every server location has sufficient server capability or server capability can be combined to this location is required. However, the server capacity at each location constrained may result in improving an increase in the execution delay at the server by contaminating the interactivity of the DIAs. Now we examine how to accommodate each propounded algorithm to deal with server quantity limitations.

- *Nearest-Server Assignment*: Client measures the network latency by itself and between all servers in network. Each client searches and select the nearest server and make the server to connect it independently. On the basis of *first-come-first-serve*, the client request is accepted by server until it is immersed. If the nearest server is immersed, the client searches for next nearest server and so on until the request send by the client is accepted by server.
- *Greedy Assignment*: In *Nearest-Server Assignment*, it first searches for saturated servers. But *Greedy assignment algorithm* searches for only unsaturated servers when selecting a pair of non-assigned client in network. The client is assigned to be unsaturated server for update of information. Thus, it starts with an empty assignment.
- *Distributed-Modify Assignment*: In each and every stage of alteration, the user-client is permitted to be reallocated to only immersed servers. Distributed-Modify Assignment goes on with allocated alterations until maximum communication path length cannot be reduced later. Finally, assignment modification is done by the coordinator in distributed server.

4 Conclusion

Here, we have found out the user-client allocation problem for interactivity improvement in endless DIAs. We need to model the completeness of interactivity under the basis of consistency and fairness. The small amount of obtainable communication time between user-clients is investigated and used as an optimization objective in continuous [13–15] client assignment problem. It is found to be NP-complete. Greedy Assignment completely exceeds Nearest-Server Assignment algorithm. Distributed-Modify Assignment algorithm deals with client participation and network latency in networks. Thus, fairness requirements and consistency are satisfied. Therefore, the objective of endless user-client allocated problem is to reduce the maximum length of communication path in networks. For heuristic algorithms, we can easily use the lengths of directed routing path between server and client without any change in algorithms. There may be a change in approximation ratios of algorithms. We allow the detailed analyses to forthcoming work.

References

1. Marzolla M, Ferretti S, D'Angelo G (2012) Dynamic resource provisioning for cloud-based gaming infrastructures. ACM Computers in Entertainment
2. Webb SD, Soh S, Lau W (2007) Enhanced mirrored servers for network games. In: Proceedings of sixth ACM SIGCOMM Workshop network and system support for games, pp 117–122

3. Raju R, Amudhavel J, Kannan N, Monisha M (2014) A bio inspired energy-aware multi objective chiropteran algorithm (EAMOCA) for hybrid cloud computing environment. International conference on green computing communication and electrical engineering (ICGCCEE) 2014, pp 1, 5, 6–8 Mar 2014. doi: 10.1109/ICGCCEE.2014.6922463
4. Amudhavel J, Vengattaraman T, Basha MSS, Dhavachelvan P Effective maintenance of Replica in distributed network environment using DST. International conference on advances in recent technologies in communication and computing (ARTCom) 2010, pp 252, 254, 16–17 Oct 2010. doi: 10.1109/ARTCom.2010.97
5. Mauve M, Vogel J, Hilt V, Effelsberg W (2004) LocalLag and timewarp: providing consistency for replicated continuous applications. IEEE Trans Multimedia 6(1):47–57
6. Dhavachelvan P, Uma GV (2003) Multi-agent based integrated Framework for intra class testing of object-oriented software. 18th ISCIS 2003, Springer Verlag—Lecture Notes in Computer Science (LNCS), vol 2869, pp 992–999. ISSN 0302-9743
7. Zhang L, Tang X (2012) Optimizing client assignment for enhancing interactivity in distributed interactive applications. IEEE/ACM Trans Net 20(6):1707–1720
8. Delaney D, Ward T, McLoone S (2006) On consistency and network latency in distributed interactive applications: a survey-part I. Presence: Teleoperators and Virtual Environ 15 (2):218–234
9. Planetlab All-Pairs-Pings (2013) http://pdos.lcs.mit.edu/strib/
10. Garey MR, Johnson DS (1979) Computers and intractability: a guide to the theory of NP-completeness. WH Freeman and Company, San Francisco
11. Raju R, Amudhavel J, Kannan N, Monisha M (2014) Interpretation and evaluation of various hybrid energy aware technologies in cloud computing environment—a detailed survey. International conference on green computing communication and electrical engineering (ICGCCEE) 2014, pp 1, 3, 6–8 Mar 2014. doi: 10.1109/ICGCCEE.2014.6922432
12. Qiu L, Padmanabhan VN, Voelker GM (2001) On the placement of web server replicas. In: Proceedings of IEEE INFOCOM '01, pp 1587–1596
13. Raju R, Amudhavel J, Pavithra M, Anuja S, Abinaya B (204) A heuristic fault tolerant MapReduce framework for minimizing makespan in hybrid cloud environment. International Conference on Green Computing Communication and Electrical Engineering (ICGCCEE) 2014, pp 1, 4, 6–8 Mar 2014. doi: 10.1109/ICGCCEE.2014.6922462
14. Jay C, Glencross M, Hubbold R (2007) Modeling the effects of delayed haptic and visual feedback in a collaborative virtual environment. ACM Trans Comput-Hum Interact 14(2):8
15. Dhavachelvan P, Uma GV (2004) Reliability enhancement in software testing: an agent-based approach for complex systems. 7th ICIT 2004, Springer Verlag—Lecture Notes in Computer Science (LNCS), vol 3356, pp 282–291. ISSN 0302-9743
16. Dhavachelvan P, Uma GV, Venkatachalapathy VSK (2006) A new approach in development of distributed framework for automated software testing using agents. Int J Knowl–Based Syst 19(4):235–247, Aug 2006
17. Brun J, Safaei F, Boustead P (2006) Managing latency and fairness in networked games. Comm ACM 49(11):46–51

Comparison and Analysis of Fuzzy Methods—TOPSIS, COPRAS-G, ANP, ELECTRE and AHP Using MATLAB

S. Kumarakrishsnan, K. Devika P. Kumar, S. Sreelakshmi, S. Sevvanthi, B. Asha, Ramachandiran and D. Arvind Kumar

Abstract Fuzzy logic is a multi-valued logic which is obtained from fuzzy set theory. It deals with human resourcing which is approximate than 'very unlikely', when compared to the traditional binary sets the truth value ranges in degree between 0 and 1. The main concept of fuzzy logic is that it provides a practical way of approach to automate the complex data analysis, fusion and inference processes which is usually performed by human experts with extensive experiences. COPRAS-G and COPRAS (Venkatesan et al. in Performance analysis of mobile agent failure recovery in e-service applications 32(1–2):38–43, 2005, Licata in Employing fuzzy logic in the diagnosis of a clinical case, 2010) decision-making methods are separately used in various applications to illustrate the approximate evaluation of attributes. This helps in selecting the best thing. It can be used to select best websites, company, project manager, etc. It can also be used to select the location for constructing warehouse, underground dam, etc. We propose the case study to compare both COPRAS and COPRAS-G (Aruldoss et al. in A survey on multi criteria decision making methods and its applications, 2013, Dhavachelvan et al. in A

S. Kumarakrishsnan (✉) · K. Devika P. Kumar · S. Sreelakshmi · S. Sevvanthi · B. Asha
D. Arvind Kumar
Department of Computer Science and Engineering, SMVEC, Pondicherry, India
e-mail: skumarakrishnan@outlook.com

K. Devika P. Kumar
e-mail: devikapkumar7@gmail.com

S. Sreelakshmi
e-mail: sree.pdy94@gmail.com

S. Sevvanthi
e-mail: sevvanthijaya@gmail.com

B. Asha
e-mail: ashababu94@gmail.com

D. Arvind Kumar
e-mail: arvind.patricks@gmail.com

Ramachandiran
Department of MCA, SMVEC, Pondicherry, India
e-mail: ramachandiran08@gmail.com

L.P. Suresh and B.K. Panigrahi (eds.), *Proceedings of the International Conference on Soft Computing Systems*, Advances in Intelligent Systems and Computing 398, DOI 10.1007/978-81-322-2674-1_53

new approach in development of distributed framework for automated software testing using agents, 2006) decision making fuzzy methods for finding the rate of diseases occurring to children, teenagers and old aged people. That is, what are the diseases may occur at certain age can be accurately predicted through this project. Two methods Simplified COPRAS and COPRAS-G are implemented with the same input to obtain the best and the worst alternative. The ranking of the Simplified COPRAS-G and COPRAS method are compared by the parameters. Each parameter is applied in both the method in order to find the better method.

Keywords COPRAS · COPRAS-G · Rating · Computation

1 Introduction

In today's technologies fuzzy logic is one of the most successful techniques for developing difficult control systems. The main reason for that is, it is very simple in working processes. Fuzzy logic resembles few applications such as human decision-making with an ability to generate the exact solutions for appropriate information. By using the mathematical approaches it fills the vacant engineering design methods like linear control design and expert system in system design. While the other approaches bring out the accurate equations in real-world behaviours, a fuzzy design can understand the human language and logic. Fuzzy logic provides both the methods for defining human terms and also automates the conversion of the system specifications into effective models.

At first they developed the fuzzy theory which was primarily industrial, such as process control for cement kilns [1, 2]. As further developments were made the fuzzy logic to be used in more useful applications. There are also various kinds of fuzzy controllers used which make the subway journey more easy and comfortable with smooth braking and smooth acceleration. Normally, the driver's work is to push the start button. It is also used in elevators in order to reduce the waiting time. After then fuzzy logic methods have been extensively used in a wide range.

Consider an example for fuzzy system which is nothing other than the fuzzy washing machine. When clothes are loaded in it and start button is pressed and then the machine starts to spin automatically choosing the best cycle. There are even many methods present here such as fuzzy microwave, place chilli, potatoes, etc., in the fuzzy oven method, add the material to be cooked or heated and then just press a single button it automatically cooks the food at appropriate time in appropriate temperature and then finishes at the right time. There is also another method in fuzzy like fuzzy car which is used to do the work of a driver by using simple verbal instructions from the driver. Here it can even stop in between when obstacles occur with the help of sensors. In practice, the exciting part of fuzzy application is its simplicity in operation. Fuzzy logic is the superset of Boolean logic [3] where the truth value lies between 'completely true' and 'completely false'. Here for fuzzy the values are not exact but are appropriate which is defined in its name.

1.1 COPRAS

The COPRAS method assumes the direct and proportional dependence and priority of alternatives on system for ranking the investigated alternatives. To transform the performance to dimensionless value COPRAS requires normalization procedure. Overview of multi-criteria and normalization procedures are discussed below. Using COPRAS method, the significance and priorities of alternatives are determined which is expressed precisely using four steps.

Stage 1 This is the first step which discusses about the normalized decision-making technique where matrix D is constructed. In MCDM methods each criteria has different units for measuring them. By using the normalization procedure the dimensionless values are obtained from the performance of chosen alternatives. The formula used for normalization is:

$$\tilde{x}_{ij} = \frac{x_{ij}}{\sum_{i=1}^{m} x_{ij}} \tag{1}$$

where x_{ij} refers the performance of the ith alternative with respect to the jth criterion, $\tilde{x}_{ij}$ is its normalized value and m is number of alternatives.

Stage 2 It involves adding the normalized criteria to evaluate the weighted normalized criteria which describes the ith alternative. For COPRAS method each and every alternative is described by summing its maximizing attributes S_{+j} and S_{-j},, i.e., minimizing the criteria and maximizing the optimization direction. To simplify the calculation in decision-making matrix, first of all place the maximum value and then place the minimum value where the criteria specified are S_{+i} and S_{-i}. Both these values are calculated using the formula:

$$S_{-i} = \sum_{j-k+1}^{n} \tilde{x}_{ij} \cdot q_{j-}. \tag{2}$$

Stage 3 The relative weight to calculate the alternatives are evaluated in this step. The relative weight Q_i can be calculated for each alternative as follows:

$$Q_i = S_{+i} + \frac{\min_i S_{-i} \sum_{i=1}^{m} S_{-i}}{S_{-i} \sum_{i=1}^{m} \frac{\min_i S_{-i}}{S_{-i}}}. \tag{3}$$

The above formula can be simplified as

$$Q_i = S_{+i} + \frac{\sum_{i=1}^{m} S_{-i}}{S_{-i} \sum_{i=1}^{m} \frac{1}{S_{-i}}}. \tag{4}$$

Stage 4 The priority order of alternatives is determined. The priority is determined on the basis of relative weight of the alternatives. The alternative with higher relative weight has higher rank. The high-ranked alternative is the most acceptable alternative.

$$A^* = \{A_i | \max_i Q_i\}. \tag{5}$$

This method can be easily applied for selecting the most efficient alternative or solution. In real world many decisions are made which are not precisely known.

There are many parameters used in COPRAS and COPRAS-G methods, they are:

- Time complexity
- Space complexity
- Rank reversal
- Sensitivity analysis.

1.2 Time Complexity

Time complexity can be defined as the amount of time taken by an algorithm to complete the execution. This algorithm runs as a function by representing string as input. It excludes the coefficients and the lower order terms, which are asymptotical. The main purpose of the time complexity is to count the number of operations that are performed by an algorithm only when it takes a lot of time. The constant factor varies for the operations performed by the algorithm and the time taken for execution also varies.

The time varies for every different inputs having the same size, the worst-case time complexity is mainly defined in the commonly used algorithm and defines the maximum amount of time taken for the input size.

1.3 Space Complexity

The space complexity algorithm is used to determine the amount of storage necessary for execution. Many space complexity algorithms are developed to work with arbitrary length as input to the algorithm. Basically, the efficiency of an algorithm is defined as the function relating to the time complexity or space complexity. In many places, the term space complexity is misused as auxiliary space. The correct definition for auxiliary space and space complexity are explained as follows. Space complexity is defined as total time required for an algorithm with respect to the size of the input. Auxiliary space is defined as the extra space used by the algorithm.

The term space complexity is actually misused for auxiliary space at many places, where the auxiliary space means the extra space or the temporary space used by an algorithm. It contains both the auxiliary space and simple space used by the input. The space complexity of an algorithm is the total time taken by the algorithm with respect to the input size.

2 Existing Work

Enormous studies are conducted for finding the best alternative selection with boundaries of criteria. In 1991, Gargano et al. combined artificial neural network and genetic algorithm for performing best alternative (hospital) selection. In this analysis, basic criteria like hospital facilities, location of the hospital, cost of operation, doctor's experience, number of operation, patient preference, pharmacy, food, special care for the patient and equipment for operations are considered. In 1993, Miller and Feinzig proposed that fuzzy set theory is the best alternative selection method. In 1994, Liang and Wang suggested an algorithm, it also uses the fuzzy set theory for its operation.

In this algorithm, subjective criteria, such as hospital facilities, doctor's experience, location of the hospital along with some objective criteria, such as, pharmacy, food were made use of Karsak et al. (2003) modelled hospital selection process by using fuzzy criteria programming and evaluates the qualitative and quantitative factors for evaluation.
A new model was developed by Capaldo and Zollo in the year 2001 for the purpose of improving the effectiveness of hospital selection process. The first and the foremost study made is to develop the decision formulations and decision samples which are based on the evaluation of the hospitals.

Second, we have built an evaluation method for hospital selection by utilizing the fuzzy logic. For the hospital selection we consider the ten criteria which are necessary for revaluating them. Hospital selection uses multi-criteria analyses which are described in the literature (Bohanec et al. 1992; Timmermans and Vlek 1992, 1996; Gardiner and Armstrong-Wright 2000; Spyridakos et al. 2001; Jessop 2004). Above methods are effectively used in evaluating the multitude of factors. For large and complicated problems, Roth and Bobko (1997) identified some issues in using multi-attribute methods. In 1998, Hooper et al. introduced an expert system named as BOARDEX. This expert systems are used by American army for its personnel selection. The personnel selection factors include grade, civilian education level, military education level, weight, height, and assignment history are provided as input to this expert system.

3 Proposed Work

The traditional MCDM methods require the precise data, i.e., its rating performance of the alternatives and the criterion weights must be precise. However, for solving many real-world problems few predictions are required and it is not possible to obtain the precise data, but it is necessary for using the traditional MCDM methods. The COPRAS-G method is an extension of the MCDM method.

The system which we created is an implementation of real-time environment. In our day-to-day life, so many decisions are being made from various criteria, so the decision can be made by providing weights to different criteria and all the weights are obtained from expert groups. It is important to determine the structure of the problem and explicitly evaluate multi-criteria. For example, in building a nuclear power plant, certain decisions have been taken based on different criteria. There are not only very complex issues involving multi-criteria, some criteria may have effect towards some problem, but overall to have an optimum solution, all the alternatives must have common criteria which clearly lead to more informed and better decisions.

This graph Fig. 1 explains about the comparison made between the COPRAS-G and COPRAS, where the comparative results show the various results displayed by applying the COPRAS-G and COPRAS algorithm. By using these two algorithms an effective way of ranking the hospitals can be done by analysing and verifying the results given by the users to the hospitals. Here various hospitals are taken for sample and the ranking system is shown. Likewise how they rank and according to which aspects they are ranged. The coding for the ranking of hospital is generated below:

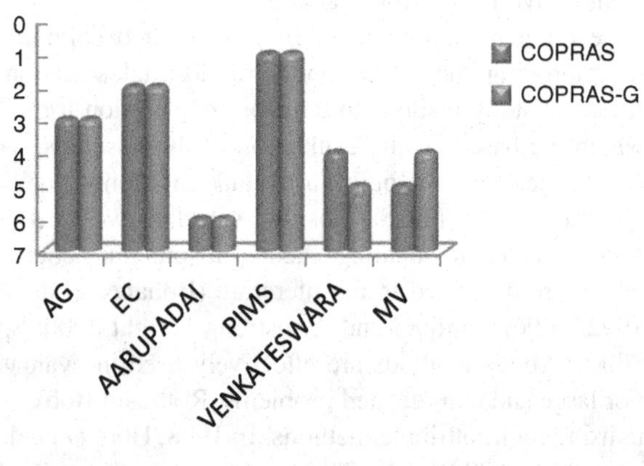

Fig. 1 Comparison between COPRAS-G and COPRAS

CODING

```
function[A,B,weight,Upper,Lower,result,result1,res,NUpper,NLower,Cex,dex,pj,rj,
pj1,rj1,nj] = COPRASg(Upper,Lower)
tic
A = [1,5,7,7,1/3,1/3,1/3,7,1/3,5; 1/5,1,1/3,1/3,5,1/3,1/3,5,5,5; 1/7,3,1,1/3,7,9,9,7,5,
1/3;1/7,3,3,1,7,5,1/3,1/3,9,7;3,1/5,1/7,1/7,1,1/3,5,9,1/3,1/3; 3,3,1/9,1/5,3,1,1/3,7,5,
5; 3,3,1/9,3,1/5,3,1,7,1/3,1/3; 1/7,1/5,1/7,3,1/9,1/7,1/7,1,9,9; 3,1/5,1/5,1/9,3,1/5,3,
1/9,1,5; 1/5,1/5,3,1/7,3,1/5,3,1/9,1/5,1];
B = bsxfun(@rdivide,A,sum(A));
A = sum(A);
B = B';
weight = sum(B)/length(B);
fprintf('\nImportance weight of each criterion:\n');
disp(weight);
fprintf('Upper limit matrix:\n');
disp(Upper);
fprintf('Lower limit matrix:\n');
disp(Lower);
result = (Lower-2*Lower);
result1 = (Upper-2*Upper);
res = 2*(Lower + Upper);
res = 0.5*res;
NLower = result./res;
NUpper = result1./res;
fprintf('Weighted Normalized Decision Matrix for Lower Limit Matrix:\n');
disp(NLower);
fprintf('Weighted Normalized Decision Matrix for Upper Limit Matrix:\n');
disp(NUpper);
disp(weight);
for ix = 1:6;
forjx = 1:10;
Cex(ix,jx) = NLower(ix,jx) * weight(1,jx);
end
end
fprintf('\nWeighted Normalized Values for Lower Limit Matrix:\n');
disp(Cex);
for ix = 1:6;
forjx = 1:10;
dex(ix,jx) = NUpper(ix,jx) * weight(1,jx);
end
end
fprintf('Weighted Normalized Values for Upper Limit Matrix:\n');
disp(dex);
rj = Cex';
```

```
pj = dex';
disp(rj);
disp(pj);
pj = sum(pj);
rj = sum(rj);
fprintf('Sum of criterion values(Smaller value):\n');
disp(rj);
fprintf('Sum of criterion values(Larger value):\n');
disp(pj);
pj1 = max(pj);
rj1 = min(rj);
qj = rdivide(rj1,rj);
qj = rdivide(rj,qj);
qj = pj1 + qj;
fprintf('Value of Qj:\n');
disp(qj);
nj = -(qj*100);
fprintf('Final Ranking:\n');
display(nj);
whos
toc
end
```

Figure 2 explains about the differences in time and space complexity done to the COPRAS-G and COPRAS. While comparing with the COPRAS the best method is COPRAS-G, where it reduces the time complexity and also improves the efficiency in ranking the hospitals.

Here in the tabular column we describe how the methods are split according to the time complexity with respect to the COPRAS and COPRAS-G.

MCDM methodology	Time complexity (sec)
COPRAS	0.007877
COPRAS-G	0.045507

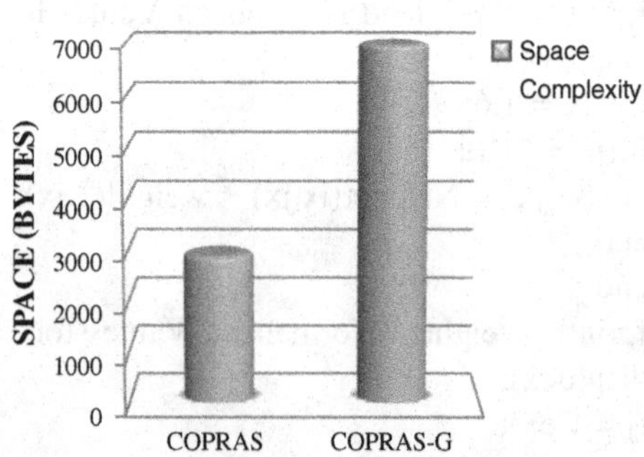

Fig. 2 Differences in time and space complexity

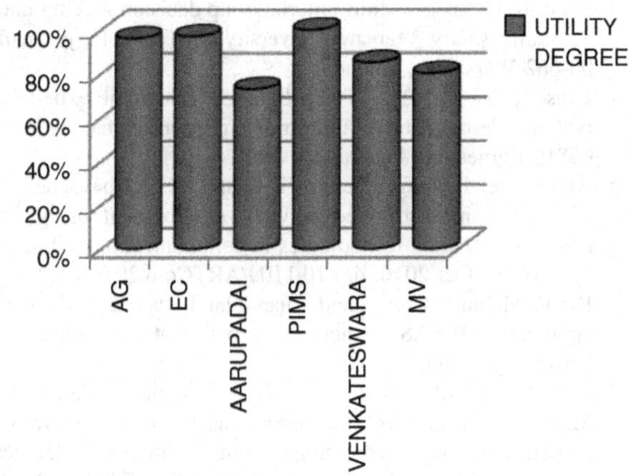

Fig. 3 Ranking chart

4 Implementation

The coding used to generate the graph using the matrix is shown above. Here we can say that there are two different types of matrix present in this, they are higher matrix and the lower matrix. The matrix is based upon the ranking criteria where the ranking method is used with the 5 ratings, where the first 3 rates have the lower matrix and the last two rates define the higher matrix. Then there is another main matrix which is said to be the decision matrix which is calculated on the basis of the formula, where the formula is based on the weights in which weights depend on the user review about the project. In order to generate the graph we must have both the lower and the matrix.

Here in Fig. 3 the ranking is done with ten criteria, they are location, operation, cost, facility, pharmacy, food, patient's suggestions, special care, equipment, and the most important and the needed criteria is the doctor's experience. These are only the few criterions used to generate the rank of the hospital. When the high ranking is done then the rate of the hospital gets increased, so by seeing the rating the patient can decide to choose the hospital accordingly with respect to the cost, facility, or doctor's experience. By doing this, patients will know the hospitals' status and the hospital can be chosen according to their willingness.

References

1. International Journal of Engineering Trends and Technology (IJETT)—Volume 4 Issue 6-June 2013 Detection of Heart Diseases using Fuzzy Logic Sanjeev Kumar, ursimranjeet Kaur Asocc. Prof., Department of EC, Punjab Technical University ACET, Amritsar, Punjab India
2. Liao H, Xu Z, Zeng X-J (2014) Distance and similarity measures for hesitant fuzzy linguistic term sets and their application in multi-criteria decision making. Info Sci 271

3. Skorupski J (2014) Multi-criteria group decision making under uncertainty with application to air traffic safety Warsaw University of Technology, faculty of transport, Koszykowa 75, 00-662 Warszawa, Poland
4. TanselIç Y (2014) A TOPSIS based design of experiment approach to assess company ranking, department of industrial engineering, faculty of engineering, Baskent University, 06810 Etimesgut, Ankara, Turkey
5. Amudhavel J, Vengattaraman T, Basha MSS, Dhavachelvan P (2010) Effective maintenance of Replica in distributed network environment using DST. International conference on advances in recent technologies in communication and computing (ARTCom) 2010, pp 252, 254, 16–17 Oct 2010. doi:10.1109/ARTCom.2010.97
6. Bindu Madhuri Ch, Anand Chandulal J, Padmaja M (2010) Selection of best web site by applying COPRAS-G method. Department of computer science and engineering, Gitam University, India
7. Raju R, Amudhavel J, Pavithra M, Anuja S, Abinaya B () A heuristic fault tolerant MapReduce framework for minimizing makespan in hybrid cloud environment. International conference on green computing communication and electrical engineering (ICGCCEE) 2014, pp 1, 4, 6–8 Mar 2014. doi:10.1109/ICGCCEE.2014.6922462
8. Junior FRL, Osiro L, Carpinetti LCR (2014) A comparison between Fuzzy AHP and Fuzzy TOPSIS methods to supplier selection, School of Engineering of São Carlos, University of São Paulo, Production Engineering Department, São Carlos, SP, Brazil. Federal University of TrianguloMineiro (UFTM), Production Engineering Department, Uberaba, MG, Brazil
9. Raju R, Amudhavel J, Kannan N, Monisha M (2014) A bio inspired energy-aware multi objective chiropteran algorithm (EAMOCA) for hybrid cloud computing environment. International conference on green computing communication and electrical engineering (ICGCCEE) 2014, pp 1, 5, 6–8 Mar 2014. doi:10.1109/ICGCCEE.2014.6922463
10. Miranda Lakshmi T, Prasanna Venkatesan V (2014) A comparison of various normalization in techniques for order performance by similarity to ideal solution (TOPSIS)
11. Raju R, Amudhavel J, Kannan N, Monisha M () Interpretation and evaluation of various hybrid energy aware technologies in cloud computing environment—a detailed survey. International conference on green computing communication and electrical engineering (ICGCCEE) 2014, pp 1, 3, 6–8 Mar 2014. doi:10.1109/ICGCCEE.2014.6922432
12. Inzinerine Ekonomika-Engineering Economics (2011) The comparative analysis of MCDA methods SAW and COPRAS' Valentinas Podvezko Vilnius Gediminas Technical University Sauletekio av. 11, LT-2040 Vilnius, Lithuania
13. Vengattaraman T, Abiramy S, Dhavachelvan P, Baskaran R (2011) An application perspective evaluation of multi-agent system in versatile environments. Int J Expert Syst Appl, Elsevier 38 (3):1405–1416
14. Tavana M, Momeni E, Rezaeiniya N, Mirhedayatian SM, Rezaeiniyab H (2012) A novel hybrid social media platform selection model using fuzzy ANP and COPRAS-G'. Business systems and analytics department, lindback distinguished chair of information systems and decision sciences, Expert Syst Appl 40
15. Venkatesan S, Dhavachelvan P, Chellapan C (2005) Performance analysis of mobile agent failure recovery in e-service applications. Int J Comput Stand Interfaces, 32(1–2):38–43. ISSN 0920-5489
16. Licata G (2010) Employing fuzzy logic in the diagnosis of a clinical case Dipartimento FIERI, University of Palermo, Viale delle Scienze, Palermo, Italy; it received 20 Oct 2009; revised 9 Dec 2009; accepted 14 Dec 2009
17. Aruldoss M, Miranda Lakshmi T, Prasanna Venkatesan V (2013) A survey on multi criteria decision making methods and its applications. Am J Info Syst
18. Dhavachelvan P, Uma GV, Venkatachalapathy VSK (2006) A new approach in development of distributed framework for automated software testing using agents. Int J Knowl–Based Syst, Elsevier, 19(4):235–247, Aug 2006

Enhanced and Secure Query Services in Cloud with Perfect Data Privacy for the Data Owner

B. Thiyagarajan, M.R. Gokul, G. Hemanth Kumar, R. Rahul Prasanth, M. Shanmugam, N. Pazhaniraja, K. Prem Kumar and V. Vijayakumar

Abstract Data mining is the process of analyzing data from different perspectives and summarizing it into useful information. It allows users to learn data from various different levels or angles, categorize it and summarize the associations identified. Increasing data intensity in cloud may also need to improve scalability and stability. Sensitive data must be preserved securely on violating the environment. This research work presents to secure the queries and to retrieve data to the client in an efficient manner. In this proposed work, two schemes, random space perturbation (RASP) and Advanced Encryption Standard (AES) are used for the purpose of initialization and to encrypt the data. Finally, Dijkstra's algorithm is applied to calculate the distance between the nearby objects in the database. Comparing to K-NN, Dijkstra's algorithm provides exact results, so its performance

---

B. Thiyagarajan (✉) · M.R. Gokul · G. Hemanth Kumar · R. Rahul Prasanth · K. Prem Kumar · V. Vijayakumar
Department of CSE, SMVEC, Pondicherry, India
e-mail: thiyagarajan0484@gmail.com

M.R. Gokul
e-mail: gokul.vasanth@gmail.com

G. Hemanth Kumar
e-mail: praveenhemanthkumar@gmail.com

R. Rahul Prasanth
e-mail: raahul.prasanth@gmail.com

K. Prem Kumar
e-mail: premkvpt@gmail.com

V. Vijayakumar
e-mail: vijayakumarv@smvec.ac.in

N. Pazhaniraja
Department of IT, SMVEC, Pondicherry, India
e-mail: pazhanibit@gmail.com

M. Shanmugam
Department of CSE, Pondicherry University, Pondicherry, India
e-mail: maddy.shan@gmail.com

L.P. Suresh and B.K. Panigrahi (eds.), *Proceedings of the International Conference on Soft Computing Systems*, Advances in Intelligent Systems and Computing 398, DOI 10.1007/978-81-322-2674-1_54

is highly accurate. In order to decrease the computational cost inherent to process the encrypted data, consider the case of incrementally updating datasets in the entire progress.

Keywords Cloud data · RASP · AES · Dijkstra's algorithm · Encryption

1 Introduction

Authentication on location-based services contributes major process of extracting queries onto the client. The name of protection in third party side violation has to be completely avoided. Finding location-based spatial data [1, 7,9, 10] with distances can be progressed through data mining technique [10, 11] (classification and regression). Classification is the process of categorizing the data and regression is the verdict relationship among the data. To address the user privacy needs, several protocols have been proposed that withhold, either partially or completely the user's location information from the LBS. For instance, the work in larger cloaking regions that is meant to prevent disclosure of exact user whereabouts. Nevertheless, the LBS can still derive sensitive information from the cloaked regions, so another line of research that uses cryptographic strength protection was started in and continued. The main idea is to extend existing private information retrieval (PIR) protocols for binary sets to the spatial domain, and to allow the LBS to return the NN to users without learning any information about users' locations. This method serves its purpose well, but it assumes that the actual data points (i.e., the points of interest) are available in plaintext to the LBS. This model is only suitable for general-interest applications such as Google Maps, where the landmarks on the map represent public information, but cannot handle scenarios where the data points must be protected from the LBS itself. More recently, a new model for data sharing emerged, where various entities generate or collect datasets of POI that cover certain niche areas of interest, such as specific segments of arts, entertainment, travel, etc. For instance, there are social media channels that focus on specific travel habits, e.g., ecotourism, experimental theater productions or underground music genres.

The content generated is often geotagged, for instance related to upcoming artistic events, shows, travel destinations, etc. However, the owners of such databases are likely to be small organizations, or even individuals, and not have the ability to host their own query processing services. This category of data owners can benefit greatly from outsourcing their search services to a cloud [1, 4] service provider. In addition, such services could also be offered as plug-in components within social media engines operated by large industry players. Due to the specificity of such data, collecting [15] and maintaining such information is an expensive process, and furthermore, some of the data may be sensitive in nature. For instance, certain activist groups may not want to release their events to the general public due to concerns that big corporations or oppressive governments may intervene and compromise their activities.

Similarly, some groups may prefer to keep their geotagged datasets confidential, and only accessible to trusted subscribed users, for the fear of backlash from more conservative population groups. It is therefore important to protect the data [9, 11] from the cloud [6, 13] service provider. In addition, due to financial considerations on behalf of the data owner, subscribing users will be billed for the service based on a pay-per-result model. For instance, a subscriber who asks for NN results will pay for specific items, and should not receive more than k results. Hence, approximate querying methods with low precision, such as existing techniques that return many false positives in addition to the actual results, are not desirable. Specifically, both the POI and the user locations must be protected from the cloud [13] provider. This model has been formulated previously in literature as "blind queries on confidential data [1]". In this context, POIs must be encrypted by the data owner, and the cloud service [4, 6] provider must perform NN processing on encrypted data.

To address this problem, previous work has proposed privacy-preserving data transformations that hide the data, while still allowing the ability to perform some geometric functions evaluation. However, such transformations lack the formal security guarantees of encryption [3, 5, 8]. Other methods employ stronger security transformations, which are used in conjunction with dataset partitioning techniques, but return a large number of false positives, which is not desirable due to the financial considerations outlined earlier. Finding location-based spatial data with distances can be progressed through data mining technique [2, 3] (classification and regression) Classification is the process of categorizing the data and regression is finding relationship among the data. Shortest path algorithm is the only solution to reduce the time for data extraction.

2 Proposed Work

This research work presents to secure the queries and to retrieve data to the client in an efficient manner (Fig. 1).

2.1 Working Steps

Step 1: Input initialization; spatial input will get initialize with storing all attributes inside the database.

Step 2: Perturbation work; extracting coordinates for finding shortest distance from one node to another node, edges and vertex gathered will lead to estimate.

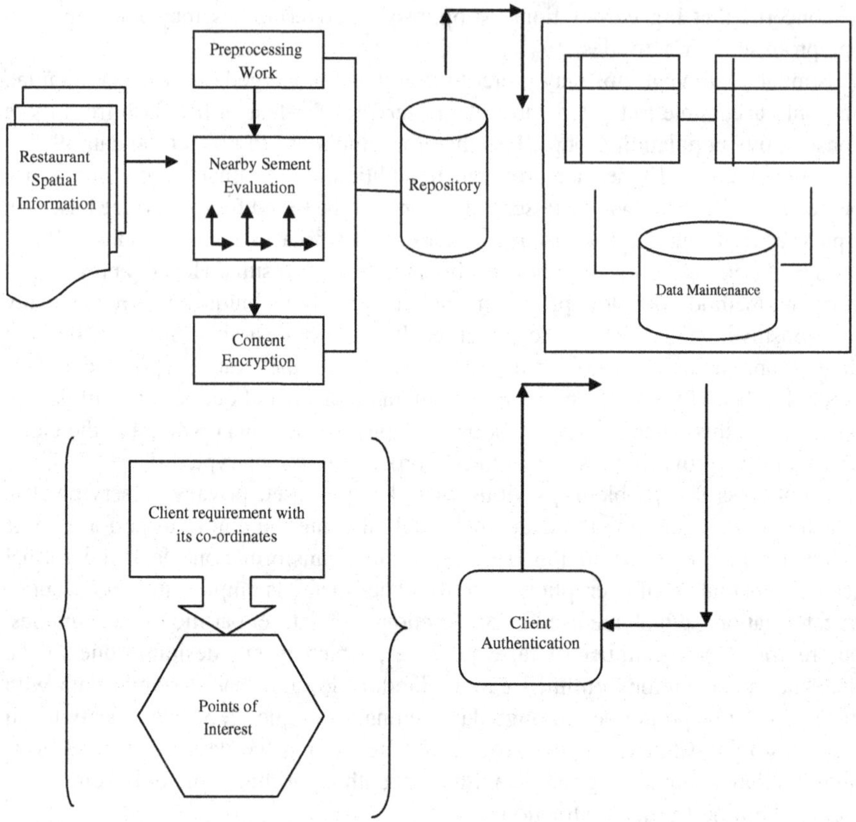

Fig. 1 System architecture

Step 3: Shortest distance values for each node will get updated into database. Entire information will be encrypted by using Advanced Encryption Standard (AES).

Encrypt(plaintext[n])
Add-Round-Key (state, round-key [0]);
For i = 1 to Nr − 1 step-size 1 do
Sub-Bytes (state);
Shift-Rows (state);
Mix-Columns (state);
Add-Round-Key (state, round-key[i]);
Update ()

Step 4: Outsourcing; encrypted values will send over to cloud side untrusted area. Cloud will maintain that information in a separate database.

Step 5: Service provision can be processed by cloud to the required client.

Step 6: After the client authentication service has been received by client, points of interest can be considered as client request which will be given as query.

Step 7: Query has been decrypted by requested client at the end.

2.2 RASP: Random Space Perturbation

Random space perturbation (RASP) is one type of multiplicative perturbation, with a novel combination of OPE, dimension expansion, random noise injection, and random projection. Let us consider the multidimensional data are numeric and in multidimensional vector space. The database has k searchable dimensions and n records, which makes a $d \times n$ matrix X. The searchable dimensions can be used in queries and thus should be indexed. Let x represent a d-dimensional record, $x \mathcal{E} \mathrm{IR}^d$. Note that in the d-dimensional vector space IR^d, the range query conditions are represented as half-space functions and a range query is translated to find the point set in corresponding polyhedron area described by the half-spaces. The RASP perturbation involves three steps. Its security is based on the existence of random invertible real-value matrix generator [12, 14, 16] and random real-value generator. In this work RASP mainly used for initialization of building a block.

2.3 Dijkstra's Algorithm

2.3.1 Basic Principles of Dijkstra's Algorithm

Currently, the best algorithm to find the shortest path between two points is publicly known as Dijkstra's algorithm which is proposed by Dijkstra in 1959. It can not only get the shortest path between the starting point and the final destination, but also can find the shortest path of each vertex from the starting point.

Step 1: Initialization.

$$\begin{aligned} V &= \{1, 2, 3, \ldots, N\}, \\ S &= \{F\}, \\ D[I] &= L[F, I], Y[I] = F, I = 1, 2, \ldots, N \end{aligned}$$

F is the path starting point. I is one of the vertexes. N is the number of all vertices in the network. V is the set of all the vertices. $L[F, I]$ is the distance between the vertex F and vertex I. S is the set of vertices. D is an array of N elements which is used to store the shortest distance from vertex F to other vertexes. Y is an array of N elements which is used to store the nearest vertex before vertex I in the shortest path.

Step 2: Find a vertex T from the $V - S$ set and make $D[T]$ be the minimum value, then add T into S.
if the $V - S$ is empty set, the algorithm is over.

Step 3: Adjust the value of array Y and array D. For the each adjoining vertex of vertex T in
the $V - S$ set,
If $D[I] > D[T] + L[I, T]$
$Y[I] = T, D[I] = D[T] + L[I, T]$ then let:

Step 4: Go to step 2.

A. ALGORITHM PSEUDO CODE

Function Dijkstra(*Graph*, *source*):
dist[*source*] ← 0
//Distance from source to source
prev[*source*] ← undefined
//Previous node in optimal path initialization
for each vertex *v* in *Graph*:
//Initialization
If *v* ≠ *source*
//Where v has not yet been removed from Q (unvisited nodes)
dist[*v*] ← infinity
//Unknown distance function from source to v
prev[*v*] ← undefined
//Previous node in optimal path from source
end if
add *v* to *Q*
//All nodes initially in Q (unvisited nodes)
end for
while *Q* is not empty:
u ← vertex in *Q* with min dist[u]
//Source node in first case
remove *u* from *Q*
for each neighbor *v* of *u*:
//where v has not yet been removed from Q.
alt ← dist[*u*] + length(*u*, *v*)
if *alt* < dist[*v*]:
//A shorter path to v has been found
dist[*v*] ← *alt*
prev[*v*] ← *u*
end if
end for
end while
return dist[], prev[]
end function

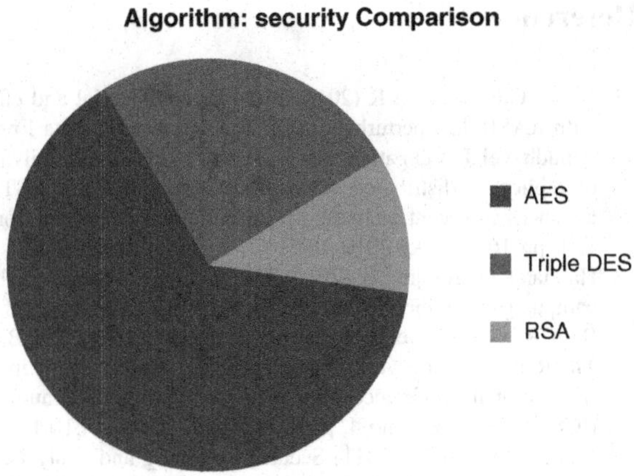

Fig. 2 Results of security comparison

2.3.2 Reward of Proposed Work

- Improve the performance of query processing for both range queries and shortest path process.
- Formally analyzing the leaked query and access patterns.
- Our proposing RASPs' possible effect on both data and query confidentiality.
- It will reduce the verification cost for serviced clients.
- Complete avoidance of security violation from provider side.

3 Experimental Result

To assess the efficiency of the proposed approach, we have made both qualitative (visual) and quantitative analysis of the experimental results (Fig. 2).

4 Conclusion

In this paper, proposed two schemes to support RASP data perturbation: Dijkstra's algorithm with encryption algorithm had proposed. They both use mutable order-preserving encoding (mOPE) as building block. Dijkstra's provides exact results, but its performance overhead may be high. K-NN only offers approximate NN results, but with better performance. In addition, the accuracy of k-NN is very close to that of the exact method. Planning to investigate ore complex secure evaluation functions on ciphertexts, such as skyline queries. And also research formal security protection guarantees against the client to prevent it from learning anything other than the received k query results.

References

1. Xu H, Guo S, Chen K (2014) Building confidential and efficient query services in the cloud with RASP data perturbation. IEEE Trans Knowl Data Eng 26(2)
2. Amudhavel J, Vengattaraman T, Basha MSS, Dhavachelvan P (2010) Effective maintenance of replica in distributed network environment using DST. In: International conference on advances in recent technologies in communication and computing (ARTCom), vol 252, no 254, pp 16–17, Oct 2010. doi:10.1109/ARTCom.2010.97
3. Haixiang D, Jingjing T (2013) The improved shortest path algorithm and its application in campus geographic information system. J Convergence Inf Technol 8(2)
4. Raju R, Amudhavel J, Pavithra M, Anuja S, Abinaya B (2014) A heuristic fault tolerant MapReduce framework for minimizing make span in hybrid cloud environment. In: International conference on Green Computing Communication and Electrical Engineering (ICGCCEE), vol 1, no 4, pp 6–8, March 2014. doi:10.1109/ICGCCEE.2014.6922462
5. Bau J, Mitchell JC (2011) Security modeling and analysis. IEEE Secur Priv 9(3):18–25
6. Raju R, Amudhavel J, Kannan N, Monisha M (2014) A bio inspired energy-aware multi objective Chiropteran algorithm (EAMOCA) for hybrid cloud computing environment. In: International conference on Green Computing Communication and Electrical Engineering (ICGCCEE), vol 1, no 5, pp 6–8, Mar 2014. doi:10.1109/ICGCCEE.2014.6922463
7. Boyd S, Vandenberghe L (2004) Convex optimization. Cambridge University Press, Cambridge
8. Cao N, Wang C, Li M, Ren K, Lou W (2011) Privacy-preserving multi-keyword ranked search over encrypted cloud data. In: Proceeding of IEEE INFOCOMM
9. Chen K, Kavuluru R, Guo S (2011) RASP: efficient multidimensional range query on attack-resilient encrypted databases. In: Proceedings of ACM conference data and application security and privacy, pp 249–260
10. Chen K, Liu L (2011) Geometric data perturbation for outsourced data mining. Knowl Inf Syst 29:657–695
11. Chen K, Liu L, Sun G (2007) Towards attack-resilient geometric data perturbation. In: Proceedings of SIAM international conference data mining
12. Chor B, Kushilevitz E, Goldreich O, Sudan M (1998) Private information retrieval. ACM Comput Surv 45(6):965–981
13. Raju R, Amudhavel J, Kannan N, Monisha M (2014) Interpretation and evaluation of various hybrid energy aware technologies in cloud computing environment—a detailed survey.In: International conference on Green Computing Communication and Electrical Engineering (ICGCCEE), vol 1, no 3, pp 6–8, Mar 2014. doi:10.1109/ICGCCEE.2014.6922432
14. Vengattaraman T, Abiramy S, Dhavachelvan P, Baskaran R (2011) An application perspective evaluation of multi-agent system in versatile environments. Int J Expert Syst Appl Elsevier 38 (3):1405–1416
15. Venkatesan S, Dhavachelvan P, Chellapan C (2005) Performance analysis of mobile agent failure recovery in e-service applications. Int J Comput Stand Interfaces Elsevier 32(1–2):38–43. ISSN:0920-5489
16. Dhavachelvan P, Uma GV, Venkatachalapathy VSK (2006) A new approach in development of distributed framework for automated software testing using agents. Int J Knowl Based Syst Elsevier 19(4):235–247

Forecasting the Stability of the Data Centre Based on Real-Time Data of Batch Workload Using Times Series Models

R. Vijay Anand, P. Bagavathi Sivakumar and Dhan V. Sagar

Abstract Forecasting has diverse range of applications in many fields like weather, stock market, etc. The main highlight of this work is to forecast the values of the given metric for near future and predict the stability of the Data Centre based on the usage of that metric. Since the parameters that are being monitored in a Data Centre are large, an accurate forecasting is essential for the Data Centre architects in order to make necessary upgrades in a server system. The major criteria that result in SLA violation and loss to a particular business are peak values in performance parameters and resource utilization; hence it is very important that the peak values in performance, resource and workload be forecasted. Here, we mainly concentrate on the metric batch workload of a real-time Data Centre. In this work, we mainly focused on forecasting the batch workload using the auto regressive integrated moving average (ARIMA) model and exponential smoothing and predicted the stability of the Data Centre for the next 6 months. Further, we have performed a comparison of ARIMA model and exponential smoothing and we arrived at the conclusion that ARIMA model outperformed the other. The best model is selected based on the ACF residual correlogram, Forecast Error histogram and the error measures like root mean square error (RMSE), mean absolute error (MAE), mean absolute scale error (MASE) and *p*-value of Ljung-Box statistics. From the above results we conclude that ARIMA model is the best model for forecasting this time series data and hence based on the ARIMA models forecast result we predicted the stability of the Data Centre for the next 6 months.

Keywords ARIMA · ACF · RMSE · MASE · MAE

R.V. Anand (✉) · P.B. Sivakumar · D.V. Sagar
Department of Computer Science and Engineering, Amrita Vishwa Vidyapeetham, Coimbatore, India
e-mail: vijay.r18@gmail.com

P.B. Sivakumar
e-mail: pbsk@cb.amrita.edu

D.V. Sagar
e-mail: dhanvsagar@gmail.com

L.P. Suresh and B.K. Panigrahi (eds.), *Proceedings of the International Conference on Soft Computing Systems*, Advances in Intelligent Systems and Computing 398, DOI 10.1007/978-81-322-2674-1_55

1 Introduction

A Data Centre processes huge amount of requests everyday which are served by many applications and also servers that host the applications. Many resources, such as CPU, disk, memory, network, etc. are utilized during the processing of requests. The Data Centre architects need to consider the factors like timely up gradation of the system, estimating the resources and determining whether the available resources are sufficient for near future, etc. Therefore, forecasting and prediction are essential for a Data Centre to maintain its working stability.

Based on the observed workload that the server has for past one year, a Data Centre architect needs a best approach for forecasting in order to accommodate the workload of the server during migration process or additional usage of the resources. Therefore, we go for selecting the best model for maintaining the workload of servers and hence avoiding bottleneck.

In this paper, we mainly concentrate on the forecasting using the auto regressive integrated moving average (ARIMA) and exponential smoothing of the time series model. ARIMA is the best model for predicting and forecasting a linear time series data. Exponential smoothing is used by many companies to forecast the time series data for shorter period of time (i.e. seasonal and trend time pattern).

In this paper, we employed two models viz. ARIMA, exponential smoothing for forecasting and we selected the best model based on the performance comparison and predicted how much workload that a server would accommodate in the future in order to maintain the stability of a Data Centre, without any failure.

2 Related Work

The paper [1] proposed a time series model for forecasting short-term load and this technique has been proved to forecast accurately the loads per hour on all days including weekdays, weekends and public holidays and it is found that this model outperformed the conventional methods like Box-Jenkins model or artificial neural network. The models viz. ARIMA and DAN2 [3] have been compared in which DAN2 outperformed ARIMA in forecasting time series and the authors proposed a hybrid model which is a combination of the above stated models which provided better results [2].

In order to calculate short-term load forecasts, a technique called exponential smoothing is used which is considered to be simple and provided more accurate forecasting results [3]. The paper [4] proposed a hierarchical model for forecasting workload of web server time series by dividing the time series into short-and long-term components; this scheme reduced the effect of changes in the trend because of the reduction in the data history for training the model.

The paper [5] suggests a capacity planning for resource pools and conducted a case study on 6 months of data for 139 enterprise applications. A new technique of forecasting workload in cloud in which a co-clustering scheme was proposed which formed the basis for designing HMM-based technique for forecasting changes in the workload trend [6].

3 Proposed System

The proposed system consists of five modules

- Preprocessing dataset
- Forecasting of data using Exponential Smoothing
- Forecasting of data using ARIMA model
- Accuracy estimate of ARIMA and exponential smoothing model
- Predicative report based on selected model

The given data is first transformed into a times series data format. The transformed time series data is made stationarity and then forecasted using the time series and ARIMA model. The transformed data is also forecasted using the Holt-Winters exponential model. The accuracy of both models is evaluated based on the various error measures like root mean square error (RMSE), mean absolute scale error (MASE) and mean absolute error (MAE) and p-value of Ljung-test. The best model is selected based on the histogram, correlogram and above-mentioned error measures and based on the forecasted value from the selected model, prediction is done and a report is generated on the workload of server for the next 6 months. The proposed system is more useful because it helps the project managers to effectively accommodate the workload growth of the server in the near future.

The proposed architecture diagram is described in Fig. 1.

4 Implementation

In this work, R Studio is used for performing all the time series predicting models. The data set that is used is the real-time batch workload data of a Data Centre. We mainly focus on the metric batch workload in this paper. Based on the value that is analyzed for past one year, we forecasted the metric for the next 6 months and according to the values that are forecasted, we generated a report indicating how the batch workload of the server will be for the next 6 months, and it also helps in planning for accommodating the workload of server if there is any abrupt change.

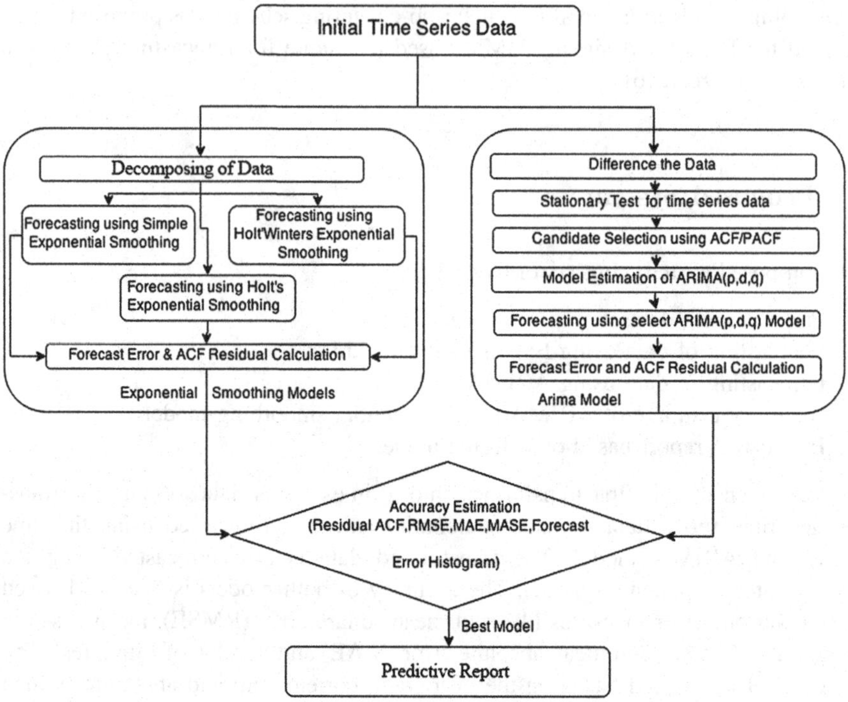

Fig. 1 Architecture diagram

4.1 Pre-processing of the Data

Initially, the real-time data is converted into a time series data and the resultant data is analyzed for the stationary, which proves that data is constant over time, so that data could be efficiently forecasted using ARIMA model. Further, we decomposed the data set and found the component under which the data falls; either in trend or seasonal or random component and we also used simple moving average (SMA) to smooth the non-seasonal data, so that it could be processed by exponential smoothing.

4.2 Forecasting of Data Using Exponential Smoothing

4.2.1 Simple Exponential Smoothing

Simple Exponential smoothing is best suited for the short-term forecasting. The data is smoothened using the Holt-Winters method and the alpha is the only parameter used to smooth the data. So the alpha value of the batch workload data is estimated keeping the beta and the gamma value constant for the exponentially smoothened data. Then, the observed data is fitted to the model and then forecasted. The accuracy of forecast is estimated by sum-of-squared error (SSE) and the auto correlation function (ACF) of the residuals is expressed using correlogram up to lag 20. Forecast error is also calculated and plotted using the histogram. In addition, Ljung-Box test is estimated and the p-value of the autocorrelation of the forecasted value for total lag is found.

4.2.2 Holt's Exponential Smoothing

Holt's exponential smoothing is used to find the level of slope in the data by smoothing using two parameters alpha and beta keeping the gamma value constant. The calculated alpha value 0.32 and the calculated beta value 0.18 both nearly tend to zero and are positive which concludes that data forms the seasonal component. Based on this, the data is forecasted and the SSE is calculated which is equal to 2060.00 and the ACF is estimated using the correlogram up to lag 20. Further, forecast error is estimated and the mean is also calculated using histogram, and the slope of the histogram was found to be normally distributed and the value tended nearly to zero. So it fits the model of forecast and error is less. In addition, Ljung-Box test is estimated and the p-value seems to be large for the model.

4.2.3 Holt-Winters Exponential Smoothing

Holt-Winters exponential smoothing is used to evaluate all the three viz. slope, seasonal and level components. In this, the log of the time series data is found and the log time series data is smoothed, based on all the three components, values of alpha, beta and gamma are estimated. The result shows that the beta value seems to be nearly 0 so it tails to a seasonal component; the alpha and beta values are 0.1 and 0.1 which are positive and also tails nearly to zero, so the pattern is positive. Then SSE is found to be 3.4143 and then the data is forecasted using the HoltWinters. The forecast error is evaluated using the histogram and mean value is also calculated. ACF of the residuals is plotted using correlogram up to lag 20.

4.3 Forecasting of Data Using ARIMA Model

The initial time series data is differentiated once to make the data stationary and then this stationarity is verified using the Dickey-Fuller (DF) test. Then, the auto correlation and Partial ACFs are estimated in order to select the candidate for the ARIMA (p, d, q) model. From the ACF and PACF, the selected model is ARIMA (0, 1, 1) model. The observed batch workload data is then fitted to the model. Using this model, the batch workload data is evaluated and then based on this model, the data is forecasted for the next 6 months. The forecast error is also calculated and plotted using the histogram and the mean is also estimated for it. The ACF of the residuals is also estimated using the correlogram up to lag 20.

4.4 Accuracy Estimate of Exponential Smoothing Models and ARIMA Model

The accuracy of the model is calculated based on RMSE, MAE and MASE measures. The three error measures of all the four models are calculated and based on that, the ARIMA model is found to have very less error measure and all the error rates are nearly zero. From Fig. 4 of the residual ACF and from Fig. 8 of the histogram, it is found that ARIMA model outperforms all the other exponential models for the batch workload data. So, we predicted the stability of the Data Centre based on the forecast done using the ARIMA (0, 1, 1) model. Based on the prediction report, the architect can visualize how the workload of the server will be for the next 6 months and also accommodate the workload growth easily.

5 Result Analysis

In this paper, we have performed forecasting and estimated the requirements of the Data Centre for the next 6 months. This was done by using the previous data collected from the Data Centre. In this paper, we performed forecasting using four models and compared their performances. Based on the performance evaluation, we selected the model which has highest accuracy and then predicted the stability of the Data Centre based on the batch workload of the server. The analysis was done on the basis of ACF correlogram, Forecast error Histogram and the error measure RMSE, MAE and MASE. Based on the result analysis, we proved that ARIMA model is best suited for forecasting when compared to other stated models (Figs. 2, 3 and 4).

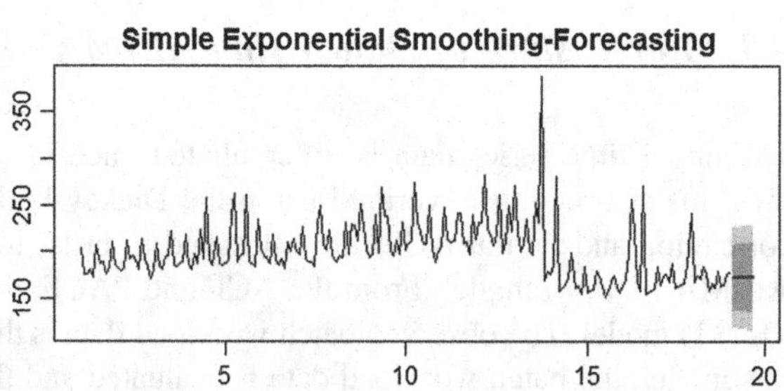

Fig. 2 Forecast of simple exponential smoothing

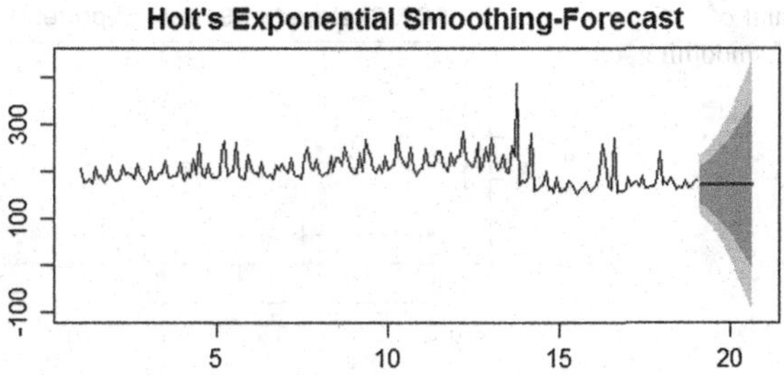

Fig. 3 Forecast of Holt's exponential smoothing

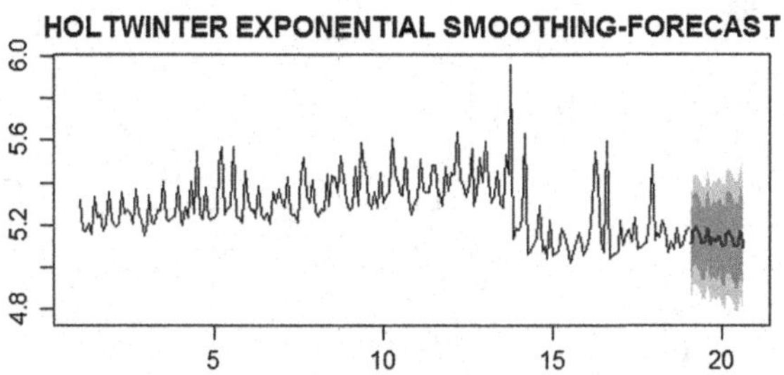

Fig. 4 Forecast of Holt-Winters exponential smoothing

Figures 5, 6, 7 and 8 present the ACF Correlogram of the residuals that is calculated for all the four models. The correlogram is estimated up to maximum lag 20. Figures 9, 10, 11 and 12 show the histogram of all the four models that is plotted to find the forecast error and from the Table 1 demonstrates the sensitivity of RMSE, MAE, MASE and *p*-value of Ljung-Box test for the four models. From the above results we conclude that ARIMA model is the best model for forecasting this time series data and hence based on ARIMA models forecast result we predicted the stability of the server for next six months (Fig. 13).

Fig. 5 Forecast of ARIMA

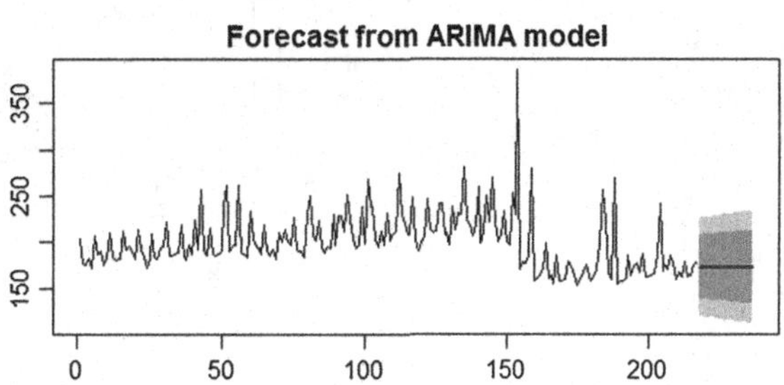

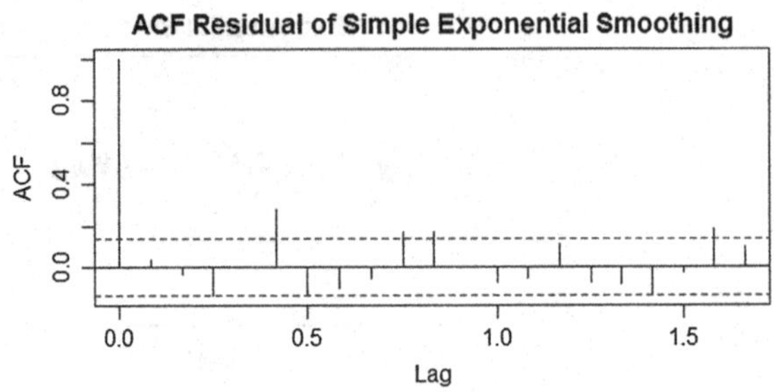

Fig. 6 ACF residual of simple exponential smoothing

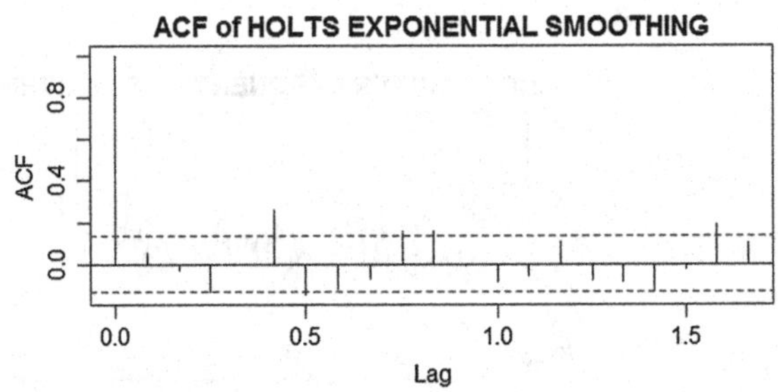

Fig. 7 ACF residual of holts exponential smoothing

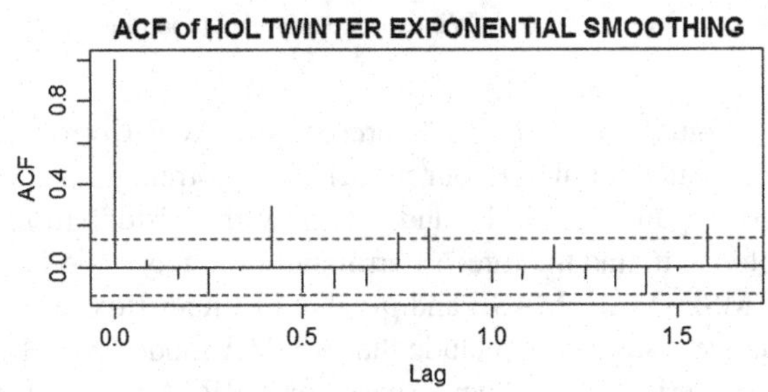

Fig. 8 ACF residual of Holt-Winters exponential smoothing

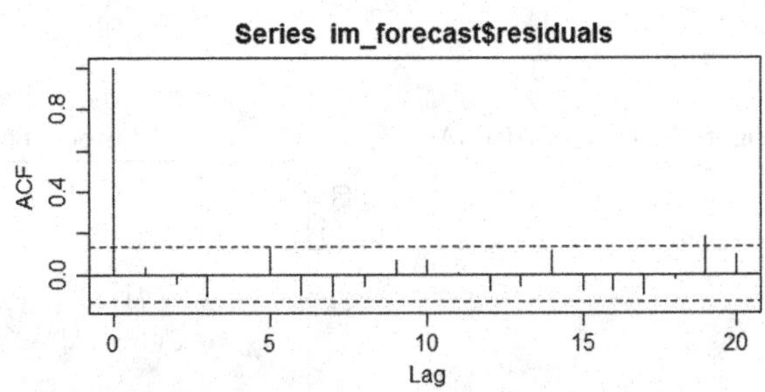

Fig. 9 ACF residual of ARIMA

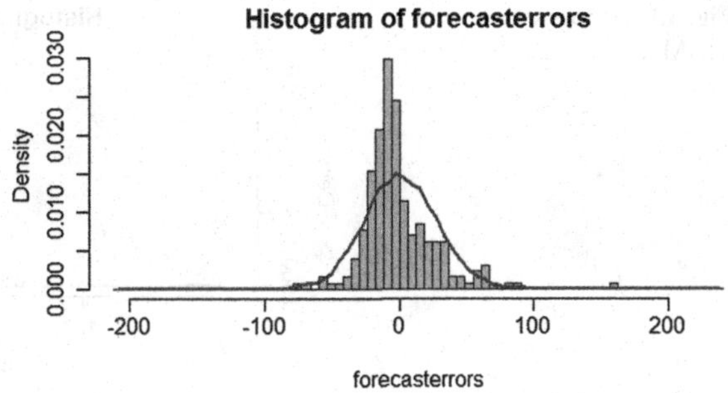

Fig. 10 Histogram of simple exponential smoothing

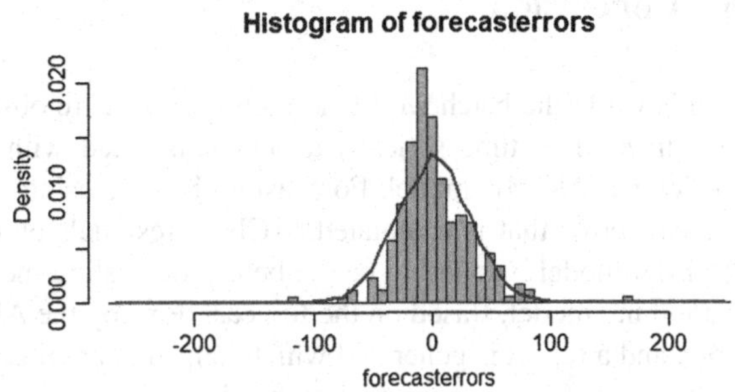

Fig. 11 Histogram of Holt's exponential smoothing

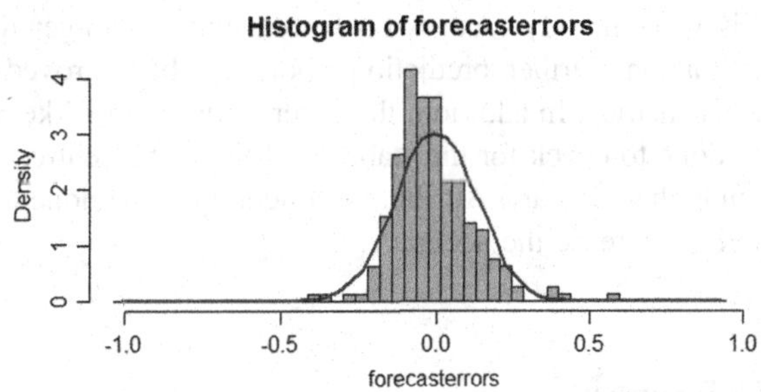

Fig. 12 Histogram of Holt-Winters exponential smoothing

Table 1 Accuracy measure

Accuracy estimation				
Model	RMSE	MAE	MASE	*p*-value
Simple exponential smoothing model	27.014	19.22	0.71	1.63
Holt's exponential smoothing model	30.954	0.098	0.75	3.28
Holt'sWinters exponential smoothing model	0.129	0.098	0.75	9.2
ARIMA model	0.092	0.097	0.69	0.014

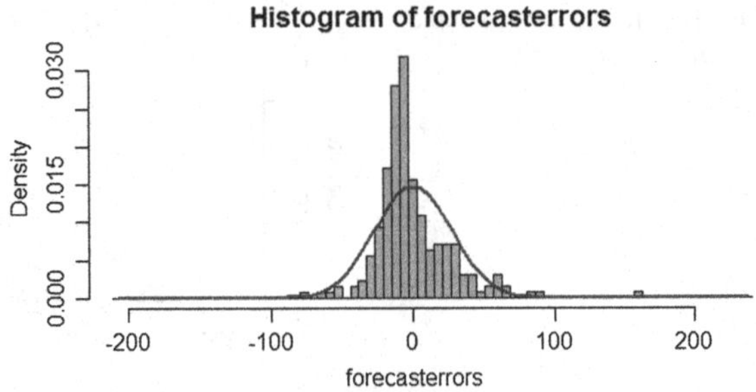

Fig. 13 Histogram of ARIMA

6 Conclusion

In this work, the batch workload metric of data are observed for past one year, data is converted to time series data and it is fitted with the Exponential smoothing model and ARIMA model. Forecasting is done using all the four models. From the forecast error that is calculated, ACF of residuals and various error measures the ARIMA model is found to be the better forecasting model compared to exponential smoothing model. Based on the forecast done by the ARIMA model, a prediction is done and a report is generated which helps the architect to know the stability of the data centre for the next 6 months. Further, it helps them to accommodate the workload in future during migration or overload of resource. The future direction of this work mainly focuses on forecasting for longer duration changing the model estimation. Further, prediction reports can be improved considering more number of other metrics. In addition, the other future works like Software fault prediction can be done to check for the stability of the Data Centre, SLA and SLO violation case can be handled and Attribute Reduction can be done to reduce the forecasting time and to increase the accuracy.

References

1. Amjady N (2001) Short-term hourly load forecasting using time-series modeling with peak load estimation capability. Power Syst IEEE Trans 16(3):498–505
2. Gomes GSS, Maia ALS, Ludermir TB, de Carvalho F, Araujo AF (2006) Hybrid model with dynamic architecture for forecasting time series. In: Neural networks, 2006 IJCNN'06, international joint conference on IEEE, pp 3742–3747
3. Tran VG, Debusschere V, Bacha S (2012). Hourly server workload forecasting up to 168 hours ahead using Seasonal ARIMA model. In: Industrial technology (ICIT), 2012 IEEE international conference on IEEE, pp 1127–1131
4. Christiaanse WR (1971) Short-term load forecasting using general exponential smoothing. Power Apparatus Sys IEEE Trans 2:900–911
5. Vercauteren T, Aggarwal P, Wang X, Li TH (2007) Hierarchical forecasting of web server workload using sequential monte carlo training. Sig Process IEEE Trans 55(4):1286–1297

6. Gmach D, Rolia J, Cherkasova L, Kemper A (2007) Workload analysis and demand prediction of enterprise data center applications. In: Workload characterization, 2007, IISWC. IEEE 10th international symposium on IEEE, pp 171–180
7. Nehinbe JO, Nehibe JI (2012) A forensic model for forecasting alerts workload and patterns of intrusions. In: Computer modelling and simulation (UKSim), 2012 UKSim 14th international conference on IEEE, pp 223–228
8. KhanA, Yan X, Tao S, Anerousis N (2012) Workload characterization and prediction in the cloud: a multiple time series approach. In: Network operations and management symposium (NOMS), 2012 IEEE, pp 1287–1294
9. Li X (2013) Comparison and analysis between holt exponential smoothing and brown exponential smoothing used for freight turnover forecasts. In: 2013 Third international conference on intelligent system design and engineering applications IEEE, pp 453–456

Active Warden Attack on Steganography Using Prewitt Filter

P.P. Amritha, K. Induja and K. Rajeev

Abstract Digital Steganography is a method used to embed secret message in digital images. Attackers make use of steganographic method for the purpose of transmitting malicious messages. In this paper, we proposed active warden method by using Prewitt filter on the input image to highlight the edge locations. Then the Discrete Spring Transform (DST) is applied on the filtered image to relocate the pixel location so that secret message cannot be recovered. This method is a generic method since it is independent of the steganography algorithms used and it does not require any training sets. We have compared our results with the existing system which used Sobel filter and curve length method. Our experimental results concluded that Prewitt was able to destroy the message to larger extent than by using Sobel filter and curve length method. Bit error rate (BER) and PSNR was used to measure the performance of our system. Our method was able to preserve the perceptual quality of the image.

Keywords Steganography · Sobel filter · Prewitt filter · Curve length method · Discrete spring transform

1 Introduction

Image steganography is used to hide and transmit secret message in an invisible way. The secret message can be embedded by altering the multimedia signal slightly. This will not cause a drastic change to the image, and hence, the human

P.P. Amritha (✉) · K. Induja · K. Rajeev
TIFAC CORE in Cyber Security, Amrita Vishwa Vidyapeetham, Coimbatore, India
e-mail: ammuviju@gmail.com

K. Induja
e-mail: indujak91@gmail.com

K. Rajeev
e-mail: rajeev.cys@gmail.com

L.P. Suresh and B.K. Panigrahi (eds.), *Proceedings of the International Conference on Soft Computing Systems*, Advances in Intelligent Systems and Computing 398, DOI 10.1007/978-81-322-2674-1_56

visual system will not understand the distortion. Security problem arises in steganography when the illegal messages also can be embedded by the terrorist. They embed their secret plans and transmit in multimedia signals. The steganography method should possess few characteristics to hide information. Two characteristics are undetectability and extracting of secret message. The steganalysis is a method to detect the secret message hidden using steganography. Because of the widespread use of steganography [1], steganalysis have been of great importance in network scenarios. Hence, how to prevent the illegal or malicious information transmission through internet became an issue. Two categories of steganalysis are targeted [2, 3] and universal [4]. These two steganalysis techniques can be built on a passive way. Passive attack is not able to actively discover the steganography information hence active attacking-based method will be more effective. An active warden method using Prewitt filter which is independent of the steganographic algorithm is proposed in this paper. This method destroys the hidden information rather than extracting the message.

In our proposed system, filtering is done at the first phase to detect the presence of edges and in the second phase discrete spring transform (DST) is applied to destroy the hidden information from the stego image. The characteristics of this method maintain the perceptual quality of the image even though it causes drastic change to the coefficients of the image. The rest of the paper is organized as follows: Sect. 2 describes about the related works. In Sect. 3 we describe the proposed system which is efficient than the existing system described in related works. Section 4 illustrates the experimental results and conclusion is drawn in Sect. 5.

2 Related Works

Sharp et al. in [5] and [6] focused on active warden attack on image. In DST, spring is altered physically but it preserves the structure and integrity of the spring. This concept of spring is applied to cover image such that whatever the modification are done by changing the coefficients of the cover image the quality of it is preserved. DST is an effective method for certain steganographic algorithms since it removes the secret message without changing the cover image. The image is interpolated into 2D window kernel. Then the image is resampled using variable sample rates.

Sharp et al. in [7] propose a novel attack method for the DST that can attack transform domains of a cover media. This technique is a stronger way to attack those steganographic methods which utilize transform domains than spatially oriented DST. For the frequency DST attack (FDST) they concentrated on the Fourier transform of the cover media. The mid-range components of the FFT are typically least affected by distortion, in fact, this is where most steganographic schemes embed information. With this premise they choose to attack the mid-range frequency components of the FFT cover media. The experiments done by them show that increasing the selection of mid range, increases the BER while decreasing the PSNR of the cover media.

Qi et al. in [8] focused on audio steganography using the transform called DST. Similar to the time scale modification, the spring transform disables the synchronization of the hidden information. Furthermore, this method has some advantages over the traditional time scale modification, therefore, the steganography method which can resist to the time scale modification still can be defeated by this method. In [7] a block-based spring transform is used to implement variable rates. The signal is divided into certain blocks and once the maximum block size is very small compared to the sampling rate the stretch in these small time slots will be negligible. The BER is more than 0.4 for the audio signal so the secret message is destroyed.

3 Proposed System

The main objective of active steganalysis is to destroy the hidden information while maintaining minimum distortion. Inspired by the paper [5] we have used Prewitt filter to destroy the steganographic content in our proposed system. Our system is divided into two phases. In the first phase, input image undergoes filtering to detect edges using Prewitt edge detector because of the fact that the edge contains more information capacity. Here, we are assuming that steganographic embedding is done more on edge areas than on smooth areas. In the second phase, the output of the filtered image is transformed using DST. In the transformed image every pixel of the input image is mapped into a new location with minimum distortion by destroying the secret information while maintaining the visual perceptual quality of the image. A block diagram for the proposed system is given in Fig. 1.

Steps for implementation

1. Take an input image I.
2. Apply Prewitt filter on I to obtain I_f (filtered image).

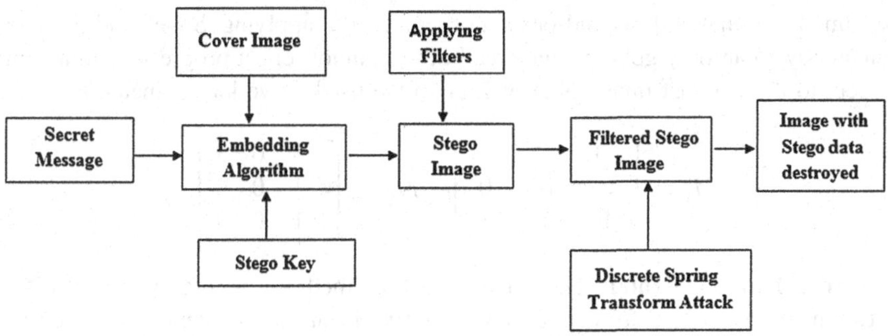

Fig. 1 Framework for active warden steganography

3. To the output image obtained from the step 2 apply DST to the highlighted edges in the I_f for pixel relocation so as to destroy the hidden information.
4. PSNR and Bit error rate (BER) is used to measure the performance of the image obtained after step 3.

Our proposed system was compared against the existing active warden approach which used Sobel filter and curve length method. For comparison we have used Sobel and curve length method instead of Prewitt using the same procedure as we have mentioned above.

Edge Detectors

Human vision system pays fewer attentions to the plain area in the image compared to the edge area in the image in terms of the content of the image. It is straightforward because the edge area contains more information capacity. On the other hand the plain area only reflects a few information because of lack of change. The following are the edge detectors used in our system.

Sobel Edge Detection: Sobel edge detection [9] is used to extract edge information of an image. The masks of the Sobel detector in horizontal and vertical directions are given below

$$M_1 = \begin{bmatrix} -1 & 0 & 1 \\ -2 & 0 & 2 \\ -1 & 0 & 1 \end{bmatrix} \quad M_2 = \begin{bmatrix} -1 & -2 & -1 \\ 0 & 0 & 0 \\ 1 & 2 & 1 \end{bmatrix}$$

The gradient of image in horizontal and vertical are calculated by performing 2D convolution of the image and Sobel detector masks in two directions by using the Eqs. (1) and (2),

$$G_1 = I \times M_1 \tag{1}$$

$$G_2 = I \times M_2 \tag{2}$$

where G_1 and G_2 are the gradients of the obtained image I.

Prewitt Edge Detection: The masks M_1 and M_2 given below can be applied to image to obtain the edge area in the horizontal and vertical direction [9]. This will be similar to first-and second-order derivative. By applying Sobel and Prewitt detectors we can only get the edge area but we cannot reflect progressive trends in respect to the contrast image pixel values so we used curve length method.

$$M_1 = \begin{bmatrix} -1 & -1 & -1 \\ 0 & 0 & 0 \\ 1 & 1 & 1 \end{bmatrix} \quad M_2 = \begin{bmatrix} -1 & 0 & 1 \\ -1 & 0 & 1 \\ -1 & 0 & 1 \end{bmatrix}$$

Curve length Method: The edge detection methods will only differentiate between the edge area from plain area. The plain areas will have shorter curve length than edge areas. The curve length method [5] is obtained by computing the interpolated signal in 2D both in horizontal and vertical directions and combining it. The curve length method is calculated using the following Eqs. (3), (4), and (5)

$$\hat{I}(x,y) = I(x,y) \times W_L(x,y) \tag{3}$$

where I is the input image $\hat{I}$ is the filtered image obtained W_L is the interpolation window kernel. The third-order Lanczos window kernel is used here. This is the curve length between $I(k, y)$ and $I(k + 1, y)$ in the horizontal direction and between $I(x, k)$ and $I(x, k + 1)$ in the vertical directions

$$d_1(k,y) = \int_{x=k}^{x=k+1} \left(\frac{d\hat{I}}{x}(x,y)^2 + 1\right)\mathrm{d}x \tag{4}$$

$$d_2(x,k) = \int_{y=k}^{y=k+1} \left(\frac{d\hat{I}}{y}(x,y)^2 + 1\right)\mathrm{d}y \tag{5}$$

3.1 Discrete Spring Transform

The concept of the DST attack is that of likening the cover media to a spring, in that it can be physically altered or manipulated in a manner of ways, while the basic structure and integrity of the spring is preserved despite these alterations. Since many steganographic schemes rely on the fact that the cover media remains somewhat stable the DST is effective in destroying the stego media for such types of steganographic algorithms. The local geometric transform will not create larger distortions in the image. The general local geometric transform is given in Cheddad et al. [1].

3.2 Analog Location Transform

Most of the steganography methods failed to survive from the print-scan process in Cheddad et al. [1] is because of the geometrical location deviation distortion. The original pixels will deviate from their original location due to aligning error of the print-scan process. Due to this kind of deviation the receiver will not able to decode the hidden information. This small deviation will be insensitive to human visual system. This pixel location can be formulated as ALT. For a $M \times N$ image $A(m, n)$, the ALT-based DST is represented in Eqs. (6) and (7)

$$I(m,n) = I'(m',n') \tag{6}$$

where I is the input image and $\hat{I}$ is the transformed image.

$$\begin{bmatrix} x' \\ y' \\ 1 \end{bmatrix} = \begin{bmatrix} 1 & 0 & \varphi_1(x,y) \\ 0 & 1 & \varphi_2(x,y) \\ 0 & 0 & 1 \end{bmatrix} \begin{bmatrix} x \\ y \\ 1 \end{bmatrix} \tag{7}$$

After applying the Sobel and Prewitt filters on the input image the following transformation in (8) and (9) is performed on the filtered image.

$$\varphi_i(x,y) = A_i\left(\frac{1}{G_i(x,y)}\right) \tag{8}$$

$$A_i = \frac{T_i}{\sum_{p=1}^{M}\sum_{q=1}^{N}\frac{1}{G_i(p,q)}} \tag{9}$$

where $T_1 = M\frac{\delta_1}{2}$ and $T_2 = N\frac{\delta_2}{2}$

After applying curve length method on the input image the image is transformed using the following Eqs. (10) and (11).

$$\varphi_i(x,y) = B_x\left(\frac{1}{d_i(x,y)}\right) \tag{10}$$

$$B_x = \frac{T_i}{\sum\frac{1}{d_i}} \tag{11}$$

The pixels of the image obtained are relocated to a new location. Next step is the reconstruction of original image from the projected image. There are two methods for reconstruction the first method is based on ALT and the second is geometrized image reconstruction.

4 Experimental Results

Experiments have been conducted on three different categories of pictures which are portrait, scenery, and still object. Steganographic algorithms used for analysis are pixel value differencing (PVD) [10] and LSB [11]. The curve length-based ALT, Sobel, and Prewitt edge detectors-based ALT were used to attack steganography. The scanning window for edge detection was chosen in the range 3–9. The BER is measured for three different categories of images using Sobel and Prewitt filters and for the curve length method. We analyzed that we were able to destroy the hidden information to larger extent by using Prewitt than Sobel and

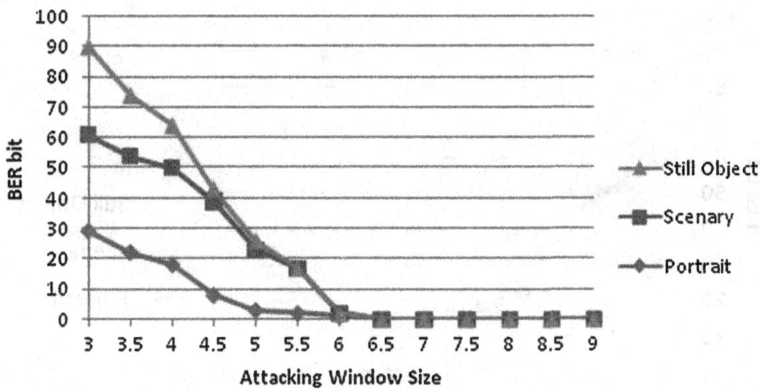

Fig. 2 Sobel edge detector-based BER

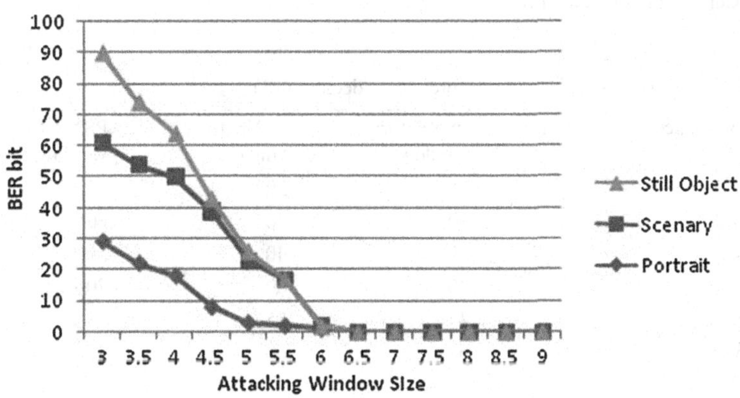

Fig. 3 Prewitt edge detector-based BER

curve length method. We found that for the portrait pictures BER was low by applying Sobel, Prewitt, and curve length method compared to the scenery and still object. Figures 2 and 3 represents the Sobel and Prewitt edge detector-based BER and Fig. 4 represents the curve length-based BER. The table below gives the PSNR measure after applying Prewitt filter to three different category of pictures. When we compared LSB and PVD, PSNR value was greater for PVD. We have also found out that the PSNR value was greater for PVD when Sobel and curve length method was applied. This implies that Prewitt was able to destroy more information but still preserve quality to some extent. This is shown in Table 1. The PSNR value was consistent for different scanning window size.

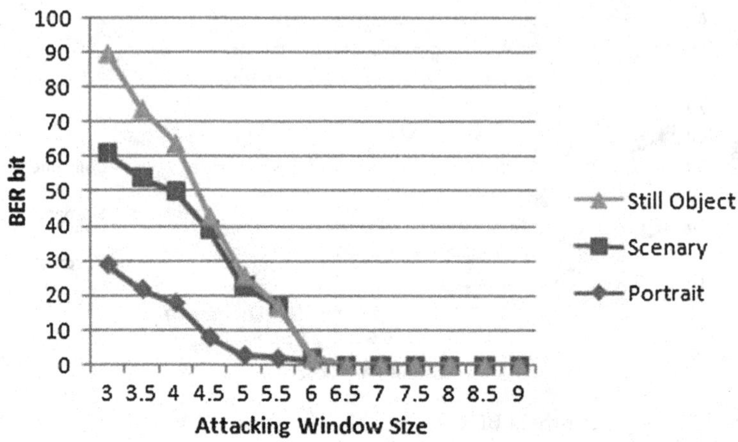

Fig. 4 Curve length-based BER

Table 1 PSNR value after DST using prewitt in decibel (dB)

Types of images (.bmp)	Algorithm	PSNR for window 3	PSNR for window 4	PSNR for window 5
Portrait	LSB	25	30	32
Scenery		30	40	60
Still object		30	40	50
Portrait	PVD	40	60	70
Scenery		45	55	65
Still object		50	60	75

5 Conclusion

We have verified an active warden attack called DST applied on three different filters namely Sobel, Prewitt, and curve length method. The Prewitt was able to destroy hidden information considerably than when other two filters were applied. By this method we could destroy the malicious information still preserving the quality of the cover image which was verified through PSNR and BER. We have done these experiments in spatial domain steganography. Further works can be done to attack steganographic algorithms done in frequency and wavelet domain.

References

1. Cheddad A, Condell J, Mc Kevitt P (2010) Digital image steganography: survey and analysis of current methods. Sig Process 90:727–752
2. Dumitrescu S, Xiaolin W, Wang Z (2003) Detection of LSB steganography via sample pair analysis. Sig Process IEEE Trans 51:1995–2007
3. Fridrich J, Goljan M, Du R (2001) Reliable detection of LSB steganography in color and grayscale images. In: Proceedings on multimedia and security: new challenges. ACM, pp 27–30
4. Fridrich J, Kodovsky J (2012) Rich models for steganalysis of digital images. Inf Forensics Secur IEEE Trans 7:868–882
5. Qi Q, Sharp A, Yang Y, Peng D, Sharif H (2014) Steganography attack based on discrete spring transform and image geometrization. In: Wireless communications and mobile computing conference (IWCMC). IEEE, pp 554–558
6. Sharp A, Qi Q, Yang Y, Peng D, Sharif H (2013) A novel active warden steganographic attack for next-generation steganography. In: 9th international conference on wireless communications and mobile computing (IWCMC). IEEE, pp 1138–1143
7. Sharp A, Qi Q, Yang Y, Peng D, Sharif H (2014) Frequency domain discrete spring transformed: a novel frequency domain steganographic attack. In: 9th International symposium on communication systems, networks & digital signal processing (CSNDSP). IEEE, pp 972–976
8. Qi Q, Sharp A, Peng D, Yang Y, Sharif H (2013) An active audio steganography attacking method using discrete spring transform. In: 24th international symposium on personal indoor and mobile radio communications (PIMRC). IEEE, pp 3456–3460
9. Gonzalez RC, Woods RE (2009) Digital image processing. Pearson Education, India
10. Wu DC, Tsai WH (2003) A steganographic method for images by pixel-value differencing. Pattern Recogn Lett 24:1613–1626
11. Provos N, Honeyman P (2003) Hide and seek: an introduction to steganography. Secur Priv IEEE 1:32–44

Similarity Scores Evaluation in Social Networking Sites

Amrita Ravindran, P.N. Kumar and P. Subathra

Abstract In today's world, social networking sites are becoming increasingly popular. Often we find suggestions for friends, from such social networking sites. These friend suggestions help us identify friends that we may have lost touch with or new friends that we may want to make. At the same time, these friend suggestions may not be that accurate. To recommend a friend, social networking sites collect information about user's social circle and then build a social network based on this information. This network is then used to recommend to a user, the people he might want to befriend. FoF algorithm is one of the traditional techniques used to recommend friends in a social network. Delta-SimRank is an algorithm used to compute the similarity between objects in a network. This algorithm is also applied on a social network to determine the similarity between users. Here, we evaluate Delta-SimRank and FoF algorithm in terms of the friend suggestion provided by them, when applied on a Facebook dataset. It is observed that Delta-SimRank provides a higher precise similarity score because it considers the entire network around a user.

Keywords Social networking sites · Friend of friend · Delta-SimRank · Similarity scores

1 Introduction

Social networking sites have become popular because they allow users to share content and offer recommendations to other users about, the people who they are most likely to befriend. For this purpose, the social networking sites collect information about users' social circle and then build a social network based on this

A. Ravindran (✉) · P.N. Kumar · P. Subathra
Department of Computer Science and Engineering, Amrita School of Engineering, Ettimadai, Coimbatore, India
e-mail: anku329@gmail.com

L.P. Suresh and B.K. Panigrahi (eds.), *Proceedings of the International Conference on Soft Computing Systems*, Advances in Intelligent Systems and Computing 398, DOI 10.1007/978-81-322-2674-1_57

information. This network is then used to recommend to a user, the people he might want to befriend. The motivation behind these recommendations is that a user might be just few steps away from a potential and desirable friend, and not realize it. Through these recommendations, rather than searching through a multitude of friends, all his potential friends are brought to the forefront.

The ever increasing popularity of social networking sites has led to an increasing interest in measuring similarities between objects. Measuring similarity between objects can be considered in a number of contexts. The similarity between network structures requires computing similarity between their nodes. There are basically two ways to compute similarity between nodes. One method is based on the content of text in the node. The other method makes use of the linking between the objects.

In this paper, we are considering the linking between the nodes in a social network. Here, we will consider two such algorithms and then compare their performance based on the friend suggestions provided by them. One such algorithm is the commonly used friends-of-friends (FoF) algorithm [1]. This algorithm calculates the count of mutual or common friends between two users. For any particular user, the user with the maximum number of common friends is recommended. A few drawbacks were observed with this approach such as, who will it recommend when there is more than one user with the same maximum common friends. In such cases, need arises for a more accurate method to determine the similarity criteria.

SimRank is a popular method to compute similarity between two objects. The limitation is that the computational cost of SimRank is very high [2]. So, an alternate method called Delta-SimRank came into use. Delta-SimRank is faster and fits better into the scenario of distributed computing [2]. SimRank computation on large networks is speeded up by computing it using Map Reduce. Through the use of Map Reduce, computation is distributed to multiple machines and memory requirement is greatly reduced. Delta-SimRank, an improvement on SimRank, provides greater speed and much less data transfer.

2 Literature Survey

Graph-based analysis, with its applications in areas like analysis of social networks, recommender engines, clustering, and classification of documents and so on, is of great interest. A graph is a representation that naturally captures objects and the relationships between the objects [3]. Objects will be represented as nodes while relationships are represented as edges [3]. There are many cases in which it would be important to answer queries to find the most similar nodes in a network. To this end, a great number of similarity measures have been proposed. There are two kinds of methods for computing the similarity between two nodes. One method is based on the content of text. The other method computes similarity between linked objects. SimRank was proposed [4] to calculate the similarity between any two objects or nodes in social network. SimRank computation suffer from the limitation that its computing cost will be very high. Optimized computation of SimRank is done using parallel algorithms for SimRank computation on Map Reduce framework [5] and more specifically its open-source implementation, Hadoop. Even though it is effective, SimRank computation is expensive in two aspects. First, time complexity of computation is huge. It takes around 46 h. to calculate SimRank measures on a synthetic network containing 10 K nodes on a single machine [6]. Second, as the size of the network gets larger, the amount of memory needed to calculate SimRank also increases which is mostly beyond the processing abilities of a single computer [2]. In [2], implementation of SimRank on Map Reduce is done. This ensures a reduction in the memory requirement and high computational cost, but it leads to lots of data transfers in a distributed system environment. A new algorithm for distributed computing was brought about, called Delta-SimRank [2], to improve the performance. Compared to SimRank, Delta-SimRank works better in a distributed computing environment, since it leads to lesser communication traffic and faster speed [2]. Social network sites such as LinkedIn and Facebook use the FoF algorithm to help users broaden their networks [1]. The FoF algorithm suggests friends that a user may know, but are not part of their immediate network [1]. With e-commerce systems, came the concept of personalized recommendation systems. Many recommender engines use the collaborative filtering technique, which has proved to be one of the most successful techniques in recommender systems in recent years [7].

3 Experiment Methodology

The Methodology followed for the course of the experiment is as shown in Fig. 1. Here, friend circle of a user is taken as input and then similarity scores are applied on such a network, which considers the user and his social circle as nodes and their relationship as edges. Two types of similarity scores, namely FoF and Delta-SimRank are applied on them and their comparison is done.

Fig. 1 The adopted methodology

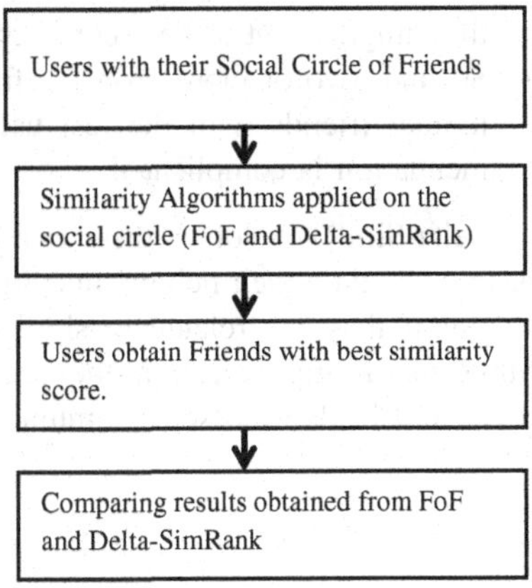

Joe	Jon	Kia	Bob	Ali	
Kia	Joe	Jim	Dee		
Dee	Kia	Ali			
Ali	Dee	Jim	Bob	Joe	Jon
Jon	Joe	Ali			
Bob	Joe	Ali	Jim		
Jim	Kia	Bob	Ali		

Fig. 2 The friend list for a given user in a social network

3.1 *Friends-of-Friends Algorithm (FoF)*

A user may want to befriend a person who may not be present in his immediate social circle but may belong in a second degree, outer network. In such cases, the FoF algorithm will use the number of common friends to recommend a friend. Consider the following example.

In Fig. 2, we can see a network of people [1].

From the above figure, it is clear that if we take a user, e.g.: Jim; then the friend list of the user gives a count of the number of common friends it has with any other user [1]. Here, the first FoF suggested for Jim is Joe. This is because Jim and Joe share three common friends namely, Kia, Bob and Ali. This count is more than that with the users Kia, Ali, Jon and Bob. The next FoF is Dee followed by Jon. In this way, even though there were a number of second degree friends, who were potential candidates to recommend for a friend, only the most relevant was recommended.

3.2 *Discussion on FoF*

The following disadvantages were observed in the traditional FoF algorithm:

- First, existing FoF algorithm will predict a friend based on the number of mutual friends. Two users may have a number of mutual friends, but that need not mean they might want to be recommended to each other.
- Second, two or more probable friends may have the same number of common mutual friends with the test user. In this case, determining the best recommendation is complicated.

Delta-SimRank solves these two disadvantages by computing similarity of two people based on their neighbours. SimRank is based on the idea that two objects are similar if they are related to similar objects [3]. Delta-SimRank is a version of SimRank, fitting better in the scenario of distributed computing environment because it leads to lesser communication traffic and faster speed [2].

3.3 Delta-SimRank

SimRank is a very easy and efficient method, which is computed iteratively. SimRank may be applied to any domain that considers object relationships and is not restricted. Although SimRank may be applied to a vast range of areas, it leads to high computational cost and also large space requirements. For large amount of data, the solution is to use Google's Map Reduce paradigm on SimRank to make it faster. Even so, the computation requirement of SimRank exceeds that of a single computer, for large data. So, a more efficient implementation of SimRank on distributed systems was introduced. This is called Delta-SimRank [2].

Let $s^i(x, y)$ denote SimRank scores at an iteration i. Then SimRank scores are initialized as,

$$s^0(x,y) = 1 \quad \text{if } x = y \text{ and } 0 \text{ if } x \neq y, \tag{1}$$

Then it was updated at each iteration, using the equation

$$s(x,y) = \left\{ \begin{matrix} 1 & \text{if } x = y \\ \frac{c}{|I(x)||I(y)|} \sum\limits_{u \in I(x), v \in I(y)} s(u,v) & \text{if } x \neq y \end{matrix} \right\} \tag{2}$$

Here x and y denote the nodes of which we need to compute the similarity. $I(x)$ and $I(y)$ denote the set of neighbours of x and y. u and v are nodes belonging in $I(x)$ and $I(y)$ respectively. This computation of SimRank leads to a time complexity of $O(N^4)$ and space complexity of $O(N^2)$. So, a Map Reduce computation of SimRank was proposed by Cao et al. in [2]. In this computation, there are map and reduce functions in a program. The map function runs parallel on the input data and produces output key value pairs. Then, each key value pair is sent to a reducer process based on their key. The reducer process will group the values with similar key and subsequently run the reduce function on them. The disadvantage with this implementation is that each mapper sends $s^i(x, y)$ similarity score from iteration i, to the reducer multiple times. On a graph consisting of N nodes, each having an average of p neighbours, the amount of data transferred from mapper to reducer is $O(p^2N^2)$. Mapper and Reducer processes usually reside in two separate machines, and so the large data transfers can slow the distributed system down and may also result in errors. To overcome this, Delta-SimRank was introduced.

$$\Delta^{i+1}(x,y) = \left\{ \begin{matrix} 0 & \text{if } x = y \\ \frac{c}{|I(x)||I(y)|} \sum\limits_{u \in I(x), v \in I(y)} \Delta^i(u,v) & \text{if } x \neq y \end{matrix} \right\} \tag{3}$$

This is the Delta-SimRank model and $\Delta^i(x,y)$ is the delta score for iteration i, between two nodes x and y. The fact that SimRank is solved by using Delta-SimRank is proved in Cao et al. [2] via the following proof.

Proof Initialise $\Delta^1(x,y) = s^1(x,y) - s^0(x,y)$,
Consider x ≠ y

$$\Delta^{i+1}(x,y) = \left\{ \frac{c}{|I(x)||I(y)|} \sum_{u \in I(x), v \in I(y)} s^i(u,v) - s^{i-1}(u,v) \right\}$$
$$= \frac{c}{|I(x)||I(y)|} \sum_{u,v} s^i(u,v) - \frac{c}{|I(x)||I(y)|} \sum_{u,v} s^{i-1}(u,v) = s^{i+1}(x,y) - s^i(x,y)$$

This holds for $x = y$ also,

$$\Delta^{i+1}(x,x) = s^{i+1}(x,x) - s^i(x,x) = 1 - 1 = 0$$

Then, we transform SimRank computation to updation of delta score.

$$\Delta^{i+1}(x,y) = \left\{ \frac{c}{|I(x)||I(y)|} \sum_{u \in I(x), v \in I(y)} \Delta^i(u,v) \right\} \text{if } x \neq y$$
$$s^{i+1}(x,y) = s^i(x,y) + \Delta^{i+1}(x,y)$$

A toy example used (Fig. 3) is taken by Cao et al. [2] to understand the differences between SimRank and Delta-SimRank. Table 1, taken from Cao et al. [2] shows the development of SimRank and Delta scores through the iterations, in the toy network between University and Prof. B.

The decay factor taken in Cao et al. [2] was $C = 0.8$. It was noted in Cao et al. [2] that small decay factors brought about faster convergence. Table 1 show that SimRank scores are unchanged in few iterations while increasing in others. Delta-SimRank scores are zero in some while SimRank scores are seen to be nonzero in all iterations. Since SimRank scores were nonzero in all iterations, all the scores need to be passed from mappers to reducers while for Delta-SimRank, only nonzero values need to be sent. Thus, the communication traffic is lowered. Also, it precomputes the value $\frac{c}{|I(u)||I(v)|}$ in map function since only nonzero delta values are being passed. The advantage of having only nonzero values passed is that, as

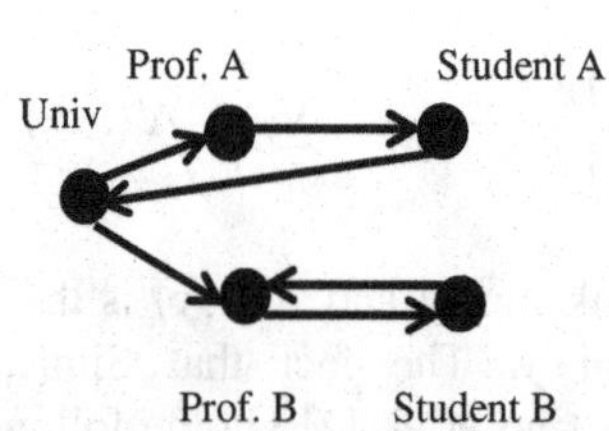

Fig. 3 Toy Example-SimRank versus Delta-SimRank

Table 1 Evolvement of Simrank and delta scores

Iterations	3	4	...	8	9	10
SimRank	0.128	0.128	...	0.128	0.132	0.132
Delta-SimRank	0.128	0	...	0	0.004	0

communication traffic decreases, the transmission errors also decrease and this leads to better system efficiency.

4 Experiments

4.1 Algorithm: Friend-of-Friend Calculation in Map Reduce [1]

```
Calculating number of common friends:
01: Map (node_name, node)
For all adjnode in node.adjacency_list do:
        Emit ((node_name, adjnode.name), 1)
        for all adj2node in node.adjacency_list do:
                If the tuple (adjnode.name, adj2node.name) hasn't already been emitted
                Then emit ((adjnode.name, adj2node.name), 2)
02: Reduce (tuple (node1.name, node2.name), [i1, i2...])
        Common_friends=0
        Already_friends=false
        For all i in counts [i1, i2 ..] do
                If i=1 then
                Already_friends=true
        Common_friends=Common_friends+1
        If Already_friends≠ true then
                emit(tuple(node1.name,node2.name), Common_friends)
Sorting the common friends:
01: Map (tuple (node1.name, node2.name), Common_friends)
        Emit (node1.name, tuple (node2.name, Common_friends))
02: Reduce (node1.name, tuple (node2.name, Common_friends))
        Sort the tuple in the decreasing order of Common_friends.
Emit (node1.name, tuple (node2.name, Common_friends))
```

4.2 Algorithm: SimRank on Map Reduce [2]

Input: Graph G, initialized s^0

For i = 0 to I-1
Mapper Function ((x, y), $s^i(x, y)$)
Find neighbours I(x) and I(y) of x, y respectively
for each u∈ I(x), v ∈ I(y)
output (u, v), $s^i(x, y)$
Reducer Function (Key = (u, v), Values = vs [])
If u = v
$s^{i+1}(u, v) = 1$
else
$s^{i+1}(u, v) = \frac{c}{len(vs)} sum(vs)$
Output (u, v), $s^{i+1}(u, v)$
Output: *updated s^i*

4.3 Algorithm: Delta-SimRank on Map Reduce [2]

Input: *Graph G, initialized Δ^i*
Mapper function ((x,y), $\Delta^i(x, y)$)
If x = y or $\Delta^i(x, y) \leq \epsilon$
Return
Find neighbours I(x) and I(y) of x, y respectively.
for each u∈ I(x), v∈ I(y)
Output (u, v), $\frac{c}{|I(u)||I(v)|}\Delta^i(x, y)$
Reducer function (Key = (u,v), Values = vs[])
If u = v
Output $\Delta^{i+1}(u, v) = 0$
Else
Output $\Delta^{i+1} = sum(vs)$
Output: *updated Δ^{i+1}*

4.4 Algorithm: SimRank Using Delta-SimRank [2]

Input: *Graph G, init SimRank s^0*
1: Using algorithm 4.1.2 update SimRank and obtain s^1.
2: Initialise Delta-SimRank by $\Delta^1 = s^1 - s^0$
3: for i = 1 to I-1
4: update Δ^{i+1} -SimRank as in Algorithm 4.1.3.
5: $s^{i+1} = s^i + \Delta^{i+1}$
Output: *updated SimRank score s^I.*

5 Implementation Details

Implementation is done on a Hadoop cluster consisting of two nodes; one master and one slave. The dataset containing the undirected graph in an edge list format is converted to a directed graph using the NetworkX package in Python. The converted dataset is then used for the similarity computation using the Delta-SimRank algorithm mentioned above. The initial dataset is also applied to FoF algorithm to determine the number of common friends.

5.1 Dataset

The dataset used in obtained from SNAP library developed by Stanford University [8]. The dataset consists of Facebook data collected from survey participants using a Facebook app [8]. The dataset includes node features (profiles), circles, and ego networks [8]. From this dataset the edge list files were used for the various modules in the project. It contains a total of 4039 nodes and 88234 edges.

6 Results and Analysis

A comparison of similarity computations was done with FoF and Delta-SimRank. Both were implemented on a Map Reduce framework on a cluster of two machines. It was observed that on a large social network dataset, the traditional FoF algorithm is not as precise as Delta-SimRank for recommendations.

6.1 FoF Output Versus Delta-SimRank Output

Here, we see that even though node 376 has many common friends (42 common friends) with 392, the similarity score of these pair of nodes is very less (0.006 whereas node 378 has 0.01) and thus 376 would not be a good recommendation, (Figs. 4 and 5). Also, we can see that in FoF, even though several nodes have the same number of common friends with node 392(all 9), Delta-SimRank provides distinct values for them, making recommendations easier, (Figs. 6 and 7). The values are 0.007 for 402, 0.008 for 408, and 0.009 for 439. This makes it evident that 439 is the best recommendation.

Fig. 4 FoF output of node 392 with 376

```
["392", "359"] 0.0008421052631490306
["392", "360"] 0.006015037593947276
["392", "361"] 0.0019138755980632313
["392", "362"] 0.009356725146062037
["392", "363"] 0.009494324045358648
["392", "364"] 0.0
["392", "365"] 0.004210526315567867
["392", "366"] 0.009881847475779368
["392", "367"] 0.004753820033915503
["392", "368"] 0.008293460924973737
["392", "369"] 0.0008421052631490306
["392", "370"] 0.008354218880499774
["392", "371"] 0.010526315788781163
["392", "372"] 0.0013157894736625694
["392", "373"] 0.007287449392687965
["392", "374"] 0.008354218880499774
["392", "375"] 0.009210526315637986
["392", "376"] 0.006804891015399243
["392", "377"] 0.0
["392", "378"] 0.010526315789420414
["392", "379"] 0.0
["392", "380"] 0.0
["392", "381"] 0.0
["392", "382"] 0.001913875598040338
["392", "383"] 0.0
["392", "384"] 0.0
["392", "385"] 0.00902255639063825
["392", "386"] 0.0
["392", "387"] 0.008421052631515633
["392", "388"] 0.00967283072539351
["392", "389"] 0.0011080332409818833
["392", "390"] 0.008421052631135734
["392", "391"] 0.011100478468846412
["392", "392"] 1
["392", "393"] 0.0
["392", "394"] 0.007017543859587566
```

6.2 *The Metrics Used*

The metrics used to arrive at these observations are:

- Common friends—measures the number of common friends between two nodes.
- SimRank score—determines the similarity measure between two nodes. Two nodes are said to be similar, if their neighbours are similar.
- Delta-SimRank score—a more optimized version of SimRank score which takes into account the SimRank scores of previous iterations.

Fig. 5 Delta-SimRank output depicting the score between 392 and 376

3919	3609	1
3919	3502	1
3919	3501	1
3919	3576	1
3919	3574	1
3919	3563	1
3919	3560	1
392	376	42
392	400	33
392	391	32
392	378	30
392	363	28
392	438	28
392	374	26
392	563	25
392	524	25
392	395	24
392	513	23
392	348	23
392	544	20
392	366	19
392	496	18
392	461	18
392	353	18
392	580	17
392	434	16
392	484	15
392	517	15
392	417	15
392	591	15
392	651	15
392	414	14
392	606	13
392	526	13
392	683	12
392	614	11

Fig. 6 FoF output for node 392

```
392      580 17
392      434 16
392      484 15
392      517 15
392      417 15
392      591 15
392      651 15
392      414 14
392      606 13
392      526 13
392      683 12
392      614 11
392      604 11
392      479 11
392      637 10
392      669 9
392      558 9
392      402 9
392      408 9
392      439 9
392      520 8
392      545 8
392      409 8
392      352 8
392      471 8
392      432 8
392      641 8
392      398 8
392      507 8
392      512 7
392      422 7
392      565 7
392      451 7
392      676 7
392      494 7
392      601 6
```

Fig. 7 Delta-SimRank output for node 392

```
["392", "397"]  0.00228832951943771
["392", "398"]  0.009210526315713731
["392", "399"]  0.0
["392", "400"]  0.01035375323550322
["392", "401"]  0.0
["392", "402"]  0.007894736842068165 <------
["392", "403"]  0.004432132963927533
["392", "404"]  0.008842105263111358
["392", "405"]  0.0021052631578393354
["392", "406"]  0.0
["392", "407"]  0.0
["392", "408"]  0.008966861598396849 <------
["392", "409"]  0.0102073365230446
["392", "410"]  0.003007518796935949
["392", "411"]  0.0
["392", "412"]  0.006937799043041455
["392", "413"]  0.007894736841975416
["392", "414"]  0.0110047846889294
["392", "415"]  0.0067368421051922445
["392", "416"]  0.0115789473682687
["392", "417"]  0.010008628127653114
["392", "439"]  0.009523809523749852 <-------
["392", "418"]  0.0066985645932213095
["392", "419"]  0.007518796992445871
["392", "420"]  0.008906882591002967
["392", "421"]  0.005263157894679133
["392", "422"]  0.003947368421020171
["392", "423"]  0.012729498163936785
["392", "424"]  0.0
["392", "425"]  0.007017543859546528
["392", "426"]  0.0024767801857201737
["392", "427"]  0.002631578947281856
["392", "428"]  0.007602339181258764
["392", "429"]  0.0
["392", "430"]  0.00967283072539351
["392", "431"]  0.009392712550569261
```

7 Conclusion

Delta-SimRank, an algorithm for computing SimRank on a distributed system using Map Reduce [2] is implemented. Delta-SimRank algorithm ensures computation of SimRank scores faster and with lesser traffic. We get up to 30 times better speed when compared with distributed SimRank algorithm [2]. The experiment was done on a social network dataset (Facebook dataset). The results were compared with the FoF algorithm applied on the same dataset. It was seen that Delta-SimRank produces more accurate scores when compared with FoF. It was observed that given the same dataset, the FoF algorithm generates common friends whereas Delta-SimRank generates a similarity score. It was seen that for a node having large number of common friends with the user, the similarity score given by Delta-SimRank was very less. This is because FoF considers only the common friends, whereas Delta-SimRank considers the whole friend circle around the user. Also, FoF will not be suitable in cases where more than one potential friend has the same number of common friends with the given user. In such cases Delta-SimRank works better since it gives a more precise score.

8 Future Enhancements

The similarity scores obtained with the Delta-SimRank algorithm can be used in a Recommender system to develop friend recommendations. In cases where relations from different sources have to be integrated, an algorithm called simfusion [9] can also be used. Delta-SimRank scores can be applied on an Epinion dataset [10] to calculate similarity between users and items and recommend similar items to users based on similar users' ratings. Epinion datasets contain trust relations and item ratings. Trust relations focus on the trust relations between users. This ensures that users receive recommendations from those users, who, they have themselves specified as their trusted circle.

References

1. Holmes A (2012) Hadoop in practice. Manning Publications Co, Connecticut
2. Cao L, Kim HD, Tsai MH, Cho B, Li Z, Gupta I (2012) Delta-SimRank computing on Map Reduce. In: BigMine12, proceedings of the 1st international workshop on big data, streams and heterogeneous source mining: algorithms, systems, programming models and applications. ACM, pp 28–35
3. Li L, Li C, Chen H, Du X (2013) Map Reduce-based SimRank computation and its application in social recommender system. In: 2013 IEEE international congress on big data, pp 133–140
4. Jeh G, Widom J (2002) SimRank. a measure of structural-context similarity. In: KDD, pp 538–543
5. Dean J, Ghemawat S (2008) Map Reduce: simplified data processing on large clusters. In: Proceedings of the 6th conference on symposium on operating systems design and implementation, vol 51, No. 1, pp 107–113
6. Lizorkin D, Velikhov P, Grinev M, Turdakov D (2008) Accuracy estimate and optimization techniques for SimRank computation. Proc VLDB J 19(1):45–66
7. Gong S (2010) A collaborative filtering recommendation algorithm based on user clustering and item clustering. J Softw 5(7):745–752
8. Stanford SNAP library. http://snap.stanford.edu/data/
9. Wang J, de Vries AP, Reinders MJT (2006) Unifying userbased and itembased collaborative filtering approaches by similarity fusion. In: SIGIR '06 Proceedings of the 29th annual international ACM SIGIR conference on research and development in information retrieval, pp 501–508
10. Meyffret S, Guillot E, Mdini L, Laforest F (2012) RED: a rich epinions dataset for recommender systems. In: Technical report at LIRIS UMR CNRS, 5205 ACM

Estimating Random Delays in Modbus Over TCP/IP Network Using Experiments and General Linear Regression Neural Networks with Genetic Algorithm Smoothing

B. Sreram, F. Bounapane, B. Subathra and Seshadhri Srinivasan

Abstract Time-varying delays adversely affect the performance of networked control systems (NCS) and in the worst case can destabilize the entire system. Therefore, modeling network delays are important for designing NCS. However, modeling time-varying delays are challenging because of their dependence on multiple parameters, such as length, contention, connected devices, protocol employed, and channel loading. Further, these multiple parameters are inherently random and delays vary in a nonlinear fashion with respect to time. This makes estimating random delays challenging. This investigation presents a methodology to model delays in NCS using experiments and general regression neural network (GRNN) due to their ability to capture nonlinear relationship. To compute the optimal smoothing parameter that computes the best estimates, genetic algorithm is used. The objective of the genetic algorithm is to compute the optimal smoothing parameter that minimizes the mean absolute percentage error (MAPE). Our results illustrate that the resulting GRNN is able to predict the delays with less than 3 % error. The proposed delay model gives a framework to design compensation schemes for NCS subjected to time-varying delays.

B. Sreram (✉)
Department of Information Technology, Kalasalingam University, Krishnankoil (via), Srivilliputtur, Tamil Nadu, India
e-mail: bsreram85@gmail.com

F. Bounapane
Department of Engineering, University of Federico II, Naples, Italy
e-mail: furio.buonopane@gmail.com

B. Subathra
Department of Instrumentation and Control Engineering, Kalasalingam University, Krishnankoil (via), Srivilliputtur, Tamil Nadu, India
e-mail: clk0602@nitt.edu; clk0602@gmail.com

S. Srinivasan
International Research Center, Kalasalingam University, Krishnankoil (via), Srivilliputtur, Tamil Nadu, India
e-mail: cpscourse@klu.ac.in

L.P. Suresh and B.K. Panigrahi (eds.), *Proceedings of the International Conference on Soft Computing Systems*, Advances in Intelligent Systems and Computing 398, DOI 10.1007/978-81-322-2674-1_58

Keywords General regression neural network · Genetic algorithm · Networked control systems · Time varying delay modelling · Modbus · TCP/IP network

1 Introduction

Networked control systems (NCS) use networks for information exchange among control components. Network proliferation into control loops has enabled many novel applications (see, [1–6] and references therein) with distinct advantages that were not realizable with traditional hard wired systems. In spite of such advantages, design, and analysis of NCS has been difficult mainly due to the time-varying delays introduced by communication channels. Such delays are potential enough to adversely affect NCS performance and in the worst case can lead to instability. Therefore, it becomes imperative to model delays and capture their influence on NCS. However, this is not straightforward due to the dependency of delays on numerous network parameters, such as length, contention ratio, connected devices, channel loading, and network protocol that are inherently random. Further, delay variations with time are nonlinear and capturing these variations with conventional models is complex. Therefore, new models that capture the influence of various factors, nonlinear and time-varying behavior of delays are required for designing NCS. Objective of this investigation is to develop one such model for estimating delays that can be used in designing NCS.

Modeling random delays in communication channels for designing NCS controllers have been investigated by researchers in the past and many approaches have been proposed. The available methods can be broadly classified into two categories: (i) deterministic and (ii) stochastic delays. The first approach tries to capture the time-varying delays to be deterministic by estimating the worst-case bounds. For instance, the investigation in [7], time-varying delays are modeled using buffers that reflect the delays in the channel. However, deterministic delay models in literature are conservative as they consider worst-case delays in modelling. Recently, stochastic models have gained significant attention. Time-varying delays have been modeled to be stochastic using empirical distribution [8], Markov chain [9], Markov Chain Monte Carlo [10] and Hidden Markov Model (HMM) [11]. Empirical models capture delays using stochastic distribution (For example, Gaussian [12, 22]) and have been used in adaptive controller design. The authors reported that the performance of controllers could be significantly improved provided an accurate estimation of delays. Except for the HMM model, the other approaches can model delays considering any one of the channel condition. Further, they can be used to design only stochastic controllers that are difficult to adapt in industries. Therefore, new models that can consider influences of multiple factors, nonlinear, and time-varying behavior of delays are required for designing controllers for NCS. To our best knowledge, current delay models cannot model delays considering the above-mentioned factors except for the method proposed in [23], wherein data-mining techniques have been used to model the delays. Further, the available delay data from time stampings is not used in these models. To overcome these research gaps, this investigation proposes to model delays using artificial neural networks due to their ability to model complex non-linear and time-varying phenomenon (For example, see [13, 14]). Further, experimental data from networks can be integrated into the model while employing the ANN-based estimation techniques. In particular, this investigation uses the general regression neural network (GRNN) due to its ability to model complex, multivariate, and time-varying process (see [15, 16] and references therein). GRNN can be a good and practical choice to classify a medical data [17]. Further, as the accuracy of estimates with GRNN depends on the smoothing parameter, this investigation uses genetic algorithm (GA) evolutionary optimization algorithm for computing the optimal smoothing parameter with the objective of reducing the mean absolute percentage error (MAPE). The resulting delay model gives more accurate estimates than the conventional GRNN.

Main contributions of this investigation are: (i) Experiments to model time-varying delays, (ii) GRNN model for time-varying delays, and (iii) improvement in GRNN performance using genetic algorithm.

The paper is organized into five sections. Section 2, presents the experimental study with MODBUS over TCP/IP network. The GRNN model and GA optimization algorithm for computing the optimal smoothing parameter is described in Sect. 3. Comparison of the actual delays and the estimated ones using GRNN with and without RCGA smoothing are presented in Sect. 4. Conclusions are drawn from the obtained results in Sect. 5.

2 Experiments to Model Time-Varying Delays

The first step to model time-varying delays in NCS is the collection of data containing the network conditions and delays. This data is the input to the GRNN for estimating the delays. The data will be used in two modes, training and testing. During the training mode, the GRNN is adjusted based on the inputs until sufficient accuracy is obtained. Once trained the GRNN produces delay estimates, when presented with the network conditions. Therefore, collection of data is pivotal to the accuracy of delay estimates. Further, the experiments should capture the various conditions envisaged during the NCS operation. This requires changing network conditions and recording delays. Then we use these conditions to generate the delay estimates.

This investigation tries to model the delays in an industrial network. Therefore, we select MODBUS over TCP/IP as the candidate network whose delays will be modeled using GRNN. The selection is motivated by the wide application of the MODBUS in industries [18]. The experimental prototype is shown in Fig. 1. It consists of a Master Controller and number of slave controllers, the Modbus over TCP/IP network is used to connect the slave controllers. The network loading, length, channel contention by varying the devices and the number of rungs of the

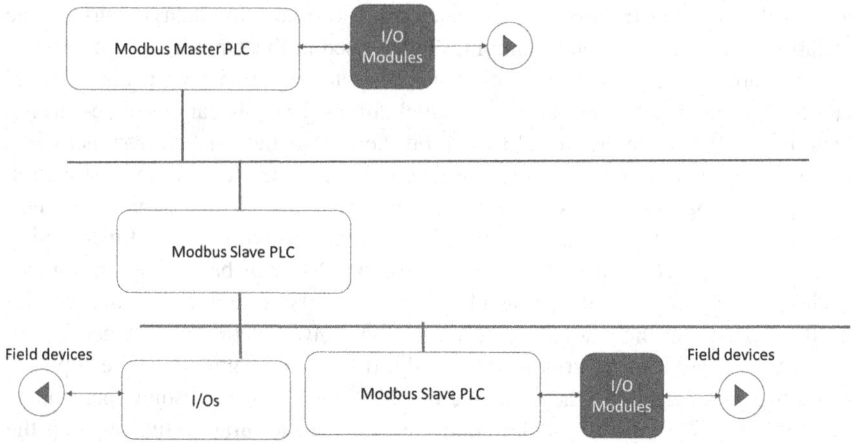

Fig. 1 Experiment on Modbus over TCP/IP

programmable logic controller (PLC) logic was changed. The delays for variations in these parameters were recorded using simple program in CoDeSys [19] following IEC 61131. The delays and the network conditions are the input to the GRNN model, while the delay estimates are the output. The GRNN uses the input samples, and estimates the time-varying delays. This experimentation procedure leads to determination of delay samples that is essential for estimating the delays.

3 Genetic Algorithm Tuned General Regression Neural Network

The delay model proposed in this investigation uses the GRNN; a feedforward neural network developed by Specht [20]. The GRNN is the neural version of Nadarya-Watson kernel regression proposed in [21]. It has its basics on well-developed statistical principles and converges asymptotically with an increasing number of samples to the optimal regression surface. The GRNN was selected for estimating the delays due to its ability to estimate nonlinear, multi-parameter, and time-varying process. A few problems with the existing approaches in using GRNN are the selection of network structure, and tuning parameters (e.g., learning rate) given by the algorithm developer leading to uncertainties and variations in the output. On the other hand, the GRNN proposed in this investigation uses the genetic algorithm to tune the spread. Therefore, ambiguities in setting up the smoothing parameter are eliminated using optimization as a decision support tool. The use of GA is motivated due to its ability to reach global optimum.

3.1 GRNN

GRNN is a nonlinear regression tool and it is preferred over BPNN as it requires less number of samples to converge. The structure of GRNN is different from ANN and it consists of four layers: (i) input, (ii) pattern, (iii) summation, and (iv) output. The schematic of GRNN is shown in Fig. 2. Unlike BPNN, GRNN employs conditional exception to calculate the regression of dependent variable y on independent variable x.

Let X is measured value of random variable x. Joint probability density function (PDF) of measurement value X and the dependent variable y is denoted by $f(X, y)$. Then the conditional mean $E[y|x]$ and the regression $Y(X)$ for a given X can be calculated by:

$$\hat{Y}(X) = E[y|x] = \frac{\int_{-\infty}^{\infty} y f(X, y)\mathrm{d}y}{\int_{-\infty}^{\infty} \mathrm{d}y f(X, y)}. \tag{1}$$

If $f(X, y)$ cannot be found, then sample observations of x and y can be used to calculate $f(X, y)$. An estimate of the nonconditional mean, denoted by

$$\hat{Y}(X, y) = \frac{\sum_{j=1}^{n} Y' \exp\left(-\frac{D^2}{2\sigma^2}\right)}{\sum_{i=1}^{n} \exp\left(-\frac{D^2}{2\sigma^2}\right)}. \tag{2}$$

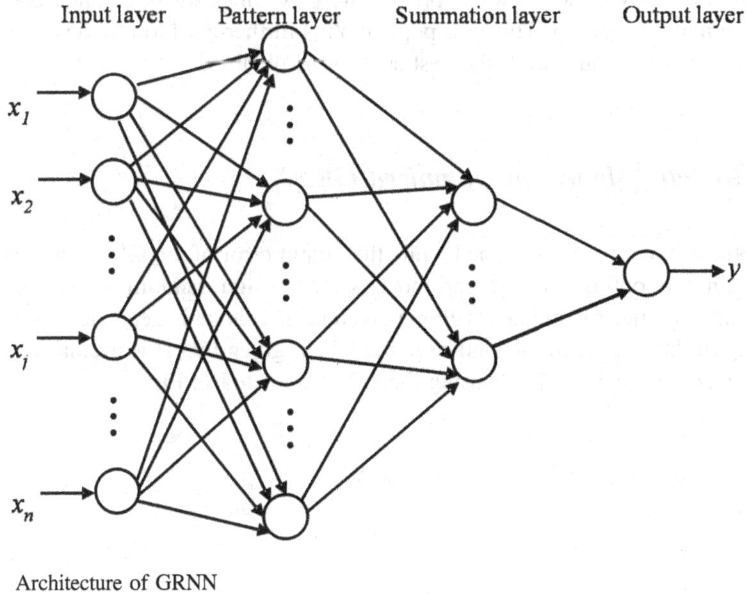

Fig. 2 Architecture of GRNN

where σ denotes the smoothing parameter of the GRNN and, D_i^2 is given by,

$$D_i^2 = (X - X')^T (X - X'). \tag{3}$$

Here, Eq. (2) can be directly used for problems with numerical data and the smoothing parameter determines the shape of the estimated density to vary between multivariate Gaussian to non-Gaussian shapes. Therefore, the GRNN performance varies considerably depending on the smoothing factor. Therefore, determining optimal value of σ is important. This study uses GA algorithm to compute the optimal parameter.

3.2 Genetic Algorithm

Genetic algorithms are evolutionary optimization algorithm that uses ideas of natural selection and genetics. A random search is used to solve an optimization problem. The algorithm starts with a random set of solution in the search space called population, solutions from one population is used to generate next set population that are better than the older one using genetic operators, such as cross-over and mutation. Initially, a random population of chromosomes are selected. Then, the fitness function of the individual chromosomes is evaluated. A selection step is then executed. Selection involves selecting two parent chromosomes with high fitness function. The crossover operation is done on the parent chromosomes to produce new offspring. The new offspring is mutated and it is placed in new population. This new population is further used to run the algorithm. This sequence continues until the best solution is obtained.

3.3 Genetic Algorithm Optimized GRNN

The optimal value of σ is obtained while the output error of the GRNN is smallest. To obtain the optimal value, chromosomes to assign random values to each smoothing parameter generated within a given spread range is generated during the first step of the GA. In the second step, the fitness given by (4) is evaluated. Then, the optimization of the GRNN using the GA is depicted in Fig. 3.

$$J = \frac{1}{\text{MAPE}} \tag{4}$$

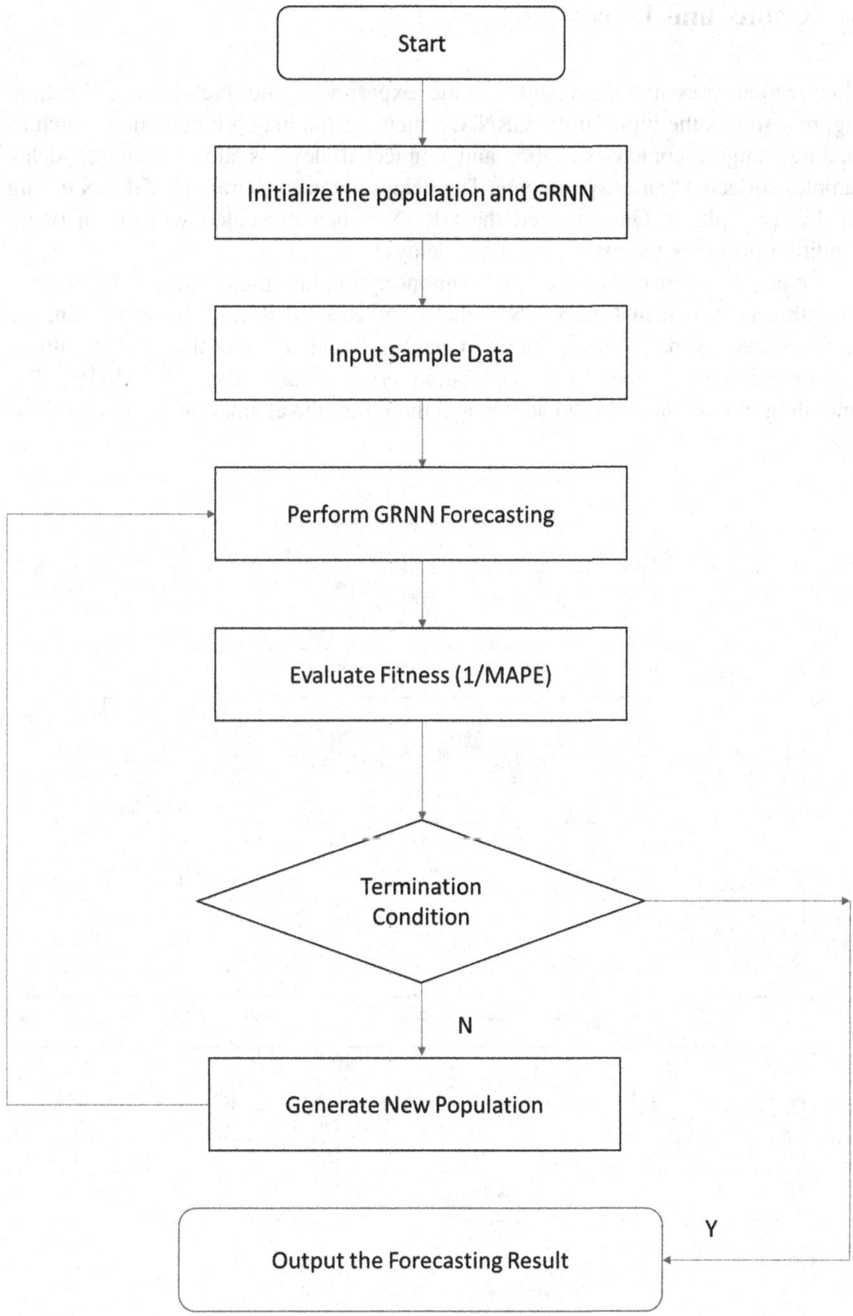

Fig. 3 Flowchart of GA optimized GRNN

4 Results and Discussions

This section presents the results of the experiment and the delay estimation. Figure 4 shows the input to the GRNN, which are the network conditions, such as loading, length, contention ratio, and connected devices along with the delay samples collected from experiments. The delays are used to train the GRNN during the learning phase. Once trained the GRNN, when presented with the network condition provides the estimates of the delays.

The delays estimated with GRNN without optimizing the network using genetic algorithm is shown in Fig. 5. The delay samples used during the validation, the delay estimates, and error are shown. Our results indicate that the GRNN without GA optimization results in an estimation error of around 16 % MAPE. The smoothing factor was selected ad hoc and therefore, the estimation accuracy is low.

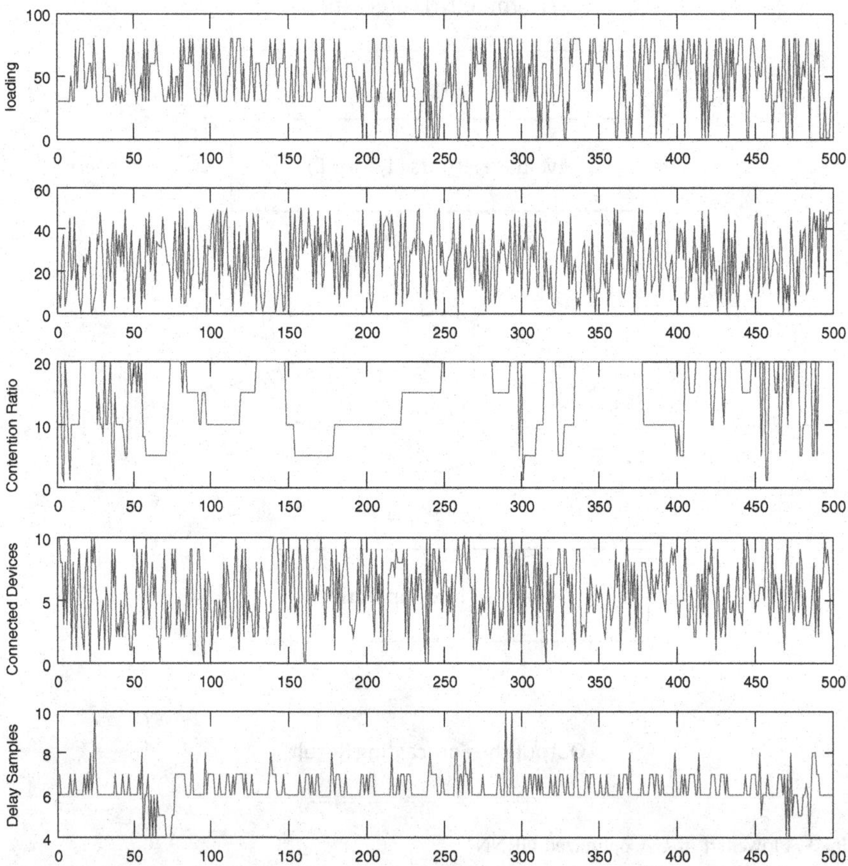

Fig. 4 Input to the GRNN network

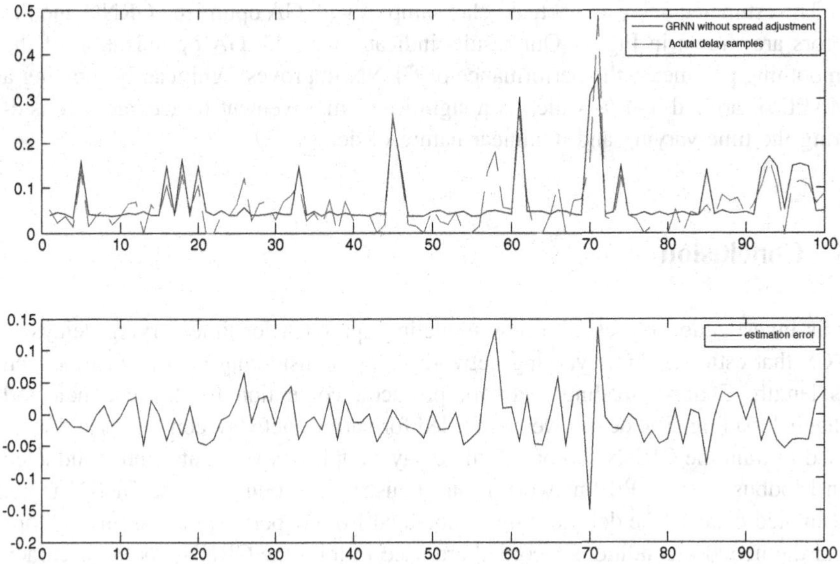

Fig. 5 Actual delay samples, delay estimates, and estimation error with GRNN

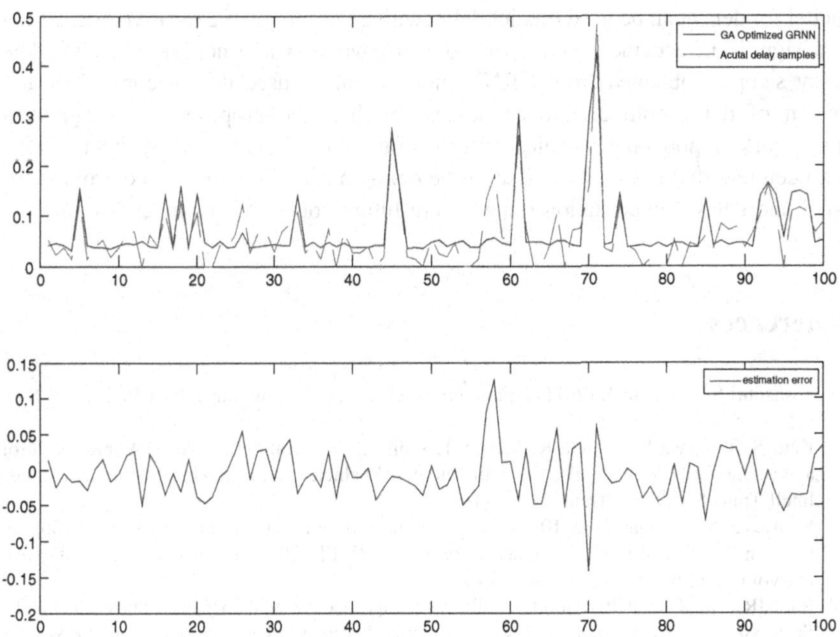

Fig. 6 Actual delay samples, delay estimates, and estimation error with GA optimized GRNN

The estimated delay and actual delay samples with GA optimized GRNN and the errors are shown in Fig. 6. Our results indicate that with GA optimization of the smoothing parameter, the performance of GRNN improves significantly showing a MAPE of around 3–4 % which is a significant improvement in accuracy, considering the time-varying and nonlinear nature of delays.

5 Conclusion

This investigation presented a new modeling approach for time-varying delays in NCS that estimated time-varying network delays considering various factors such as length. Channel loading, network protocol, contention for the channel, and channel loading. The delays were recorded for various network conditions and were used to train the GRNN. To obtain the delay samples experiments were conducted on Modbus over TCP/IP network in an industry. The output of the GRNN is the estimated delays. The delays samples obtained from experiment using time-stamps and the network conditions recorded are used to train the GRNN. As the accuracy of the GRNN depends on the smoothing parameter, GA was used to obtain its optimal value that reduces the MAPE. Once trained, GRNN produced delay estimates based on observed network conditions. Our results show that the GRNN model for delay can be used predict delay with an accuracy of 2–3 % MAPE, which is a significant accuracy considering the time-varying and nonlinear delays. The delay samples obtained from GRNN models can be used design controllers and design of delay compensation schemes. Further, the proposed model gives a framework to model delays considering various channel conditions, such a model is not been reported in literature. Use of the delay models in designing controllers for NCS and delay compensation schemes are future course of this investigation.

References

1. Seshadhri S, Ayyagari R (2011) Platooning over packet-dropping links. Int J Veh Auton Syst 9(1):46–62
2. Kato S, Tsugawa S, Tokuda K, Matsui T, Fujii H, Shin Kato et al (2002) Vehicle control algorithms for cooperative driving with automated vehicles and intervehicle communications. Intell Trans Syst IEEE Trans 33:155–161
3. Srinivasan S, Ayyagari R (2010) Consensus algorithm for robotic agents over packet dropping links, in Biomedical Engineering and Informatics (BMEI), 2010 3rd International Conference on, vol 6, no., pp 2636–2640, 16–18 Oct
4. Raol JR, Gopal A (2010) Mobile intelligent autonomous systems. Defence Sci J 60(1):3–4
5. Srinivasan S, Ayyagari R (2014) Advanced driver assistance system for AHS over communication links with random packet dropouts. Mech Syst Sig Process 49(1–2), 20 December, 53–62, ISSN 0888-3270

6. Perumal DG, Saravanakumar G, Subathra B, Seshadhri S, Ramaswmay S (2014) Nonlinear state estimation based predictive path planning algorithm using infrastructure-to-vehicle (I2V) communication for intelligent vehicle. In: Proceedings of the Second international conference on emerging research in computing, information, communication and applications (ERCICA)
7. Luck R, Ray A (1990) An observer based compensator design for distributed delays. Automatica 26(5):903–908
8. Srinivasan S, Vallabhan M, Ramaswamy S, Kotta U (2013) Adaptive LQR controller for networked control systems subjected to random communication delays. In: American Control Conference (ACC), pp 783–787, 17–19 June
9. Nilsson J (1998) Real-time control systems with delays. Ph.D. thesis, Lund Institute of Technology, Lund
10. Seshadhri S, Ayyagari R (2011) Dynamic controller for Network Control Systems with random communication delay. Int J Syst Control Commun 3:178–193
11. Cong S, Ge Y, Chen Q, Jiang M, Shang W (2010) DTHMM based delay modeling and prediction for networked control systems. J Syst Eng Electron 21(6):1014–1024
12. Srinivasan S, Vallabhan M, Ramaswamy S, Kotta U (2013) Adaptive regulator for networked control systems: MATLAB and true time implementation. In: Chinese Control and decision conference (CCDC), 25th Chinese. IEEE, pp 2551–2555
13. Seshadhri S, Subathra B (2015) A comparitive analysis of neuro fuzzy and recurrent neuro fuzzy model based controllers for real-time industrial process. Syst Sci Control Eng 3:412–426
14. Zeng J, Qiao W (2011) Short-term solar power prediction using an RBF neural network. In: Power and energy society general meeting. IEEE, pp 1–8
15. Leung MT, Chen AS, Daouk H (2000) Forecasting exchange rates using general regression neural networks. Comput Oper Res 27(11):1093–1110
16. Ben-Nakhi AE, Mahmoud MA (2004) Cooling load prediction for buildings using general regression neural networks. Energy Convers Manag 45(13):2127–2141
17. Kayaer K, Yıldırım T (2003) Medical diagnosis on Pima Indian diabetes using general regression neural networks. In: Proceedings of the international conference on artificial neural networks and neural information processing (ICANN/ICONIP), pp 181–184
18. http://new.abb.com/plc
19. http://www.codesys.com/
20. Specht DF (1991) A general regression neural network. Neural Networks IEEE Trans 2 (6):568–576
21. Nadaraya EA (1964) On estimating regression. Theor Probab Appl 9(1):141–142
22. Vallabhan M, Seshadhri S, Ashok S, Ramaswmay S, Ayyagari R (2012) An analytical framework for analysis and design of networked control systems with random delays and packet losses. In: Proceedings of the 25th Canadian conference on electrical and computer engineering (CCECE)
23. Srinivasan S, Buonopane F Subathra B, Ramaswamy S (2015) Modelling of random delays in networked automation systems with heterogeneous networks using data-mining techniques. In: Proceedings of the 2015 conference on automation science and engineering (CASE 2015), Gothernberg, Sweden, pp. 362–368

Opinion Mining of User Reviews Using Machine Learning Techniques and Ranking of Products Based on Features

P. Venkata Rajeev and V. Smrithi Rekha

Abstract Online shopping websites and the people using the Online shopping websites are proliferating every day. The widely available internet resources are letting the users to shop any products anywhere, anytime at any cost. With the brisk development in the 3G and 4G we can expect a tremendous development in the area of M-commerce and E-commerce. In this paper, we have presented our work which is an extension to our earlier work which is the comparison of two mobile products based on predefined score and features of the Mobile. Therefore, we have shown in this paper the ranking of products, ranking of products based on features, comparison of websites Flipkart and Amazon, comparison of algorithms Naive Bayes classifier, decision tree classifier and Maximum Entropy classifier based on accuracy which is used in the classification of reviews. Finally, we have shown these rankings in a graphical user interface (GUI) to recommend the user the best product.

Keywords Naive Bayes classifier · Maximum entropy classifier · Decision tree classifier · Opinion mining · GUI

1 Introduction

There are a number of Online shopping websites that are available on the internet, such as Amazon, Flipkart, Snapdeal, Jabong, Myntra, Paytm, Zovi, etc. These websites allow the users to buy products with ease. A lot of attractive and day-to-day useful products like books, electronic goods, home appliances, clothing,

P.V. Rajeev (✉)
CSE Department, Amrita Vishwa Vidyapeetham, Coimbatore, India
e-mail: rajivanindian@gmail.com

V.S. Rekha
Center for Research in Advanced Technologies for Education, Amrita Vishwa Vidyapeetham, Coimbatore, India
e-mail: smrithirekha@gmail.com

L.P. Suresh and B.K. Panigrahi (eds.), *Proceedings of the International Conference on Soft Computing Systems*, Advances in Intelligent Systems and Computing 398, DOI 10.1007/978-81-322-2674-1_59

footwear are sold from these sites. These websites provide an option to the customers to write their review about their product that they buy from these sites. These reviews are very helpful to the users, manufacturers of the product as well as the developers of the website. The users who are in dilemma to buy a product can read the reviews about the particular product from these websites so that they can have a view about their product before buying it. Potential buyers can make decisions based on the reviews of customers who have purchased and experienced the product. The manufacturers of the product will be able to know the minor or major drawbacks of product from the reviews which helps the manufacturers to get a chance to release the updated version of the product which satisfies the reviews that are mentioned in the websites. Hence online reviews play a significant role in understanding the "customer's voice." Sentiment analysis and opinion mining through machine learning algorithms offer a great possibility in automating the process of gathering, processing, and making sense of the data. In our paper, we present our feature-based opinion mining of reviews. The reviews are of mobile phones on popular Indian E-commerce sites.

2 Related Work

Opinion mining or Sentimental analysis is playing a vital role in the current era. Many works have been done on this topic. This paper is an extension to our previous work [1] in which we have created a prototype web-based system for recommending and comparing products sold online. We have extracted user reviews of two mobile products and we have shown comparison of both the products by displaying product's score and features. The score is calculated by using function which contains four parameters namely review text, rating of the review, age of the review, and helpfulness votes. Features are extracted by using the tool mallet. Naïve Bayes Classifier is used for classifying a review into positive or negative. Peter D. Turney had proposed a system that classifies reviews into *recommended* or *not recommended* by using Pointwise Mutual Information (PMI) and Information Retrieval (IR) algorithm [2]. Ranking products based on reviews is presented by Zhang et al. in paper [3], in which they have taken customer reviews from Amazon.com and their product ranking score contains three parameters namely polarity of the review which is the difference of positive and negative review, helpfulness votes and age of the review. Aspect-based opinion mining, i.e., classifying a review for each feature is shown in the paper [4], where it displays a feature as well as the related positive and negative reviews. Feature-based product ranking technique is given by Zhang et al. [5], they evaluated the relative importance of each product by its feature using the pRank algorithm. Liu et al. [6] have created a prototype system called Opinion Observer; they have done a Feature-by-Feature comparison of different products based on consumer opinions on websites, where it displays the strengths and weaknesses of each product. Abbasi et al. [7] have developed entropy weighted genetic algorithm (EWGA) for efficient feature selection in order to improve accuracy and identify key features. Pang et al. [8] have applied machine learning algorithms, such as Naïve Bayes, maximum entropy, and support vector machines (SVMs) to the sentimental classification problem. Bo Pang and Lillian Lee have removed the objective sentences or the irrelevant sentences from the document and then applied support vector machines (SVMs) classifier on the subjective or the relevant sentences in the document. The main aim of this subjective and objective separation is to prevent the classifier from classifying an irrelevant sentence which does not contain any opinion of the reviewer [9]. Ye et al. [10] used Naïve Bayes classifier, SVM classifier, and dynamic language model classifier for classification of reviews; they have shown that all the three classifiers reached an accuracy of 80 % when they are applied on large training data set.

3 Proposed System

The steps in our proposed system contain selection of website, selection of product, calculation of positive and negative features, calculation of polarity, product-based ranking and feature-based ranking.

3.1 Selection of the Online Shopping Website

In this step, we display the websites and let the user to select any one of the websites. If the user selects a particular website like Flipkart, all the product-based and feature-based ranking is based on the reviews from Flipkart. To start with, we have provided two websites named Flipkart and Amazon from which user can select one of them.

3.2 Selection of Particular Product

In this step we display the list of product categories like mobiles, tablets, etc. Later, we let the user to select any one of the product categories. If the user selects a particular product category like mobile then we show the user the list of mobiles present in the category. We have provided two categories of products named mobiles and tablets. In the mobile category, we have provided four mobiles namely

(i) Asus Zenfone 5, (ii) Apple iphone 5s, (iii) Nokia Lumia 520, and (iv) Samsung Galaxy Duo S2. Similarly in the tablets category, we have provided four tablets namely (i) Apple 16 GB iPad, (ii) Asus Fonepad 7, (iii) HP Slate 7 VoiceTab, and (iv) Lenovo A7-30.

3.3 Calculation of Positive and Negative Features

A particular feature is rated high if a positive word like 'good,' or 'excellent' occurs closer to the word in a specific sentence. If it is closer to a negative word, then it is ranked lower. We count the number of times the pair feature and opinion word occurs together in all the reviews. If the count of the pair feature and positive opinion word is more than the count of feature and negative opinion word then we say that the feature is positively oriented or else the feature is negatively oriented. The features we have considered for the mobile and tablet are speed, display, flash, memory, speaker, processor, camera, software, ram, screen, battery, sound, and speaker (Table 1).

3.4 Summary of the Product

When the user selects a particular product, and then ,we show the user the list of positive features and negative features of the product (Table 2).

Table 1 Sample procedure for identification of feature and opinion word

Sample review	'The camera is amazing in this fone but ram is worst'
After eliminating stop words	'camera amazing fone ram worst'
Feature, positive opinion word	camera amazing
Feature, negative opinion word	'ram worst'

Table 2 Sample summary display of a product

Categories	Mobiles, tablets
When user clicks 'Mobiles'	Asus Zenfone 5 Apple iphone 5s Nokia Lumia 520 Samsung Galaxy Duo S2
When user clicks 'Asus Zenfone 5'	Positively oriented features: negative oriented features Screen = 6 Screen = 2 Ram = 5 Ram = 3 Memory = 1 Memory = 4 Speaker = 2 Speaker = 5

3.5 *Calculation of Polarity of Reviews Based on Accuracy of Machine Learning Algorithms*

We have calculated the polarity of each review, i.e., classifying a review into a positive review or a negative review. This classification is based on accuracy of machine learning algorithms like Naive Bayes classifier, maximum entropy classifier, and decision tree classifier, etc. Initially, we calculate accuracy of different algorithms on different training sets then we choose the best accuracy algorithm, later on using that algorithm we will find polarity of the review.

3.6 *Product-Based Ranking*

Every product is ranked based on the score given by our score function which is calculated based on four parameters, i.e., star rating of the review, actual review given by user, helpfulness votes given to the review, and age of the review. We will let the user to select one of the products then we display the rank of that product when compared to the other products.

3.7 *Feature Based Ranking*

Each product has some good features and some bad features. Certain products may have an overall good reputation but it may have one or two features which adds to the product's disadvantages. Each product may have one or two features which are best features when compared to the other products. So, we will let the user to select a feature from a product category and then display the product in which that features is best reputed (Fig. 1).

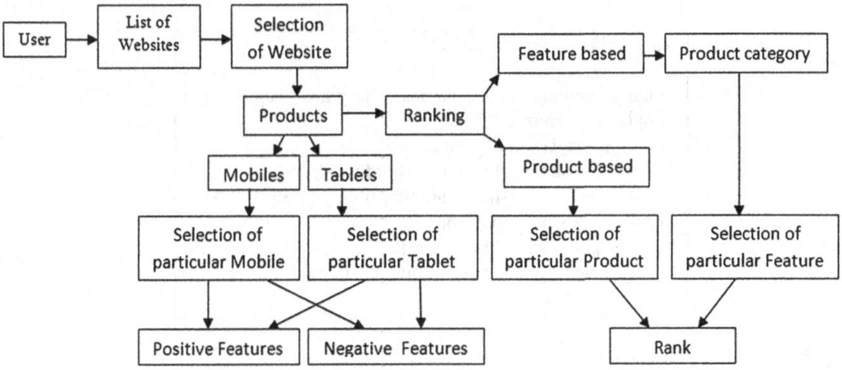

Fig. 1 The product ranking system

4 Implementation

The step for finding the best accuracy algorithms is as follows:
Input: Algorithm = ['NaiveBayes Classifier', 'DecisionTree Classifier', 'Maximum Entropy Classifier']

Output: Best Accuracy algorithm.

1. Take a training dataset of N reviews (Fig. 2).

```
TrainingDataSet =
[
('camera is good in this phone ', 'pos'),
('perfect phone for price ', 'pos'),
('very nice phone beautiful specifications', 'pos'),
('excellent phone ', 'pos'),
('great and smart phone','pos')
]
```

Fig. 2 Sample training data set

2. Apply Algorithm [0] on the training data set which is taken in step 1.
3. Find the polarity of all the reviews based on Algorithm [0] (Fig. 3).

```
('hanging while playing games getting hot ),
('battery is amazing'),
('only fools would buy this '),
('i cannot use this phone getting shutdown '),
(' apps are fast, i installed number of apps'),
('camera is very good'),
('worst phone ever seen life time ')
```

Fig. 3 Sample review data set

4. Testing data set is the 'reviews and its polarity' which is found in step 3 (Fig. 4).

```
Testing Dataset =
[
 ('hanging while playing games getting hot ','neg'),
 ('battery is amazing' ,'pos'),
 ('only fools would buy this ','neg'),
 ('i cannot use this phone gets shutdown ','neg'),
 ('apps are fast, i installed number of apps', 'pos'),
 ('camera is very good', ' pos'),
 ('worst phone ever seen life time ', ' neg'),
 ('heating problem is very frequent', ' neg')
 ]
```

Fig. 4 Sample testing dataset

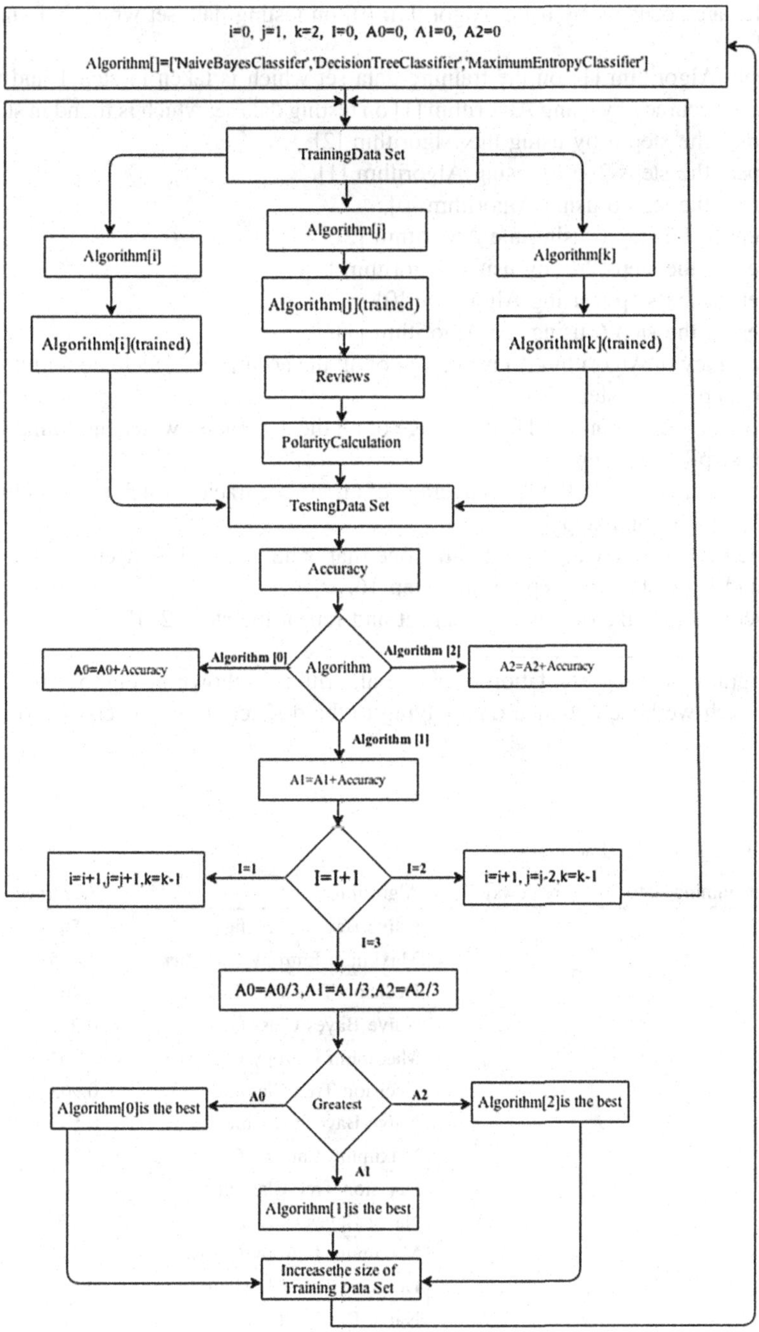

Fig. 5 Best accuracy algorithm

5. Calculate accuracy by using Algorithm [0] on testing data set which is found in step 4.
6. Apply Algorithm [1] on the training data set which is taken in step 1 and then calculate accuracy by using Algorithm [1] on testing data set which is found in step 4.
7. Repeat the step 6 by using the Algorithm [2].
8. Repeat the steps 2–5 by using Algorithm [1].
9. Repeat the step 6 using Algorithm [0].
10. Repeat the step 6 using the Algorithm [2].
11. Repeat the steps 2–5 by using Algorithm [2].
12. Repeat the step 6 using Algorithm [0].
13. Repeat the step 6 using the Algorithm [1].
14. Accuracy of Algorithm [0] is average of all the accuracies which are found from step 5, step 9, and step 12.
15. Accuracy of Algorithm [1] is average of all the accuracies which are found from step 6, step 8, and step 13.
16. Accuracy of Algorithm [2] is average of all the accuracies which are found from step 7, step 10, and step 11.
17. The best accuracy algorithm is the one that is having highest accuracy that can be found from step 14, step 15, and step 16.
18. Increase the size of training data set and repeat the steps 2–17.

Diagrammatic representation of above algorithm is shown in Fig. 5. The accuracy which we have obtained on applying to the different training sets is shown in Table 3.

Table 3 Algorithm and accuracy

Size of training data set (in reviews)	Algorithm	Accuracy
30	Naïve Bayes Classifier	0.56
	Maximum Entropy Classifier	0.415
	Decision Tree Classifier	0.461
56	Naïve Bayes Classifier	0.9725
	Maximum Entropy Classifier	0.9645
	Decision Tree Classifier	0.968
73	Naïve Bayes Classifier	0.5655
	Maximum Entropy Classifier	0.4174
	Decision Tree Classifier	0.4658
97	Naïve Bayes Classifier	0.6708
	Maximum Entropy Classifier	0.5122
	Decision Tree Classifier	0.5678
120	Naïve Bayes Classifier	0.6708
	Maximum Entropy Classifier	0.5122
	Decision Tree Classifier	0.5678

5 Result Analysis

Based on the results of Algorithms and Accuracy which are depicted in the Table 3, we have observed that the accuracy which we have got from different algorithms varies according to the size training dataset. We have also observed that this accuracy depends on type of the training dataset which we have taken. The results which we got from the python graphical user interface (GUI) is as follows (Figs. 6, 7, 8, 9, 10 and 11).

Fig. 6 App in Flipkart mode

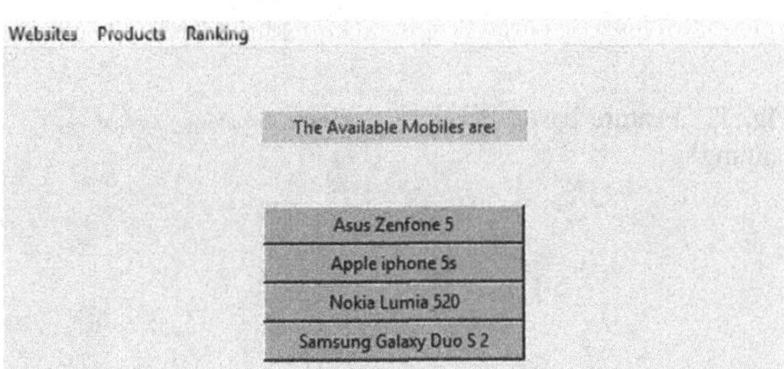

Fig. 7 Available mobiles

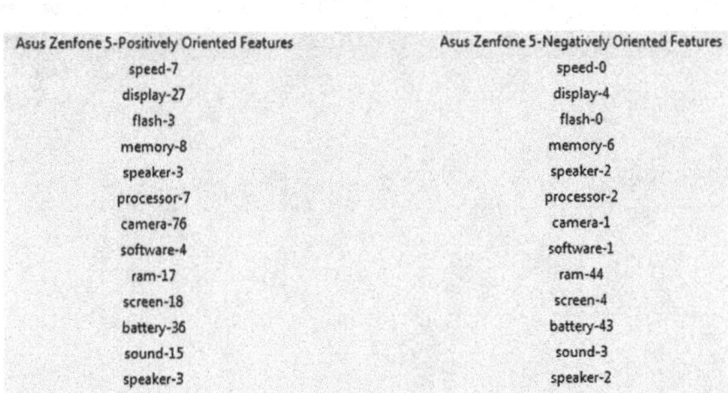

Asus Zenfone 5-Positively Oriented Features
speed-7
display-27
flash-3
memory-8
speaker-3
processor-7
camera-76
software-4
ram-17
screen-18
battery-36
sound-15
speaker-3
Asus Zenfone 5-Negatively Oriented Features
speed-0
display-4
flash-0
memory-6
speaker-2
processor-2
camera-1
software-1
ram-44
screen-4
battery-43
sound-3
speaker-2

Fig. 8 Features based on Flikart reviews

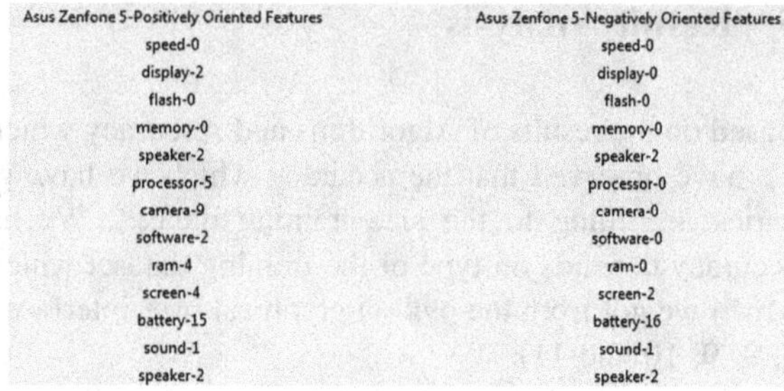

Asus Zenfone 5-Positively Oriented Features
speed-0
display-2
flash-0
memory-0
speaker-2
processor-5
camera-9
software-2
ram-4
screen-4
battery-15
sound-1
speaker-2
Asus Zenfone 5-Negatively Oriented Features
speed-0
display-0
flash-0
memory-0
speaker-2
processor-0
camera-0
software-0
ram-0
screen-2
battery-16
sound-1
speaker-2

Fig. 9 Features based on Amazon reviews

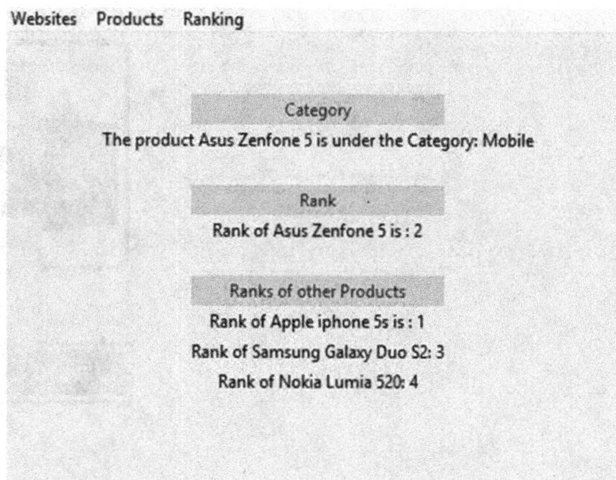

Fig. 10 Product based ranking

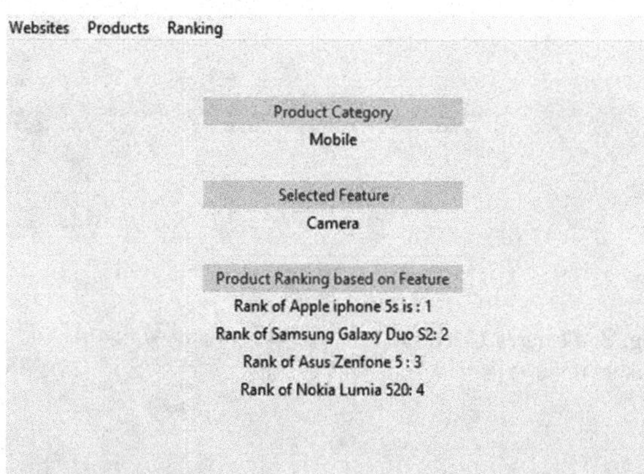

Fig. 11 Feature based ranking

6 Conclusion

In this paper, we have shown our work on opinion mining of online customer reviews of mobiles and tablets. We have shown the ranking of products, ranking of products based on features, comparison of accuracy of algorithms like Naive Bayes classifier, maximum entropy classifier, and decision tree classifier. We have also shown the comparison of two websites Amazon and Flipkart. We have shown all our results on a python GUI. The aim of our proposed system is to help the user to select the best product he needs. As a future work, we want to increase the number of product categories, increase the number of products in each product category, and also we want to increase the number of websites to collect the customer reviews.

References

1. Rajeev PV, Rekha VS (2015) Recommending products to customers using opinion mining of online product and features. ICCPCT
2. Turney PD (2002) Thumbs up or thumbs down?: semantic orientation applied to unsupervised classification of reviews. In: Proceedings of the 40th annual meeting on association for computational linguistics. Association for computational linguistics
3. Zhang K, Cheng Y, Liao WK, Choudhary A (2011) Mining millions of reviews: a technique to rank products based on importance of reviews. In: Proceedings of the 13th international conference on electronic commerce. ACM, p 12
4. Sharma R, Nigam S, Jain R (2014) Mining of product reviews at aspect level. arXiv preprint arXiv:1406.3714
5. Zhang K, Narayanan R, Choudhary A (2010) Voice of the customers: mining online customer reviews for product feature-based ranking. In: Proceedings of the 3rd conference on online social networks, pp 11–11
6. Liu B, Hu M, Cheng J (2005) Opinion observer: analyzing and comparing opinions on the web. In: Proceedings of the 14th international conference on World Wide Web. ACM, pp 342–351
7. Abbasi A, Chen H, Salem A (2008) Sentiment analysis in multiple languages: Feature selection for opinion classification in Web forums. ACM Trans Inf Syst 26(3):12
8. Pang B, Lee L, Vaithyanathan S (2002) Thumbs up?: sentiment classification using machine learning techniques. In: Proceedings of the ACL-02 conference on Empirical methods in natural language processing, vol 10. Association for Computational Linguistics, pp 79–86
9. Pang B, Lee L (2004) A sentimental education: sentiment analysis using subjectivity summarization based on minimum cuts. In: Proceedings of the 42nd annual meeting on association for computational linguistics. Association for Computational Linguistics, p 271
10. Ye Q, Zhang Z, Law R (2009) Sentiment classification of online reviews to travel destinations by supervised machine learning approaches. Expert Syst Appl 36(3):6527–6535

Identification of Attacks Using Proficient Data Interested Decision Tree Algorithm in Data Mining

M. Divya Shree, J. Visumathi and P. Jesu Jayarin

Abstract The key feature of today's networks is being open, communication between any pair of Internet end points is easier. This leads to various types of intrusions which are actions that threaten the confidentially of the network and lack of effective network infrastructures for distinguishing and dropping malicious traffics. This approach of intrusion detection with data mining concepts involving the KDD cup dataset that generates rules for the detection which works well for new as well as unknown attacks. Data mining is the process of identifying valid understandable patterns in data. It can help learn the traffic through supervised and unsupervised learning we have applied here the semi supervised way. To classify the given data resourcefully, the Proficient Data Interested Decision Tree (PDIDT) algorithm is functioned. We have concentrated on mitigating the Distributed Denial of service (DDos) attacks and in reducing the false alarm rate (FAR) with a global network monitor which can observe and control every flow between any pair of hosts.

Keywords Intrusion detection · Data mining · Classification algorithm · Denial of service attack

1 Introduction

Intrusion is the process of secretly trying to access the shared resources available on the network and it occurs quickly in this modern era. Intellectual IDS could differentiate intrusion and non-intrusion activities. These existing system of Intrusion detection is usually accomplished through either signature or anomaly detection. With signature detection, attacks are analyzed to generate unique descriptions. This allows for accurate detections of known attacks. However, attacks that have not been analyzed cannot be detected. This approach does not work well with new or

M.D. Shree (✉) · J. Visumathi · P.J. Jayarin
Department of Computer Science and Engineering, Jeppiaar Engineering College, Chennai, India
e-mail: divya.sri991@gmail.com

L.P. Suresh and B.K. Panigrahi (eds.), *Proceedings of the International Conference on Soft Computing Systems*, Advances in Intelligent Systems and Computing 398, DOI 10.1007/978-81-322-2674-1_60

unknown threats. With anomaly detection, a model is built to describe normal behavior. Significant deviations from the model are marked as anomalies, allowing for the detection of novel threats. Since not all anomalous activity is malicious, false alarms become a issue.

Intrusion occurs in a few seconds in this fast changing modern era. Intruders easily modify the commands and erase their prints in review and record files. Prosperous IDS intellectually differentiate between intrusive and nonintrusive records. Utmost existing systems have security breaks that make them easily vulnerable and could not be solved. Extensive research is going on IDS that are yet undeveloped tool in contradiction of intrusion. It is a most challenging task for network admin and security authorities to protect the resources on network which are defenseless. Hence this has to be interchanged with more secure system.

IDS grounded with Data mining efficiently recognize data of user interest also prophesies the result that can be used in the near future. Data mining or knowledge discovery in databases has multiplied a great pact of consideration in Information Technology industry as well in the society. To analyze the handy information from large volumes of data the concept of data mining is been involved which is placed centrally to seizure all the incoming packets that are communicated over the network.

Elicited data are sent for pre-processing to remove the noise, irrelevant and missing attributes that are replaced. Then the pre-processed data are analyzed and classified according to their severity measures. If the record is normal, then it is considered not require any change else it send for report generation to raise alarm. If the attack identified from a user, the admin is given an alert as process level attack detected. Once the admin finds such a message displayed blocks the user from further processing these are logged into the server machine. This proceeded as the user request is certain to server.

2 Related Work

The techniques to create a intrusion detection system various literature papers are involved. Mohammad et al. (has analyzed 10 % of KDD cup'99 training dataset based on intrusion detection. They have focused on establishing a relationship between the attack types and the protocol used by the hacker. ODM tool is been used to build thousand clusters which worked fine in accomplishing the task. K-means clustering algorithm using ODM has not yet been applied for analyzing the KDD Cup 99 Dataset.

Bhuyan et al. [1] have proposed a tree based clustering technique for finding clusters in network intrusion data. Also for sensing unknown attacks without using any labeled traffic or signatures or training. This technique is larger when compared to existing unsupervised network anomaly detection practices. Works on a faster, incremental version of Tree CLUSTER is underway for both numeric and mixed type network invasion data.

Kavitha et al. have anticipated port hopping technique to support multiple clients without synchronization. An algorithm for a server to support port hopping with many clients is presented. Conclusion derived that it is possible to employ the port hopping method in multiparty applications in a scalable way. Still the adaptive method can work under timing uncertainty and specifically fixed clock drifts.

Warusia et al. given an integrated machine learning algorithm across K-Means clustering and Naïve Bayes Classifier called KMC + NBC to overcome the drawbacks of existing paper. K-Means clustering is applied to labeling the entire data into corresponding cluster sets. Naïve Bayes Classifier is smeared to reorder the misclassified clustered data into correct categories. The essential resolution is to split the data among the conceivable attack and normal data into different clusters in earlier module. One disadvantage of labeled clustered data are later re-classified into specific classes which tedious.

Ganapathy et al., have surveyed on intelligent techniques for feature selection and classification for intrusion detection in networks based on intelligent software agents, neural networks and particle swarm intelligence. These techniques have been useful for identifying and for preventing network intrusions to provide security. Here the quality of service has been enhanced. Two new algorithms namely intelligent rule-based attribute selection algorithm and intelligent rule-based enhanced multiclass support vector machine have been proposed.

Uttam et al. have aimed at making improvements on existing work in three perspectives. Firstly, the input traffic pattern is pre-processed and redundant instances are removed. Next, a wrapper based feature selection algorithm is adapted which has a greater impact on minimizing the computational complexity of the classifier. Finally, a neuro tree model is employed as the classification engine which will improve detection rate.

Zhang Fu et al. have proposed effective Distributed Denial of Service (DDos) defense mechanism that assures legitimate users access to an Internet service. It detects threat and discard malicious packets in an online fashion. Given that emerging data streaming technology can enable mitigation in an effective manner STONE, a stream based DDoS defense framework data streaming technology.

Anshu et al. came up with an Anomaly-Detection based Intrusion Detection System. The main objective was to try various data mining approaches and analyze the results of the same when used for Anomaly Detection. Schemes such as Outlier Detection for Network based IDS and Prediction of system calls for Host based IDS, were vexed out. They have been able to detect attacks using both Network and Host-based IDS, which is the hybrid technique which is suggested for perfect IDS.

Mrutyunjaya et al. has investigated some novel hybrid intelligent decision technologies using data filtering by adding supervised or un-supervised methods along with a classifier to make intelligent decisions in order to detect network intrusions. They used a variant of KDD Cup 1999 dataset as NSL-KDD to build the proposed IDS. The performance comparison amongst different hybrid classifiers was made in order to comprehend the effectiveness of proposed system in terms of various performance measures.

Gisung et al. aimed of optimizing rule's quality individually, without directly taking into account the interaction with other rules. A less greedy, but possibly more expensive way to approach the problem was to comrade a particle with an entire rule set and then to consider the quality of it when evaluating a particle. It is known as the Pittsburgh approach in the evolutionary algorithm, is a stimulating research direction. Also the nominal part of the rule is discovered first which is advantageous to a co-evolved approach.

Fatima proposed a PSO-SVM technique to optimize the performance of SVM classifier. 10-fold cross validation is applied in order to legalize and evaluate the results. The outcomes found shows this approach gives a improved classification in terms of accuracy even though when the execution time is increased. Involvement consists of adapting an evolutionary method for optimization of this factor. They chose the method PSO which makes it possible to optimize the performance of classifier SVM.

Marina et al. proposed SIEVE that drape nodes to form a lightweight distributed filtering system against DDoS attacks. SIEVE uses source addresses as lightweight authenticators to filter malicious traffic and to save magnitude of computing rule compared with common message authentication algorithms. They presented that SIEVE can resourcefully filter attack during traffic also save server and router bandwidth for authentic traffic flow.

3 System Architecture

Figure 1 shows the architecture of the proposed system is the hybrid IDS developed for testing the intrusion detection system where client possess a valid username and password and sends request message to collect data from server if authorized user gets access else considered intruder and defended.

Fig. 1 System architecture

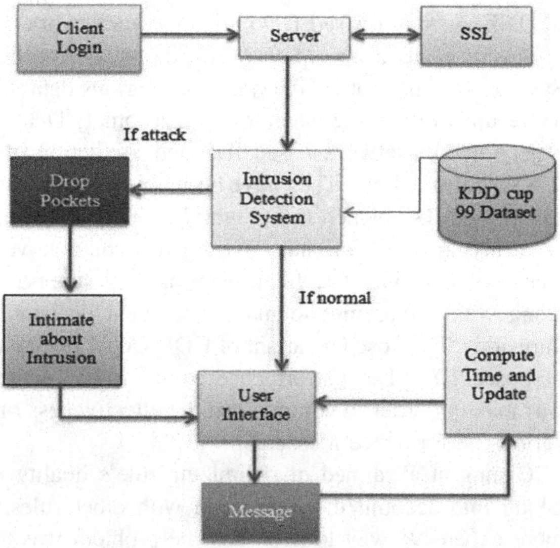

Main server module which is always on the go for intrusion contains various layers that sense the client, detects if it's an attack alerts the user on finding any intrusion addition data log is present where all alerts and data for future references. In experimentation result, the real system work is shown. Now login is created for any type of user at first then certain privileges are provided to admin. Once the packets are authorized sent to the module of Pseudorandom Function with Clock drift Technique (PFC) which detects particular intrusion if any intimate the user through gadgets else the packets are dropped making the system secured.

4 Implementation

4.1 Proficient Data Interested Decision Tree Algorithm (PDIDT)

The proposed PDIDT algorithm utilizes technique that identifies the local as well global best values for n number of iterations to obtain the optimal solution. The best solution is attained by manipulative the regular value and by finding the exact efficient features from the given training data set which are trained based on this classification algorithm. The normalized information gain for each attribute and decision node forms a best attribute.

Step 1: For each attribute a, unique values are selected.
Step 2: If n unique values fit to the same class label, these are divided into m intervals, and m must be less than n.
Step 3: If the unique values go to different c class label, check the probability of the value belongs to same class.
Step 4: If it is found then change the class label of values with the class label of highest probability.
Step 5: Split the unique values as c interval then repeat Checking of unique values in the class label for all values in the data set.

4.2 Hybrid Intrusion Detection System

Swing in java is installed to seizure the network packets in real time also KDD Cup 99 dataset is used. KDD Cup 99 data set contains 23 attack types and their names are revealed and its features are grouped as, basic feature, traffic feature, and content feature. For all 23 attacks labeled in the KDD cup dataset, the associated features are intended by aiding the threshold value. If the attribute satisfies the specified constraints then the attribute is chosen as the related features of particular attack. In semi supervised approach the dataset is divided into training and testing data. Training data comprises of both the labeled data and unlabeled data. By means of the labeled data the unlabeled data is labeled.

4.3 Pseudorandom Function with Clock Drift Technique (PFC)

To mitigate the efficiency of DDos Attack pseudorandom function with clock drift (PFC) technique is used. This tactic has a variable clock drift process to avoid the client waiting time for server and at the same time message loss is avoided significantly. It includes the features, contact initiation and data transmission part.

Step 1: Client initiates a request to the server for connection.
Step 2: Once the server receives the contact initiation message it sends the varying clock value of the client and the server clock. t1... tn Vh(t1) = {low-rate(L),mediumrate(M),high rate(H)}
Step 3: It would be stored by the client to estimate the variable clock drift.
Step 4: The client waits for the Acknowledgment from the server for A specific period of time, say 2l + L.
Step 5: On over waiting chooses another interval and starts sending messages

It may take any a number of trials to get access to the server. In our proposed algorithm, the number of trails made by the client in contact initiation part has been minimized so as to improve the reliability of the application. Once the message is received, the server waits for the port to open. As soon as the port is open server sends the acknowledgment with pseudorandom function, index value and varying time to the client that is used for further communication.

5 Experimental Results

5.1 Training and Testing Data

The Third International Knowledge Discovery and Data Mining Tools Competition (KDD Cup 99) dataset are made used in these experimental result. There are a total of 41 attributes defined in each connection record. The intrusion detection is really a challenging assignment owing to the list of attributes that consists of both continuous-type and discrete type variables, with statistical distributions varying significantly from the former. The KDD dataset contains five million network connection records viz. land attack, Neptune attack, password guess, port scan, etc. The 22 categories of attacks are described these following four major classes: DoS, R2L, U2R and Probe. This basic information of 41 features describe about the network packet, network traffic, host traffic and content information. The five class labels are contained in each record such as normal, probe, DOS, R2L and U2R. It also has 391458 DOS attack records, 52 U2R attack records, 4107 Probe attack record, 1126 R2L attack records and 97278 normal records shown only in this 10 % of this data set.

5.2 Results

Table 1 compares the C4.5, SVM, proposed PDIDT classification based algorithms. Figure 2 specifies the corresponding chart for the result obtained in Table 1. The chart illustrates the build time of, SVM, C4.5 and Improved PDIDT algorithm which provides higher accuracy percentage than the other two classification algorithms.

According to Table 1; Fig. 3 shows the performance of existing and proposed PDIDT algorithms based on false alarm rate (FAR). Thus the proposed PDIDT Algorithm effectively detects attack with less computational time and FAR.

Table 1 Accuracy based on attacks

Algorithms	Accuracy	Computational Time (CT)	False Alarm Rate (FAR)
C4.5	8	4	1.6
SVM	4.5	2	3.2
SVM + PSO	9.1	12	1.94
Proposed PDIDT	9.5	8	0.18

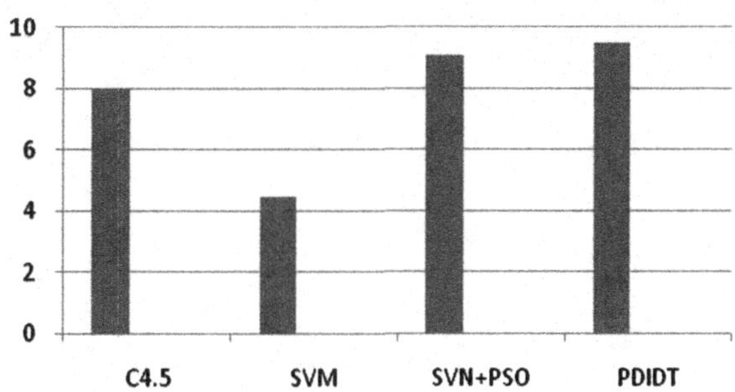

Fig. 2 Results comparison built on accuracy

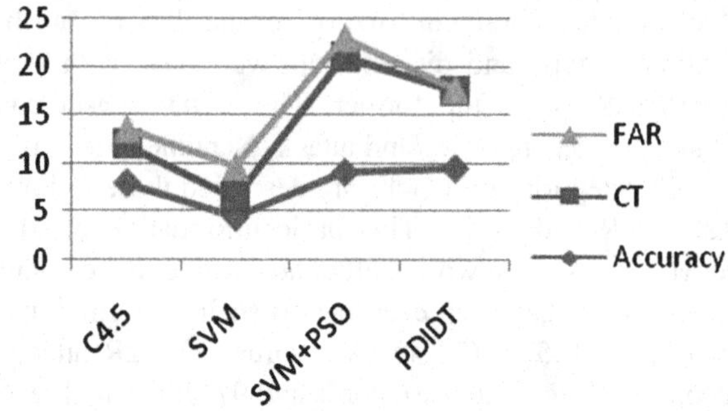

Fig. 3 Performance based on CT, FAR, accuracy

6 Conclusion

The authentic size of the dataset is reduced that assist the admin to analyze the attacks efficiently with less FAR. The semi supervised approach is applied to resolve the irresistible problem of supervised and unsupervised methods. Various typed attacks are identified by the IDS proposed which are immediately elucidated using the alert layer in the admin end. Mitigation of DDoS attack is worked based on the pseudorandom function with varying clock drift technique. The mitigation effort is identified by the admin as well the user through monitoring of the network. The information flow to user is done by toolkit application. Hence the intrusion system generated is operative with enhanced accuracy and performance of the system.

Reference

1. Bhuyan MH, Bhattacharyya DK, Kalita JK (2012) An effective unsupervised network anomaly detection method. In: International conference on advances in computing, communications and informatics, no. 1, pp 533–9

NEBULA: Deployment of Cloud Using Open Source

Mulimani Manjunath, Bhandary Anjali, N. Ashwini, K. Nagadeeksha and S. Jayashree

Abstract Cloud computing is a neoteric model which facilitates suitable, on-demand network access can be quickly supplied and released with minimal management effect or service provider interaction to a shared pool of configurable computing resources. Here, our objective is to use physical resources in a cost-effective manner. Use of different applications in different OS on dedicated physical machine for each OS leads to wastage of resources. The virtualization is the best solution for the use of resources. It allows all the users to run their likely OS virtual machines on single system using KVM for virtualization. In this paper, we provide a detailed review of a deployment of cloud computing using open source.

Keywords Cloud computing · Deployment · KVM · Virtual machine · Virtualization

---

Manjunath Mulimani, Anjali, Ashwini N, Jayashree, Nagadeeksha, Nebula: Deployment of cloud using Open Source, 2015.

M. Manjunath · B. Anjali (✉) · N. Ashwini · K. Nagadeeksha · S. Jayashree
Department of Computer Science & Engineering, Sahyadri College of Engineering and Management, Adyar, Mangaluru 575007, Karnataka, India
e-mail: anjalibhandary04@gmail.com

M. Manjunath
e-mail: manjunath.gec@gmail.com

N. Ashwini
e-mail: ashwinirai.cse@gmail.com

K. Nagadeeksha
e-mail: ndeekshad@gmail.com

S. Jayashree
e-mail: jayashreesalian6@gmail.com

L.P. Suresh and B.K. Panigrahi (eds.), *Proceedings of the International Conference on Soft Computing Systems*, Advances in Intelligent Systems and Computing 398, DOI 10.1007/978-81-322-2674-1_61

1 Introduction

Cloud computing has become the widely adopted computing service in the modern era. Cloud computing is the use of computing resource that are delivered as a service over a network. Deployment models and service models (SAAS, PAAS, and IAAS) are the two types of cloud computing. Cloud resources are dynamically reallocated per demand and shared by multiple users. The colossal amount of resources in the cloud needs proper utilization of the physical machines and its related virtual machines. The main criteria in the cloud is to allocate the resources for user job requests toward attaining optimum performance with least energy consumption and thereby, leading to low cost service. Virtualization is a technique which facilitates an abstract version of the physical resources in the form of VMs [1–3]. In recent years, operating system virtualization has enticed substantial interest, peculiarly from the data center and cluster computing communities. Using virtualization, it is possible to easily adjust the resource allocations by changing the number of running VMs and perform load balancing by migration of VMs across the physical machines (PMs) without disrupting any active network connections of the environment. A useful tool for administrators of data centers and clusters are migrating operating system instances across different physical hosts are: It enables a clean separation between software and hardware. And it also provides load balancing, fault management, and low-level system maintenance [4].

2 Background

Cloud computing makes extensive use of virtual machines (VMs) because they permit workloads to be kept far away from one another and for the resource usage to be somewhat controlled. However, the extra levels of abstraction involved in virtualization reduce workload performance, which is passed on to customers as worse price/performance. Newer advances in container-based virtualization simplify the deployment of applications while continuing to permit control of the resources allocated to different applications [5]. Since 1960s Virtual machine concept was in existence. It was first developed by IBM to provide concurrent, interactive access to a mainframe computer. All the VM used to be an instance of the physical machine. The users wanted to access physical machine directly. So it gave users mirage of accessing the physical machine directly. It was an artistic and transparent way to allow resource-sharing and time-sharing on the highly expensive hardware [6]. It was early recognized that for testing and resource utilization purposes the idea of using a computer system to imitate another computer system was useful. IBM led the way with their Virtual Machine system like the many other computer technologies. VMware's software only virtual machine monitor has been quite successful till date. Recently, the Xen [xen] open-source hypervisor brought virtualization to the open-source world. First, it brought with a variant termed as paravirtualization. And later as hardware became available, full virtualization was also provided [7]. Kernel virtual machine (KVM) [8] is a feature of Linux that allows Linux to act as a type 1 hypervisor [9], running an unmodified guest operating system (OS) inside a Linux process. KVM uses hardware virtualization features in recent processors to reduce complexity and overhead; for example, Intel VT-x hardware eliminates the need for complex ring compression schemes that were pioneered by earlier hypervisors like Xen [10] and VMware [11]. KVM supports both emulated I/O devices through QEMU [12] and paravirtual I/O devices using virtio [13]. The combination of hardware acceleration and paravirtual I/O is designed to reduce virtualization overhead to very low levels [14]. KVM supports live migration, allowing physical servers or even whole data centers to be evacuated for maintenance without disrupting the guest OS [15]. KVM is also easy to use via management tools such as Libvirt [16]. The Collective project [17] has previously traversed VM migration as a tool to facilitate mobility to users. The users will be working on different physical hosts at different times in different places. For example, when the user completes his work at one place and goes to his house and he will want to transfer an OS instance to his home computer to work on it. Their work aims to optimize for slow (e.g., ADSL) links and longer time spans. So stops OS execution for the duration of the transfer, with a set of enhancements to reduce the transmitted image size.

3 Problem Statement

When we need to use different applications in different OS on dedicated physical machine for each OS leads to wastage of resources and maximum power consumption. To run multiple operating systems on single machine by making partition leads to resource (hard disk, RAM, etc.) wastage. In other way to separate physical machine for each OS leads to maximum power consumption [18–20].

4 Proposed System

A complete cloud laboratory will be set up. Laboratory is highly necessary for all students for their research on cloud. From our cloud initially we are providing infrastructure as service using open-source KVM (Kernel-based virtual machine). Later, we can build cloud to provide PAAS and SAAS services. For an efficient resource utilization virtualization is the best solution. Virtualization allows us to run different virtual machines on single server system. It is possible for all the users to create virtual machines with specified requirements and their likely OS. Users can access their virtual machines over internet whenever he wants. No question of resource wastage or power consumption of user system. We are proposing a toolkit named Nebula, it provides nice GUI for users to create VM's and manage them on server through internet. Nebula is described as an integrated work flow system for request and commissioning of virtual machines. It provides facilities like suspend, resume, shutdown, power off, and power on. Nebula also provides live migration of VM's from one physical host to the other. Dynamic resource scheduling, resource utilization monitoring, and power management is possible.

5 System Design

Nebula will have graphical user interface (Fig. 1). The clients first have to register to Nebula, and then he can login to it any time. His details will be stored in the database. Once he logins he can request for virtual machines with the required specifications and manage or list his virtual machines if any previously available. The administrator grants permission and creates virtual machines for the requested client. Users of the administrator privilege can switch role to access the administrators interface. The user can start, stop, suspend, pause, resume, and destroy his virtual machines. It also provides live migration (Fig. 4).

A virtual machine is logical container, like a file system, that is stored as set of files in its own directory in a database, which hides the specifics of each storage device and provides a uniform model for storing virtual machine files. ISO images, virtual machine templates, and floppy images can also be stored in data stores (Fig. 2). They can be backed by network file system. Cloud storage contains several virtual machines and each virtual machine has required OS and corresponding applications as shown in Fig. 3. When we analyze further, the benefit enabled by virtualization: it is live OS migration. When the user is using any virtual machine,

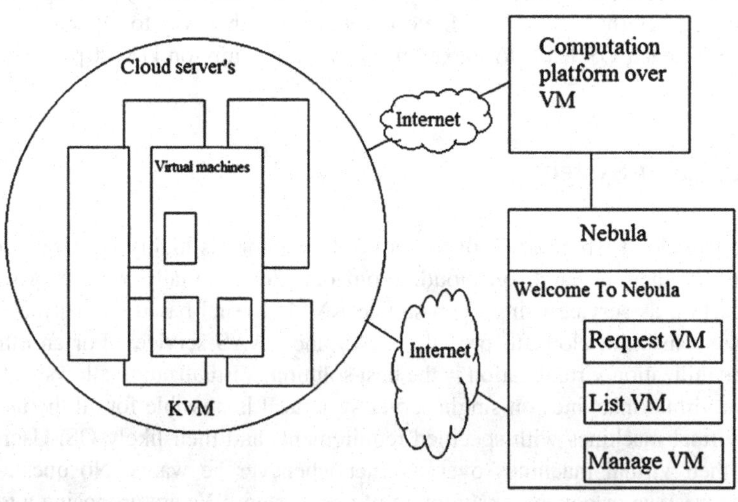

Fig. 1 Architecture of Nebula

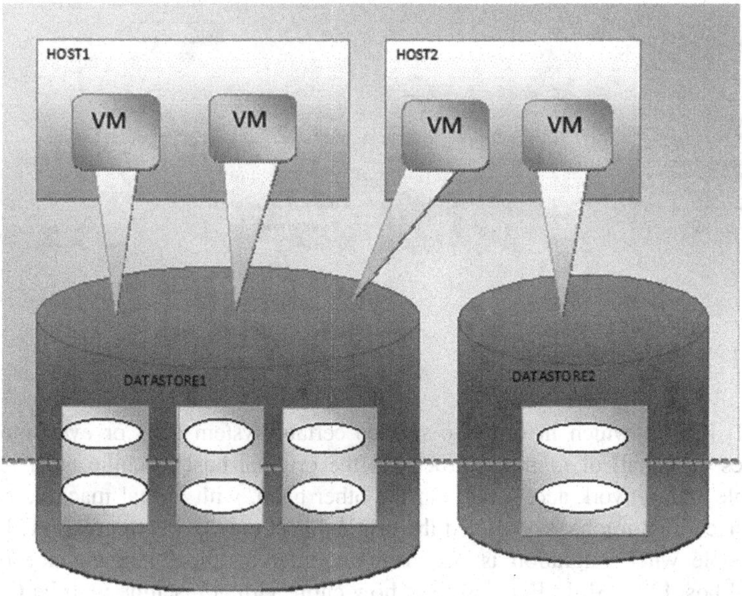

Fig. 2 Database storage of Nebula [20]

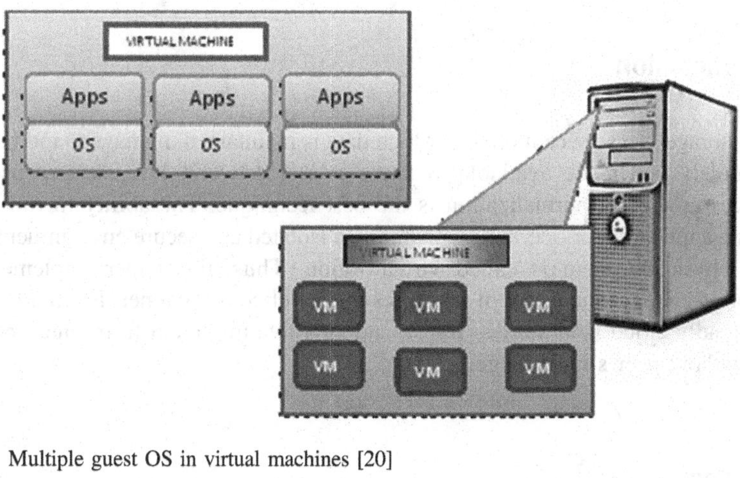

Fig. 3 Multiple guest OS in virtual machines [20]

live migration of that VM is possible. Live migration moves running virtual machines from one physical server to another without any impact on virtual machine availability to the users. We can avoid many of the difficulties faced by process-level migration approaches by migrating an entire OS and all of its applications as one unit. It is easy avoid the problem of 'residual dependencies' by the narrow interface between a virtualized OS and the virtual machine monitor

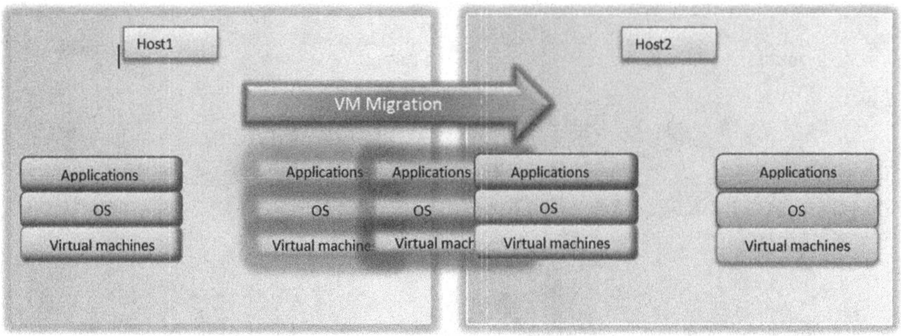

Fig. 4 Live VM migration [20]

(VMM) [21] in which in order to service certain system calls or even memory accesses on behalf of migrated processes the original host machine must remain available and network accessible. On the other hand, with virtual machine migration, once migration has completed the original host may be decommissioned. This is valuable when migration is occurring in order to allow maintenance of the original host [22, 23]. In Fig. 4 we see how entire virtual machine with its OS and applications is migrated from one physical machine (host 1) to another physical machine (host 2).

6 Conclusion

Cloud storage is a service model in which data is maintained, managed, and backed up remotely and made available to users over a network. For users to use the resources efficiently, virtualization is the best technique. The ability to run more than one application or operating system in an isolated and secure environment on a single physical system is called virtualization. Thus, this paper implements a solution to enhance the usage of resources in cost-effective manner. It provides nice user-friendly cloud laboratories. It also supports live migration from one server to other without user's knowledge.

References

1. Chen S, Mulgrew B, Grant PM (1993) A clustering technique for digital communications channel equalization using radial basis function networks. IEEE Trans Neural Netw 4:570–578
2. Duncombe JU (1959) Infrared navigation—Part I: an assessment of feasibility. IEEE Trans Electron Devices ED-11:34–39
3. Lin CY, Wu M, Bloom JA, Cox IJ, Miller M (2001) Rotation, scale, and translation resilient public watermarking for images. IEEE Trans Image Process 10(5):767–782

4. Clark C, Fraser K, Hand S, Hansen JG, Jul E, Limpach C, Pratt I, Warfield A (2005) University of Cambridge Computer Laboratory, Cambridge
5. Felter W, Ferreira A, Rajamony R, Rubio J (2014) An updated performance comparison of virtual machines and linux containers. IBM Research, Austin
6. Nanda S, Chiueh T (2005) A survey on virtualization technologies. Department of Computer Science, SUNY at Stony Brook, Stony Brook
7. Kivity A, Kamay Y, Laor D, Lublin U, Liguori A (2007) kvm: the Virtual Machine Monitor
8. Kivity A, Kamay Y, Laor D, Lublin U, Liguori A (2007) KVM: the Linux virtual machine monitor. In: Proceedings of the linux symposium, vol 1. Ottawa, Ontario, Canada, pp 225–230
9. Popek GJ, Goldberg RP (1974) Formal requirements for virtualizable third generation architectures. Commun ACM 17(7):412–421
10. Virtualization overview. http://www.vmware.com/pdf/virtualization.pdf
11. Xen Project Software Overview. http://wiki.xen.org/wiki/Xen Overview
12. Bellard F (2005) QEMU, a fast and portable dynamic translator. In: Proceedings of the annual conference on USENIX annual technical conference, ATEC '05. USENIX Association, Berkeley, CA, USA, p 41
13. Russell Rusty (2008) Virtio: towards a de-facto standard for virtual I/O devices. SIGOPS Oper Syst Rev 42(5):95–103
14. McDougall R, Anderson J (2010) Virtualization performance: perspectives and challenges ahead. SIGOPS Oper Syst Rev 44(4):40–56
15. Balogh A (2013) Google Compute Engine is now generally available with expanded OS support, transparent maintenance, and lower prices. http://googledevelopers.blogspot.com/2013/12/google-compute-engine-is-now-generally.html, Dec 2013
16. Bolte M, Sievers M, Birkenheuer G, Niehörster O, Brinkmann A (2010) Non-intrusive virtualization management using Libvirt. In: Proceedings of the conference on design, automation and test in Europe, DATE '10, European Design and Automation Association, pp 574–579
17. Sapuntzakis CP, Chandra R, Pfaff B, Chow J, Lam MS, Rosenblum M (2002) Optimizing the migration of virtual computers. In: Proceedings of the 5th symposium on operating systems design and implementation (OSDI-02), Dec 2002
18. Laor D, Kivity A, Kamay Y, Lublin U, Liguori A (2007) KVM: the linux virtual machine monitor. Virtualization technology for directed I/O. Intel Technol J 10.225–230
19. Ubuntu enterprise cloud—overview. http://www.ubuntu.com/business/cloud/overview
20. VMWare vCloud Director—deliver infrastructure as a service without compromise. http://www.vmware.com/products/vclouddirector/features.html
21. Milojicic D, Douglis F, Paindaveine Y, Wheeler R, Zhou S (2000) Process migration. ACM Comput Surv 32(3):241–299
22. Stage A, Setzer T (2009) Network-aware migration control and scheduling of differentiated virtual machine workloads. Proceedings of the 2009 ICSE workshop on software engineering challenges of cloud computing, pp 9–14
23. Lloyd W, Pallickara S, David O (2011) Migration of multi-tier applications to infrastructure-as-a-service clouds: an investigation using kernel-based virtual machines. IEEE/ACM 12th international conference on grid computing, (GRID), pp 137–144

An Efficient Task Scheduling Scheme in Cloud Computing Using Graph Theory

S. Sujan and R. Kanniga Devi

Abstract Cloud computing is an archetype that consists of standard concepts and technologies like distributed computing, grid computing, utility computing, Virtualization, and Internet technologies in large scale. Incorporating various technologies and domains into one single concept, cloud computing delivers it resources as a service to the end users. Being an emerging technology in the current computing scenario, the cloud Computing faces various issues that include resource provisioning, security, energy management and reliability. Considering the context of scheduling from the resource provisioning issue, this work focuses on an efficient task scheduling scheme. Adapting the concept of graph theory this paper comes out with a theoretical proposal on scheduling tasks in the Cloud environment. In this paper, bandwidth is being considered as the metric and an efficient bandwidth utilization concept is proposed to schedule the task using graph theory. This theoretical proposal is being proposed considering all the attributes like virtualmachine (VM), host and the task that constitutes the element of Cloud environment.

Keywords Cloud Computing · Scheduling · Graph theory

1 Introduction

Cloud Computing is one of the emerging technologies which has got its issues and researches perennial among the other technical environments and domains. Standing unique among all the other domains in modern computing scenarios like grid computing, utility computing, and Soft computing, the cloud computing has

S. Sujan (✉) · R. Kanniga Devi
Department of Computer Science and Engineering, Kalasalingam University,
Anand Nagar, Krishnankoil, Srivilliputhur, Tamil Nadu 626126, India
e-mail: sujansuresh7@gmail.com

R. Kanniga Devi
e-mail: rkannigadevi@gmail.com

L.P. Suresh and B.K. Panigrahi (eds.), *Proceedings of the International Conference on Soft Computing Systems*, Advances in Intelligent Systems and Computing 398, DOI 10.1007/978-81-322-2674-1_62

got both the developers as well as the end users actively use this environment. This makes the environment a lot more versatile to the user and vibrantly challenging to the developers to render a reliable and a quality service to the user who are associated with the system. Having more number of users incoming and outgoing in a dynamic environment, each and every task by the user has to be reliably serviced by the system and hence scheduling the tasks may be considered as one of the foremost constraints that has to be taken into account. On the other hand, there is a steady increase in users who get into the cloud computing environment every day. Hence scheduling the tasks in an efficient way becomes mandatory.

Adhering the concepts like heterogeneity among the users who are accessing the system, maintaining various levels of data abstractions in the system and inheriting its core behavior from conventional computing concepts like distributed and utility computing the archetype of cloud environment stands wide and complex. Incorporating the virtualization concept in the conventional data centers, the classical data center (CDC) and converting it into be a virtualized data center (VDC) from which Cloud services are offered. As Internet plays a main role in establishing the Cloud environment, pairing the Internet technology with a VDC gives out the Cloud Data Center. Each and every task that is initiated by the user gets into the Cloud Data Center. As Cloud Data Center consists of thousands of networked nodes, the entire Cloud Data Center can be represented in a graph structure. A weighted graph consists of set of nodes which can be taken as a host and the vertices establish the connection between the nodes that are present in the environment.

Imagining the entire Cloud environment into an undirected weighted graph, it extends like a road map where each and every city that is marked in the map is represented as vertices and the edge connections are obtained from a city to another city. Cloud Computing too is such a dynamic environment with varied number of hosts and virtual machines that are paired with one another to service the request that enters into the system. Efficient utilization of bandwidth becomes mandatory when it comes to huge set of tasks. At the same time mapping the appropriate task to the machine that is present in the system decides the efficient utilization of the resources. Considering an undirected weighted graph as a Cloud environment the weight assigned for the edges is taken as the bandwidth that is being provided for those particular vertices. Hence the task traveling from one virtual node to the other virtual node is supposed to be done under that provided bandwidth. Representing in a graph-based model makes it a lot more flexible in analyzing the behavior of the system in some enterprise-based organizations that rely on Cloud networks. They enable the application of proven graph algorithms to solve enterprise topology research problems in general and Cloud research problems in particular [8]. As the user uses the Cloud environment with a service level agreement (SLA) with the Cloud service provider and pays for what they use, it becomes necessary to provide the user a highly available service. In such cases, no path caching between the nodes is taken. This is done based on the widest path algorithm. This concept works on the Dijkstra's-based approach. Graph traversal of task is being done with the route which provides higher bandwidth so that a reliable and a quality network path is obtained in order to provide a measured service to the end user without any lagging or distortion.

2 Related Work

With an uncountable number of dynamic network nodes, a Cloud environment extends as a pool of computational scenario which has got huge set of tasks incoming, initiated by the user, huge set of resources present in the system, which are being mapped to the tasks to get executed. In order to maintain this environment, computation scenario available to the end user who could access the system anytime, scheduling the task to the resources becomes mandatory. Many research papers and techniques have been proposed so far in the concept of task scheduling in the Cloud computing. Incorporating the concept of graph theory into it has also led to the evolution of ideas into various levels and it has also been made practically possible.

Mezmaz et al. [1] proposed a technique on bi-objective genetic algorithm for that takes into account, not only makespan, but also energy consumption using the NP hardness problem. Minimization of energy consumption is done based on dynamic voltage scaling (DVS).

Frîncu [2] considering the job scheduling mechanism, Scheduling highly available applications on the Cloud environment has been proposed as a technique. A multi-objective scheduling mechanism that holds with the probability of scheduling highly available environments, has been proposed, inspiring the property of previously used application scaling property.

Abrishami et al. [3] an algorithm for deadline constraint workflow scheduling has been proposed considering the workflow scheduling mechanism. This model is taken for Infrastructure as a service Cloud and an algorithm is proposed on an ICPCP (Infrastructure Cloud Partial Critical Paths) derived from the conventional PCP algorithm. He has obtained results considering the deadline factor and normalized cost.

Liu et al. [4] considered Cloud services with data access awareness. A profit driven scheduling for Cloud services with data access awareness has been proposed, taking cost as the metric of work. He has proposed a new algorithm called max profit scheduling algorithm new optimization algorithm for profit-driven service request scheduling based on dynamic reuse, using the service request scheduling algorithm, which takes account of the personalized SLA characteristics of user requests and current system workload.

Wu et al. [5] an energy efficient scheduling mechanism for the Cloud centers has been proposed, considering the job scheduling technique using the conventional job scheduling mechanism. A scheduling algorithm for the Cloud datacenter with a dynamic voltage frequency scaling technique, where the scheduling algorithm can efficiently increase the resource utilization, has been proposed; hence, it can decrease the energy consumption for executing jobs.

Malik et al. [6] Latency-based group discovery algorithm for network aware Cloud scheduling is proposed considering the time as the metric. A model for the grouping of nodes with respect to network latency has been proposed as the technique. The application scheduling is done on the basis of network latency.

Xu et al. [7] a model for dynamic scheduling has been proposed. The resource allocation using the fairness constraint and the justice function has been compared, understanding the conventional Cloud scheduling mechanism inspired by the economic model. The first constraint is to classify user tasks by QoS preferences, and establish the general expectation function. This is done in accordance with the classification of tasks to restrain the fairness of the resources in selection process. The second constraint is to define resource fairness justice function to judge the fairness of the resources allocation.

Tobias et al. [8] to bring formalization and provability in Cloud environment and enterprise topology scenario is being handled using a graph model enterprise topology graph (ETG). The experimental cost which is incurred in the IT consolidation is reduced by simplifying the charges on enterprise topology, using the power of ETG approach is illustrated.

Speitkamp et al. [9] a decision models to optimally allocate source servers to physical target servers while considering real-world constraints, is being presented. Besides an exact solution method, a heuristic is presented to address large-scale server consolidation projects. A preprocessing method for server load data is introduced allowing for the consideration of QoS levels.

Jayasinghe et al. [10] considering an IaaS cloud environment, an idea about how to improve performance and availability of services hosted is presented. A structural constraint aware virtual machine placement (SCAVP) is being proposed as the system where a hierarchical placement approach with four approximation algorithms that efficiently solves the SCAVP problem for large problem sizes is being designed and the efficiency and importance of the proposed approach is discussed.

Chan et al. [11] a computing Cloud is being formulated as a graph in which computing resource such as services or intellectual property access rights as an attribute of a graph node, and the use of the resource as a predicate on an edge of the graph. Algorithms that are used to compose Cloud computations and model-based testing criteria to test Cloud applications are being presented.

Out of all the surveys made for this model, none of the above-mentioned work uses graph theory, to represent Cloud infrastructure, and hence, it is a motivational factor for incorporating the graph theory into Cloud Computing.

3 Proposed Work

Each and every user who enters into the Cloud environment comes in with set of tasks which has to be serviced by the resources present in the Cloud environment. Considering a real time Cloud environment with huge set of users with plethora task the system must be able to handle them. In a Cloud computing environment, each and every task is being serviced by the virtual machines present in the physical machine. A physical machine is nothing but the host which holds the set of processing elements that are required to process the task that is provided by the user into the system. A Cloud environment can have any number of physical machine deployed, depending upon the amount of requests that it handles regularly. A physical machine is capable of holding humongous amount of virtual machines in them. A task is being mapped to the host based on the broker policy in the Cloud environment. The Broker is the place where policy creations for the scheduling are being made for the task to get scheduled to the particular host.

Imagining the entire Cloud infrastructure as a graph-based model, the task first enters the Cloud environment. As soon as it enters the Cloud environment, the task reaches the broker. The broker receives the tasks and maps the set of generalized tasks to the hosts that are present in the system. The concept of mapping is done based on a policy that is created and stored in the broker, which differs from environment to environment. This policy is created by the organization that would like to establish the Cloud environment based on their ideas and priorities. So the policy with which the task is mapped from the broker to the system is purely dynamic and an organizational-based criteria. As the task gets mapped from the broker to the host, the execution of the task is being carried out based on the theoretical proposal which extends this way. Each and every host has a lot of virtual machines which are taken in a graph-based representation.

An undirected weighted graph where each and every vertices taken as nodes are connected by the respective weighted edges in which the weight represents the bandwidth that is allocated for that particular edge. The task which enters the host is mapped to a VM-based on the widest path algorithm using the Dijkstra's approach. The destination VM is being reached from the host through the graph using the edge which has got the maximum bandwidth that is being allocated to it. This is a process which is carried out for a single task that is entering the system. In real time for a Cloud computing environment which has got huge set of tasks entering and exiting the system, the same theoretical proposal is extended where the consecutive tasks that enter the host are mapped to the next nearest virtual machine with the edge bearing the highest cost. This conceptual methodology dynamically extends with the set of host that is present in the system (Fig. 1).

Each and every host that is present in the Cloud environment is treated as a network node in the virtual data center.

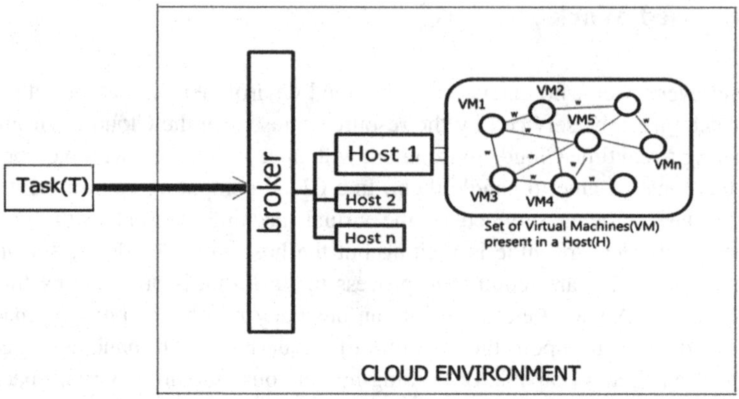

Fig. 1 Proposed system model

Relating the vertices and edges in the graph the host and the virtual machine present in the Cloud environment are represented as $C = \{\text{Host, VM}\}$ where Host represents set of physical machines and VM represents set of Virtual machines.

Having a huge set of networked nodes in a single Cloud environment the entire Cloud environment can be represented as an undirected weighted graph.

The graph-based representation for the physical machine is given as $G_i = (V_i, E_i)$, where

$V_i = \{v_1, v_2, v_3\}$ be the set of physical machines in the Cloud environment.

$E_i = \{e_1, e_2, e_3\}$ be the set of edges connecting physical nodes.

Similarly, the graph-based representation for the set of virtual machines that are present in the physical machine is given as

$G_j = (V_j, E_j)$, where $V_j = \{v_1, v_2, v_3, v_4, v_5\}$ represents the set of virtual machines and $E_j = \{e_1, e_2, e_3, e_4, e_5\}$ represents the set of edges of the respective virtual nodes.

Hence from the above graph-based representation, the Cloud environment C can be represented as

$C = G_i \cup G_j$ that is,

$$C = \{V_i \cup V_j\} \ \& \ \{E_i \cup E_j\}$$

In such a setup, the task must reach the VM which is of higher cost/bandwidth.

Let $T = (T_1, T_2, T_3, \ldots, T_n)$ be the set of tasks that are entering the system, $H = (H_1, H_2, H_3, \ldots, H_n)$ be the number of hosts present in the system.

Hence the VM mapped to the host is represented as,

$$\begin{aligned} H(\text{VM}) = \{ & H_1(\text{VM}_1^w, \text{VM}_2^w, \text{VM}_3^w, \ldots, \text{VM}_n^w), H_2(\text{VM}_1^w, \text{VM}_2^w, \text{VM}_3^w, \ldots, \text{VM}_n^w), \\ & H_3(\text{VM}_1^w, \text{VM}_2^w, \text{VM}_3^w, \ldots, \text{VM}_n^w), \ldots, H_n(\text{VM}_1^w, \text{VM}_2^w, \text{VM}_3^w, \ldots, \text{VM}_n^w)\} \end{aligned}$$

where '*w*' represents the weight which is considered as the bandwidth that is allocated to every VM present in a host.

Hence, a task T_i allocated to a VM is represented as follows

$T_i \rightarrow \{H_j(\mathrm{VM}_k^w)/W(\mathrm{VM}_k) > [H_j(\mathrm{VM}^w) - W(\mathrm{VM}_k)]\}$, that is, for a task T_i entering the Cloud environment is mapped to the virtual machine VM_k^w present in the host H_j, where, the weight of the VM_k^w is greater than the weights of all the other virtual machines present in the host H_j.

4 Conclusion

The model presented in this paper proposes a scheduling mechanism for the Cloud Computing environments. Incorporating the concept of graph theory the proposed theoretical model comes out with an efficient solution for allocation of resources considering bandwidth as the constraint. Efficient utilization of bandwidth along all the virtual machines connected in a host provides an efficient way of servicing the incoming tasks. Scheduling the incoming task to a virtual machine of higher bandwidth makes the task to get executed faster and without any lagging when compared to other scheduling concepts. The proposed model can further be improved considering some real time organization based needs by incorporating various scheduling algorithms among the graph of virtual machines present in a host.

References

1. Mezmaz M, Melab N, Kessaci Y, Lee YC, Talbi E-G, Zomaya AY, Tuyttens J (2011) A parallel bi-objective hybrid metaheuristic for energy-aware scheduling for Cloud computing systems. Parallel Distrib Comput 71:1497–1508
2. Frîncu ME (2014) Scheduling highly available applications on Cloud environments. Future Gener Comput Syst 32:138–153
3. Abrishami S, Naghibzadeh M, Epema DHJ (2013) Deadline-constrained workflow scheduling algorithms for Infrastructure as a Service Clouds. Future Gener Comput Syst 29:158–169
4. Liu Z, Sun Q, Wang S, Zou H, Yang F (2012) Profit-driven cloud service request scheduling under SLA constraints. J Inf Comput Sci 9(14):4065–4073
5. Wu C-M, Chang R-S, Chan H-Y (2014) A green energy-efficient scheduling algorithm using the DVFS technique for Cloud datacenters. Future Gener Comput Syst 37:141–147
6. Malika S, Huet F, Caromel D (2014) Latency based group discovery algorithm for network aware Cloud scheduling. Future Gener Comput Syst 31:28–39
7. Xu B, Zhao C, Hu E, Hu B (2011) Job scheduling algorithm based on Berger model in Cloud environment. Adv Eng Softw 42:419–425
8. Binz T, Fehling C, Leymann F, Nowak A, Schumm D (2012) Formalizing the Cloud through enterprise topology graphs. Institute of Architecture of Application Systems, University of Stuttgart, Germany
9. Speitkamp B, Bichler M (2010) A mathematical programming approach for server consolidation problems in virtualized data centres. IEEE Trans Serv Comput 3

10. Jayasinghe D, Pu C, Eilam T, Steinder M, Whalley I, Snible E (2011) Improving performance and availability of services hosted on IaaS Clouds with structural constraint-aware virtual machine placement. Georgia Institute of Technology, USA
11. Chan WK, Mei L, Zhang Z (2009) Modeling and testing of cloud applications. In: Proceedings on IEEE APSCC 2009. City University of Hong Kong, Singapore, 7–11 Dec 2009

A Stochastic Modelling Approach for the Performance Analysis of an Intrusion Detection System

Ethala Kamalanaban and R. Seshadri

Abstract In this paper, an architecture of an intrusion detection system (IDS) based on a two-server queueing model is considered and studied. By using an integral equation approach, an explicit analytical solution is found for the steady-state probabilities of the states of the IDS. Some of the performance measures of the IDS such as the throughput, queueing delay, system utilization and packet loss are also obtained.

Keywords Intrusion detection system · Interconnected two-server queue · Throughput · Packet waiting time · Queueing delay · CPU utilization · Packet loss

1 Introduction

A wide range of malicious attacks like worms, trojans and port scans against active networks poses numerous issues such as wastage of network resources and degradation of network traffic. Usage of authenticated security measures has become an important activity for preventing such malwares from damaging the network performance. On the other hand, extensive usage of security measures causes overutilization of network resources and undesirable degradation in the performance of networks. A balanced viewpoint is necessitated in combining network security and network performance together. Accordingly, intrusion detection systems (IDS) have become an integral part of any active network for detecting

E. Kamalanaban (✉)
Department of Computer Science and Engineering, Vel Tech Rangarajan Dr. Sagunthala R&D Institute of Science and Technology, Avadi, Chennai 600062, India
e-mail: kamalanaban2009@gmail.com

R. Seshadri
Department of Computer Science and Engineering, Sri Venkateswara University, Tirupathi 517502, Andhra Pradesh, India
e-mail: ravalaseshadri@svuniversity.ac.in

L.P. Suresh and B.K. Panigrahi (eds.), *Proceedings of the International Conference on Soft Computing Systems*, Advances in Intelligent Systems and Computing 398, DOI 10.1007/978-81-322-2674-1_63

malicious signatures. A huge amount of research has been carried out in the past in the design and implementation of IDS targeting to detect malicious packets and reject them (see, for example, Debar et al. [1], Dreger et al. [2], Alsubhi et al. [3], Mitchell and Chen [4]). Shin et al. [5] have proposed an advanced probabilistic approach for network intrusion forecasting and detection. In their approach, they defined the network states by using K-means clustering algorithm and used Markov chain model to measure the level of the malicious nature of the incoming data. Representing the level of attacks in stages, they achieve improved detection performance and also show that their approach is robust to training datasets and the number of states in the Markov model. However, intruders manoeuver the architecture of any IDS and come out with new strategies to create powerful malwares to damage the performance of the network. This naturally increases the size of the signature database of IDS in a server and the processing time of IDS per packet is increased. Furthermore, memory space of the system in which IDS is installed becomes limited and packets are queued resulting in longer waiting times before being processed for transmission. Consequently, configuration of IDS is contemplated to achieve increase in throughput and decrease in (a) waiting times, (b) packet delay, (c) packet loss. Keeping this as the principal objective, Alsubhi et al. [6] have proposed an analytical queuing model based on the embedded Markov chain which analyses the performance of the IDS and evaluates its impact on performance. Their model achieves simultaneously trade-off between security enforcement levels on one side and network Quality of Service (QoS) requirements on the other. However, they have not provided explicit expressions for the steady-state probabilities for the states of the IDS. In this paper, we fill the gap by obtaining explicit expressions for the state probabilities and system performance measures.

The paper is organized as follows:

In Sect. 2, the queueing model of the IDS is presented. Section 3 derives the integral equations for the state probabilities in real time. The steady-state probabilities for the states of the IDS are obtained in Sect. 4. Some performance measures are obtained in Sect. 5. Section 6 provides the conclusion of the results of the paper.

2 Two-Server Queueing Model of IDS

We consider an IDS which has two stages connected in tandem. The first stage does the header analysis for the incoming packets and the second stage performs the content analysis for those packets which have been assigned by the first stage. IDS is said to be idle if both servers are idle. Packets arrive to the IDS according to a Poisson process with rate λ. We assume that the IDS has a finite capacity L. In other words, the IDS has a finite buffer of capability $L - 1$. The IDS is modelled as a First-Come-First-Served finite queue with two servers working in tandem. We shall call the server in the first stage as server 1 and the server in the second stage as server 2. We assume that the mean service time of server 1 is $1/\mu$ and that of server

2 is $1/\alpha$. An arriving packet to an idle IDS enters into the server 1 for header analysis. If the server 1 is busy at a time doing the header analysis of a packet, an arriving packet at that time joins the queue in the buffer. After completing the header analysis, the packet either leaves the system with probability p or routed to server 2 with probability $1-p$ where the content analysis is made. After completing the content analysis, the packet leaves the system. When the sever 2 is busy, no packet can enter for header analysis even if server 1 is idle. A server can process only one packet at a time. This assumption leads to the situation that the IDS can be in three stages, namely the stage 0 stands for the idle stage when both servers are idle (no packets in the IDS), the stage 1 stands for the situation when server 1 alone is busy, and the stage 2 stands for the situation when the server 2 alone is busy. Let $N(t)$ be the number of packets in the system at time t. Let $S(t)$ be the state of the IDS. We set

$$S(t) = \begin{cases} 0 & \text{if the IDS is idle,} \\ 1 & \text{if the IDS is performing the} \\ & \text{packet processing task,} \\ 2 & \text{if the IDS is performing the} \\ & \text{rule–checking process.} \end{cases}$$

Then $(N(t), S(t))$ is Markov whose state space is given by

$$\mathcal{U} = \{(0,0); (1,1), \ldots, (L-1,1), (L,1); (1,2), \ldots, (L,2)\}.$$

We assume that at time $t = 0$, the IDS is idle. Define the state probabilities

$$p_{m,n}(t) = \Pr\{N(t) = m, S(t) = n\},$$

where $m = 0, 1, 2, \ldots, L; n = 0, 1, 2$. The state transition diagram is as shown in Fig. 1. In the next section, we derive the integral equations for $p_{m,n}(t)$.

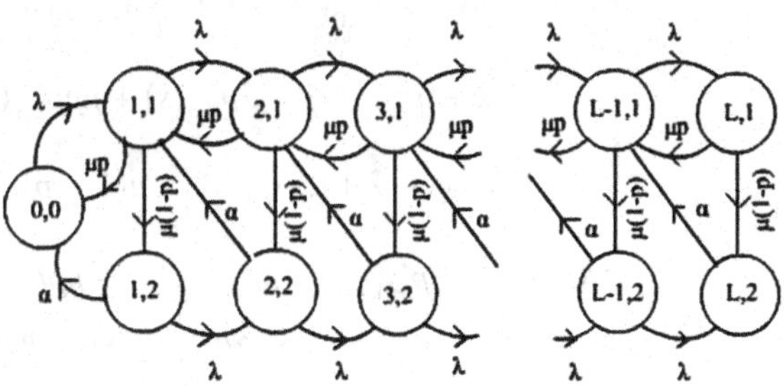

Fig. 1 State transition diagram for proposed model

3 Integral Equations for $p_{m,n}(t)$

Using probability laws, we get

$$p_{0,0}(t) = e^{-\lambda t} + p_{1,1}(t)\mu p \copyright e^{-\lambda t} + p_{1,2}(t)\alpha \copyright e^{-\lambda t}, \tag{1}$$

$$\begin{aligned} p_{1,1}(t) = p_{0,0}(t)\lambda \copyright e^{-(\lambda+\mu)t} + p_{2,1}(t)\mu p \copyright e^{-(\lambda+\mu)t} \\ + p_{2,2}(t)\alpha \copyright e^{-(\lambda+\mu)t}, \end{aligned} \tag{2}$$

$$p_{1,2}(t) = p_{1,1}(t)\mu(1-p) \copyright e^{-(\lambda+\alpha)t}, \tag{3}$$

$$\begin{aligned} p_{m,1}(t) = p_{m-1,1}(t)\lambda \copyright e^{-(\lambda+\mu)t} + p_{m+1,1}(t)\mu p \copyright e^{-(\lambda+\mu)t} \\ + p_{m+1,2}(t)\alpha \copyright e^{-(\lambda+\mu)t}, \quad m = 2, \ldots, L-1, \end{aligned} \tag{4}$$

$$\begin{aligned} p_{m,2}(t) = p_{m-1,2}(t)\lambda \copyright e^{-(\lambda+\alpha)t} \\ + p_{m,1}(t)\mu(1-p) \copyright e^{-(\lambda+\alpha)t}, \quad m = 2, \ldots, L-1, \end{aligned} \tag{5}$$

$$p_{L,1}(t) = p_{L-1,1}(t)\lambda \copyright e^{-\mu t}, \tag{6}$$

$$\begin{aligned} p_{L,2}(t) = p_{L,1}(t)\mu(1-p) \copyright e^{-\alpha t} \\ + p_{L-1,2}(t)\lambda \copyright e^{-\alpha t}, \end{aligned} \tag{7}$$

where we have used the copyright notation for

$$u(t) \copyright v(t) = \int_0^t u(\tau)v(t-\tau)\mathrm{d}\tau.$$

Denoting the Laplace transform of $p_{m,n}(t)$ by $p^*_{m,n}(s)$, Eqs. (1)–(7) yield

$$(s+\lambda)p^*_{0,0}(s) = 1 + \mu p p^*_{1,1}(s) + \alpha p^*_{1,2}(s), \tag{8}$$

$$(s+\lambda+\mu)p^*_{1,1}(s) = \lambda p^*_{0,0}(s) + \mu p p^*_{2,1}(s) + \alpha p^*_{2,2}(s), \tag{9}$$

$$(s+\lambda+\alpha)p^*_{1,2}(s) = \mu(1-p)p^*_{1,1}(s), \tag{10}$$

$$\begin{aligned} (s+\lambda+\mu)p^*_{m,1}(s) = \lambda p^*_{m-1,1}(s) + \mu p p^*_{m+1,1}(s) \\ + \alpha p^*_{m+1,2}(s), \quad m = 2, \ldots, L-1, \end{aligned} \tag{11}$$

$$\begin{aligned} (s+\lambda+\alpha)p^*_{m,2}(s) = \lambda p^*_{m-1,2}(s) + \mu(1-p)p^*_{m,1}(s), \\ m = 2, 3, \ldots, L-1, \end{aligned} \tag{12}$$

$$(s+\mu)p^*_{L,1}(s) = \lambda p^*_{L-1,1}(s), \tag{13}$$

$$(s+\alpha)p^*_{L,2}(s) = \mu(1-p)p^*_{L,1}(s) + \lambda p^*_{L-1,2}(s). \tag{14}$$

We define the steady-state probabilities

$$\pi_{i,j} = \lim_{t\to\infty} \Pr\{N(t) = i, S(t) = j\}, \quad (i,j) \in \mathcal{U}.$$

Using the final value theorem of Laplace transform, we get

$$\pi_{i,j} = \lim_{s\to 0} sp^*_{i,j}(s).$$

In the next section, we obtain $\{\pi_{i,j}, (i,j) \in \mathcal{U}\}$.

4 Steady-State Probabilities $\pi_{i,j}$

Using (8)–(14), we obtain

$$\lambda\pi_{0,0} = \mu p\pi_{1,1} + \alpha\pi_{1,2}, \tag{15}$$

$$(\lambda+\mu)\pi_{1,1} = \lambda\pi_{0,0} + \mu p\pi_{2,1} + \alpha\pi_{2,2}, \tag{16}$$

$$(\lambda+\alpha)\pi_{1,2} = \mu(1-p)\pi_{1,1}, \tag{17}$$

$$\mu\pi_{L,1} = \lambda\pi_{L-1,1}, \tag{18}$$

$$\alpha\pi_{L,2} = \mu(1-p)\pi_{L,1} + \lambda\pi_{L-1,2}, \tag{19}$$

$$(\lambda+\mu)\pi_{i,1} = \lambda\pi_{i-1,1} + \mu p\pi_{i+1,1} + \alpha\pi_{i+1,2}, \quad i = 2,\ldots,L-1, \tag{20}$$

$$(\lambda+\alpha)\pi_{i,2} = \lambda\pi_{i-1,2} + \mu(1-p)\pi_{i,1}, \quad i = 2,\ldots,L-1. \tag{21}$$

To solve the above system of equations, we define

$$G_1(s) = \sum_{i=1}^{L} \pi_{i,1}s^i, \quad G_2(s) = \sum_{i=1}^{L} \pi_{i,2}s^i. \tag{22}$$

Using (15), (18) and (20), we obtain

$$[(\lambda+\mu)s - \lambda s^2 - \mu p]G_1(s) - \alpha G_2(s) = \lambda s(1-s)[\pi_{L,1}s^L - \pi_{0,0}]. \tag{23}$$

Similarly, using (17), (19) and (21), we obtain

$$-(1-p)\mu G_1(s) + [(\lambda+\alpha) - \lambda s]G_2(s) = \lambda \pi_{L,2} s^L (1-s). \tag{24}$$

Solving (23) and (24), we get

$$G_1(s) = \frac{\lambda}{D(s)} \{ s(\lambda+\alpha-\lambda s)(\pi_{L,1} s^L - \pi_{0,0}) + \alpha \pi_{L,2} s^L \}, \tag{25}$$

$$\begin{aligned} G_2(s) = \frac{\lambda}{D(s)} [&\{ (\lambda+\mu)s - \lambda s^2 - \mu p \} \pi_{L,2} s^L \\ &+ s(\pi_{L,1} s^L - \pi_{0,0})(1-p)\mu], \end{aligned} \tag{26}$$

where

$$D(s) = -\lambda^2 s^2 + s(\lambda^2 + \lambda\mu + \lambda\alpha) - \mu(\alpha + \lambda p).$$

Let ω_1 and ω_2 be the roots of the quadratic equation

$$-\lambda^2 s^2 + s(\lambda^2 + \lambda\mu + \lambda\alpha) - \mu(\alpha + \lambda p) = 0. \tag{27}$$

The discriminant of (27) is

$$\Delta = \lambda^2 \left[(\lambda - \mu + \alpha)^2 + 4\lambda\mu(1-p) \right] > 0.$$

Consequently, both ω_1 and ω_2 are real, positive and distinct. These roots have the following properties:

$$\omega_1 = \frac{\lambda + \mu + \alpha - \sqrt{(\lambda - \mu + \alpha)^2 + 4\lambda\mu(1-p)}}{2\lambda}, \tag{28}$$

$$\omega_2 = \frac{\lambda + \mu + \alpha + \sqrt{(\lambda - \mu + \alpha)^2 + 4\lambda\mu(1-p)}}{2\lambda}, \tag{29}$$

$$\omega_1 \omega_2 = \frac{\mu(\alpha + \lambda p)}{\lambda^2}, \tag{30}$$

$$-\lambda^2 s^2 + s(\lambda^2 + \lambda\mu + \lambda\alpha) - \mu(\alpha + \lambda p) = -\lambda^2 (s - \omega_1)(s - \omega_2), \tag{31}$$

$$(1-\omega_1)(1-\omega_2) = \frac{\mu\alpha}{\lambda^2} \left[1 - \frac{\lambda}{\mu} \left\{ 1 + \frac{\mu(1-p)}{\alpha} \right\} \right]. \tag{32}$$

Now, (25) and (26) give

$$\lambda(s-\omega_1)(s-\omega_2)G_1(s) = -s(\lambda+\alpha-\lambda s)(\pi_{L,1}s^L - \pi_{0,0}) - \alpha\pi_{L,2}s^L, \tag{33}$$

$$\begin{aligned}\lambda(s-\omega_1)(s-\omega_2)G_2(s) = &-\{(\lambda+\mu)s - \lambda s^2 - \mu p\}\pi_{L,2}s^L \\ &- s(\pi_{L,1}s^L - \pi_{0,0})(1-p)\mu.\end{aligned} \tag{34}$$

Since both $G_1(s)$ and $G_2(s)$ are polynomials of degree L, we get

$$G_1(\omega_1) \neq 0, \quad G_1(\omega_2) \neq 0, \quad G_2(\omega_1) \neq 0, \quad G_2(\omega_2) \neq 0.$$

Consequently, we obtain

$$\omega_1(\lambda+\alpha-\lambda\omega_1)(\pi_{L,1}\omega_1^L - \pi_{0,0}) + \alpha\pi_{L,2}\omega_1^L = 0, \tag{35}$$

$$\omega_2(\lambda+\alpha-\lambda\omega_2)(\pi_{L,1}\omega_2^L - \pi_{0,0}) + \alpha\pi_{L,2}\omega_2^L = 0, \tag{36}$$

$$\{-(\lambda+\mu)\omega_1 + \lambda\omega_1^2 + \mu p\}\pi_{L,2}\omega_1^L - \omega_1(\pi_{L,1}\omega_1^L - \pi_{0,0})(1-p)\mu = 0, \tag{37}$$

$$\{-(\lambda+\mu)\omega_2 + \lambda\omega_2^2 + \mu p\}\pi_{L,2}\omega_2^L - \omega_2(\pi_{L,1}\omega_2^L - \pi_{0,0})(1-p)\mu = 0. \tag{38}$$

We note that

$$\lambda\omega_1^2 - \omega_1(\lambda+\mu) + \mu p = -\frac{\mu\alpha}{\lambda} + \omega_1\alpha, \tag{39}$$

$$\lambda\omega_2^2 - \omega_2(\lambda+\mu) + \mu p = -\frac{\mu\alpha}{\lambda} + \omega_2\alpha. \tag{40}$$

Substituting (39) and (40) in (37) and (38) respectively, we get

$$(\mu - \lambda\omega_1)\alpha\pi_{L,2}\omega_1^L + \omega_1(\pi_{L,1}\omega_1^L - \pi_{0,0})(1-p)\lambda\mu = 0, \tag{41}$$

$$(\mu - \lambda\omega_2)\alpha\pi_{L,2}\omega_2^L + \omega_2(\pi_{L,1}\omega_2^L - \pi_{0,0})(1-p)\lambda\mu = 0. \tag{42}$$

Solving (41) and (42), we get

$$\pi_{L,1} = \frac{\pi_{0,0}}{\mu}\left[\lambda\sum_{j=1}^{L}\frac{1}{\omega_1^{L-j}\omega_2^{j-1}} - \mu\sum_{j=1}^{L-1}\frac{1}{\omega_1^{L-j}\omega_2^{j}}\right], \tag{43}$$

$$\pi_{L,2} = \frac{\lambda(1-p)\pi_{0,0}}{\alpha}\sum_{j=1}^{L}\frac{1}{\omega_1^{L-j}\omega_2^{j-1}}. \tag{44}$$

To get $\pi_{0,0}$, we make use of the total probability law:

$$\pi_{0,0} + G_1(1) + G_2(1) = 1. \tag{45}$$

Consequently, we get

$$\pi_{0,0} = \frac{1 - \gamma}{1 - \gamma(\gamma A - B)}, \tag{46}$$

where

$$\gamma = \frac{\lambda}{\mu}\left\{1 + \frac{\mu(1 - p)}{\alpha}\right\}, \tag{47}$$

$$A = \sum_{j=1}^{L} \frac{1}{\omega_1^{L-j}\omega_2^{j-1}}, \tag{48}$$

$$B = \sum_{j=1}^{L-1} \frac{1}{\omega_1^{L-j}\omega_2^{j}}. \tag{49}$$

Substituting (43) and (44) in (33) and equating the coefficients, we get

$$\pi_{L,1} = \frac{\lambda\pi_{0,0}}{\beta(\omega_1, \omega_2)}\left[\lambda\left(\omega_2^L - \omega_1^L\right) - \mu\left(\omega_2^{L-1} - \omega_1^{L-1}\right)\right], \tag{50}$$

$$\pi_{L-1,1} = \frac{\mu\pi_{0,0}}{\beta(\omega_1, \omega_2)}\left[\lambda\left(\omega_2^L - \omega_1^L\right) - \mu\left(\omega_2^{L-1} - \omega_1^{L-1}\right)\right], \tag{51}$$

$$\begin{aligned}\pi_{k,1} = \frac{\pi_{0,0}}{\beta(\omega_1, \omega_2)}&[\{(\lambda + \alpha)\psi(L - 1 - k) - \lambda\psi(L - k)\}\\ &\times \left\{\mu\left(\omega_2^{L-1} - \omega_1^{L-1}\right) - \lambda\left(\omega_2^L - \omega_1^L\right)\right\}\\ &- \psi(L - 2 - k)\left\{-\lambda^2\omega_1\omega_2\left(\omega_2^L - \omega_1^L\right) + \lambda\mu\left(\omega_2^{L+1} - \omega_1^{L+1}\right)\right.\\ &\left.+ \lambda\mu\omega_1\omega_2\left(\omega_2^{L-1} - \omega_1^{L-1}\right) - \mu^2\left(\omega_2^L - \omega_1^L\right)\right\}], \quad 1 \le k \le L - 2,\end{aligned} \tag{52}$$

where

$$\beta(\omega_1, \omega_2) = \lambda\mu\omega_1^{L-1}\omega_2^{L-1}(\omega_2 - \omega_1),$$

$$\psi(n) = \sum_{i=0}^{n} \omega_2^{n-i}\omega_1^{i} = \frac{\omega_2^{n+1} - \omega_1^{n+1}}{\omega_2 - \omega_1}.$$

We observe that (50) is same as (43). Similarly, substituting (43) and (44) in (34) and equating the coefficients, we get

$$\pi_{L,2} = \frac{\lambda^2\mu(1-p)\pi_{0,0}}{\alpha\beta(\omega_1,\omega_2)}\left(\omega_2^L - \omega_1^L\right), \tag{53}$$

$$\pi_{L-1,2} = \frac{\mu^2(1-p)\pi_{0,0}}{\beta(\omega_1,\omega_2)}\left(\omega_2^{L-1} - \omega_1^{L-1}\right), \tag{54}$$

$$\begin{aligned}\pi_{k,2} = &\frac{(1-p)\pi_{0,0}}{\alpha\omega_1^{L-1}\omega_2^{L-1}}\{\lambda\psi(L-1)\psi(L-k)\\ &-\lambda\psi(L)\psi(L-1-k) - \mu p\psi(L-2)\psi(L-1-k)\\ &+\mu p\psi(L-1)\psi(L-2-k)\},\quad 1\leq k\leq L-2.\end{aligned} \tag{55}$$

The explicit expressions (43), (44), (46), (51), (52), (54) and (55) for the steady-state probabilities form the major contribution of the present paper. Using these steady-state probabilities, we obtain the performance measures of the IDS in the next section.

5 Performance Measures of the IDS

Performance of a network system is characterized by the metrics such as average time spent in the IDS per packet, the packet loss ratio at the IDS level, the average system throughput, the mean number of packets in the system and average waiting delay of a packet.

(a) **Average number of packets leaving the IDS per second.**

Let η denote the average number of packets leaving the IDS per second. We note that a packet can leave either from the first stage or from the second stage. Further, there must be at least one packet in the system so that a packet can leave from the IDS. Therefore, we get

$$\eta = \sum_{j=1}^{L}\pi_{j,1}\mu p + \sum_{j=1}^{L}\pi_{j,2}\alpha. \tag{56}$$

(b) **Packet Loss probability of the IDS.**

Let p_{Loss} denote the packet loss probability of the IDS. An arriving packet is not allowed to join the IDS when the IDS is full. In this situation, such packets are called lost packets. Therefore, we get

$$p_{\text{Loss}} = \pi_{L,1} + \pi_{L,2}. \tag{57}$$

(c) **Mean number of packets in the IDS.**

Let $\overline{X}$ be the average number of packets in the IDS. In the long run, the IDS can be either in the first stage or in the second stage and the number of packets in the IDS can be $j, j = 0, 1, 2, \ldots, L$. Therefore, we get

$$\overline{X} = \sum_{j=1}^{L} j(\pi_{j,1} + \pi_{j,2}). \tag{58}$$

(d) **Average time spent by a packet in the IDS.**

Let W_s denote the average time spent by a packet in the system. Since η is the throughput (average number of packets leaving the IDS per second) and $\overline{X}$ is the average number of packets in the IDS, we get

$$W_s = \frac{\overline{X}}{\eta}. \tag{59}$$

(e) **Average service time rendered to a packet in the IDS.**

Let W_τ be the average service time rendered to a packet in the IDS. The service is completed either from stage 1 or from stage 2. Therefore,

$$W_\tau = \frac{p}{\mu} + (1-p)\left(\frac{1}{\mu} + \frac{1}{\alpha}\right) = \frac{1}{\mu} + \frac{1-p}{\alpha}. \tag{60}$$

(f) **Average time spent by a packet in the buffer content of the IDS.**

Let W_q be the average time spent by a packet in the buffer content of the IDS. Using W_s and W_τ, we get

$$W_q = W_s - W_\tau. \tag{61}$$

6 Conclusion

The intention of employing proposed IDS in active networks is to detect intrusions and sophisticated attacks to breach into the security system of an organization or its networking premises. However, deploying these kinds of intelligent IDS compromises the performance of the active running host. This leads to significant delay in packet transmission and increases packet loss ratio. In this paper the above-stated problems are taken into consideration and the input metrics taken for the performance analysis decide various security levels that can be implemented.

Acknowledgments The authors would like to thank S. Sibi Chakkaravarthy, Research Scholar, Department of Computer Science and Engineering, Anna University (MIT Campus), Chennai 600044, India and S. Udayabaskaran, Department of Mathematics, Vel Tech Rangarajan Dr. Sagunthala R&D Institute of Science and Technology, Avadi, Chennai 600062, India for their help in carrying out the computations.

References

1. Debar H, Dacier M, Wespi A (2000) A revised taxonomy for intrusion-detection systems. Ann Telecommun 55(7–8):361–378
2. Dreger H, Kreibich C, Paxson V, Sommer R (2005) Enhancing the accuracy of network-based intrusion detection with host-based context. Detection of intrusions and malware, and vulnerability assessment, second international conference, DIMVA 2005, Vienna, Austria, 7–8 July 2005. Proceedings. Lecture Notes in Computer Science, vol 3548. Springer, Berlin, pp 206–221
3. Alsubhi K, Bouabdallah N, Boutaba R (2011) Performance analysis in intrusion detection and prevention systems. In: IFIP/IEEE integrated network management symposium (IM)
4. Mitchell R, Chen I-R (2004) A survey of intrusion detection in wireless network applications. Comput Commun 42:1–23
5. Shin S, Lee S, Kim H, Kim S (2013) Advanced probabilistic approach for network intrusion forecasting and detection. Expert Syst Appl Int J 40:315–322
6. Alsubhi K, Alhazmi Y, Bouabdallah N, Boutaba R (2012) Security configuration management in intrusion detection and prevention systems. Int J Secur Netw 7(1):30–39

Towards Modelling a Trusted and Secured Centralised Reputation System for VANET's

T. Thenmozhi and R.M. Somasundaram

Abstract Vehicular Networks facilitate communication among vehicles to notify and exchange road-related information, and thereby, ensure road safety. In VANETs', the network infrastructure provides a facility to generate the messages. But all such messages need not be reliable. Therefore, in order to build reliability on the message, the vehicle in which the message was generated can be evaluated based on a reputation score that it has earned during its prior transmissions. This paper aims to design and analyse a reputation system for VANET's which aids the receiving vehicle to decide the reliability on the message based on the score that has been earned by the transmitting vehicle.

Keywords VANET's · Security · Message reliability · Centralised architecture · Reputation systems

1 Introduction

The development of technological innovations has led to the communication enabled between vehicles with the aid of V2V communication and V2I communication. The VANET is characterised by the mobile and self-organising nodes. VANET's provide timely updates on safety-related information, entertainment updates, etc., which are broadcasted across the network. The neighbouring vehicles are dependent on the message transmitted by the nodes. Therefore, a valid secured infrastructure is required to authenticate the messages sent in the network [1].

Without security the network is open for attacks like suppression of messages, faulty message propagation, etc., Messages related to the safety of vehicles are safety-related

T. Thenmozhi (✉)
Avinashilingam Institute for Home Science and Higher Education for Women, Coimbatore, India
e-mail: thenmozhi74@yahoo.co.in

R.M. Somasundaram
SNS College of Engineeering and Technology, Coimbatore, India

L.P. Suresh and B.K. Panigrahi (eds.), *Proceedings of the International Conference on Soft Computing Systems*, Advances in Intelligent Systems and Computing 398, DOI 10.1007/978-81-322-2674-1_64

messages and an announcement scheme is followed by the vehicle for generating and broadcasting these messages. Therefore, proper authentication of messages is required in the infrastructure to build belief about the sending node. Various cryptographic techniques are used for proving the authentication of the messages. Even then, only if the vehicles are reliable, the messages benefit the receiving nodes. If the sending vehicle is reliable then so is the message too. The consequences of an unreliable message are very high. These unreliable messages may be generated by a faulty sensor in the vehicle or may even be generated intentionally too. Therefore, any message generated by the vehicle should be evaluated for its reliability [2].

In general, all the participating vehicles do not have a major trust on the vehicles from which it receives the message. Therefore, when a message is received, the level to which the message can be depended upon is a serious issue. Therefore, it becomes mandatory to for a reputation system so that communication becomes very much trustworthy. A Reputation system aids in building a trust value for every vehicle in the network. These values help the other vehicle to determine which vehicles they should rely on. Resnicke and Zeckhauser [3] defines operational objective of a reputation systems as (a) To provide information that helps in distinguishing a reliable and unreliable vehicle (b) To encourage vehicles' to behave in a trustworthy way (c) To discourage suspected vehicles from not participating in this system.

Initially, the reputation of the vehicle is void. When the vehicle starts transmitting warnings and when other vehicles find which other vehicles find it valid. Reputation of a vehicle is the measure of belief which other vehicles have about the sender based on the reliability of earlier sent messages [4]. Usually, the belief is represented as a numerical value. With time, the participating vehicles rate the vehicle with a score. As the vehicle becomes reliable among the neighbours, it scores a positive feedback and a positive value, else and the score earned decreases.

2 Related Work

The reliability of the transmitted messages is based on the validity of the messages that have been transmitted by the vehicle. Various schemes have been proposed to implement security, reliability, authentication of messages in VANETs. Digital signatures [5] provide integrity and authentication of the transmitted messages. The Threshold method [6] verifies if same message is received from a certain "*n*" no of vehicles, but consumes a long time to check for the message validity. Network modelling [7] allows detection and correction of malicious nodes in the vehicle. But possessing the knowledge of all participating vehicles is highly infeasible and unpractical and imposes storage constraints. In a Decentralised infrastructure, reputation-based models have been proposed, but does not guarantee Robustness. In [8], Opinion piggybacking, a method named vehicle appends its opinion to the already received opinions, but does not explore on the computational burden on the vehicles, initialisation, and updating of scores. In [9], Vehicle behaviour analysis is proposed, but has a limitation that the neighbouring vehicles are in a position to react immediately. A simple, practical model of a reputation system-based announcement scheme has been proposed in [10] which analyses a secure and efficient announcement scheme. A method to improve the reception rates of the messages has been proposed in [11], which uses an adaptive broadcast protocol. In order to increase the level of reliability for safety applications in VANET's, a sublayer has been suggested in [12].

3 Proposed System

In order to scale the reputation system to a very large area, thereby benefitting many vehicles, we proposed to design and analysed a reputation system for VANET's. The proposed system analyses the drawbacks of the earlier approaches as below:

1. The existing centralised infrastructure can be utilised to a greater extent in designing and establishing a reputation system.
2. The Reputation system evaluates and disseminates the reliability score of the vehicle which assists the other vehicles in the network.
3. Broad casted messages are transmitted to the vehicles are received by the vehicles in a small area. A large number of vehicles can benefit if the technique can be extended to a greater area. A centralised reputation server which collectively groups the scores from certain number of regional reputation servers, extends the service to a larger area.
4. Reputation scores help the vehicle to decide on either accepting the message or not. The score earned by the vehicle indicates the level to which the vehicle had involved in reliable transmission.

4 Network Modelling and Simulation

4.1 System Components

In order to develop a centralised reputation system, a three-tier architecture is proposed as shown in the Fig. 1. Comprising of the following components:

(a) **Centralised Reputation Server (CRS)**: The **Centralised reputation server** is the topmost entity in the hierarchy. The CRS covers a large area and controls a certain number of RSU's. In this model, CRS is the trusted authority. The CRS aggregates the scores received from various regional reputation servers and calculates the score for a period of time. These scores can further be sent to the reputation servers on a query.
(b) **Regional Reputation Server (RRS)**: Admission and revocation of the vehicles are monitored by the regional reputation server. The RRS plays the role of receiving the scores from other neighbours, aggregating them and sends them to the CRS.

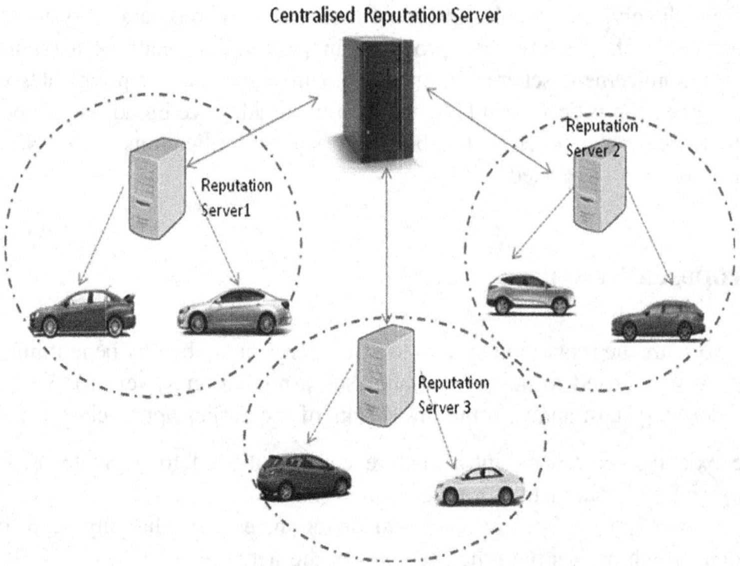

Fig. 1 3-Tier archíture

(c) **Access Points (AP)**: Wireless communication devices facilitate connection between the reputation server and the vehicles. These access points can be installed in frequently visited points. The number of access points decides the area covered.
(d) **Vehicle (V)**: Vehicles broadcast and receive messages from their neighbours. On experiencing the road-related messages, the vehicles compose a feedback and sends to its regional reputation server.

4.2 System Settings and Simulation

The basic assumptions for the design, the components of the network, the algorithmic components and the operation of the system are discussed in detail. The Network model is simulated in Ns-2 with the following assumptions.

(i) Assumptions:

(a) Vehicles move at random speeds on the roads.
(b) Traffic jam and other road conditions are simulated to occur randomly lasting for few seconds.
(c) Vehicles are comparatively closer to the occurring event.
(d) Any vehicle message received is evaluated for reliability based on a reputation threshold parameter and a time discount function.

(e) When the receiving vehicle is experiencing the event that was informed earlier, the vehicle reports to the reputation server and assumed that all experiences are reported immediately.
(f) Assuming all vehicles are in communication range and thus latest reputation certificates are received and reports are sent without delay.
(g) When a feedback is received for a vehicle, the existing reputation score is further updated and stored. Based on the new score, the certificate is generated accordingly.
(h) It is assumed that the vehicles in the network do not have any earlier trust built among them.
(i) It is assumed that all the vehicles, servers, RSU's have a synchronised clock settings.

(ii) Algorithmic components

The algorithm uses the following components:

(a) Aggr—an aggregation algorithm that calculates the feedback obtained by the vehicle. The Reputation score is computed based on the feedback obtained.
(b) Time Discount (TD)—Time discount function—When a score of a vehicle is received, it need not be accepted as such, because with time, the score might have either increased or decreased after this received value. Therefore, a time discount function can be used. Based on the time when the score is received, the reputation value is offset by multiplying score with some value in the range of [0, 1] to discount the reputation value.
(c) Digital Signature Schemes: Message integrity can be verified using Digital Signatures. Here, two schemes namely DS_1, DS_2 such that $DS_1 = (KG_1, Sign_1, Verify_1)$ $DS_2 = (KG_2, Sign_2, Verify_2)$ can be used.
(d) Hash Function H, message authentication code algorithm MAC.
(e) Three configurable parameters Th_{rs}, Th_t, Th_{cert} and T such that.
    (i) Th_{rs}—a threshold value to determine whether another vehicle is reputable, usually a value between 0 and 1
    (ii) Th_{time}—a threshold used to determine whether a message tuple is sufficiently fresh for feedback reporting.
    (iii) T—a large time interval during which the vehicles report feedback
    (iv) Th_{cert}—time period for which the certificate is valid

5 Initialisation of the System Components

(a) *Initialisation of the Centralised Reputation Server*

The centralised reputation server (CRS) is initialised with a set of public and private key pairs that are to be assigned for the regional reputation servers. The CRS is geographically placed such that it covers a set of RRS covering a larger area. This enables the reputation scores of a smaller area to be aggregated by the

CRS and then transmitted to a larger number vehicles. Any RRS that registers with the CRS receives a pair of keys for further communication. The three-tier architecture is as shown in Fig. 1.

(i) The CRS receives the aggregated reputation scores that are collected by the RRS at specific intervals.
(ii) The scores are then segregated and stored as per the identity of the vehicles.
(iii) The feedback ratings are the calculated for the individual vehicles and using continuous feedback rating algorithm the ratings are prepared and stored locally.
(iv) Any RRS can further enquire the CRS for obtaining the scores of the vehicles that are not within the range.

(b) *Initialisation of the Regional Reputation Server (RRS)*

The regional reputation server is initialized as follows.

(i) The regional reputation server registers itself with the centralised reputation ng a public, private key pair (PU_{RRS1}, PR_{RRS1})
(ii) The RRS receives the scores from all the vehicles within its range and creates a local database for storing the details of the vehicles, such as the vehicle's identity, Public Key, MAC Key, current reputation score of the vehicle and a feedback value.
(iii) The reputation score of a vehicle is aggregated using the algorithm Aggr

(c) *Installation of the Access Points:*

The access point is installed in the system to facilitate a communication between the Vehicles and the RRS for which a communication channel needs to be established.

6 Operation of the Reputation System

The CRS, RRS, access points and the vehicles are initialised and installed with certain algorithms for their operation. Some of the basic terminologies used in here are as below:

Notation	Purpose
Aggr	Reputation aggregation algorithm
MAC	Message authentication code algorithm
KG_1,KG_2	Key generation algorithm
DS_1, DS_2	Digital signature schemes
TD	Time discount function
$Verify_1$, $Verify_2$	Verification algorithms
Th_{rs}	Reputation threshold (range 0–1)

(continued)

(continued)

Notation	Purpose
Th_{time}	Threshold to determine the freshness of a message
Th_{cert}	Certificate validity time
id_{V_1}, id_{V_2}	Identity of the vehicles
$(pu_{V_1}, pk_{V_1}), (pu_{V_2} pk_{V_2})$	Public–private key pair of vehicles
(PU_{RRS1}, PR_{RRS1})	Public private key pair of regional reputation server
t_1	Certificate generation time
t_2	Message broadcast time
t_3	Message reception time
rs_{V_1}	Reputation score of vehicle V_1
$H(m)$	Hash of the message "m"
F_r	Feedback rating in the range {0, 1}
mk_{V_1}, mk_{V_2}	MAC key of the vehicles V_1 and V_2
t_{RRS}	Time when the consolidated score is sent by RRS to CRS

Once the System components have been installed, the stage wise process by which the CRS, RRS ,and the vehicles work collaboratively to establish and maintain a reputation system is as below.

(a) ***Vehicle Registration and Requisition for Reputation Certificate***

The registered vehicle requests for its Reputation Certificate from the Regional Reputation Server as below:

(i) The vehicle sends its identity id_{V_1} to the RRS.
(ii) On receiving a request the RRS creates a new record in the database for the requesting vehicle with the identity id_{V_1}.
(iii) If the requesting vehicle had earlier registered with the RRS, then the Certificate can be retrieved locally. Otherwise, the request for the Certificate is sent to CRS. The query for the vehicle id_{V_1} is then sent as $Q = \left(\left(id_{V_1}\right)_{pr_{RRS_1}}\right)_{pu_{CRS}}$ and receives a reply as $R = \left((id_{V_1}, rs_{V_1})_{pu_{RRS_1}}\right)_{pr_{CRS}}$ from which the Certificate can be generated. The Certificate, C for the requesting vehicle is then generated as $C = (id_{V_1}, pu_{V_1}, t_1, rs_{V_1}, \alpha)$, which holds the identity of the vehicle id_{V_1}, the public key of the vehicle pu_{V_1} the time t_1, when the certificate was generated and the reputation score of the vehicle as rs_{V_1}, which it has earned at time t_1. Here, α *is* Digital Signature using the algorithm $Sign_1$ such that

$$\alpha = Sign_1\left(id_{V_1}, pk_{V_1}, t_1, rs_{V_1}\right)_{pr_{RRS1}}.$$

(iv) The Certificate is then sent to the requesting vehicle, which is further stored by the vehicle locally. The Certificate remains valid for the defined time interval, $\mathrm{Th}_{\mathrm{cert}}$.

(b) ***Road-related warnings generated and Broadcasted by the vehicle***

On obtaining the certificate C from the RRS, the vehicle now generates the message and broadcasts to its neighbours.

(i) A message "*m*" could be any information composed by the driver or generated from the sensors of the vehicle. The Hash of this message is calculated as $H(m)$.
(ii) At the receiving time t_2, a time stamped Signature is generated by the Vehicle as Θ, which is $\Theta = \mathrm{Sign}_2(t_2, H(m))_{\mathrm{pr}_{V_1}}$.

The vehicle composes the message $M = (m, t_2, \Theta, C)$ and broadcasts to all the nodes in the network.

(c) ***Reliability of the Message is evaluated***

When a vehicle V_2 receives the message $M = (m, t_2, \Theta, C)$ from the sender at time t_3, the message is retrieved as below:

(i) V_2 checks if the reputation score is acceptable, i.e. $\mathrm{rs}_{V_1} \cdot \mathrm{TD}(t_2 - t_1) \geq \mathrm{Th}_{\mathrm{rs}}$
(ii) Checks if the message received is also fresh, $t_3 - t_2 \leq \mathrm{Th}_t$
(iii) The verification algorithm Verify_1 is used to check if $\alpha \in C$ by using the public key of the reputation Server $\mathrm{PU}_{\mathrm{RRS}}$.
(iv) A check on the validity of the message received by V_2 is performed using the verification algorithm Verify_2 and the public key of the Vehicle, pu_{V_1}, that is extracted from C.
(v) Once the validity of the message is verified, the vehicle from which the message was received is considered reliable. The message "*m*" is therefore considered and a feedback is computed for the vehicle. This feedback is further stored for future reporting. If the vehicle is not a reputable one, then further messages from the vehicle is not considered.

(d) ***Generation and Reporting of Feedback***

On receiving the message m at the time t_3, the vehicle stores the message and waits to experience the warning received. Once the vehicle V_2 experiences the event that was described by *m*, the reliability of the message received can be justified. If the vehicle V_2 wishes to participate in reporting the feedback about the vehicle to the RRS, then the feedback is generated, which may be either 1 (if true) or 0 (if false) and is calculated as below:

(i) V_2 generates a feedback $F_r \in \{0, 1\}$, where 1 indicates a reliable message and 0 indicates an unreliable message.
(ii) V_2 submits $(\mathrm{id}_{V_2}, \mathrm{id}_{V_1}, F_r, t_2, t_3, H(m), \Theta)$ to the trusted hardware.

(iii) The trusted hardware computes the message authentication code "D" from t_2, Θ and its MAC key mk_{V_2} as $D = \mathrm{MAC}(\mathrm{id}_{V_1}, \mathrm{id}_{V_2}, F_r, t_3, t_2, H(m), \Theta)_{\mathrm{mk}_{V_2}}$

(iv) V_2 generates the feedback tuple F as $F = (\mathrm{id}_{V_1}, \mathrm{id}_{V_2}, F_r, t_3, t_2, H(m), \Theta, D)$. If the value of F_r is 1, it is a positive feedback else if its value is 0, it is a negative feedback.

(e) ***Aggregation of Reputation Score at the RRS***

The RRS checks the following:

(i) RRS receives the feedback score from the set of registered vehicles.

(ii) RRS performs the set of following tasks.

  (a) Whether $t_3 - t_2 <= \mathrm{Th}_{\mathrm{time}}$

  (b) Calculates D by calculating MAC from the tuple (id_{V_1}, id_{V_2}, f_r, t_2, t_3, $H(m)$, Θ) using mk_{V_2}

  (c) Checks if Θ is valid, using the algorithm Verify_2 and pr_{V_2}.

(iii) For a vehicle with id_V, the scores received by the vehicle for a time period say "t_{start}" to "t_{end}" are aggregated. If any of the above check fails, then the message F is discarded.

(iv) If the message is found valid, then the feedback tuple "F" is stored in the database.

(v) The RRS applies the Aggregation algorithm "*Aggr*" for a specific Vehicle V_x on all the received feedback messages and replaces the new score with the already available score rs_{V_1}.

(vi) This aggregated Reputation score "S" is further composed into messages and sent to the CRS, at time t_{RRS}.

$$S = (\mathrm{id}_{V_1}, \mathrm{rs}_{V_1}, t_{\mathrm{RRS1}}, \mathrm{Th}_{\mathrm{cert}})\mathrm{pr}_{\mathrm{RRS1}}$$

(vi) ***Reputation Aggregation Algorithm***

(i) For a specific vehicle V, the algorithm selects all the feedback that have been reported from the start time t_{start} until the present time t_{end}, from the available database in the RRS, as:

$$S = \{F : (\mathrm{id}_{V_1} = \mathrm{id}_V) \;\&\; (t_{\mathrm{start}} < t_3 < t_{\mathrm{end}})\}$$

(ii) Multiple Feedbacks reported for a vehicle is then aggregated into a single value, by averaging, and denoted as "r_{V_i}."

(vii) ***Vehicle Revocation***

A belief parameter r_{belief} is configured for a node, (say) 70 %. The vehicle should have earned at least 70 % scoring. For a set of (say), 10 transmissions, the vehicle should have earned a reputation score of value "1", at least for seven transmissions. If a vehicle does not satisfy this constraint, it cannot be issued a Reputation Certificate for further communication.

7 Network Simulation

The Simulation is performed using Ns-2 with the Parameters as shown in Table 1. The configurable parameters Th_{rs}, Th_t, Th_{cert} and T are set as 0.7, 30, 30 and 10 min respectively. These minimal values help to visualise the effect of these parameters within the simulation time.

Three Regions with quite a geographical distance between them is considered for this simulation. Vehicles with ID's 1, 2, 3, 4, 5, 11 are configured under Region 1, Vehicles with IDs' 6, 7, 8, 9, 10, 12 under Region 2 and ID's 13, 14, 15, 16, 17, 18, 19, 20 in Region 3. The three locally configured RRS's collect the scores from the vehicles within their geographical domain, aggregate them and further send the scores to the centralised reputation server. The entralised reputation server accumulates all the scores and stores these values for further queries. Any RRS can query the CRS to obtain the scores for a far away Vehicle. For instance, when RRS1 queries the CRS for the score of Vehicle with ID 11, that does not belong to its geographical domain, the value so far aggregated for the vehicle is sent by the CRS. Thus, the reputation of the vehicle earned so far can be distributed to vehicles in Larger area (Tables 2, 3, 4 and 5).

The CRS identifies the minimum and maximum scores obtained by the Vehicles. When the vehicles decide the next forwarding vehicle based on the reputation score it had earned, there is a considerable better performance and the throughput is found to increase as in Fig. 1.

Compared to the previous schemes, the time taken by the vehicles in the current scheme to update the scores has been comparatively reduced as shown in Fig. 2.

Therefore, as in Figs. 1 and 2. The suggested scheme is from Fig. 3, it can be seen that the time taken to share the reputation scores in the three-tier architecture is better than the earlier schemes, which does not support a centralised reputation system.

Table 1 Simulation parameters

1	No. of nodes	60
2	Total simulation time	30 min
3	Channel	Wireless channel
4	Propagation	Two ray ground
5	Net info	Phy/Wireless Phy
6	MAC	MAC/802_11
7	Ifq	Queue/DropTail/PriQueue
8	Antenna	Antenna/Omni Antenna
9	Ifqlen	150
10	Routing protocol	AODV

Table 2 Reputation scores of the vehicles in region 1 stored at RRS1

Veh. ID	Time t_1	Time t_2	Time t_3	Time t_4	Time t_5	Time t_6	Time t_7	Time t_8	Time t_9	Aggr score $t_5 < t < t_9$
1	7	1	6	0	4	8	8	1	7	28
2	2	9	5	5	3	1	3	0	8	15
3	4	6	4	6	1	9	7	9	7	33
4	6	9	9	3	4	8	4	9	7	32
5	8	7	0	3	3	8	5	7	3	26
11	7	2	4	2	1	1	6	1	6	15

Table 3 Reputation scores at RRS2

Veh. ID	Time t_1	Time t_2	Time t_3	Time t_4	Time t_5	Time t_6	Time t_7	Time t_8	Time t_9	Aggr score $t_5 < t < t_9$
6	5	0	2	2	6	7	0	4	5	22
7	8	8	6	3	2	2	4	1	0	9
8	6	2	2	0	2	2	5	2	2	13
9	4	3	3	4	9	5	6	3	9	32
10	9	6	3	7	1	9	4	4	7	25
12	9	2	1	6	1	9	9	2	7	28

Table 4 Reputation scores at RRS3

Veh. ID	Time t_1	Time t_2	Time t_3	Time t_4	Time t_5	Time t_6	Time t_7	Time t_8	Time t_9	Aggr score $t_5 < t < t_9$
13	7	2	1	2	4	1	2	4	0	11
14	6	2	0	7	0	6	3	2	2	13
15	4	8	9	0	2	5	2	1	8	18
16	0	1	5	3	0	8	2	4	4	18
17	3	8	0	8	1	4	2	3	5	15
18	6	0	2	9	3	1	6	7	0	17
19	2	4	1	0	5	8	4	2	5	24
20	5	5	0	0	6	9	1	2	8	26

Table 5 Aggregated scores at CRS at time t_9

ID	1	2	3	4	5	6	7	8	9	10	11	12	13	14	15	16	17	18	19	20
Score	28	15	33	32	26	22	9 (Min)	13	32	25	15	28	11	13	18	18	15	17	24	36 (Max)

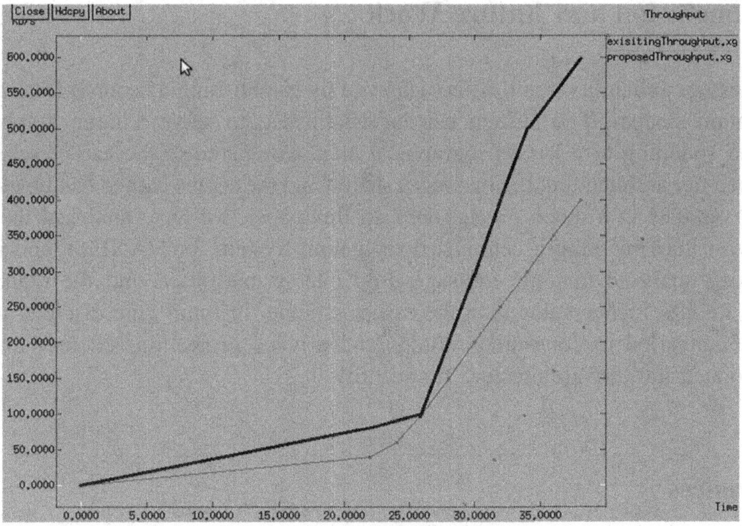

Fig. 2 Better throughput obtained when vehicles use reputation scores

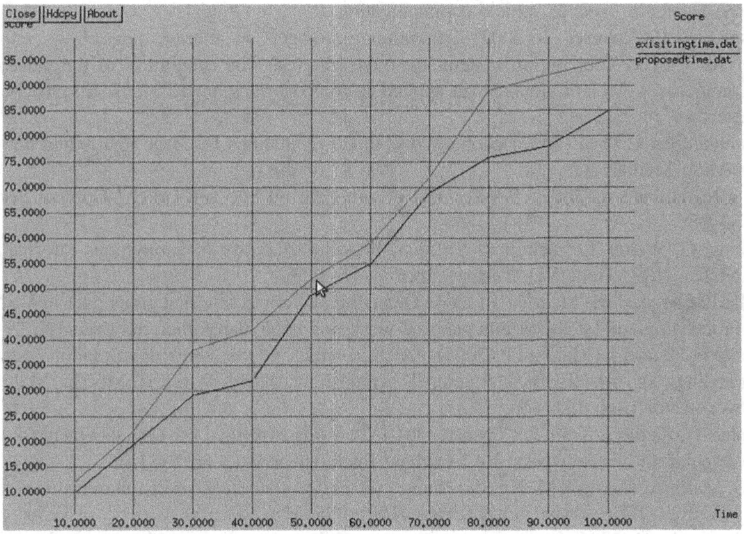

Fig. 3 Less time consumptioned to update the scores

8 Conclusion and Future Work

The message reliability can thus be achieved by establishing a secured Centralised Reputation system. The system can be established to serve a large number of vehicles spanning to a large geographical area. Compared to the earlier schemes, this three-tier architecture, the messages shared and the scores earned can be utilised by the vehicles in a much greater area. In this paper, we have analysed the possibility of implementing a centralised reputation System for VANETs' and it has also been analysed that the message drop rate is minimised and the reputation scores are at a higher value than the earlier scheme. In future, the discrete ratings can be converted to continuous ratings and privacy protection schemes may be incorporated into the architecture for security.

References

1. Luo J, Hubaux JP (2004) A survey of inter-vehicle communication. Tech Rep IC/2004/24. EPFL, Lausanne, Switzerland
2. Leinmüller T, Buttyan L, Hubaux JP, Kargl F, Kroh R, Papadimitratos P, Raya M, Schoch E (2006) Sevecom—secure vehicle communication
3. Resnick P, Zeckhauser R (2002) Trust among strangers in internet transactions: empirical analysis of eBay's reputation system. In: Baye MR (ed) The economics of the internet and E-Commerce, vol 11 of advances in applied microeconomics, pp 127–157. Elsevier Science, Amsterdam
4. Swamynathan G et al (2007) Globally decoupled reputations for large distributed networks. Adv Multi Media 1:12
5. Raya M, Hubaux J (2007) The security of vehicular ad hoc networks. J Comput Secur 15 (1):39–68
6. Kounga G, Walter T, Lachmund S (2006) Proving reliability of anonymous information in VANET's. IEEE Trans Veh Technol 56(6):3442–3456
7. Golle P, Greene DH, Staddon J (2004) Detecting and correcting malicious data in VANETs. In: Proceedings of 1st ACM international workshop vehicular Adhoc networks, pp 29–37
8. Dötzer F, Fischer L, Magiera P (2005) VARS: a vehicle ad hoc network reputation system. In: Proceedings of 6th IEEE international symposium World Wireless Mobile Multimedia Networks, vol 1, pp 454–456
9. Minhas U, Zhang J, Tran T, Cohen R (2010) Towards expanded trust management for agents in vehicular ad hoc networks. Int J Comput Intell Theor Pract 5(1):3–15
10. Li Q, Malip A, Martin KM, Ng SL, Zhang J (2012) A reputation-based announcement scheme for VANETs. IEEE Trans Veh Technol 61(9):4095–4108
11. Balon N, Guo J (2006) Increasing broadcast reliability in vehicular ad hoc networks, VANET '06. In: Proceedings of the 3rd international workshop on vehicular ad hoc networks, pp 104–105
12. Hassanabadi B, Valaee S (2014) Reliable periodic safety message broadcasting in VANETs using network coding. IEEE Trans Wireless Commun 13(3):1284–1297

A Survey on Trusted Platform Module for Data Remanence in Cloud

M. Arun Fera and M. Saravana Priya

Abstract Cloud computing is the process of storing data in a common place rather than positioning the data in a computer or server. Though cloud computing offers various security features, various security issues has been in place. This paper aims to address one of the unidentified issue or inconsiderable issue in the cloud. One such issue is data remanence. All knows that dragging a file into the recycle bin and then emptying it, does not completely delete the file. Instead, the file remains in the drive, which paves the way for an attacker to get access to that file and make use of it. This paper deals with a survey of trusted platform technologies used for overcoming the data remanence problem in cloud.

Keywords Cloud computing · Data remanence · Trusted platform module · Virtualization

1 Introduction

Cloud is a vast and wide network which is present in remote location. Cloud computing is the process of storing data in a common place rather than positioning the data in a computer or server. Besides the benefit of cloud computing, there exist various complex issues. One such issue is data remanence. It is the residual representation of data that exists even after data has been erased from its storage. Trusted platform module is one of the identified solutions for resolving the data remanence issue. Section 2 gives the overview of trusted platform module functionalities. Section 3 explains the various challenges analyzed in implementing

M.A. Fera (✉) · M.S. Priya
Computer Science and Information Security, Department of Information Technology, Thiagarajar College of Engineering, Madurai, Tamil Nadu, India
e-mail: fera26@gmail.com

M.S. Priya
e-mail: Saravanapriya1991@gmail.com

L.P. Suresh and B.K. Panigrahi (eds.), *Proceedings of the International Conference on Soft Computing Systems*, Advances in Intelligent Systems and Computing 398, DOI 10.1007/978-81-322-2674-1_65

trusted platform module. Section 4 explains the existing solutions for data remanence issue and Sect. 5 concludes the paper.

2 Survey on Trusted Platform Module

A trusted platform module provides secure asymmetric key generation. This paper [1] describes the use of a secure key generating authority in Shamir identity-based signature scheme implementation. They proposed an idea of identity-based asymmetric cryptosystems (IBC) together with an identity-based asymmetric signature. The proposed IBS scheme in this paper has itself proven secure against forgery under chosen message attacks. This paper also proposed a new concept that assigns TPM as key generating authority and list out the various benefits of implementing it.

The paper [2] initially identifies the challenges for establishing the trust in the cloud and then proposes a secure framework which helps in addressing those identified challenges. This paper is actually an extension of their previous work. In their previous work, they proposed a unique framework for establishing trust in the cloud environment. By extending their previous work, the current paper addresses those issue; it clearly covers applications data and their integration with infrastructure management data. The proposed framework [2] has four types of software agents, each run on trusted devices. The paper also explains about the controlled content sharing between devices.

In [3], security is ensured using C-code-like formal modeling at the application level. As a result of this approach, security of the protocol is ensured not only at the abstract level of protocol l, but also at the concrete level. In [4], the authors propose the virtualization of trusted platform module, so that not only single machine can use the TPM but also any number of virtual machines can also use the TPM; doing so will support higher level services like remote attestation and so on. This paper [4] proposes that the full TPM has been implemented in the form of software and integrate into hypervisor to make the TPM available to virtual machines also. In this environment, virtual TPM helps to establish trust using remote attestation and sealing capabilities. Establishing trust in computer platform is purely dependent upon validation. Validation allows external entity to keep up their trust on their platform based upon the configuration of platform. This paper [5] proposes a unique validation method to validate tree-formed data platform. This paper also uses Merkle hash tree to protect the integrity of the secure start up process of a trusted platform.

In [6], a survey is done about the various security issues in cloud. This paper [6] initially clearly explains about what are the security issues that are present in the various levels of cloud and suggest suitable countermeasures for resolving those issues. This also addresses some open issues and researches in cloud [6].

TPM usually contains a unique identity to provide security functions. This paper [7] proposes a new method of using TPM-enabled computer as client and server to detect anti-forensics. This paper presents specifications and analysis of an anti-forensic system constructed by utilizing TPM-enabled security on a client-server system. It extends the basic system specifications presented and provides detailed analysis of its anti-forensic capabilities. The system design considers various vectors in which forensic examination can be conducted. Security analysis of the system showed how each component contributes to the overall objective of the system being anti-forensics capable. An important note is that the system is designed to hinder forensics, not prevent it. Therefore, as a hinderer, it works as it should. However, the system does not completely prevent forensics since human factor comes into play [7].

In [8], a new mechanism is proposed for rooting trust in cloud environment called trusted virtual environment module. This paper introduces the high-level system architecture and design concepts of a necessarily somewhat TVEM system. In this paper [8], the TVEM protects information and conveys ownership in the cloud through the TEK generation process, which creates a dual rooted trust for the virtual environment and finally, when compared with other cloud computing security technologies such as private virtual infrastructure and locator bots, TVEMs enable a powerful solution to protecting information in cloud computing [8]. Trusted cloud computing platform provides a confidential execution of virtual machines [9]. Before launching their virtual machines, they allow the users to ensure whether the service is secure [10].

The trusted computing group (TCG) claims the technology with TPMs has now reached about more than 600 million PCs [11]. TPM is tamper-resistant security chip that can be used for machine authentication, machine attestation, and data protection. Though there exist a number of applications that makes use of TPM, there are number of problems that needs to be solved before we can fulfill the grand vision of trusted computing [12]. In [13], detailed explanation of TPM is given. It explains about the root of trust for storage (RTS), which is used for secure data storage implemented as hierarchy of keys [13].

TCG software stack (TSS) [14] is the supporting software on the platform supporting the platform's TPM.TCG mainly explains with protected storage and protected capabilities. Since TPM is very expensive, the resources within the TPM should be kept in a restrictive manner [15]. The integration of various computing technologies into virtualized computing environments enables the protection of hardware [16]. Here, they addressed the problem of enabling secure migration in private clouds [17]. The cloud networking (CloNe) infrastructure provides various services to virtualized network resources [18]. The framework supports AAI (authentication, authorization, and identity management) of entities in its infrastructure [19].

3 Challenges Analyzed in Trusted Platform Module

Table 1 addresses various security issues and challenges are addressed in their proposal [20].

Table 1 Comparison of various TPM techniques

A framework for establishing trust in the cloud	This paper identifies the related challenges for establishing trust in the cloud and then proposes a foundation framework for identifying those challenges. Mainly focuses on IaaS. The framework presented in this paper is not enough by itself and it requires further extension as establishing trust in the cloud. Cloud provenance is not covered in this paper. It will be addressed in their future work
Design and implementation of a trusted monitoring framework for cloud platforms	In this paper, they have designed a trusted monitoring framework, which provides a chain of trust and have implemented this framework on Xen and integrate it with open nebula to improve the performance. But, this monitoring VM could crash under some circumstance. Therefore, recovering the monitoring functionality is something which is needed to be taken into consideration
A hijacker's guide to communication interfaces of the trusted platform module	They proposed the some attacks in hijacking the trusted platform module. They have proposed active attack and implemented it. Though active attacks perform various activities, it does not allow direct retrieval of TPM protected data, like private parts of nonmigratable keys. To extract this kind of information it is still necessary to resort to invasive high-effort method which directly targets the TPM chip [30]
Security in cloud computing: Opportunities and challenges	This survey paper presented the security issues that arise due to the shared, virtualized, and public nature of the cloud computing paradigm. Subsequently, the counter measures presented in the literature are presented
Fine-grained refinement on TPM-based protocol applications	In this paper they formalize parts of the interfaces of TPM. Thus in their future work they try to expand our refinement framework to more general applications by formalizing all the interfaces of TPM

4 Existing Solution for Data Remanence

4.1 Existing solution

There are various solutions existing for addressing data remanence problem. But, these solutions do not completely solve the remanence issue. The various existing solutions are

4.2 Encryption

Encryption is the best solution offered for data remanence problem till now. Always ensure that all the data in the cloud are encrypted. We have to manage our keys by ourself, instead of storing them in the cloud environment. In this way, not only are our data confidential, but all we have to do to securely delete our data is to delete the key [21].

4.3 Limitations in Encryption

Though encryption is the best solution, it has some major limitation. There is simply not much that we can do with encrypted data unless we decrypt them. At decryption stage, again data remanence problem will occur. So, if we want to process our data in the cloud, the encryption approach is insufficient [21].

4.4 Virtual private storage

Another approach in addressing this problem is the virtual private storage or VPS. With the VPS, both encryption and decryption takes place in a transparent manner that discusses all interactions with the cloud. From the cloud perspective, none of our data are unencrypted [22].

4.5 Limitations in VPS

Unfortunately, VPS is not a complete solution because it limits what we do in the cloud environment. Hence, this approach is also insufficient [22].

5 Conclusion

With the help of cloud computing applications over, the Internet can be accessed anytime [23]. Though cloud computing provides flexible solutions, various security threats has been addressed [24]. Cloud computing security should ensure all security aspects related to customer's data with necessary regulations [25]. One such issue is what we have discussed earlier is data remanence [26]. Though various solutions have been proposed for addressing the data remanence issue [27], they are not able to provide permanent solution for addressing this issue. Building security to the computers is provided by the TCG. Trusted platform module (TPM), which is in hardware-assisted environment provides a root of trust to the user [28]. Trusted monitoring framework ensures the integrity of the monitoring environment [29].

References

1. Goh W, Yeo CK (2013) Teaching an old trusted platform module: repurposing a tpm for an identity-based signature scheme. Apr 2013
2. Abbadi Imad M, Alawneh Muntaha (2012) A framework for establishing trust in cloud. Comput Electr Eng 38:1073–1087
3. Huang W, Xiong Y, Miao F, Wang X, Wu C, Lu Q, Xudong G (2013) Fine-grained refinement on tpm-based protocol applications. IEEE Trans Info Forensics Sec, 8(6), June 2013
4. Ramon S, Caceres R, Kenneth A, Goldman R, Sailer P, Leendert R vTPM: Virtualizing the Trusted Platform Module. USENIX Association, Security'06:15th USENIX Security
5. Schmidt Andreas U, Leicher A, Brett A, Shah Y, Cha I (2013) Tree-formed verification data for trusted platforms. Comput Secur 32:19–35
6. Ali M, Khan SU, Vasilakos AV (2015) Security in cloud computing: opportunities and challenges. Inf Sci 305:357–383
7. Goh W, Leong PC, Yeo CK (2011) The plausibly-deniable, practical trusted platform module based anti-forensics client-server system. IEEE J Sel Areas Comm, Aug 2011
8. John Krautheim F, Phatak DS, Sherman AT (2010) Introducing trusted virtual environment module: a new mechanism for rooting trust in cloud computing
9. Santos N, Gummadi KP, Rodrigues R (2009) Towards trusted cloud computing. In: Proceedings of the conference on cloud computing. Berkeley, CA USENIX association
10. VMware. VMware vCenter Server; 2010. http://www.vmware.com/products/vcenter-server/
11. Ashford W (2012) Will this be the year TPM comes of age?" [Online]. Available: http://www.computerweekly.com/news/2240157874/Analysis-2012-Will-this-be-the-year-TPM-finally-comes-of-age. Accessed: 18 June 2012
12. ISO/IEC-11889:2009, Information technology – Trusted Platform Module
13. Sadeghi A-R Trusted platform module, lecture slides for secure, trusted and trustworthy computing, Technische Universität Darmstadt, Germany. [Online]. Available: http://www.trust.informatik.darmstadt.de/fileadmin/user_upload/Group_TRUST/LectureSlides/STC-WS2011/Chap3_-_Trusted_Platform_Module.pdf.pdf
14. The TCG Software Stack (TSS) Specification-version 1.20 Errata A Golden Candidate 2, Trusted Computing Group
15. BitLocker Drive Encryption Technical Overview, Microsoft TechNet. [Online]. Available: http://technet.microsoft.com/en-sus/library/cc732774(v=ws.10).aspx. Accessed: 3 June 2012
16. Chen D, Zhao H (2012) The data security and privacy protection issues in cloud computing. In: international conference on computer science and electron ics engineering (ICCSEE, IEEE)
17. Danev B, Masti RJ, Karame GO, Capkun S (2011) Enabling secure VM migration in private clouds. In: Proceedings of the ACM 27th annual computer security applications conference, 2011, pp 187–196
18. Dhungana RD, Mohammad A, Sharma A, Schoen I (2013) Identity management framework for cloud networking infrastructure. In: IEEe International conference on innovations in information technology (IIT), 2013, pp 13–17
19. See also: for Data Remanence solutions http://fas.org/irp/nsa/rainbow/tg025-2.htm

20. Fan K, Mao D, Lu Z, Wu J (2013) OPS: offine patching scheme for the images management in a secure cloud environment. In: IEEe International conference on services computing (SCC), 2013, pp587–594
21. Cloud Computing—https://zapthink.com/2011/05/19/data-remanence-cloud-computing-shell-game
22. Cloud Storage—http://securosis.com/blog/securing-cloud-data-withvirtual-private-storage/
23. Cloud Computing—http://www.cornwallcloudservices.co.uk/index.php/easyblog/categories/listings/theinternetofthings
24. Cloud Computing Security—http://searchcompliance.techtarget.com/definition/cloud-computing-security
25. Cloud Computing Security—http://www.forbes.com/sites/netapp/2012/12/12/cloud-security-1/
26. Data Remanence—http://www.itrenew.com/what-is-data-remanence
27. Data Remanence—http://en.wikipedia.org/wiki/Data_remanence
28. Zou Deqing, Zhang Wenrong, Qiang Weizhong, Xiang Guofu, Laurence TianruoYang, KanHu HaiJin (2013) Design and implementation of a trusted monitoring framework for cloud platforms. Future Gener Comput Syst 29:2092–2102
29. Data Remanence—http://elastic-security.com/2010/01/07/data-remanence-in-the-cloud/
30. Winter Johannes, Dietrich Kurt (2013) A hijacker guide to communication interfaces of the trusted platform module. Comput Math Appl 65:748–761

Multimodal Fuzzy Ontology Creation and Knowledge Information Retrieval

G. Nagarajan and R.I. Minu

Abstract The main objective of this paper is to design an information retrieval system for both text and image data using the concept of ontology. The focus of this paper is to improve information retrieval for sports events using Ontologies. For this purpose, an integration of domain knowledge with images using fuzzy ontology technique was implemented. In this work, the domain of basketball event is considered for creating low-level visual ontology for certain sport event images. The created multimodal ontology definition will provide a wide domain applicability, which allows the user to construct an ontology for any sport event. The domain ontologies are compared with the non-ontological information retrieval system to show the effectiveness of the ontological system.

Keywords Ontology · Knowledge representation · Semantic analysis · OWL · Machine learning · Computer vision

1 Introduction

In the past decades, the path taken by the image retrieval was from text-based retrieval to content-based retrieval [1]. CBIR is a prominent area in image processing due to its diverse applications in Internet, multimedia, medical image archives and crime prevention. Demand for improved image database has increased the need to store and retrieve digital image. Extraction of visual feature, viz. colour,

G. Nagarajan (✉)
EEE Department, Sathyabama University, Chennai, India
e-mail: nagarajanme@yahoo.co.in

R.I. Minu
CSE Department, Jerusalem College of Engineering, Chennai, India
e-mail: r_i_minu@yahoo.co.in

L.P. Suresh and B.K. Panigrahi (eds.), *Proceedings of the International Conference on Soft Computing Systems*, Advances in Intelligent Systems and Computing 398, DOI 10.1007/978-81-322-2674-1_66

texture and shape is an important component of CBIR [2]. The fundamental difference between content-based and text-based retrieval systems is that the human interaction is an indispensable part of the latter system. Humans tend to use high-level features (concepts) such as keywords, text descriptors, to interpret images and measure their similarity. While the features automatically extracted using computer vision techniques are mostly low-level features (colour, texture, shape, spatial layout, etc.). In general, there is no direct link between the high-level concepts and the low-level features [3]. Though many sophisticated algorithms have been designed to describe colour, shape and texture features, these algorithms cannot adequately model image semantics and have many limitations when dealing with broad content image databases. Extensive experiments on CBIR systems show that low-level contents often fail to describe the high-level semantic concepts in user's mind. Therefore, the performance of CBIR is still far from user's expectations.

In [4–6] content representation and support automatic content extraction for modelling ontology have not been used. Whereas in our paper we used Action, Item and Cognitive spatial relations component classes for the creation of multimodal ontology. For the semantic content representation, ontology introduces fuzzy classes [7] and properties. Cognitive Spatial Relation Component [8], Event Definition, Similarity, Item Composed of Relation and Concept Component classes are fuzzy classes as they aim to having fuzzy definitions [9]. Cognitive Spatial relation calculations return fuzzy results [10] and Spatial Relation Component instances are extracted with fuzzy membership values [11].

The challenges faced while designing an effective Semantic Information Retrieval System are grouped into three categories. The challenges with respect to the images are the ambiguities in selecting the competent visual feature. For this both the local [12] and global [13] features are to be studied. The selected visual features are used to create a visual word by underlying the technique of bag of visual word [14]. The values of the generated visual words are embedded as instance in the created sports event domain-specific low-level ontology. For text-based information search, the knowledge base is created as high-level ontology by extracting the required information from the Web. Here the challenges are the extracting of required entities (class) from the HTML pages and merging the entities as instance in the created sports event domain-specific high-level ontology. Once the instances of low-level and high-level ontologies are created, they have to be integrated. To integrate, the fuzzy ontology technique [15] is employed. The challenges in this integration are the identification of appropriate fuzzy membership function for integration of ontology instance and rule generation for extracting the entities. Each identified challenge is recovered by analyzing the existing algorithm and formulizing a new one in each stage of the work.

2 Visual Ontology Creation

Visual ontology is used to provide semantic to the generated visual word. In this paper the domain of basket ball game images were considered. Ontology with respect to the image feature and domain knowledge is created. For the creation of ontology fuzzy concept is incorporated. Mainly the created ontology consists of three main classes: Item, Action and Cognitive spatial relationship. In item class the general living and non-living objects such as player name, referee name, team name, ball, basket hop, free throw line, etc. were listed. In action class the general action of basketball games Assist, Dunk, Freethrow, Jumpball, Pass, Rebound, and shot were listed. In this ontology the concept of cognitive spatial relationship is used to find the relationship between ball, body, hand, basket, and free throw line in the image. To measure the distance and to identify the cognitive spatial relationship fuzzy membership function were used.

To generate fuzzy basketball ontology, two types of ontology have to be created and mapped using the A-box and T-Box techniques. The creation of high-level ontology is explained briefly in [16, 17]. For the creation of low-level ontology, the semantic visual word creation is briefed in above section. The A-box and T-box mapping is also explained in our previous papers.

The entities of the fuzzy basketball ontology are shown in Fig. 1. For creating the low-level ontology, the images with different actions are considered. From the images the visual word is substituted in the created bag of visual word data property, then form the image the ball, basket, human, head are all segmented using standard watershed algorithm [18].

The ontology consists of two types of data properties: data property and object property. In this fuzzy basketball ontology the data property is declared for the names of the player, referee and team, then the low-level feature of the image is

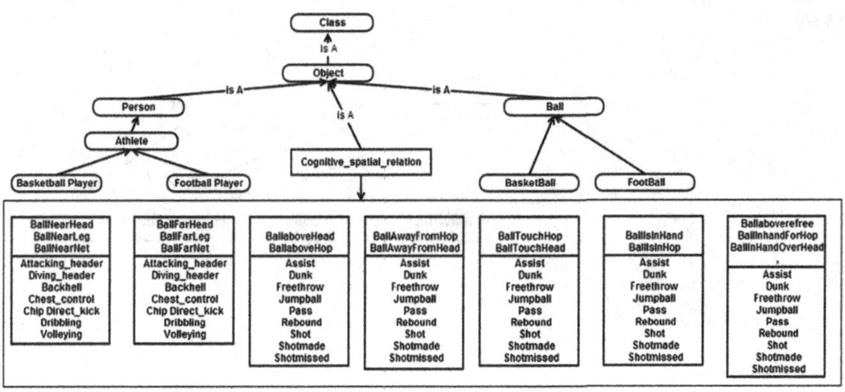

Fig. 1 Identified entities

given according to the type of image. Figure 2 elaborately shows the dunk and attacking header action and the cognitive spatial relation between the ball and the player.

For an action Dunk to decide which cognitive spatial relation will suit is decided by the distance between the ball with hoop, hand and player. From Fig. 3 the distance between the ball head is 20–60 pixel value. As we are working with user specific images the value would differ in images. To solve this hypothetical uncertainty fuzzy system was introduced to decide the entities in the ontology.

As shown in Fig. 3 the value of the distance has two sets of value, to solve this Type-2 fuzzy set is used. The Type-1 fuzzy set is good in handling crisp set of values where there is vagueness. But when it comes to handling uncertain data, where the agent is not aware on the whole truth, table Type-1 fuzzy set does not give the exact rules as it would be difficult to fix upon an absolute member function. So, there comes the Type-2 fuzzy set which will solve this problem effectively. The membership function can be categorized into lower membership function (LMF) and upper membership function (UMF). Thus, the distance similarity can be fixed between the values of 20–60 in case of ball and hoop relation.

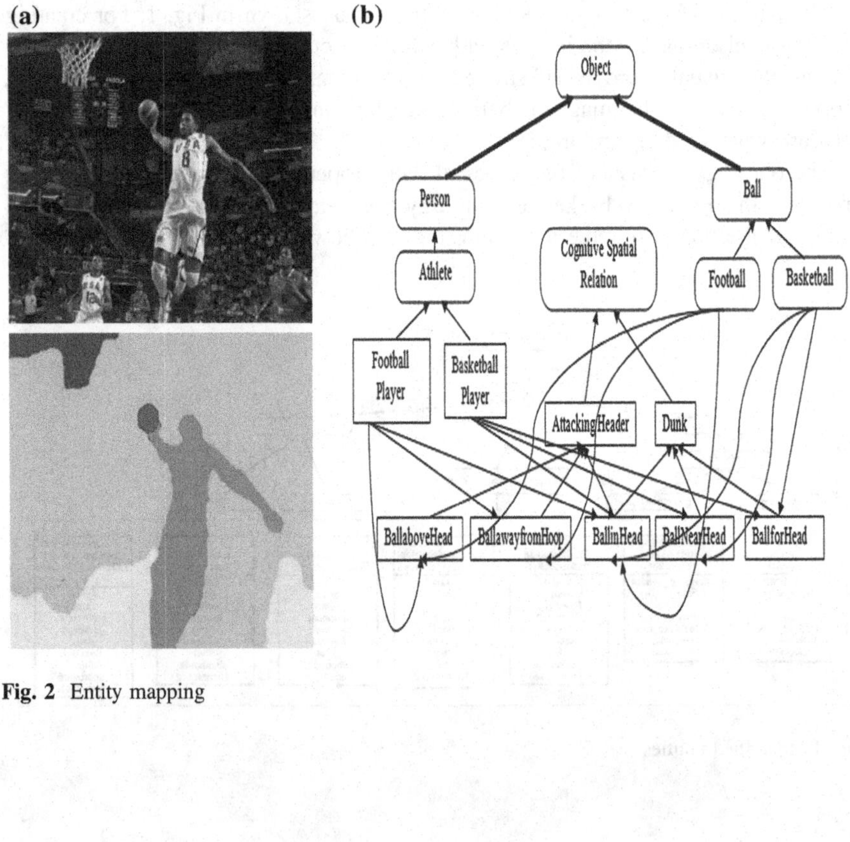

Fig. 2 Entity mapping

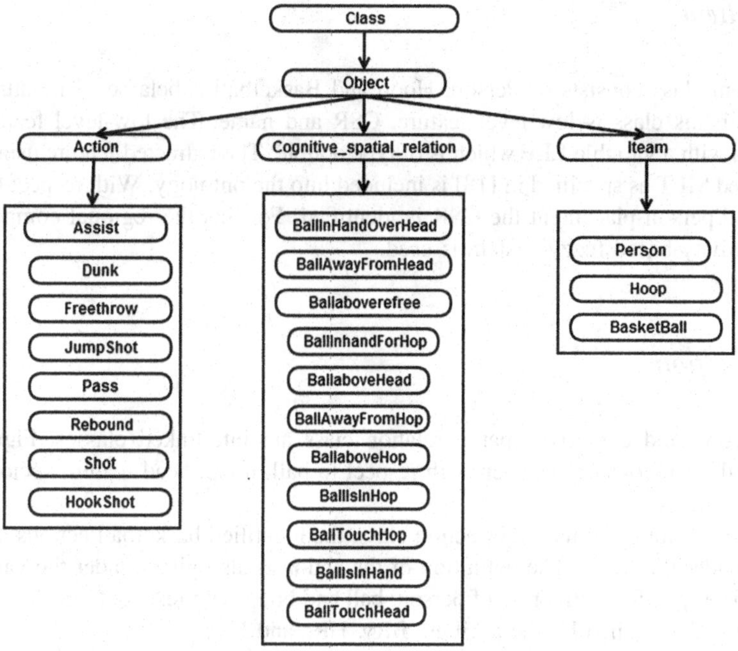

Fig. 3 Entities of fuzzy ontology

3 Component

The class of any ontology consists of object. The objects can be called either resources or component. The components identified are item, action and cognitive_spatial_relationship. The instance of this component is shown in Eq. (1), where I_x, A_y and C_z represent the component Item, Action and Cognitive spatial relation, respectively. The attribute used to measure similarity is cognitive spatial relation (CSR) which is the fuzzy membership between the ball and the player. This membership value is derived by implementing RCC8 related fuzzy variation to it.

$$\text{Component:} \begin{cases} [\text{type} \Rightarrow \{I_x, A_y, C_z\}, \text{sim} \Rightarrow \{\text{CSR}_m\}, \text{synname} \Rightarrow [\text{String}]] \\ \text{where} \\ \text{ind}(I_x, \text{Iteam}), \text{ind}\big(A_y, \text{Action}\big), \text{ind}(C_z, \text{Cognitive\_spatial\_relation}), \\ \text{ind}(\text{CSR}_m, \text{Similarity}), \\ x, y, z > 0 \end{cases} \tag{1}$$

3.1 Item

The Item class consists of Person, Hoop and Basketball subclasses. The attribute used for this class is low-level feature, CSR and name. The low-level feature is defined with a variable LLF which is of type double. The extracted feature using the modified SIFT as specified in [18] is included into the ontology. With respect to the ball and person placement the CSR is identified. For this the regional component connectivity-based fuzzy system is used.

3.2 Action

The action and cognitive_spatial_relation class are interlinked classes. Figure 4 shows the relationship between with respect to ball, hoop, head feature vector and CSR value.

In the attribute formation of action class, the identified basketball actions are all listed under this class. The definition of the action is also given under the variable AD*x*. The specific spatial role of person, ball and hoop with respect to the low-level feature is determined by the attribute BR*y*. HR*z* and PR*g*.

An action can have several definitions where each definition describes the action with a certainty degree. In other words, each action definition has a membership value for the action it defines and it denotes the clarity of description. Action definitions contain individuals of Cognitives Spatial Change, CSR Component.

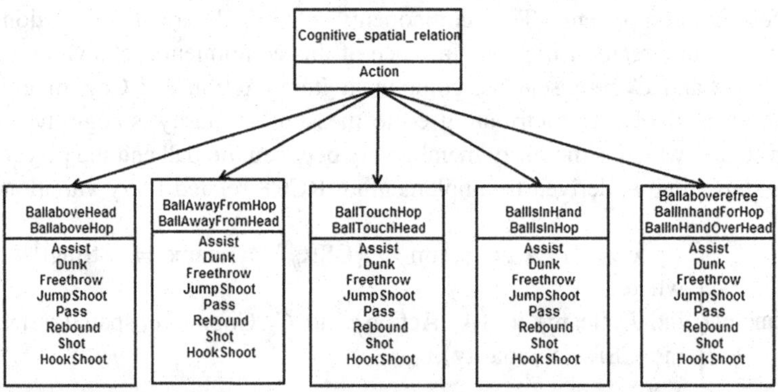

Fig. 4 Cognitive_Spatial_relation and action class relationship

3.3 *Cognitive_Spatial_Relation*

The regional connecting calculus (RCC) has totally eight relations [19]: connected, disconnected, partially connected, tangentially partially connected, not tangentially partially connected, tangentially partial part inverse connected, not tangentially partial part inverse connected and equally connected. This can be categorized in two different ways either classified as connected, not connected and membership connected or connected, disconnected and partially connected. For action recognition these RCC8 connection calculus are categorized into three spatial connections as shown in Fig. 6. Equation 5 shows the conversion of actual RCC component to the connection calculus {Touch, disjoint, overlap}. To justify the combination of regional connection calculus into three categories, Table 1 specifies the logical definition of identified cognitive spatial relation.

To identify the action Type-2 based [20] fuzzy membership function is used. The Type-1 fuzzy set is good in handling crisp set of values where there is vagueness. But to handle uncertain data, where the agent is not aware on the whole truth table Type-1 fuzzy set does not give exact rules as it would be difficult to fix upon an absolute member function. So there comes the Type-2 fuzzy set, the symbolic representation for Type-2 fuzzy set is the Type-2 fuzzy set which requires two sets of membership function values.

3.4 *Cognitive Spatial Relation Component*

Cognitive Spatial Relation Component classes are used to represent cognitive spatial relations between item individuals. It takes two item individuals from the

Table 1 Logical definition of cognitive spatial relation

RCC-8 spatial relation	Interpretation	Spatial relation definition	Cognitive spatial relation
C(A,B)	Connected	C(A,B) ∧ O(A,B)	Touch(A,B)
DC(A,B)	Disconnected from	C(A,B)	Disjoint(A,B)
PO(A,B)	Partially overlaps with	O(A,B) ∧ O(A, B) ∧ O(B,A)	Overlap(A,B)
TPP(A,B)	Tangential proper part of	PP(A,B) ∧ NTP(A, B)	
NTPP(A,B)	Non-tangential proper part of	P(B,A) ∧ NTP(A, B)	
TPP-1(A,B)	Tangentially partially Part(A,B) inverse		
NTPP-1(A,B)	Not tangentially partially Part(A,B) inverse		
EQ(A,B)	Equal to	P(B,A) ∧ P(B,A)	

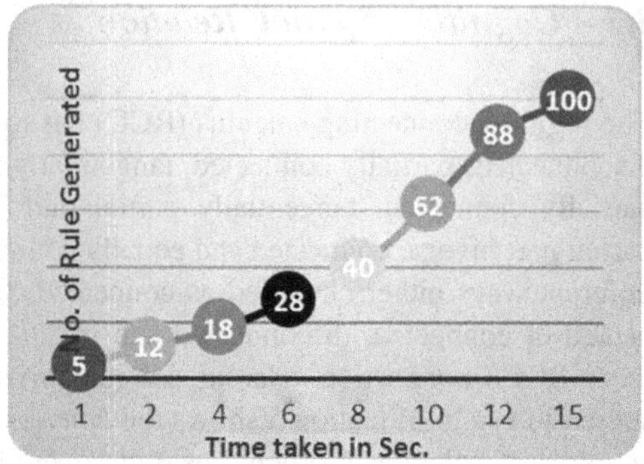

Fig. 5 Evaluation graph

respective classes and its subsequent spatial relationship would be identified. These classes are utilized in cognitive spatial change and action definition modelling. It is possible to define accurate relationship by specifying the membership value for the cognitive spatial relation individual used in its definition. For the basketball domain, Player under Hoop is an example of Spatial Relation Component class individuals.

3.5 Semantic Similarity

Similarity class is used to represent the relevance of a component to another component class using fuzzy membership function. Whenever a component which has a similarity relation with another component is extracted, the semantically related component is automatically extracted using this similarity relation as specified in [21, 22]. Initially, the created basketball ontology is inserted into the system to extract the different classes separately. The number of rules is generated and its evaluation is shown. The evaluation graph with respect to the number of rules generated and time taken is shown in Fig. 5.

4 Evaluation of the Proposed System

The entire procedure was conceived as an interface using Matlab and implemented on Stanford University sport event datasets. To measure the classification performance quantitatively a precision–recall curve is computed and its average precision–recall is found from the graph. Precision is the fraction of the images classified, that are relevant to the user's query image. Recall is the fraction of the images classified, that are relevant to the queries that are successfully classified. From the precision–recall plot, the area under the precision–recall curve gives the

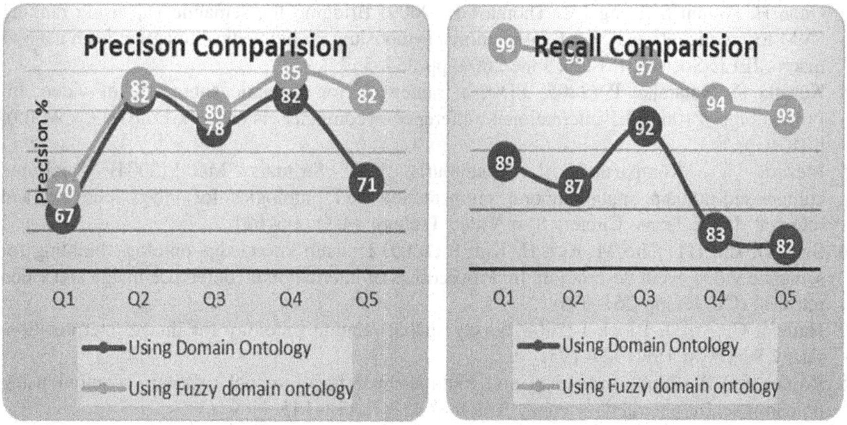

Fig. 6 Precision and recall

average precision–recall. The AP provides an accuracy of 83.5 % for the given basketball event image. The precision and recall for action image retrieval and textual retrieval are analyzed as shown in Fig. 6.

5 Conclusion

The use of an ontological knowledge model can capture users' interests more effectively. The fact that the statistical technique relies solely on numeric data can result in a failure to understand the meaning of users' interests. Use of a statistical model alone, however, fails to capture the context of the visual content in which he is interested. This feature is not supported by usage mining techniques, but through using a knowledge-based model. In this research work, the domain of basketball sports event is considered and analyzed thoroughly to create an action-based fuzzy ontology. This idea can be applied in live video stream to identify different types of actions happened. This domain ontology can be created for different kinds of sports event images such as football, cricket to determine the action of the player effectively and to merge the player detail with the action if we have a live feed of Web information.

References

1. Liu Y, Zhanga D, Lu G, Mab W-Y (2007) A survey of content based image retrieval with high level semantics. Pattern Recog (Elsevier) 40:262–282
2. Smeulders AWM, Worring M, Satini S, Gupta A, Jain R (2000) Content-based image retrieval at the end of the early years. IEEE Trans Pattern Anal Mach Intell (IEEE Society) 22 (12):1349–1380

3. Guan H, Antani S, Long LR, Thoma GR (2009) Bridging the semantic gap using ranking SVM for image retrieval. IEEE international symposium on biomedical imaging: from nano to macro. IEEE Society, Boston, June 2009, pp 354–357
4. Nevatia R, Natarajan P (2005) EDF: a framework for semantic annotation of video. In: Proceedings of 10th IEEE international conference on computer vision workshops (ICCVW '05), p 1876
5. Mezaris V, Kompatsiaris I, Boulgouris NV, Strintzis MG (2004) Real-time compressed-domain spatiotemporal segmentation and ontologies for video indexing and retrieval. IEEE Trans. Circuits Syst Video Technol 14(5):606–621
6. Song D, Liu HT, Cho M, Kim H, Kim P (2005) Domain knowledge ontology building for semantic video event description. In: Proceedings of international conference image and video retrieval (CIVR), pp 267–275
7. Hudelot C, Atif J, Bloch I (2008) Fuzzy spatial relation ontology for image interpretation. Fuzzy Sets Syst 159:1929–1951
8. Robinson VB (2000) Individual and multipersonal fuzzy spatial relations acquired using human-machine interaction. Fuzzy Sets Syst 113(1):133–145
9. Li Y, Li S (2004) A fuzzy sets theoretic approach to approximate spatial reasoning. IEEE Trans Fuzzy Syst 12(6):745–754
10. Schockaert S, Cornelis C, De Cock M, Kerre EE (2006) Fuzzy spatial relations between vague regions. In: Proceedings of the 3rd IEEE conference on intelligent systems, pp 221–226
11. Liu K, Shi W (2006) Computing the fuzzy topological relations of spatial objects based on induced fuzzy topology. Int J Geogr Inf Sci 20(8):857–883
12. Lee Chang-Shing, Jian Zhi-Wei, Huang Lin-Kai (2005) A fuzzy ontology and its application to news summarization. IEEE Trans Syst Man Cybern B Cybern 35(5):859–880
13. Yan B et al (2013) An Improved image corner matching approach. Intell Comput Theor 7995:472–481
14. Lowe David G (2004) Distinctive image feature from scale invariant keypoints. Int J Comput Vis (Springer) 2(60):91–110
15. Bellavia F, Cipolla M, Tegolo D, Valenti C (2009) An evolution of the nonparametric Harris affine corner detector: a distributed approach. In: Proceedings of international conference on parallel and distributed computing, applications and technologies, pp 18–25
16. Cimiano P, Haase P et al (2008) Reasoning with large A-Boxes in fuzzy description logics using DL reasoner: an experimental evaluation. In: Proceedings of the ESWC workshop on advancing reasoning on the web: scalability and commonsense
17. Minu RI, Thyagharajan KK (2014) Semantic rule based image visual feature ontology creation. Int J Autom Comput (Springer Publication) 11(5):489–499
18. Nagarajan G, Thyagaharajan KK (2014) Rule-based semantic content extraction in image using fuzzy ontology. Int Rev Comput Softw 9(2):266–277
19. Kong H, Hwang M, Kim P (2006) The study on the semantic image retrieval based on the personalized ontology. Int J Inf Technol 12(2)
20. Lee C-S, Wang M-H, Hagras H (2010) A Type-2 fuzzy ontology and its application to personal diabetic-diet recommendation. IEEE Trans Fuzzy Syst 13(2):374–395
21. Yildirim Y, Yilmaz T, Yazici A (2007) Ontology-supported object and event extraction with a genetic algorithms approach for object classification. In: Proceedings of sixth ACM international conference image and video retrieval (CIVR '07), pp 202–209
22. Yildirim Y, Yazici A, Yilmaz T (2013) Automatic semantic content extraction in video using a fuzzy ontology and rule based model. IEEE Trans Knowl Data Eng 25(1):47–61

A Hybrid Cloud Architecture for Secure Service—Measures Against Poodle Vulnerability

V. Sabapathi, P. Visu and K.A. Varun kumar

Abstract Cloud computing and its services are growing day by day in a real world scenario. In hybrid cloud architecture also, we have same service security threats like poodle (Padding Oracle on Downgraded Legacy Encryption) attack that will affect the SSL-based communication between client and server. If the connection between client and server is compromised, then it will be a serious issue for cloud users, which gives hackers more privilege and cause effect. **POODLE** will disconnect the SSL connections. In cloud, it is a open connectivity over the network from which we can access the resources for user requirement. Connection setup, recently everywhere used SSL. So, **POODLE (This vulnerability discovered by Google Team at September 2014)** will crack. So far, strong authentication in connection setup, server-side authentication should be in cloud. For sever-side keystone which is in OPENSTACK, for sever-side authentication. So, in this paper, mainly for SAAS (Secure as a Service), model for cloud environment.

Keywords Cloud security · POODLE · Openstack · ECC · DH algorithm

1 Introduction

Cloud: is a resource centric technology, so that we can access the resources over the Internet. If the resource that may be the application, software, storage as the services like PAAS (Platform As A Service) SAAS (Software As A Service), IAAS (Infrastructure As A Service) respective service. So, as the number of user increase

V. Sabapathi (✉) · P. Visu · K.A. Varun kumar
Department of Computer Science and Engineering, Vel Tech University,
Chennai 600062, Tamil Nadu, India
e-mail: sabapathi2000@gmail.com

P. Visu
e-mail: pandu.visu@gmail.com

K.A. Varun kumar
e-mail: varun.kumar300@gmail.com

L.P. Suresh and B.K. Panigrahi (eds.), *Proceedings of the International Conference on Soft Computing Systems*, Advances in Intelligent Systems and Computing 398, DOI 10.1007/978-81-322-2674-1_67

automatically number of security issues also arises. So, security is the most important thing for cloud development and using the cloud in real-time and long-term usage. Once the technology wants to become very popular and people accessing them mostly then confidential, believes should be in that technology. Cloud is the emerging technology, so security is most important thing for that confidentiality, availability of data anywhere, and affordability of resources at low cost.

In this paper, we deploy the secure connection setup against POODLE VULNERABLE and keystone.

Authentication at server-side and elliptic curve cryptography for key generating, Diffie–Hellman Key exchange protocol was used for secure key exchange. But actually, cloud computing is the basic concept of separating everything like applications, software, and even the infrastructure from the hardware you are working on. Ex. Google Doc is a classic web application, Google spreadsheet, Zen, Quick Books, and many more. According to NIST definition of cloud computing, Cloud computing is a model for enabling ubiquitous, convenient, and on-demand network access to a shared pool of configurable computing resources (e.g., networks, servers, storage, applications, and services) that can be rapidly provisioned and released with minimal management effort or service provider interaction [1]. The idea behind having cloud computing is that if somehow your system gets crash or your windows got corrupted or some other fatal damage is done to your hardware, then the software, even the application on it gets affected and one is left with nothing in hand. One way is that you can have entire backup of your system. But that is too costly, not everyone can afford. External hard disk are very expensive. The other way out is one can buy some storage from cloud service providers and can store their data.

2 Related Works

In this section, to discuss about hybrid cloud architecture and Poodle vulnerability [2] using a lot of open source cloud solutions to build a private cloud with IaaS cloud service layer. Voras et al. [3] devise a set of criteria to evaluate and compare most common open source IaaS cloud solutions. Mahjoub et al. [4] compare the open source technologies to help customers to choose the best cloud offer of open source technologies. Most common open source cloud computing platforms are scalable, provide IaaS, support dynamic platform, Xen virtualization technology, linux operating system, and Java [5]. However, they have different purposes. For example, Eucalyptus [6] fits well to build a public cloud services (IaaS) with homogeneous pool of hypervisor, while OpenNebula [7] fits well for building private/hybrid cloud with heterogeneous virtualization platforms. CloudVisor [8] uses nested virtualization to deal with the compromise of the hypervisor. In this technique, a secure hypervisor is introduced below the traditional hypervisor and the interactions between the traditional VMM and virtual machines are monitored by the secure hypervisor. However, since the resource management is still performed by the traditional VMM, the compromise of VMM can impact the operation of the virtual machines. Compared to CloudVisor, the main focus of our work is securing the network interactions of tenant virtual machines. The technique proposed in [9] allocates a separate privileged domain for each tenant. The tenants can use this for the enforcement of VMM-based security on their virtual machines. However, the model can become more complex as different tenant virtual machines can be hosted on the same physical server. Furthermore, such models cannot deal with the case of malicious tenants that misuse the cloud resources to generate attacks on other hosts. Our architecture considers the case of malicious cloud administrators and malicious tenants. There have also been some prior works addressing privacy-related issues in the cloud domain (Sdom0) and privileged client domains. The techniques proposed in [10] consider making the cloud services scalable to the dynamic changes in the runtime environment.

3 System and Models

Our system model includes cloud system administrators, tenant, and full-fledge secure model. When registering new clients means that new client information should be more peculiar and secure. Administrators (or operators) who manage the tenant virtual machines, and tenant users (or tenant's customers) who use the applications and services running in the tenant virtual machines. Cloud providers are entities such as Amazon EC2 and Microsoft Azure who have a vested interest in protecting their reputations. The cloud system administrators are individuals from these corporations entrusted with system tasks and maintaining cloud infrastructures, who will have access to privileged domains. We assume that as cloud providers have a vested interest in protecting their reputations and resources, the adversaries form following modules (Fig. 1).

HTTPS/SSLV3 for initially used in cloud setup for connection establishment. Hence, SSL 3.0 will disable by the POODLE Hacker, he can hijack or intercept the data, so Man-In-Middle time attack will happen. Hence, the user confidentiality will be gone. So, we have to use TLS 1.2 20. Symmetric-key cryptography relies on both the client and server having to share some sort of secret information before

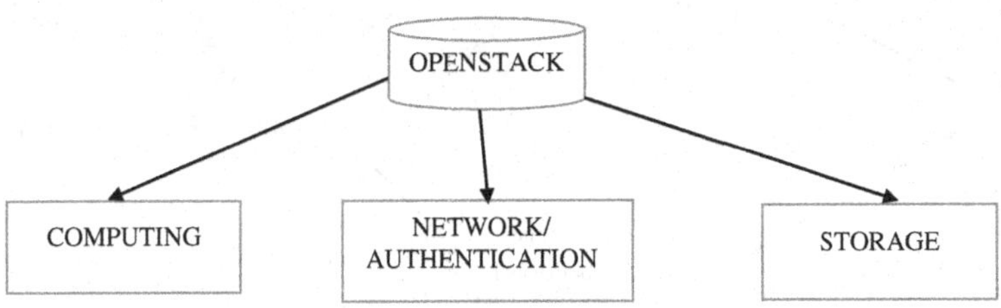

Fig. 1 Cloud interaction and network authentication

authentication can take place. It also relies on the server to securely store the client secret information in some sort of database. Here, the process will need to be entered for each new client in the database itself.

4 Proposed Scheme

In this section, we discuss the implementation and analysis of our security architecture. Section A presents the implementation setup. The implementation of our security architecture and main contribution in this paper is a security architecture that provides a flexible security as a service. Our security as a service model offers a measures against Poodle vulnerability (Fig. 2).

4.1 Methodologies

In this paper, mainly concentrate on multi-security way over the network. So that

(i) Connection setup using HTTPS/TLS against POODLE vulnerability
(ii) Account creation form—collecting peculiar details from individual user for if he had forgotten the password or hacker tries for hacking password, so that he can make login details little bit tougher.
(iii) In server-side authentication keystone authentication tool.
(iv) Openstack—cloud manager act as a local server, in backend of this local cloud server elliptic curve cryptography-based key generation (for speedy computation's) Diffie–Hellman key exchange protocol for secure connection establishment to the user OPENSTACK components were installed for the cloud platform. Hence, keystone (Authentication) service is used for authentication that will help for more secure connectivity

1. **Compute (Nova)**
2. **Object Storage (Swift)**

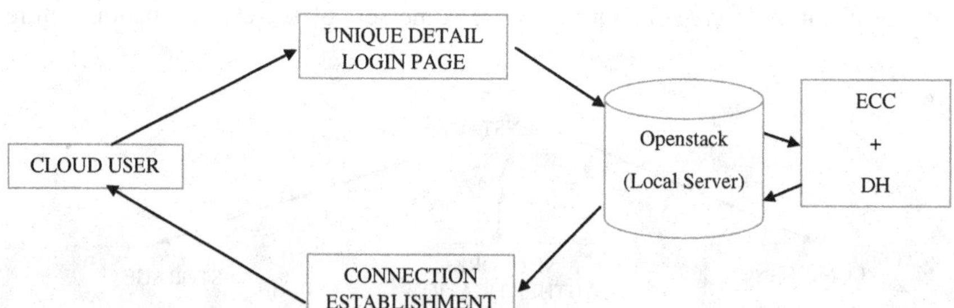

Fig. 2 Secure service model for cloud

3. **Block Storage (Cinder)**
4. **Networking (Neutron)**
5. **Identity Service (Keystone)**
6. **Image Service (Glance)**
7. **Orchestration (Heat)**
8. **Database (Trove)**

4.2 Modules

In this paper, we define five modules as following.

4.2.1 Login Module

If the user wants to login, he should be registered with every detail; in that register form he should fill up his ID proof and he have to fill some of his USER'S PSYCHE IDENTITY personal interest and USER'S PHYSICAL IDENTITY of his appearance (like a MOLE on his body), they have entered. If some ID will loss hacker may try, but these may be little stronger compared to normal login page with basic details.

4.2.2 Connection Setup

HTTPS/SSLV3 is initially used in cloud setup for connection establishment. Hence, SSL 3.0 will disable by the POODLE [2] Hacker, he can hijack or intercept the data, so Man-In-Middle time attack will happen. Hence, the user confidentiality will gone. So we have to use TLS 1.2 (Fig. 3).

4.2.3 Unique ID Generating

Using elliptic curve cryptography, we can generate the around 160 bits providing same security level as 1024 bits [11]. So that computation speed is high, less memory, and long-term battery life. So ECC will generate key efficiently. As NIST recommends shown in Table 1 so their way we choose ECC.

4.2.4 Server Authentication

In this project, we deploy Openstack act as local server. Openstack is a collection of software that can manage the cloud environment. Openstack act as a local server.

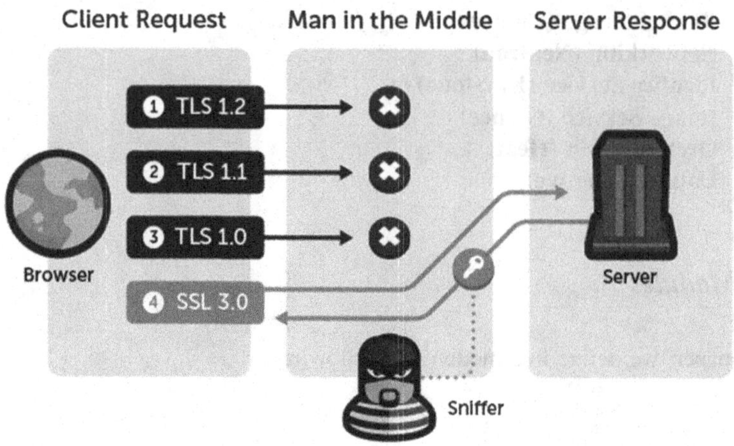

Fig. 3 Middle and packet sniffing

Table 1 NIST recommended key sizes

Symmetric-key size (bits)	RSA and Diffie–Hellman key size (bits)	Elliptic curve key size (bits)
80	1024	160
112	2048	224
128	3072	256
192	7680	384
256	15,360	521

4.2.5 Client-Authentication Protocol

Diffie–Hellman (DH) key exchange algorithm was used for key exchange protocol. It is a public key cryptography. It uses two keys, for sending message to server using this private key and send this public key. Receiver side using this private key for decryption and response using this public key. DH for Handshaking, connection establish supports

5 Conclusion

In this paper, we propose as SAAS (Secure as a Service) model using TLS 1.2 channel for connection establishment. Registration login form, we make as much as unique details from the user because if user forget the password or lost his ID proof hacker may attack. In server-side, we provide keystone authentication technology and ECC algorithm for small and speedy unique ID and key is generated. DH for secure handshaking key transfer between user and server. The paper described the design of the security architecture and discussed how different types of attacks are counteracted by the proposed architecture.

References

1. NIST National Institute of Standards and Technology U.S. Department of Commerce
2. /POODLE%20-%20Wikipedia,%20the%20free%20encyclopedia.html#References
3. Voras I, Mihaljevic B, Orlic M (2011) Criteria for evaluation of open source cloud computing solutions. In: Proceedings of the ITI 2011 33rd international conference on Information Technology Interfaces (ITI), June 2011, pp 137–142
4. Mahjoub M, Mdhaffar A, Halima RB, Jmaiel M (2011) A comparative study of the current cloud computing technology-gies and offers. In: Proceedings of the 2011 first international symposium on network cloud computing and applications, NCCA'11. IEEE Compute Society, Washington, DC, pp 131–134
5. Peng J, Zhang X, Lei Z, Zhang B, Zhang W, Li Q (2009) Comparison of several cloud computing platforms. In: Proceedings of the 2009 second international symposium on information science and engineering, ser. ISISE'09. IEEE Computer Society, Washington, DC, pp 23–27
6. Eucalyptus. Eucalyptus cloud. Retrieved March 2013(Online). Available: http://www.eucalyptus.com/
7. Open Nebula. Open nebula cloud software. Retrieved March 2013 (Online). Available: http://Opennebula.org
8. Ng C-H, Ma M, Wong T-Y, Lee P, Lui J (2011) Live deduplication storage of virtual machine images in an open-Source cloud. In: Proceedings of the 12th ACM/IFIP/USENIX international conference on Middleware, ser. Middleware'11. Springer, Berlin, pp 81–100
9. KVM, "Kernel based virtual machine". Retrieved March 2013 (Online). Available http://www.linux-kvm.org/page/Main%20Page
10. UML, "User-mode Linux kernel". Retrieved March 2013 (Online). Available http://user-mode-linux.sourceforge.net/
11. Modares H, Shahgoli MT, Keshavarz H (2012) Recommendation for KeyManagement, Special Publication 800-57 Part1 Rev 3,NIST,07/2012

An Efficient Authentication System for Data Forwarding Under Cloud Environment

S.V. Divya and R.S. Shaji

Abstract Cloud offers numerous services and applications to its customers and those resources can be accessed by the customers anywhere and at any time without any software installation. Apart from the countless benefits it provides some issues such as privacy, data confidentiality, data leakage, data integrity, and security, which are still threatening the cloud. In all the prior works, the user enters into the owners cloud and requests access to the data needed by him, but this paper focus that the user challenge the owner and the requested data are stored in the users' cloud instead of the owners' cloud so that high security is maintained by hiding all the other confidential data from the users. We also proposed a secure cloud storage framework with an efficient authentication mechanism that also supports data forwarding functionality. Apart from data forwarding functionality, our method guarantees security with respect to smaller key size and storage consumption and also sends an online alert message if any unauthorized access happens.

Keywords Storage system · Authentication · ECC · Forwarding · Online alert · Unauthorized access

1 Introduction

With the usage of Internet everywhere and the development of various technologies, a unique computing technology called 'cloud' have emerged. Unlike the traditional models, the data in the cloud reside in the remote location. This brings cloud a security challenge. The users find it more convenient to move their data into the cloud as they do not have to care about the complexities of direct hardware

S.V. Divya (✉) · R.S. Shaji
Department of IT, Noorul Islam University, Kumaracoil, India
e-mail: divyasadasivam@gmail.com

R.S. Shaji
e-mail: shajiswaram@yahoo.com

L.P. Suresh and B.K. Panigrahi (eds.), *Proceedings of the International Conference on Soft Computing Systems*, Advances in Intelligent Systems and Computing 398, DOI 10.1007/978-81-322-2674-1_68

management. Since the cloud is a distributed environment, the resources in the cloud are shared across multiple servers in multiple locations for redundancy purposes. This allows for efficient computing by centralized storage, memory, processing.

Various security challenges [1, 2] such as integrity, confidentiality, privacy are the major obstacles which affect the overall performance of the cloud. Based upon the customers need, the services are offered to the customers by the appropriate service providers and they are metered on pay-as-you basis. Each service providers have “locked-in” with a particular vendor and service-level agreements. The SLAs are not able to provide the same level of security to all their customers. Hence it is very difficult for the enterprise to swift from one service provider to another unless they are given huge fees.

Since the data is handed over to a third party in the cloud, the unused data may be deleted sometimes because of the dishonest third party auditor (TPA) and also to reuse the storage space, the resources are deleted. Hence it is necessary to check the storage correctness and the data integrity, auditing is performed. In order to enable public auditability, users use a TPA to check the integrity of outsourced data. Wang et al. [3] proposed a secure cloud storage system that supports privacy-preserving public auditing scheme to check the integrity of outsourced data. They enabled the TPA to perform auditing for multiple users simultaneously and efficiently by means of Public Ke-based Homomorphic Linear authenticator technique. Their method guarantees data confidentiality and data integrity issues.

As data is handed to third parties concern, there is a chance of altering the data without the owner’s consent. Hence authenticity of data is also important. Cloud also suffers from data leakage problem. When data is moved into the cloud, some standard encryption methods were used to store the data and to secure the operations. Various cryptographic algorithms such as homomorphic encryption, attribute-based encryption, identitybased encryption, and threshold proxy re-encryption have been used. In order to overcome the problem of traditional cryptosystems of greater complexity and large key size, elliptic curve cryptographic system comes into practice in cloud. The most important issue is to check whether the data integrity and confidentiality is attained while the data is stored in the cloud systems. In order to ensure the data storage correctness, provable data possession [4, 5, 6] and proof of storage or retrievability [7] was introduced.

Various researches have been done on providing security to cloud environment using different mechanisms; but still no proper mechanism have been found for providing better security in cloud.

2 Background Works

Storing data in a third-party cloud system causes serious concern over data confidentiality. Users just use services without being concerned about how computation is performed and storage is managed. Lin and Tzeng [8] had focused on designing a secure cloud storage system for robustness, confidentiality, and functionality. Their work focuses on forwarding the data to another user directly under the command of the data owner. They proposed a threshold proxy re-encryption scheme with multiplicative homomorphic property. Their method combines encrypting, encoding, and forwarding technique which makes the storage efficient and fulfill the requirements of data confidentiality, data robustness, and data forwarding.

Due to data outsourcing, there requires an auditing service to check the integrity of the data in the cloud. Some existing methods were applied only to static archive data and cannot be applied for dynamic data. An efficient and secure dynamic auditing protocol was proposed by Yang and Jia [9] which guarantees the data storage correctness. Their protocol supports confidentiality, batch auditing, and dynamic auditing. The main challenge in their design was data privacy problem. To overcome that, an encrypted proof with challenge stamp was generated by means of bilinearity property so that the auditor can verify the correctness of the proof without decrypting it. Later, Wang and Li [10] constructed a public auditing mechanism specifically for cloud data so that the confidential data of the users are not exposed to the TPA in auditing. One of the most important issues in auditing is "identity-privacy" when a group of users share a common file in the cloud. In their method, the file is divided into independent blocks and each block is separately signed by the users and if any modification is performed by the user, he needs to sign the modified block with his private/public key and informed to the TPA about his identity.

A PDP scheme was introduced to guarantee the data storage correctness in the cloud. But this also poses new challenges. Hence in order to overcome the drawbacks of PDP system, a cooperative PDP technique was proposed by Zhu and Hu [6]. Their scheme was initially developed for distributed systems to check the integrity and availability of their data in all the CSPs (cloud service providers), and they highlighted the data migration problem. Homomorphic verifiable response and hash index hierarchy were the methods used. Their work supports dynamic data operations and minimizes the computation costs of both the clients and storage servers. The major pitfalls of their method were for large files, this method leads to complexity.

An identity-based data storage system was introduced by De Capitani di Vimercati et al. [11] to overcome the collusion attacks and to support intra and inters domain queries.

As security threats get increases everyday, it is necessary to encrypt the data before storing it in the cloud. In order to protect the data from the various network attacks and attackers, several encryption algorithms such as identity-based encryption, attribute encryption, proxy re-encryption were used in the cloud. All these traditional algorithms increase the key size and therefore ECC was used in the cloud environment. Because of the lesser key size and network bandwidth, the ECC system was popularly used in all the environments such as wireless systems, mobile networks, and grid and in cloud environments. The ECC system was compared with the RSA for wireless environment was proposed by Lauter [14].

All the prior methods of data forwarding system lack security in terms of the threshold limit delay. Moreover, all those methods increase the computation and the communication cost rapidly.

Maintaining the keys during data sharing is a difficult job since the encrypted data takes different forms in different stages. Unless the owner gets compromise with the attacker or a hacker they may not able to acquire those keys. Hence we proposed an efficient method for secure authentication and data forwarding using ECC with mobile alert is used.

3 Modeling and Architecture

3.1 System Model

The proposed cloud storage system is constructed in such a way that it consists of distributed servers. The architecture of the proposed authentication system is given in Fig. 1.

The cloud storage system consists of four phases: system setup, data storage, data forwarding, and data retrieval. The system setup phase is used for setting up the parameters and generating the keys. The data storage and data forwarding phase are used for storing the encrypted data and forwarding the data respectively. The data retrieval phase deals with how the data can be retrieved correctly by the owner.

Moreover our proposed method includes two stages of authentication and hence it increases the security level.

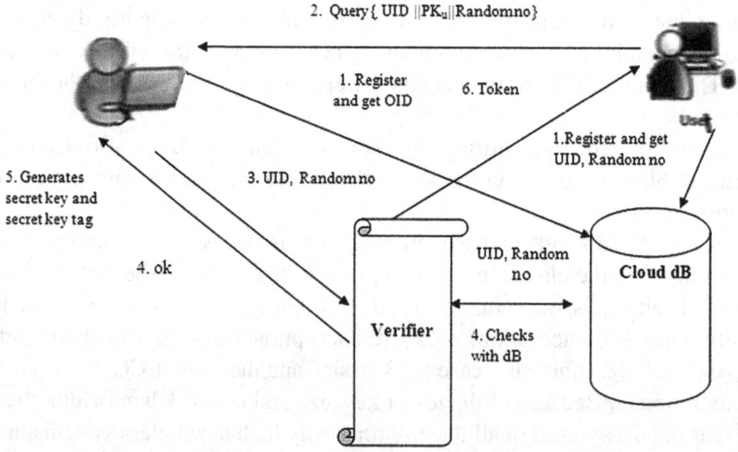

Fig. 1 Proposed authentication model

3.2 Phases in the Proposed System

Our proposed system consists of 4 phases. They are:

1. **System Setup**: In this phase, the system manager chooses system parameters and publishes to all. The data owner generates a private key PK_o. Similarly each user, generates a private key PK_u; where $u \in \{1,2...n\}$. The key servers are responsible for holding the private keys.
2. **Data Storage**: When the data owner needs to store the data, the data or the message m is divided into **k** blocks m_1, m_2, $m_{3,}...m_k$. The data is encrypted using the elliptic curve cryptography Enc() algorithm and store it in the cloud. The storage servers hold the encrypted data inside the cloud network. Elliptic curve cryptosystem is used in our proposed system [13] for encryption purpose.
3. **Data Forwarding**: This phase deals with how the data is forwarded between the servers.
4. **Data Retrieval**: This deals with how the data is retrieved by the user. The user requests the data owner for accessing the data. Upon checking the authenticity of the user, the owner sends the secret key tag to and a one-time key to retrieve a message from storage servers.

3.3 Steps for Proposed Method

Step1: The data owner and the user register with the cloud for accessing its services. After registering with the cloud, the data owner gets an OID and the user gets an UID and a random number.

First Stage of Authentication:

Step2: The user first sends query to prove his authenticity with the data owner. The user query includes {UID $||PK_u||$Random no}; where UID is the user identity; PK_u is the private key of the user and k be the random number.

Step3: The owner in turn sends this UID and the random number to the verifier (step 4) and if it is successful, then the owner generates a secret key and secret key tag by means of Secret_KeyGen() and chall() algorithm respectively.

Step4: The verifier checks:

If Received UID and the Random no = UID and
Random number already stored in the cloud db,
then
the user is authenticated and the verifier sends a token to him
else
the user is unauthenticated.

The token includes {UID || Session time/Expiry time ||token no}; where session time or Expiry time mentions the time period the user can access the data and token no is a number given to each user when a request is made.

Step 5: The owner now encrypt the data by means of elliptic curve cryptography with the help of the Enc() algorithm and store the encrypted data in the owners' cloud storage.

Authentication and Forwarding Stage:

Step 6: The user again request the owner for accessing the data. The request contains {Token || MID||DSize}; where MID is the Mobile ID of the user; and DSize is the size of the data which he needs.

Step 7: The owner checks the token with the verifier. If the token is valid, then

(a) The owner generates a one-time key and sends it to the users' mobile and
(b) Simultaneously forward the encrypted data to the cloud server 2 i.e. the user cloud storage.
else
Access is denied and notified to owner's mobile

Step 8: The user decrypts the data using the one-time key which is obtained from the owner.

Step 9: The user can access the data from the users' local server within that stipulated time mentioned in the token; else the data is locked and the user is not able to retrieve the data. Hence go to step 1.

Step 10: Using a different key, the data is again encrypted by the owner and stored in the owner's cloud.

The second stage of authentication and the data forwarding phase is shown in Fig. 2.

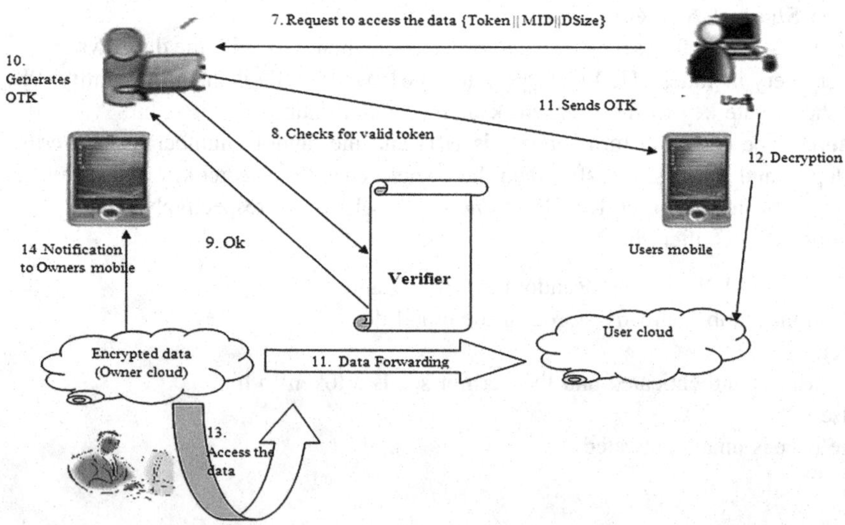

Fig. 2 Proposed forwarding model

If a valid user, the data owner provides the user a one-time pad through mobile and using this one-time pad, the user can decrypt the data which is encrypted already by the ECC and he can access the data. If the data is requested by another user, the data will be forwarded to him. Otherwise, access to the data is denied and the owner sends an alert message to the user through mobile.

3.4 Methodology

3.4.1 Elliptic Curve Cryptography

In order to overcome the drawbacks of public key cryptosystems of larger key size and more bandwidth, an elliptic curve cryptography is evolved. In our proposed method, elliptic curve cryptography is used for encryption in secure data forwarding. ECC has the unique property that it takes two points on a specific curve, and when added together, get a third point on the same curve. An elliptic curve E over a real numbers is explained by,

$$E: y^2 + a_1 xy + a_3 y = x^3 + a_2 x^2 + a_4 x + a_6$$

where a_1, a_2, a_3, a_4, $a_5 \in k$ and Δ is the discriminant of E and the value of $\Delta \neq 0$. The values are:

The elliptic curve over F_p is defined as,

$$\begin{aligned} & y^2 \bmod p = \left(x^3 + ax + b\right) \bmod p; \text{where} \left(4a^3 + 27b^2\right) \bmod p \neq 0 \\ & x, y, a, b \in [0, p - 1] \end{aligned}$$

The points in the curve E can be given as, $E(F_p) = (x, y)$; where $x, y \in F_p$ which satisfies the equation $y^2 = x^3 + ax + b$. Point addition and Point multiplication are the operations performed on elliptic curves.

In our proposed method, the ECC cryptographic method is used for encrypting and decrypting the data by means of Enc() and Dec() algorithm.

3.4.2 One-Time Key (OTK)

In our proposed method, one-time key mechanism is used to decrypt the data from the user cloud storage by the user. When the user requests the owner for accessing the data, the owner checks for a token and if it is successful, then he sends a one-time key to the user. The one-time Key is generated by means of the Current time/Timestamp—the time at which the user requests the data from the owner after verification by the verifier.

Keys used for Decryption: OTK + SK_t

The data is decrypted by means of 2 keys: secret key tag—SK_t and a one-time key (OTK) in our method and hence if any attacker or malicious hacker gets any one of the keys, it is not possible for him to decrypt the data because the OTK is generated at the time of request of the user and it is sent only to the legal and authenticated user by checking their token and sent through mobile. Hence our method provides an efficient way of forwarding the data from the owners' cloud to the users' cloud.

3.5 Algorithms of Our Proposed Method

The key generation, encryption and the decryption process is as follows,

Algorithm for Key Generation: KeyGen()

Input: Private keys of the owner PK_o, the private key of the user PKu and the random number k.
Output: Generates the secret key for the owner.
Step 1: Select a curve parameter p such that $p \in [1, n-1]$.
Step 2: The public key is $PUK_o = PK_o{*}p$; where PK_o is the private key and p –a point in the elliptic curve say as affine point.
Step 3: Return (PUK_o, PK_o)
Step 4: Calculate secret key of the owner by $SK_o = \{PUK_o + \{k{*}p\}\}_{PKo}$;

Encryption Process: In order to meet the security requirements of the conventional public key cryptography, elliptic curve cryptography is used in our proposed system for encryption purpose. In the encryption process, each character in the message m is converted into an ASCII value say m_a and multiplied with the random number and the cipher texts are calculated.

Algorithm for Encryption: Enc()

Input: The Message m, a private key PK_o, secret key $\mathbf{SK_o}$ and curve point k.
Output: Generates the cipher text C1 and C2.
E ($M, SK_o, PK_o p$)->C1,C2
Step 1: Represent the entire message M as m1 and m2.
Step 2: Select a point a in the elliptic curve domain [1,n-1]
Step 3: Compute $C1 = m1 + \{k{*}\ PK_o\}SK_o$ and $C2 = C1 + p{*}m2$
Both the cipher texts C1 and C2 are calculated by the above equations.
Step 4: Return (C1, C2).

Chall(): The user sends a challenge message to the owner indicating his identity for data retrieval.

Algorithm for Challenge: Chall()
Input: Private key of the user PK_u and the random number k.
Output: Generates the secret key tag and a token for the user
Step 1: The user sends the PK_u and k to the owner.
Step 2: Owner computes the secret key tag SK_t by $SK_t = PK_u{*}SK_o$ (Fig. 3)

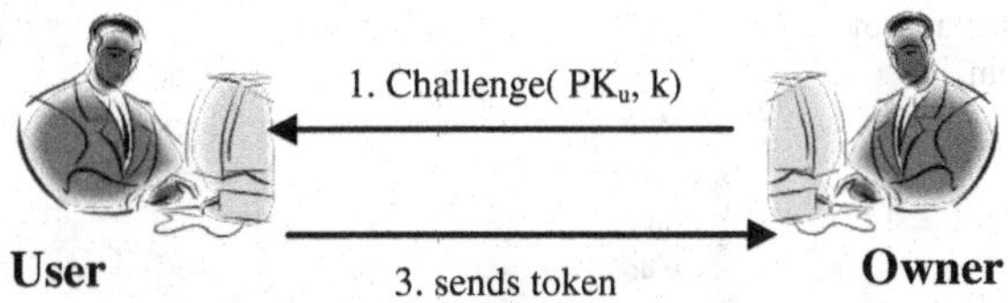

Fig. 3 Challenge-response scenario

Step 3: The owner after generating the secret key tag, sends a token to the user which includes {UID || Session time/Expiry time ||token no}.
Step 4: The Session time is calculated at when the user request is made to the owner.

Decryption Process: The decryption process explains about how the data is retrieved by the user.

Algorithm for Decryption: Dec()

Dec (C1,C2, SK_t, OTK) ->M
Inputs: The cipher texts C1, C2 and the secret key tag SK_t and the One-Time Key (OTK).
Output: The Message M
Step 1: Extract the values of the SK_o and PK_o
Step 2: Calculate the message m from the equation C1- {k* PK_o }- SK_o = m1;and C2-C1/p = m2;
Step 3: Combine the value of m1 and m2 to extract the original message M.

4 Results and Discussion

Even though the benefits of cloud were tremendous, there are a lot of security challenges that need to be solved [1, 2].

The proposed method is compared with that of the existing system with key size and storage and computation cost (Table 1).

Figure 4 shows the simulation graph of the existing system with that of proposed system.

Table 1 Comparison chart of existing system with proposed system w.r.t. key size and cost

Key size (kB)	Storage and computation cost (threshold proxy re-encryption)	Storage and computation cost (ECC)
115	1250	1090
210	2250	2100
275	4230	2250

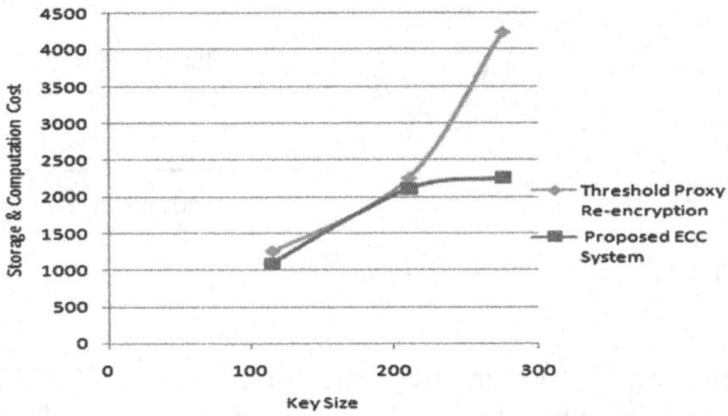

Fig. 4 Simulation graph of the proposed system

Since two-stage authentication is used, our proposed method attains greater security when compared to the previous methods and it also guarantees a security with respect to two-phase authentication as well as data forwarding.

5 Conclusion

Our proposed method yields a multilevel security for data forwarding in the cloud environment by means of:

1. Two-stage authentication
2. ECC encryption
3. Key confidentiality by means of OTK
4. Mobile Alert

The focus is also being done on the size of the encrypted message, number of packets to be transmitted, time complexity for executing the algorithm. It is clear, that our proposed scheme is superior in terms of security, performance, confidentiality when compared to existing methods.

References

1. Rong C, Nguyen ST, Jaatun MG (2012) Beyond lightning: a survey on security challenges in cloud computing. Elsevier Computers and Electrical Engineering
2. Gohring N (2008) Amazon's S3 down for several hours. Online Amazon.com, "Amazon Web Services (AWS). Online at http://aws.amazon.com
3. Wang C, Chow SSM, Wang Q (2012) Privacy-preserving public auditing for secure cloud storage. IEEE conference, pp 525–533
4. Ateniese G, Burns RC, Curtmola R, Herring J, Kissner L, Peterson ZNJ, Song DX (2007) Provable data possession at untrusted stores. In: Ning P, De Capitani di Vimercati S, Syverson PF (eds) ACM conference on computer and communications security. ACM, pp 598–609

5. Erway CC, Küpçü A, Papamanthou C, Tamassia R (2009) Dynamic provable data possession. In: Al-Shaer E, Jha S, Keromytis AD (eds) ACM conference on computer and communications security. ACM, pp 213–222
6. Wang Q, Wang C, Ren K, Lou W, Li J (2011) Enabling public auditability and data dynamics for storage security in cloud computing. IEEE Trans Parallel Distrib Syst 22(5):847–859
7. Shacham H, Waters B (2008) Compact proofs of retrievability. In: ASIACRYPT. Lecture notes in computer science
8. Lin H-Y, Tzeng W-G (2012) A secure erasure code based cloud storage system with secure data forwarding. IEEE Trans Parallel Distrib Syst 23(6)
9. Yang K, Jia X (2012) An efficient and secure dynamic auditing protocol for data storage in cloud computing. IEEE Trans Parallel Distrib Syst
10. Wang B, Li B (201X) Privacy-preserving public auditing for shared data in the cloud. IEEE Trans XXXXXX, vol X, no. X, pp XXXX
11. De Capitani di Vimercati S, Foresti S, Jajodia S, Paraboschi S, Samarati P (2013) Integrity for join queries in the cloud. IEEE Trans Cloud Comput 1(X):XXXXXXX
12. Zhu Y, Hu H (2011) Cooperative provable data possession for integrity verification in multi-cloud storage. IEEE Trans Parallel Distrib Syst 23(12):1–14
13. Koblitz N (1998) An elliptic curve implementation of the finite field digital signature algorithm. In: Advances in cryptology (CRYPTO1998). Lecture notes in computer science, vol 1462. Springer, Berlin, pp 327–337
14. Lauter K (2004) The advantages of elliptic curve cryptography for wireless security. IEEE Wirel Commun 11(1):62–67

Survey on Data Mining Techniques with Data Structures

Y. Jeya Sheela, S.H. Krishnaveni and S. Vinila Jinny

Abstract In the current technology evolution of various applications of data mining requires efficiency which can be achieved in various ways by improving the steps of the data mining techniques. Current research shows a wide opening in the step that uses data structure of data mining methodology. Data mining deals with large database of various formats. Processing of this data requires multiple scans which increases computational time. Using data structures, the computation operations involved in the data mining techniques can be improved. Efficient data structures make a data mining methodology more effective. We have presented a survey paper by reviewing standard data structures and various data mining techniques. We have concentrated on the concept of standard data structures as well as data structures in research area. Also we illustrate some data structure with examples. Thus after the survey we identified the current issues in the data structure when associated with data mining methodologies.

Keywords Data mining · Data structure · Suffix tree · FP-tree · DOM tree

Y. Jeya Sheela (✉) · S.H. Krishnaveni
Department of Information Technology, Noorul Islam University, Kumaracoil, India
e-mail: minisheela@ymail.com

S.H. Krishnaveni
e-mail: shkrishnaveni@gmail.com

S. Vinila Jinny
Department of Computer Science and Engineering, Noorul Islam University, Kumaracoil, India
e-mail: vinijini@gmail.com

L.P. Suresh and B.K. Panigrahi (eds.), *Proceedings of the International Conference on Soft Computing Systems*, Advances in Intelligent Systems and Computing 398, DOI 10.1007/978-81-322-2674-1_69

1 Introduction

Data mining is a well-known process of mining unknown patterns from large database. It is also known as knowledge discovery. It also deals with various forms of data, methods for managing large data, and ways for retrieving patterns from huge database. Data mining is a proficient process that abstracts some novel and important data hidden in huge records. The aim of data mining process is to determine concealed patterns, unforeseen trends or other understated relationships in the data in large database. Today data mining is found to be a new discipline that finds significant applications in various commercial, technical, and industrial circumstances. For example, there are various commerce applications which store different types of confidential and financially viable information about the contender. Researchers in the field of data mining found many issues and challenges that can be overcome by applying data structures in the apt place. Issues like memory requirement while handling huge data, complex computation for pattern recognition, exhausted executed time, etc., can be dealt efficiently by incorporating data structures with data mining. Data structure is defined as a way of organizing data for easy access and representation. Here we have made a study on various existing data structures and data structures on research. Section 2 deals with existing data structures which gives basic idea about organization of various data. Section 3 explains the data structures applied on various data mining techniques. Section 4 gives the conclusion and suggestions for future work.

2 Existing Data Structure

2.1 DOM Tree

The Document Object Model (DOM) [1] is a standard application programming interface that can be used in various applications. DOM finds a significant usage in HTML and XML applications. With DOM, the objects in HTML and XML documents can be represented effectively and interacted in an efficient way. It defines the coherent structure of documents and tells how a document can be accessed and manipulated. With the DOM, programmers can create documents, navigate their structure, add, or remove and modify elements in the documents. HTML or XML document can be accessed, changed, added, or deleted using the DOM. The DOM can be used with any programming language. For e.g., consider the following table, taken from an HTML document:

```
<TABLE>
<THEAD>
<TR>
<TD>WELL</TD>
<TD>NICE</TD>
```

Fig. 1 DOM for table

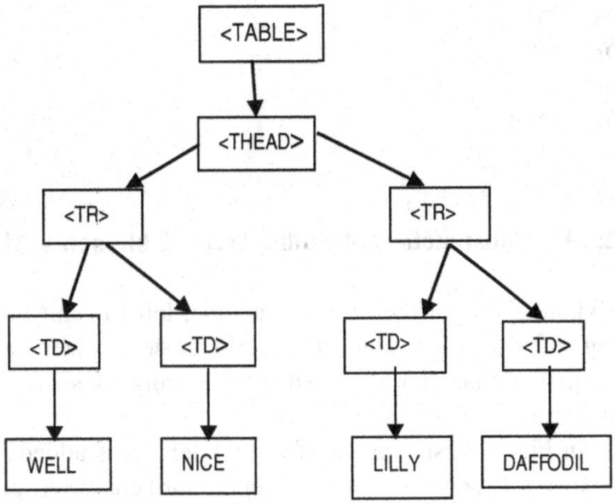

</TR>
<TR>
<TD>LILLY</TD>
<TD>DAFFODIL</TD>
</TR>
</THEAD>
</TABLE>

Figure 1 represents the DOM for the above HTML table.

2.2 *Suffix Tree*

A suffix tree [2] is also called PAT tree. A suffix tree is a data structure that solves many problems related to strings. A string is a sequence of characters. A suffix tree for a string S of size n is a rooted directed tree with n leaves (numbered as $1 - n$). All internal nodes except the root have at least two children and edges are labeled with a sub string of S. No two out coming edges of a node can have the labels with same first character. For any leaf i, joining the edge labels on the path from the root to leaf i spells the suffix of S that begins at position i. That is, $S[1 \ldots n]$. Suffix tree is used to find substring of size m in a string of size n.

If str = $s_1 s_2 \ldots s_i \ldots s_n$ is a string, then $S_i = s_i s_{i+1} \ldots s_n$ is the suffix of str that begins at position i.

For e.g., [3]

$S_1 = xyzxyw = \text{str}$
$S_2 = yzxyw$
$S_3 = zxyw$

$S_4 = xyw$
$S_5 = yw$
$S_6 = w$
$S_7 =$ (empty)

2.2.1 Construction of Suffix Tree—Ukkonen's Algorithm

Ukkonen's algorithm is a linear time, left to right online algorithm for constructing suffix tree. The algorithm works in step by step basis from left to right. In each step, one character is added to the string of size n which increases the size of substring to $n + 1$.

In Fig. 2, first character from the left (x) is added into the suffix tree. In Fig. 3, next character (y) is added. In Fig. 4, third character (z) is added. In Fig. 5, character x is added. In Figs. 6 and 7 characters y and w are added.

2.3 *FP-Tree*

Data mining is a process of extracting hidden predictive information from large databases. Data mining deals with large data. Finding frequent item sets in large databases is an important task. Han has proposed an effective, capable, extensible, and modular algorithm named FP-Growth algorithm [4] for discovery of frequent

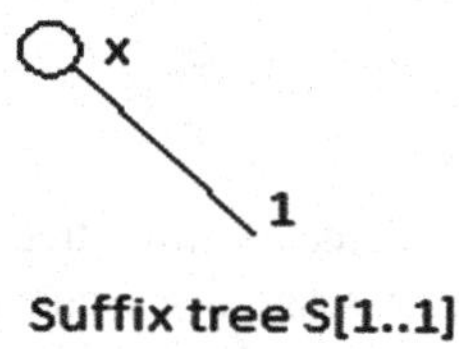

Fig. 2 x is added

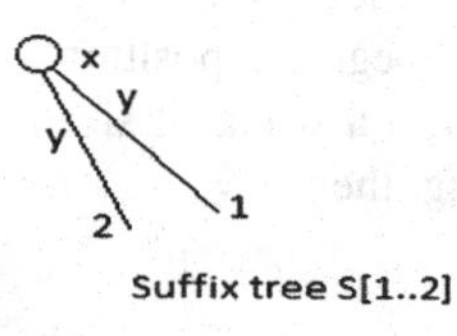

Fig. 3 y is added

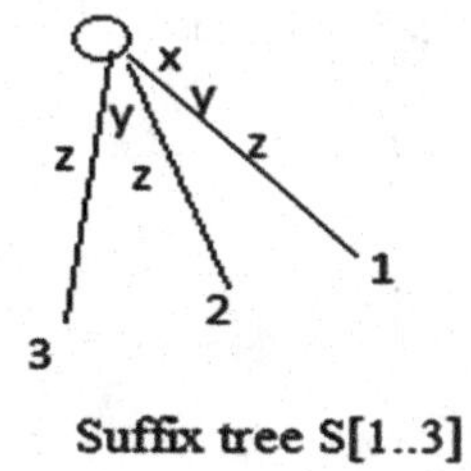

Fig. 4 *z* is added

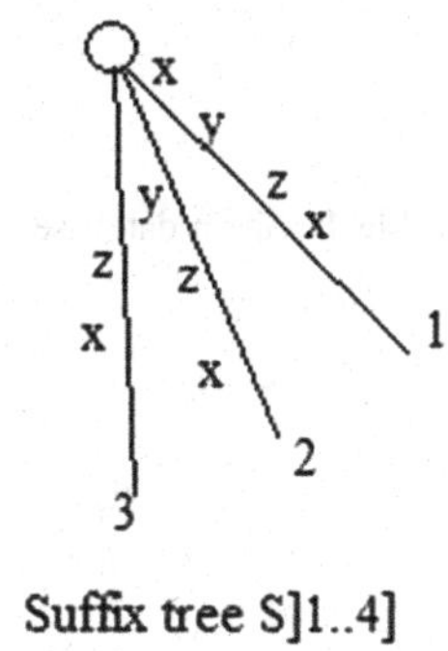

Fig. 5 *x* is added

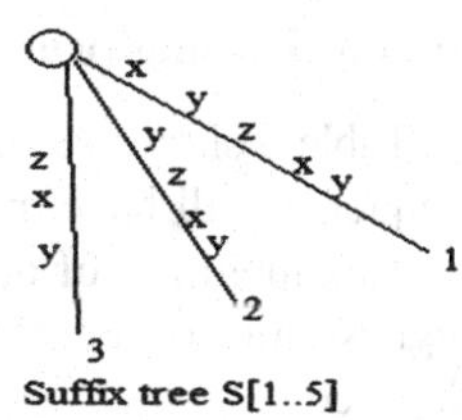

Fig. 6 *y* is added

item sets from large dataset. This algorithm uses a data structure named frequent pattern tree which maintains the item set association information.

Example, find all frequent itemsets [5] in the database given in Table 1, taking minimum support as 30 %.

Step 1: Calculation of Minimum support

$$\begin{aligned}\text{Minimum support count} &= 30/100 * 8\\ &= 2.4\\ &= 3\end{aligned}$$

Step 2: Finding frequency of occurrence

Table 2 shows the frequency of occurrence of items.

For e.g., item X has occurred in row 1, row 2, and row 3. Totally it has occurred three times. In the same manner frequencies of remaining items are found.

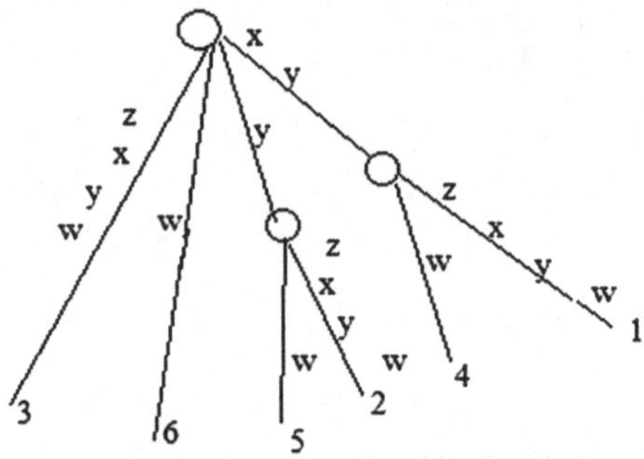

Suffix tree S[1..6]

Fig. 7 *w* is added

Table 1 Given database

TID	Items
1	X, A, C, B
2	C, A, Y, X, B
3	Y, A, B, X
4	B, A, C
5	C
6	C, B

Step 3: Finding priority of items

Table 3 shows the priority of items. Items which have the maximum number of frequency will have the highest priority.

Maximum no. of occurrences for A, B, C, X, Y are 4, 5, 5, 3, 2, 5. B has the highest priority. Items that does not satisfy minimum support requirement can be dropped.

Step 4: Ordering of Item based on priority

Table 4 shows the ordering of items based on priority. Item B has occurred the maximum number of times, and Item Y has minimum number of times. Hence Item B has highest priority and item Y has the lowest priority. Order of items are B, C, A, X, Y.

Step 5: Construction of FP-Tree

Figure 8 shows the FP-tree constructed for items in Row 1 of Table 4. Root node is a NULL node which is drawn first and then items in row 1 (B, C, A, X) are attached one by one.

Figure 9 shows the FP-tree constructed for items in Row 2 of Table 4. Items in row 2 are: B, C, A, X, Y.

Figure 10 shows the FP-tree constructed for items in Row 3 of Table 4. Items in row 3 are: B, A, X, Y.

Table 2 Frequency of occurrence

ITEM	Frequency
A	4
B	5
C	5
X	3
Y	2

Table 3 Priority of items

Item	Priority
B	1
C	2
A	3
X	4
Y	5

Table 4 Ordering of items

TID	Items	Ordered items
1	X, A, C, B	B, C, A, X
2	C, A, Y, X, B	B, C, A, X, Y
3	Y, A, B, X	B, A, X, Y
4	B, A, C	B, C, A
5	C	C
6	C, B	B, C

Figure 11 shows the FP-tree constructed for items in Row 4 of Table 4. Items in row 4 are: B, C, A.

Figure 12 shows the FP-tree constructed for items in Row 5 of Table 4. Items in row 5 are: C.

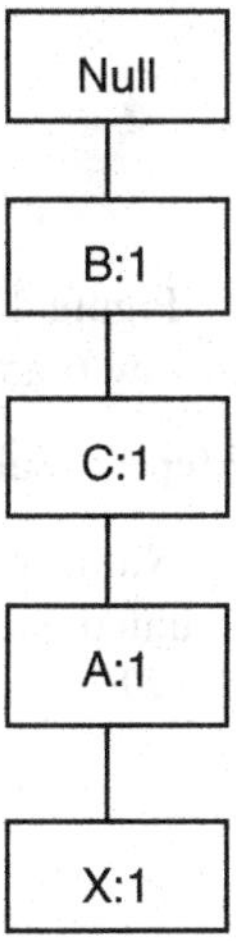

Fig. 8 FP-tree for row 1 of Table 4

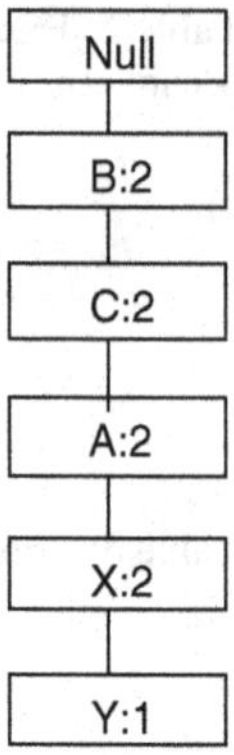

Fig. 9 FP-Tree for row 2 of Table 4

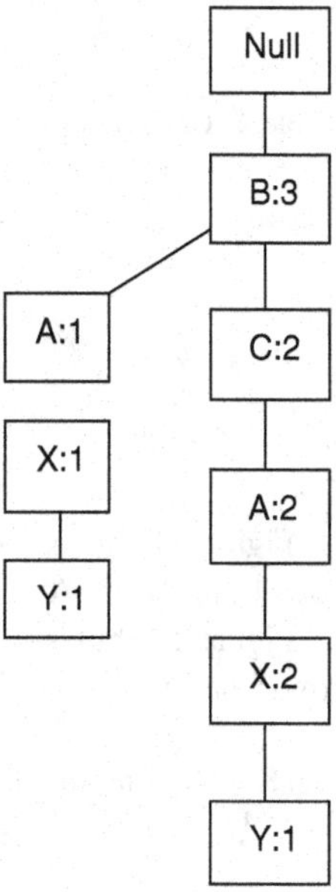

Fig. 10 FP-tree for row 3 of Table 4

Figure 13 shows the FP-tree constructed for items in Row 6 of Table 4. Items in row 6 are: B, C.

Step 6: Validation

Number of frequency of occurrences of items in the FP-Tree in Fig. 13 is counted and compared with the values in Table 2. If both the values matches, then the FP-tree constructed is correct. Here all the values are matched.

Fig. 11 FP-tree for row 4 of table

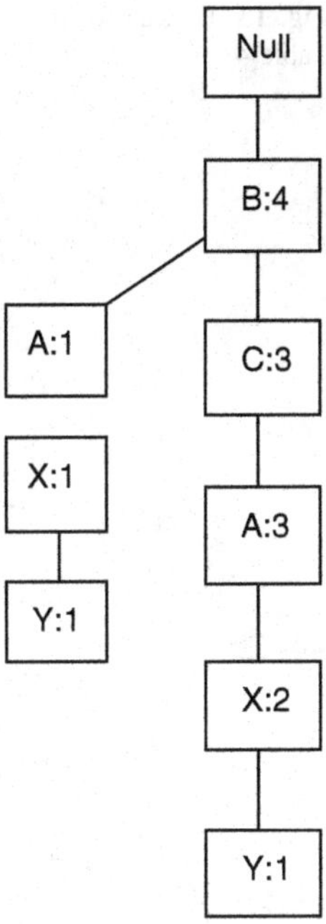

Fig. 12 FP-tree for row 5 of Table 4

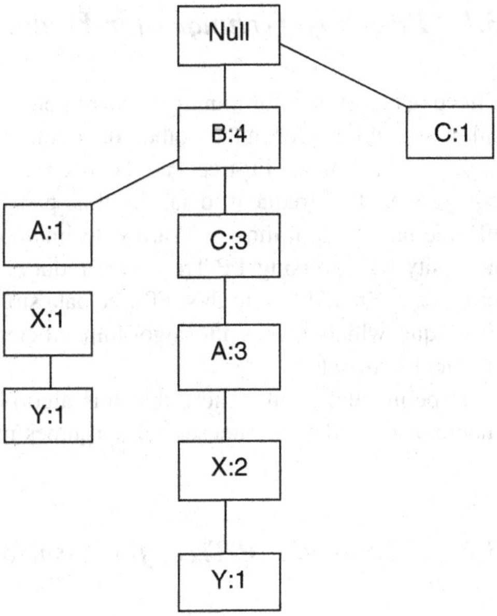

Fig. 13 FP-tree for row 6 of Table 4

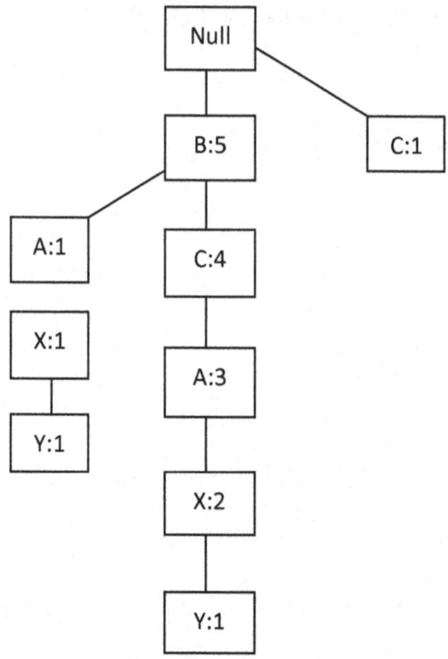

3 Research Method

3.1 FP-Array Technique for Finding Frequent Item Sets

Effective algorithms for mining frequent item sets are critical for mining association rules as well as for many other data mining tasks. A well-known prefix-tree structure known as FP-tree can be used for mining frequent item sets which compresses the frequent data. In this paper, to improve the performance of FP-tree-based algorithms, FP-array technique [6] is used which decreases the necessity of traversing FP-Tree. This reduces the computational time needed for processing. In addition to this, FP-tree data structure is combined with the FP-array technique which makes the algorithm efficient for mining all, maximal, closed frequent item sets.

Experimental results show that this algorithm is very fast and consumes less memory when data is intense, but consumes much memory when data is bare.

3.2 T-Trees and P-Trees for Association Rule Mining

Association rule learning is a well-known and well-examined method for finding inspiring relations between variables in large databases. It is anticipated to discover strong rules found in databases using diverse events of interestingness. Two new data structures T-Tree and P-Tree [7] are together used with associated algorithms which make the algorithm more efficient and improve the storage and execution time.

Experimental results are compared with various other data structures like Hash tree approach and FP-tree.

Results show that T-Tree provides less generation time and storage requirements compared to hash tree structures and the P-Tree gives important preprocessing advantages in terms of generation time and storage requirements compared to the FP-tree.

3.3 Evolutionary Programming Technique for Detecting-Repeated Patterns from Graph

A proficient technique like evolutionary programming (EP) technique is used for finding replicating patterns in a graph. The searching ability of evolutionary programming [8] is used for this purpose. The method used in this correspondence is tree-like structure. If a pattern is found on a particular level of the tree, then the graph is compacted using it, and the substructure discovery algorithm is reiterated with the compacted graph. This permit one to find replicating substructures within huge structures. The proposed technique is useful for mining information from data records that can be easily expressed as graphs. Experimental results reveal that best and improved substructures can be found successfully to compact the data.

3.4 CMTreeMiner for Mining Closed and Maximal Frequent Subtrees

Tree structures are used widely in field of computational biology, pattern recognition, XML databases, computer networks, etc. The significant problem in mining the records of trees is detecting repeatedly appearing subtrees. Because of the combinable detonation, the number of recurring subtrees grows regularly with the length of recurring subtrees. So, mining all recurring subtrees from large trees is futile. In this paper, an effectual and enumerating method named CMTreeMiner [9] is used, which finds closed and maximal frequent sub trees from a database of labeled rooted ordered trees and labeled rooted unordered trees. The algorithm first uses various pruning and interrogative techniques to decrease the search space and to get better computational efficacy and then mines closed and maximal frequent subtrees. Closed and maximal frequent sub trees are identified by visiting the computational tree methodically. Performance of the algorithm is studied through wide-ranging experiments, through artificial and real-world datasets. According to the experimental results this algorithm is very proficient in decreasing the search space and rapidly finds all closed and maximal frequent subtrees

3.5 Geometric Hashing Technique for Finding Patterns in 3D-Graph

In this paper, a method for finding patterns in 3D graph is given. The nodes in the graph are labeled and non-decomposable. The link between the nodes is called edges. Patterns are solid substructures that occur in a graph after undergoing random number of complete structure rotations, translations, and a small number of edit operations. The various edit operations are node relabeling, inserting, and deleting a node. In this paper, an efficient technique called geometric hashing technique [10] is used, which hashes three nodes of the graph at a time, into a 3D table, and compacts the label in table. This technique is used to find recurring substructures in a set of 3D graphs. A frame work for structural pattern discovery is developed and it is used in the field of Biochemistry for classifying proteins and clustering compounds. Experimental results illustrate good performance of the algorithm and better utility in pattern discovery.

4 Conclusion

In this paper, we have made detailed study on various data structures and applying data structure on data mining algorithms. The standardized data structures shows promising way of organizing data of various forms for easy access. In the field of data mining, various issues like memory storage problem on handling huge data and complex computational operations are dealt effectually, by incorporating efficient data structures with data mining. Data Structures under Research gives best results when appropriately selected based on the requirement of the application.

References

1. http://en.wikipedia.org/w/index.php?title=Document_Object_Model&action=history
2. http://en.wikipedia.org/w/index.php?title=Suffix_tree&oldid=649776788
3. http://www.geeksforgeeks.org/
4. en.wikibooks.org/wiki/Data_mining_Algorithms_In_R/Frequent_Pattern_Mining/The%20FP_Growth_Algorithm
5. hareenlaks.blogspot.com/2011/06/
6. Grahne G, Zhu J (2005) Fast algorithms for frequent item set mining using FP-Trees. IEEE Trans Knowl Data Eng 17(10)

7. Coenen F, Leng P, Ahmed S (2004) Data structure for association rule mining: T-trees and P-trees. IEEE Trans Knowl Data Eng 16(6)
8. Maulik U (2008) Hierarchical pattern discovery in graphs. IEEE Trans Syst Man Cybern Part C: Appl Rev 38(6)
9. Chi Y, Xia Y, Yang Y, Muntz RR (2005) Mining closed and maximal frequent sub trees from databases of labeled rooted trees. IEEE Trans Knowl Data Eng 17(2)
10. Wang X, Wang JTL, Shasha D, Shapiro BA, Rigoutsos I, Zhang K (2002) Finding patterns in three-dimensional graphs: algorithms and applications to scientific data mining. IEEE Trans Knowl Data Eng 14(4)

Author Index

L.P. Suresh and B.K. Panigrahi (eds.), *Proceedings of the International Conference on Soft Computing Systems*, Advances in Intelligent Systems and Computing 398, DOI 10.1007/978-81-322-2674-1

Printed in the United States
By Bookmasters